GIS *for the* URBAN ENVIRONMENT

Juliana Maantay and John Ziegler

Foreword by John Pickles

ESRI PRESS

REDLANDS, CALIFORNIA

ESRI Press, 380 New York Street, Redlands, California 92373-8100

All rights reserved. First edition 2006
10 09 08 07 06   1 2 3 4 5 6 7 8 9 10

Printed in the United States of America

Library of Congress Cataloging-in-Publication Data
Maantay, Juliana, 1951–
    GIS for the urban environment / Juliana Maantay, John Ziegler.
      p.   cm.
    Included index.
    ISBN 1-58948-082-1 (hardcover : alk. paper)
    1. Cities and towns—Geographic information systems.
    2. Geographic information systems. I. Ziegler, John, 1931–. II. Title.
    HT153.M314 2006
    307.760285—dc22   2006013778

ISBN-13: 978-1-58948-082-7
ISBN-10: 1-58948-082-1

Ask for ESRI Press titles at your local bookstore or order by calling 1-800-447-9778. You can also shop online at www.esri.com/esripress. Outside the United States, contact your local ESRI distributor.

ESRI Press titles are distributed to the trade by the following:

In North America, South America, Asia, and Australia:
Independent Publishers Group (IPG)
Telephone (United States): 1-800-888-4741
Telephone (international): 312-337-0747
E-mail: frontdesk@ipgbook.com

In the United Kingdom, Europe, and the Middle East:
Transatlantic Publishers Group Ltd.
Telephone: 44 20 7373 2515
Fax: 44 20 7244 1018
E-mail: richard@tpgltd.co.uk

Cover design by Doug Huibregtse
Book design and production by Jennifer Galloway

Table of Contents

Preface   XI

Foreword   XVII

Acknowledgments   XXI

Part 1

Chapter One   **Basics of Mapping and GIS**   **3**
What is a map?
Historic and cultural perspectives
Maps, technology, and power
What is a GIS?
Why planners use GIS
GIS for the urban environment
Layers
Using GIS in an urban context
Basic GIS functionality
Typical GIS tasks
Written Exercise 1: GIS on the Internet

Chapter Two   **Modeling Spatial Data and Basic Mapping Concepts**   **25**
Modeling
The vector model
The raster model
Combining raster and vector
Continuous vs. discrete data
Scale
Generalization
Geographic extent, geographic unit of analysis, and data aggregation
Map projections and coordinate systems
Common coordinate systems
Written Exercise 2: Maps, scales, and coordinate systems

Chapter Three   **Thematic Mapping**   57
Maps: Tangible vs. mental
Maps: Reference vs. thematic
Components of the thematic map
Quantitative vs. qualitative thematic maps
Types of thematic maps
Multivariate maps
Written Exercise 3: Mental mapping

Chapter Four   **Data Classification Methods and Data Exploration**   93
Quantitative vs. qualitative information
Measurement scales
Statistics and mapping
Summarizing data with statistics
Methods of classification
Issues in classification
Data exploration
Written Exercise 4: Thematic map interpretation

Chapter Five   **Data Visualization and Map Design**   125
Types of charts and graphs
Data visualization
Map design and composition
Main map elements
Graphic design
Color
Typography
Written Exercise 5: Map design critique

Chapter Six   **Sources of Urban Data**   157
A wealth of data
Choosing the appropriate data
Level of detail
Quality
Map accuracy
National map accuracy standards
Attribute data accuracy and currency
Metadata
The U.S. Decennial Census
U.S. Census TIGER/Line files
U.S. Postal zone mapping
United States Geological Survey (USGS) mapping
GIS data produced by state and local government
Commercial sources
The Internet
Data clearinghouses
Access to urban GIS data
Written Exercise 6: Evaluation of an urban dataset

Chapter Seven   **Mapping Databases**   **181**
Point geocoding
Unmapped attribute tables
Creating vector mapping
GPS
Written Exercise 7: Data acquisition

Chapter Eight   **Attribute Data and Relational Database Management Systems**   **191**
Attribute database tables
Creating database tables
Working directly with external database management systems
Managing database tables
Tables and views
Grouping data
Relational databases
Adding the spatial dimension
RDBMS and data integration
Written Exercise 8: Integrating data using an RDBMS

Chapter Nine   **Methods of Spatial Data Analysis**   **209**
Types of analysis
Simple query: Data search and analysis
Reclassification
Geoprocessing
Optimal location/suitability analysis
Boolean operators
Modeling
Expert systems and rule-based reasoning
Interpolation
Written Exercise 9: Using GIS for problem solving

Chapter Ten   **GIS Project Development and Institutional Issues**   **247**
Driving forces
Proposing a GIS
Initiating a project
Project development
Development cycle
Management perspective
Written Exercise 10: Develop a project flow chart

Chapter Eleven   **Ethical Issues in GIS and Urban Planning   267**
A multitude of issues
The subjectivity of maps
Maps as propaganda
Participatory GIS and the democratization of GIS
GIS for advocacy purposes: Counter-mapping
Technology and the use of local knowledge bases
Access to hardware, software, and skills
Vernacularization of GIS
Data access
Data ownership
Data liability
Privacy, confidentiality, and surveillance issues
Written Exercise 11: Participatory community-based planning with GIS

Chapter Twelve   **Other Geotechnologies and Recent Developments in GIS   293**
Three-dimensional GIS
Advanced GIS analysis
Remote sensing
Global Positioning System (GPS)
Lidar technology
Mobile mapping
Internet mapping software
Wide area networks
Conclusion
Written Exercise 12: Project documentation and policy recommendation report

Part 2

Case Study One **Nonprofit Organization GIS Strategic Planning and Public Outreach** 323
Public outreach: Imagine New York
The Lower Manhattan Preservation Fund
The campaign for community-based planning
Olympics 2012

Case Study Two **GIS Use by Urban Nonprofit Organizations for Housing Initiatives and Urban Services** 335
Fordham Bedford Housing Corporation, University Neighborhood Housing Program, and the Enterprise Foundation

Case Study Three **Delivering Health-care Services to an Urban Population** 341
Access to health care in Greenpoint–Williamsburg, Brooklyn, New York

Case Study Four **Natural Habitat and Open Space Assessment** 349
Long Island Sound Stewardship System

Case Study Five **Urban Environmental Planning** 357
Land-use classification for extensive vegetative roof acreage potential in the Bronx

Case Study Six **Emergency Management and Disaster Response** 367
Using GIS at Ground Zero

Case Study Seven **Infrastructure Mapping for Planning and Maintenance** 377
GIS mapping of the New York City sewer and storm drain system

Case Study Eight **Archaeology and Historic Preservation with GIS** 383
Archaeology GIS project for the New York City Landmark Preservation Commission

Case Study Nine **Health and Environmental Justice** 393
Mapping environmental hazards and asthma in the Bronx

Case Study Ten **Crime Pattern Analysis** 409
Exploring Bronx auto thefts using GIS

Case Study Eleven **Community-based Planning** 415
Graffiti in Community District Two, Queens, New York: Private expression in the public realm

Case Study Twelve **Advocacy Planning and Public Information** 425
Community Mapping Assistance Project for the Metropolitan Waterfront Alliance and Open Accessible Space Information Systems Internet map site

Part 3

Lab Exercise One   **Making Maps of Urban Data   439**

Lab Exercise Two   **Exploring Basic GIS Functionality   443**

Lab Exercise Three   **Thematic Mapping: Dot Density Maps   459**

Lab Exercise Four   **Thematic Mapping: Choropleth Maps   471**

Lab Exercise Five   **Integrating Graphs and Maps, and Designing Map Layouts   483**

Lab Exercise Six   **Developing an Attribute Database from an Internet Source   499**

Lab Exercise Seven   **Geocoding and Labeling   505**

Lab Exercise Eight   **Working with Relational Databases   519**

Lab Exercise Nine   **Generating Buffers and Using Selection for Proximity Analysis   531**

Lab Exercise Ten   **Geoprocessing Operations and Joining Tables   541**

Lab Exercise Eleven   **Data Exploration and Geostatistical Analysis   549**

Lab Exercise Twelve   **Advanced Layout Techniques   555**

Afterword   567

About the authors   569

Data credits   571

Data license agreement   575

Installing the data   579

Index   581

Preface

This book focuses on urban planning applications and urban problem solving using geographic information systems (GIS) technology. It is intended to serve as an introduction to these subjects for students and professionals in the fields of urban geography, urban studies, urban planning, urban public health, urban environmental assessment, and hazard and emergency management. One of its main objectives is to minimize technical jargon while still describing urban GIS concepts in a comprehensive fashion, enabling the reader to gain a thorough understanding of GIS techniques and so to make full use of them. A second objective is to give students a taste of the real-world activities that are often required in urban GIS projects but rarely included in prepared lab work—tasks such as data acquisition, integration of data into the GIS, and manipulation of real data. A third objective is to demonstrate project design and analysis methodologies through presentation of real examples—case studies—of urban GIS projects. Because a solid understanding of project design is crucial to urban planners using GIS, these case studies form an important part of the book.

Scope

This book is designed for three main purposes and audiences: as a textbook for an undergraduate- or graduate-level course in GIS applications for urban planning; as a self-study guide to basic GIS mapping for urban planners and GIS staff; and as an introduction to the uses of GIS for urban decision makers. To satisfy such a wide and diverse readership, the book is necessarily comprehensive. It encompasses the fundamentals of GIS theory, skills, and practice; introduces more advanced concepts of modeling, simulation, and other geotechnologies related to GIS; and throughout, focuses specifically on how GIS is used for urban issues. In this way, we intend for it to be useful not only for students, but also for those with GIS experience and for professional urban decision makers.

Organization

Part 1 consists of twelve chapters detailing basic GIS and mapping concepts, in increasing degrees of complexity. Maps and other graphics illustrating the concepts are interspersed throughout the text, and these are taken primarily from actual urban GIS projects created by the authors. GIS terminology that is likely to be new to students is highlighted in bold text, with an explanation immediately following, or close by. Students may also want to consult the ESRI GIS Dictionary located at *support.esri.com/gisdictionary.* At the conclusion of each chapter, there is an extensive reference and suggested reading list and a written assignment designed to give students further practical experience using the topics presented in that chapter.

Part 1 is an integral component of the book, a collection of twelve case studies covering a wide range of urban planning applications of GIS. Studying these will help the reader understand some of the important questions that must be asked (and answered) in every GIS project: how was a study designed, what datasets were needed, where and how were they obtained, what types of analyses were used, what were the results? By dissecting the case studies, we believe readers will learn much about how to put together their own projects, will be better able to understand how GIS works, and will be able to do so more quickly.

The case studies were selected with an eye toward representing as many of the applications of GIS in urban planning as possible, including affordable housing development, crime-pattern analysis, environmental assessment, land-use planning, emergency management, community-based planning, historic preservation, infrastructure operations and maintenance, and health and environmental justice studies.

We tried to obtain case studies from a wide range of organizations that use GIS for planning, including federal agencies, municipal planning departments, not-for-profit groups, regional authorities, community-based planning organizations, private consulting firms, and academic centers.

Part 3, the lab exercises, correspond roughly to the content of each chapter and reinforce the GIS knowledge presented in that chapter, through the use of ArcGIS desktop technology, and with the datasets that come with the book on CD. The exception is lab exercise 1, which presents a variety of urban applications of GIS and is intended as an introduction and demonstration only. The exercises in the first half of part 3 allow the reader to become familiar with the way ArcView software is structured—its graphic user interface (GUI), the menu bars, the different windows, and basic display—and the way it functions, through tasks such as adding data, zooming, identifying, panning, selecting, querying, sorting the attribute table, labeling, and symbolizing. The later lab exercises develop the reader's GIS skills further, through the building of an urban planning project and exploring the connections among demographics, land use, environment, and health. The GIS tasks in the lab exercises progress for the reader in order of complexity—starting with the making of dot density maps, choropleth maps, and graphs for data exploration and visualization, then preparing maps and data for presentation, downloading a dataset from an Internet site, geocoding, creating a database, and finally, geoprocessing, spatial analysis, and advanced cartographic design.

Classroom use

Because of the potential for student error (and hence, student frustration) we strongly recommend some level of instructor supervision or guidance be available for part 3. In fact, the exercises are written with the expectation that an instructor well versed in ArcGIS software will be available to readers to explain concepts and techniques that can be touched on only briefly in this work. Only the geographic data to do the exercises is provided on the CD; software is not provided. The exercises assume that the reader has access to ArcView 9 software, either at home or (more likely) in a computer lab at the school where the course is being taught. In addition, lab exercise 11 requires the use of the ArcGIS software extension Geostatistical Analyst (and for one task, Spatial Analyst). A spreadsheet program such as Microsoft Excel will be needed for lab exercises 6, 8, and 10.

This book is intended to fill a need for a GIS college-level textbook focused on urban planning and urban geography, a need that we, as teachers of these subjects, have felt acutely. It can be used in either a one-year undergraduate course of study, or in a more accelerated one-term graduate or undergraduate course in these general topics. It is not expected that students will have had any prior experience with GIS before using this book, although some degree of familiarity with PCs and the Microsoft Windows operating system is strongly recommended, and previous experience with ArcView will be helpful. Within the context of urban planning coursework, we would recommend that students take an introductory GIS course using this text near the beginning of graduate study, or no later than the junior year of their undergraduate work. This will enable students to get the full benefit of applying their GIS knowledge to a significant part of their other planning coursework, and allow GIS and planning courses to be mutually reinforcing. It will also encourage students (and instructors) to more fully integrate GIS analytical and graphic capabilities within their planning coursework.

In a one-semester course, each week could be devoted to one of the twelve chapters, including the written and lab assignments. The remaining weeks could then be devoted to exams, guest speakers, student presentations, or team projects. Case-study readings could be assigned as the instructor sees fit, and incorporated into class discussion, written exercises, or exams. By the end of the term, it is anticipated that students would be able to start their own GIS projects, incorporate GIS analysis and graphic presentations into their thesis or capstone projects, or continue on to a more advanced GIS course covering complex spatial analysis, modeling, simulation, three-dimensional analysis, network analysis, geostatistics, or animation.

For a one-year course, it is envisioned that the first term would break at the end of chapter 6, covering basic GIS and cartographic concepts. The second term, covering data management, spatial analysis, and project design, would consist of chapters 7 through 12. The case studies might be interspersed throughout or alternated with the other topics, and the written exercises could be done by small teams (or individually) within the class period or outside of it. There would be ample time for in-class student presentations of the various team or individual assignments, and in-class critiques of student lab assignments. This would allow for a more leisurely paced course, and an increased opportunity for students to assimilate their GIS experience into other planning coursework.

Geographic extent

Much of the content of this book—case studies, datasets, and maps—use New York City as its focus. As the country's largest urban area, New York City is uniquely positioned for studying urban planning issues, and for extrapolation to other urban locales. New York's urban geography is highly diverse, encompassing skyscrapers and urban canyons; inner city ghettos; tenements and public housing; urban renewal areas; luxury residential highrises; affluent gated, planned, and suburban bedroom communities from an earlier era; and more contemporary suburban developments, all within New York's five boroughs. The city's population and economy are equally varied. In addition, New York embodies many of the problems found in most cities, and is so large and diverse that almost every type of land-use, transportation, housing, social, environmental, and health-planning issue can be found here. In our view, New York provides an unusually inclusive example of the range of urban planning issues found around the world and how GIS can be used to deal with them. It would be difficult to replicate elsewhere the range of planning issues that exists in New York City today, but most of the individual issues that are presented in the case studies, lab exercises, maps, and examples of GIS analyses are illustrative of the types of issues common to urban planning everywhere.

Goals

Urban planning professionals and students do not necessarily have a background in geography, although most have worked with maps, plans, and other geographic data before. The authors assume that readers will be interested in learning enough about basic geographic concepts to enable them to use GIS fully for urban planning purposes. However, we do not attempt to be a substitute for a comprehensive geography course or textbook. Neither is the book a comprehensive instructional manual for using or learning to use ArcView 9, and so we have made a conscious effort to minimize technical software terminology; we think that most planners aspire to be users of GIS or to be GIS analysts, rather than GIS software developers. (Readers interested in further developing their ArcView or ArcGIS skills should consult the *Getting to Know* workbook series from ESRI Press, or ESRI Virtual Campus online course offerings.) Our aim here is to create a textbook and a course of study that is a practical and feasible means for planning professionals and students to acquire a working knowledge of GIS particularly geared toward planning applications.

The book, like much of planners' work, is project-oriented, and both the case studies and the lab exercises exemplify this. The project that is the focus of the lab exercises pertains to one particular geographic location and makes use of several datasets to examine existing demographic, land-use, environmental, and health conditions that would be exemplary background to a policy recommendation report. Since so much of acquiring competence in GIS depends on knowing where and how to get data, and how to incorporate it into a GIS project, we felt it was important to give the reader experience with those activities in the lab exercises, activities which are often missing from other prepared lab exercises in this subject area, and which prevent the student from getting a true perspective on what it takes to build a GIS project. Even so, the lab exercises cannot cover every contingency or problem one might encounter in a real GIS project, and it would be virtually impossible to create such exercises. Similarly, GIS technology now in use is quite complex, and it is doubtful that any one person knows every available function. However, we hope that the lab exercises will be a stepping-stone for an ongoing GIS learning process, and will provide enough problem-solving techniques to engender confidence and independence in the GIS beginner.

One has to be a bit fearless in attempting GIS. Our one piece of advice is: relax, and don't be afraid of making mistakes. In making mistakes, we learn how to learn, to figure things out when there is no recipe to follow. This approach works well for GIS because learning GIS is more than just knowing the right keys to press. In many ways, it's a completely new and unique approach to the process of planning and thinking about urban spaces. We hope that this book will allow readers to figure out the tools they need to go off on their own.

On the other hand, we don't want to be too theoretical and insufficiently technical. Too many people take GIS courses and learn much about GIS theory, but don't know how to make a map when they're finished—or worse, don't understand the limits of GIS. In short, we want to give a well-rounded and representational view of how GIS is used in the urban environment, shed some light on its potential, and in the end, encourage more people to jump in and get involved with GIS.

Although we encourage readers to experiment with the technology, and to do their own exploring, the GIS community is now too large to allow isolation. The community of GIS users is continually connecting and learning from each other, and this is one of the great pleasures of working with GIS.

GIS is an adventure, one that we hope you will also find enjoyable as well as enlightening.

Juliana Maantay
John Ziegler

Foreword

In the 2003 National Research Council (NRC) report *GIS for Housing and Urban Development,* the authors and committee members suggested that nowhere is the rapid entry of technology into our daily lives more evident than in the development and use of GIS. As the NRC report pointed out, "Urban, metropolitan, and state governments throughout the country have already turned to GIS as a means to deal with their own burgeoning demands for more effective and efficient service," as have nongovernmental agencies and community organizations.

GIS clearly has much to offer to researchers and policy makers who deal with the challenges of the urban environment, and it has become an essential tool for many who deal with urban planning, transport logistics, community development, environmental hazards, health care, or a host of other crucial concerns in the contemporary city. As costs of GIS have declined and software has become easier to use, individuals have also seen spatial data technologies become ever more present in their daily lives. These include GPS units in cars and boats; mapping systems for locating hotel rooms and navigating cities; integrated facilities planning by universities and other organizations; electoral polling and constituency management in politics; neighborhood targeting by commercial marketing organizations; and in the community itself, where people strive for a cleaner environment. The range of available data and the experience of citizens with digital spatial data and its uses has grown significantly. The urban environment and the practices of city life are being reshaped by these new spatial data handling and imaging technologies and capacities.

In the context of this GIS growth, and at a time of increasing accessibility of software and georeferenced data, questions of how we use GIS, how we train users, and to what uses we put GIS (and could or should put GIS), have become increasingly urgent. The need to reflect on the ways in which new data structures and infrastructures can support democratic practices, recognized by the NRC report, was one of the reasons that I published *Ground Truth* in 1995, and it remains one of the abiding challenges for all teachers and practitioners of GIS. It is the very power and reach that GIS permits—the power to predict spatial patterns of voting or consuming behavior, the ability to shape urban movement in the interests of spatial control and visibility, the ability to create new spaces of accessibility for the disabled, or the ability to develop institutional practices for the more rational and efficient location–allocation models for health care, policing, emergency response systems, business applications, or community action—that makes it attractive and important. But it is this very power and reach that require instructors and users of GIS to think about how to use GIS in various contexts and how it (and any technology) reshapes lives and landscapes.

GIS for the Urban Environment is a welcome addition to this literature. The book focuses on GIS in urban planning and problem solving, and is aimed particularly at large cities and metropolitan areas. It will be of interest to all students and practitioners who want to learn more about how they might apply GIS to their daily practices, be they in urban planning, public health, urban environmental monitoring, hazard and emergency management, geographical analysis, or community development. It will become an essential text to get started on what the authors call "the GIS adventure."

For a number of years, Juliana Maantay has been drawing our attention to the ways in which GIS can contribute in all sorts of ways to this adventure. Drawing on a background in urban planning, geography, and GIS, and having worked in the New York City Department of Planning, she knows how GIS can be used to address issues of environmental hazard, risk, and health in disadvantaged neighborhoods and among impacted communities of color. At Lehman College she has trained students and community groups to map and analyze the effects of urban zoning on the geographical distribution and redistribution of hazard and risk. By working with local groups and training students from the community, she has shown how the deployment of GIS contributes to the ways in which individuals and local communities are able to participate in shaping their living conditions. Urban zoning and the necessity of dealing with toxic waste materials produce geographies of transport, storage, and disposal that highlight the fractured geographies of race, class, and locality in the city. Some of these community groups have been able to use GIS to respond to the uneven impacts of toxic exposure by mobilizing community resources and local knowledge. The results are powerful diagnostic analyses of local patterns of toxicity and the identification of hotspots in the community, such as transport routes and overnight parking facilities where toxic waste trucks concentrate, often in close proximity to schools, playgrounds, and nursing homes. In all of these examples, Maantay has been able to show the complex interweaving of GIS practices in the city and how GIS can be used in ways that support and sustain important conversations about the choices facing those involved in shaping urban and metropolitan futures.

GIS for the Urban Environment focuses on teaching students and practitioners how to learn GIS in terms of project design and analysis methodologies that speak to this wider range of urban practitioners. The book includes twelve chapters in part 1 that introduce the elements and issues involved in using GIS, and it does so with a wide range of useful maps and other graphics. Each chapter provides readers with a carefully chosen set of further readings, along with written exercises, and in part 3, laboratory exercises. Students and practitioners will enjoy working through the individual chapters and exercises, and will find the twelve case studies of part 2 to be particularly valuable. The exercises and case studies illustrate GIS use across a wide range of contexts, providing readers with clear examples of actual applications dealing with important issues in contemporary cities. The authors have been able to link a rigorous presentation of GIS tools and approaches with a text that is accessible, clearly written, and interesting to read, and with exercises to help students and practitioners learn and develop their skills and applications. The examples and exercises are designed to serve as conceptual and technical introduction, training manual, exercise book, and rich analysis of the possibilities for using GIS in urban settings. They introduce the student to GIS practices, sources, and social issues. They include issues for nonprofit public sector groups, transport operations, emergency management and recovery, habitat and open space assessment, community planning, environmental planning, infrastructural management, community education, health and environmental justice, real estate, land-use zoning, affordable housing, industry and economic development, and historic preservation. This is an exciting GIS book that will be of great interest and value to students and practitioners alike. It covers a rich array of technical details, effectively presented. And the entire work remains refreshingly open to a multiplicity of potential users.

It is particularly refreshing to see a book on GIS that speaks to urban planning, community GIS, and new experimental applications of public participation GIS. It does so in ways that will help practitioners learn how to deal with the crucial interfaces of technical detail, practical application, and the ethics of GIS practice in their work. Many of the case studies draw on New York City. These alone are worth the price of admission, providing a wonderful introduction not only to GIS urban applications, but to the complex and uneven urban and socio-spatial dynamics of New York as an exemplar of metropolitan infrastructures and built environment.

John Pickles
Earl N. Phillips Distinguished Professor of International Studies and Geography
University of North Carolina, Chapel Hill
January 2006

References

Maantay, J. A. 2001. Race and waste: options for equity planning in New York City. *Planners Network,* Jan/Feb. *http://www.plannersnetwork.org/htm/pub/archives/145/maantay.html*

Maantay, J. A. 2002a. Mapping Environmental Injustices: Pitfalls and Potential of Geographic Information Systems (GIS) in Assessing Environmental Health and Equity. The journal *Environmental Health Perspectives,* April, 2002; Volume 110, Supplement 2, pp. 161–171. (Special issue *Advancing Environmental Justice Through Community-Based Participatory Planning.*) *http://ehpnet1.niehs.nih.gov/members/2002/suppl-2/161-171maantay/EHP110s2p161PDF.PDF*

Maantay, J. A. 2002b. Zoning Law, Health, and Environmental Justice: What's the Connection? *Journal of Law, Medicine, and Ethics,* Fall, 2002; pp. 572–593. (Special issue *Health, Law, and Human Rights.*)

Maantay, J. A. 2002c. Industrial Zoning Changes in New York City and Environmental Justice: A Case Study in "Expulsive" Zoning. *Projections: the Planning Journal of Massachusetts Institute of Technology (MIT),* Winter 2002, pp. 63–108. (Special issue *Planning for Environmental Justice.*)

National Research Council. 2003. *GIS for Housing and Urban Development.* Washington D.C.: National Academies of Sciences.

Pickles, J. 1995. *Ground Truth: The Social Implications of Geographical Information Systems.* New York: The Guilford Press.

Pickles, J. 2004. *A History of Spaces: Cartographic Reason, Mapping, and the Geo-Coded World.* London and New York: Routledge.

Acknowledgments

The initial impetus for this book stemmed mainly from our experiences in trying to teach GIS for urban planning, which gave us the kernel of the topics outline and the concept of how to present the material in a logical sequence. The book evolved substantially over the course of its preparation, and benefited immensely from the contributions of many people who read preliminary drafts, discussed ideas with us, provided constructive criticism, and produced written or graphic material.

Juliana Maantay would especially like to thank her students, who over the last few years have provided excellent feedback and suggestions for improving the teaching of GIS theory and practical methods. Many of her students gave the lab exercises a test drive, and helped iron out glitches and point out potential problems. Several students made direct and important contributions to this book, including Karen Kaplan, Andrew Maroko, and Greg Studwell, all in the graduate Geographic Information Science program at Lehman College, City University of New York (CUNY); as did Holly Porter-Morgan, PhD candidate at the CUNY Graduate Center; and Juan Carlos Saborio and Dellis Stanberry, geography majors at Lehman. Additional thanks are due for the tireless efforts of Gene Laper, the interlibrary loan specialist at Lehman College Library, for his valuable assistance with obtaining research materials and references, no matter how obscure or rare; and to William Bosworth, professor emeritus at Lehman College. Dr. Bosworth directs the Bronx Data Center, and he provided much of the data for the lab exercises, and guidance in the use of Census data for GIS research. John Ziegler would like to thank Sean Fitzpatrick of ESRI New York for reviewing parts of the manuscript.

One of the distinguishing aspects of this textbook is the wealth of urban GIS case studies in part 1. These would not have been possible without the significant labor of more than a dozen urban planning and GIS professionals who submitted to us their materials, both written and graphic. These case studies are intended to demonstrate the breadth and scope of the uses of GIS in planning, and if the book succeeds in meeting that objective, it is due in large measure to the time and expertise generously given by these skilled project directors and GIS analysts.

As in most books with more than one author, each of us took the lead for various sections and chapters, based on our expertise and familiarity with those topics. Juliana Maantay was responsible for chapters 3, 4, 5, 9, 11, the lab exercises, and the written exercises at the end of each chapter, while John Ziegler was primarily responsible for chapters 6, 7, 8, 10, and 12. Chapters 1 and 2 were a combined effort.

The graphics, maps, and illustrations were created by the authors or drawn from a wide variety of recent GIS and urban geography literature and from specialized atlases. Credits and sources for illustrations and graphics created from third-party sources are given with each; those without specific attribution are listed in the credits section in the back of the book.

Although parts of the book were written individually, our goal was to make this an integrated textbook expressing our shared vision of GIS in planning. GIS for urban planning is an exciting field, and we are pleased that ESRI Press was sufficiently enthusiastic and farsighted to support our efforts during the two years it took to produce the book. We would like to acknowledge the crucial roles that former ESRI Press Publisher Christian Harder and the book's editors, R. W. Greene and Mike Kataoka, played in bringing this book to fruition. Thanks are also due to these indispensable members of the ESRI Press team: Jennifer Galloway for her dedicated work in book design and production, and in achieving consistency in the graphic elements; Amy Collins, for carefully honing the lab exercises; and Carmen Fye, for marshalling the myriad permissions. Also thanks to Doug Huibregtse of ESRI Graphics for his wonderful cover design and to Brian Parr, ESRI Educational Services project manager, for his patience and attention to detail on the technical aspects of organizing the lab assignments.

We, of course, take full responsibility for any shortcomings, but also recognize that all positive aspects of the book are due to the fact that we stand on the shoulders of giants.

Juliana Maantay
John Ziegler
New York City

Part 1

Basics of Mapping *and* GIS

One

Mapping and GIS are two separate but closely related disciplines,

with common roots in geographic visualization and spatial analysis.

It is impossible to fully understand and use GIS without having an

adequate background in cartographic terminology and concepts, and

it is becoming equally impossible, at the beginning of the twenty-first

century, to realize the full potential of cartography and mapping without

also understanding the basic principles and uses of GIS.

What is a map?

Maps are at the heart of GIS, and are used as both the raw materials and the final products of GIS projects. Maps are intended to convey information, and are abstractions, simplifications, and representations of reality—"a representation, usually on a flat surface, of the features of an area of ground, a portion of the heavens, and so on, showing them in their correct forms, sizes, and relationships, according to some convention of representation," as defined by the *Random House Dictionary of the English Language*. A map may also be defined as a unique and complex combination of image, language, and mathematics (Woodward 1998). These definitions may still be too narrow. One of the most authoritative thinkers on this subject, Mark Monmonier, says that "maps are scale models of reality," and this broader definition probably most accurately reflects the scope of mapping today.

In fact, the subject matter treated by mapping has been expanded in recent years well beyond the earth-bound spaces that maps have traditionally depicted. We now map human-scale entities such as the human brain; submicroscopic-scale entities such as the genome; and macro-scale astronomical entities including galaxies and, indeed, the known extent of the universe (Hall 1993). We are no longer mapping only earth's lands and seas, but a variety of other intangible and unseen places, heretofore viewed only in our mind's eye. Maps are increasingly employed to delineate abstract concepts such as the distribution of a nation's average household income or its rates of crime. While mapping is an important component of GIS, GIS is much more: it combines mapping with information technology, and thereby transfers control of the mapping process from the cartographer to the user.

unknown or legendary
lands beyond the ocean

Figure 1.1   **Some scholars consider this the oldest map of the world we have found. It shows the homeland of its creators, the Babylonians, as the center of the world.** © 1998. From *New Found Lands: Maps in the History of Exploration*, by Peter Whitfield. Reproduced by permission of Routledge/Taylor & Francis Books, Inc.

Historic and cultural perspectives

Virtually every human society makes and uses maps to some degree, often only to communicate to others where some physical or spiritual resource is located, sometimes to accomplish much more. Maps are not always composed of ink and paper, but may be made of wood, stone, or other durable material—for example, the wonderfully accurate and innovative navigational maps created by Polynesian oceanic explorers. These maps, made of bamboo sticks and shells, mimic and thus identify the patterns made by waves from the open ocean as they break in unique and distinctive formations on each South Pacific island. Maps are not necessarily permanent; they may be as transitory as the map scratched in the dirt and later obliterated by rain or footsteps. Maps can also be enshrined in memory, with unique expressions of geographic information transmitted in song, dance, and oral traditions—such as the songlines of the indigenous peoples of Australia.

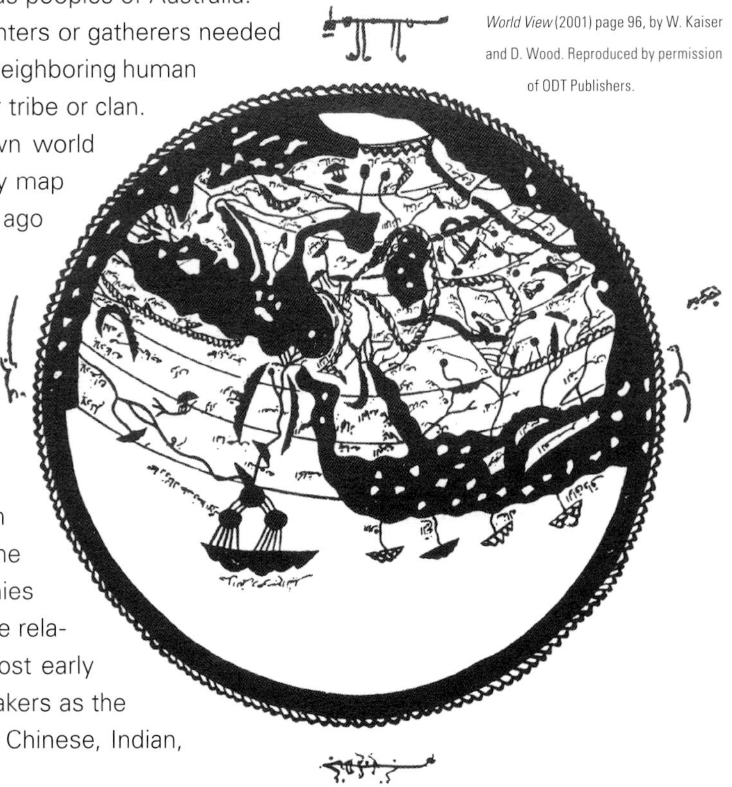

Figure 1.2 Islamic cartographer Al Idrisi created this map in 1154. He placed Mecca, the holy city of Islam, at the center. From *Seeing Through Maps: The Power of Images to Shape Our World View* (2001) page 96, by W. Kaiser and D. Wood. Reproduced by permission of ODT Publishers.

Mapping likely started out in prehistoric times when hunters or gatherers needed to communicate the whereabouts of food, water, animals, neighboring human groups, or dangerous conditions to other members of their tribe or clan. These were strictly local maps. Maps of the entire known world were apparently not attempted until about 500 BCE. A tiny map of the world, produced in Mesopotamia over 2,500 years ago on a baked clay tablet, is the earliest surviving world map in existence (figure 1.1). There are two lines running down the center of the flat, round earth, probably representing the Tigris and the Euphrates rivers. In the central circle are Babylon and other important cities. Surrounding these is the Bitter River, beyond which lived all sorts of legendary beasts (Berthon and Robinson 1991).

Many of the first maps attempted to explain more than just physical geography, but also to show graphically how the world is organized in accordance with religious philosophies and beliefs. Early mapmakers also tried to contend with the relationship of earth to the moon, planets, sun, and stars. Most early maps of the known world show the location of the mapmakers as the center of the world—no matter whether the society was Chinese, Indian, Roman, Aztec, or Babylonian (figure 1.2).

This geocentricity persists to modern times: wherever "we," the mapmakers, happen to live is portrayed as the center of the world on the map. One only has to look at modern maps of the world in which the Prime Meridian, 0° longitude, runs through the Royal Observatory in Greenwich, United Kingdom—the institution where the latitude and longitude coordinate system that we use today was codified in the eighteenth century—to see that mapmaking is still a profoundly ego- and ethnocentric activity.

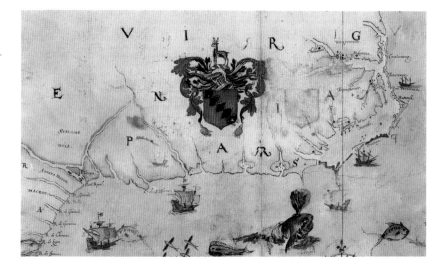

Figure 1.3   Early map of Virginia, by John White, 1585, with Sir Walter Raleigh's coat of arms displayed prominently, a stamp of English possession imposed on the existing indigenous population. © 1998. From *New Found Lands: Maps in the History of Exploration*, by Peter Whitfield. Reproduced by permission of Routledge/Taylor & Francis Books, Inc.

Historically, one of the most important purposes in the development of maps was for political control. Those who could describe a place graphically and understand its location in relation to other places could put a claim to it and control it: if you were the one who could define boundaries and map an area, it belonged to you. Maps helped the powerful consolidate political power and establish jurisdiction over conquered territories (figure 1.3).

Maps also helped promote the ascendancy of centralized governments and bureaucracies. If the emperor knew the exact layout of his land, it could be ruled more efficiently (Dorling and Fairbairn 1997).

Related to the role of political control is the role that maps played in trade and exploration, especially important in light of the globalization of markets and resources that began in the fifteenth century. Maps were enormously expensive to produce, requiring knowledge that cost vast sums to obtain, and rare technical skills to record them on paper. This effectively limited map production, ownership, and use to the wealthy few who could afford to procure the knowledge and skills required to make them. Those who obtained geographic knowledge and its two-dimensional representations could amass further fortunes in trade and exploitation of "new" lands. In Europe, for a time—roughly from the fifteenth through eighteenth centuries—maps were highly secret documents; death sentences awaited those caught sharing navigational maps with rival powers. Some maps and sensitive geographic information are still classified as secret by commercial or governmental interests today. Maps are obviously of vital importance in times of war and military conflicts, and having an accurate map of an enemy territory or battleground has often proven to be a decisive factor in warfare. Today both hard-copy maps of certain areas and access to the most accurate satellite data are often withheld from the public for reasons of national security.

Another important purpose of mapping was and continues to be the assessment and collection of taxes, duties, and revenues. Julius Caesar in fact ordered that the "whole world" be measured, in part so that it could be taxed. To this day, a consistently rich source of map data is to be found in local property (cadastral) maps, usually produced and maintained by county or municipal tax assessment offices (figure 1.4).

Although political power always provided a significant and pervasive impetus in the development of maps through history, it was not the only one. Maps have also been a method by which a society conceived of its place in the universe and the earth's place in the larger scheme of things. Maps in certain societies during certain periods have had symbolic and religious significance, such as the *mappae mundi* of the mediaeval European world; the cosmological maps of the Jains and Buddhists of India, showing heaven, hell, and earth; or the ancient Egyptian route maps to the afterlife (figure 1.5).

These maps were intended to portray not only the real world that we experience with our senses, but also a spiritual realm, to help guide our lives for the future beyond this world and to make sense of the past. In some ways, the uses that urban planners make of maps serve those same purposes, both the mundane and the idealistic: to portray what exists now, as well as what could (or should) be.

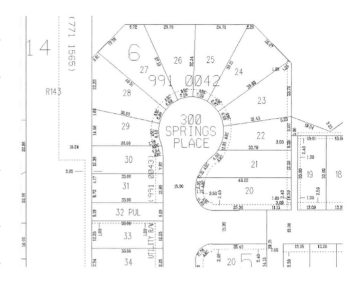

Figure 1.4 A cadastral map showing lot boundaries, lot numbers, and dimensions. Cadastral maps have been used for millennia to track property ownership and tax assessment.

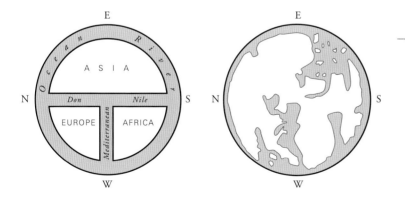

Figure 1.5 Left: Schematic *mappa mundi,* also called a T-in-O map because of the distinct shape formed by the major rivers and the encircling ocean. *Mappae mundi* typically depicted the world known to medieval Europeans, with the Garden of Eden at the top (east) of the map and north at the left of the map. Jerusalem, the spiritual center of the Christian world, would be located in the center of the map. These maps were often very elaborate and intricately detailed, and sometimes Christ's head and feet were shown at the cardinal directions, encompassing the known extent of the world. Right: Later *mappae mundi* from the fourteenth and fifteenth centuries show a more realistic depiction of the known world, as influenced by the newly rediscovered geographic writings of Ptolmey and advances in cartography and navigation. The symbolic tri-part division of the world, however, still formed the basis of these maps, just as it did in earlier, more religiously oriented *mappae mundi.* Redrawn from *History of Cartography* (1988), Vol. 1, figure 18.4, edited by J. B. Harley and D. Woodward. Reproduced by permission of The University of Chicago Press.

Maps, technology, and power

Throughout history, advances in mapping have paralleled advances in scientific knowledge and in technology. Astronomical observations, the understanding of the spherical nature of the earth, the development of geometry (literally, "earth measurement"), the compass, surveying equipment and techniques, meteorological and geological discoveries, paper manufacture, woodblock printing, the printing press, the chronometer, hot air balloons, statistics, aerial photography, computers, satellites and remotely sensed images, the Global Positioning System (GPS), and other developments have all contributed to the advancement of mapping and cartography. The modern conception and practice of mapping have depended on ever more accurate measurement techniques and ever more sophisticated equipment. This link with science and technology has contributed to the perception of mapping as a tool of the wealthy and elite, since historically science has generally been in the employ of the powerful.

The growth of GIS has not attenuated the criticism that the power of maps is concentrated in the hands of those who are already powerful. GIS mapping tends to be resource-intensive, and still requires a relatively high level of technical expertise and equipment, putting it out of the reach of many. However, advances in mapping software, the diversity of options for obtaining geographic data, and the increased distribution of maps and mapping capabilities through the Internet have allowed GIS and mapping to become accessible, if not to everyone, then to a far broader spectrum of people than ever before. This has thus democratized mapmaking. Interactive GIS maps, available to anyone with an Internet connection, enable people to find the nearest retail store or to map out a travel route. In organizations, GIS is becoming available to nontechnical people at all levels because of front-end user interfaces that are targeted to specific information needs and are programmed for ease of use. The democratization of mapping is a topic we will return to in chapter 11: ethical issues in GIS.

What is GIS?

There are probably as many definitions of GIS as there are GIS practitioners, and it is difficult to find a single definition that encompasses the multiplicity of GIS uses. It is perhaps easiest to think of GIS as an integrated system of components: information about the real world that has been abstracted and simplified into a digital database of spatial and nonspatial features, which, in conjunction with specialized software and computer hardware, and coupled with the expert judgment of the GIS user or analyst, produces solutions to spatial problems.

Many definitions of GIS either downplay or omit the role of the GIS user, but we believe the GIS user is a critical part of the GIS system. At every step of the way in a GIS analysis, the user is making decisions about what data to collect, how to integrate that data into the system, how to analyze it, what assumptions to make, how to classify the data, and how to present and display the data. GIS obviously depends on data, but it is also driven by the decisions of the user and analyst. The computer and the software cannot make sense of the data without the expertise of the user (figure 1.6).

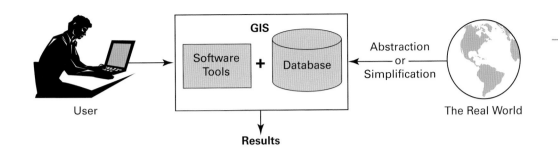

Figure 1.6   The components of a GIS: the real world, abstracted and simplified into a database, which in conjunction with specialized software and computer hardware, and the expert judgment of the GIS user/analyst, produces results such as solutions to a problem. ESRI.

This fact is apparent in studying the real-world cases in part 2 and will also become obvious when you begin your own projects.

Why planners use GIS

In our increasingly global twenty-first century, maps (and geographic knowledge) have become ever more important. A wide variety of professions require the use of geographic data, and so these professionals need to know how to make and use maps without specialized cartographic training. Disciplines and industries that depend on spatially-oriented information include public health, environmental science, political science, marketing, public administration, demography, ecology, conservation, tourism, asset management, facility management, real estate, utility and energy management, transportation, and, of course, urban planning, among many others (figure 1.7).

Urban planning requires an understanding not only of the physical space where we live, but also the underlying social, cultural, political, and economic landscape. Maps, being graphic representations of various aspects of reality, are indispensable to this effort of understanding and visualizing the existing urban environment, and the complex task of envisioning future conditions. The profession of planning could not exist without maps.

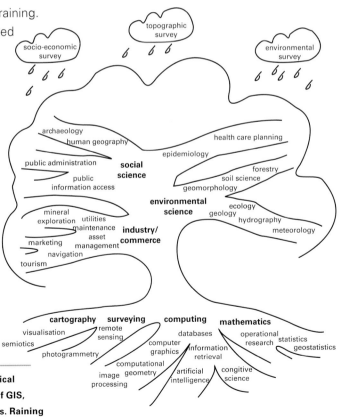

Figure 1.7   The GIS Tree. The roots represent the technical foundations of GIS. The branches represent applications of GIS, the results and requirements of which feed back to the roots. Raining down on the tree are the data sources on which the individual applications depend. Redrawn from Jones, C., *Geographical Information Systems and Computer Cartography*, 1st Edition, © 1997, pp.5, 274. Reprinted by permission of Pearson Education, Inc., Upper Saddle River, NJ.

Geographic location, spatial patterns, and distribution of features or events across a landscape inform many of the decisions that planners either make or help others to make. Planners must be intimately familiar with the physical layout and locations of, as well as relationships among, such spatial features as commercial or residential zoning districts, public schools, hospitals, transportation routes, and land uses. They must also be familiar with conditions or events such as crime patterns, accident locations, and patterns of movement. One of the most important ways GIS helps the practice of urban planning is in enhancing its decision-making capabilities.

Planning functions that use GIS to increase efficiency and flexibility include:

+ water supply and wastewater infrastructure maintenance and development
+ solid waste management
+ affordable housing specification and operation
+ zoning and land-use planning
+ regulation of the built environment
+ urban renewal and economic development
+ crime analysis and prevention
+ traffic analysis
+ facility management
+ issuing of permits
+ code inspection and enforcement
+ natural resource management and environmental assessment
+ demographic analysis
+ emergency management planning and disaster recovery
+ public health and health-care access
+ tax assessment and collection
+ municipal service allocation
+ transportation and public transit planning
+ historic preservation
+ parks and open-space planning
+ utilities maintenance and planning
+ many other aspects of municipal operations and public administration

In addition to government agencies, nonprofit organizations also engage in planning and can also benefit from using GIS technology, as can other nongovernmental organizations that seek to influence how their local environments are planned. Most notable in this regard are Participatory GIS (PGIS) or Public Participation GIS (PPGIS), usually undertaken for the purposes of advocacy or the empowerment of local peoples. Community-based planning groups engaged in these efforts are increasingly using GIS technology to prepare alternative plans, analyze official proposals, and present to the community, to elected officials, and to other decision makers a vision for their own environment.

The private sector—especially engineering and environmental consulting firms and the real estate industry—also uses GIS for planning purposes; much of the built environment in the United States is not driven by local government planners but rather develops through private market forces.

The use of GIS differs from the use of traditional, static paper maps primarily because GIS can deal with complex and constantly changing data and geographic information, and enables decision makers to respond rapidly to changing conditions.

GIS for the urban environment

The word urban connotes density—density of people, and also of the multiple support systems such as public transportation, infrastructure, and services that people need and expect. In an urban environment there is a high level of interaction among these systems because they coexist in a relatively compressed space. GIS, with its ability to manage and display information about many aspects of the same geographic area, has a special role to play in urban planning and operations.

The ability of GIS to communicate information visually in ways that can be understood by everyone is vital to the urban decision-making process, both because decision makers require clear presentations of conditions and options, and because urban decisions typically involve a high degree of public participation. Urban planning decisions, and their effects, are more visible to people than regional planning decisions. What happens within walking distance is more important to the individual than what happens in the next county, and neighborhood political activism plays an important role in the urban decision process.

Secondly, the multiplicity of data—spatial and nonspatial—that describes an urban area and its built and natural environment, is exceedingly difficult to understand and manage without GIS. In a dense urban environment, more events occur in smaller spaces than in a suburban or rural setting. But traditional maps combine and display limited sets of spatial and nonspatial information. A road map, for example, displays mapped information about a particular area that interests only drivers of automobiles, including road names, street locations, highway designations, water body areas, and city names. An urban planner looking at the same area may need different information, such as census tract boundaries or median household income within the city. This will require a different map. As a practical matter, it is impossible anymore for traditional maps to meet the demands of urban decision makers, who must have the ability to see the interaction of different combinations of elements within the same space simultaneously. GIS technology can provide this and enable GIS users to produce composite maps on demand, combining information as needed.

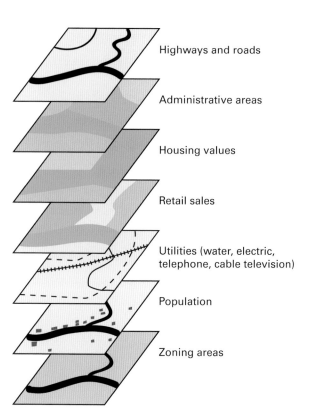

Highways and roads

Administrative areas

Housing values

Retail sales

Utilities (water, electric, telephone, cable television)

Population

Zoning areas

Figure 1.8 The concept of information layers in a GIS: Information about the real world is disaggregated into its constituent layers, each representing a different set of geographic information. Individual layers can then be recombined in GIS and relationships between different aspects of geographic reality become clear. Redrawn from *Urban Geography,* by David Kaplan, James O. Wheeler and Steven R. Holloway. © 2003. Reprinted by permission of John Wiley and Sons.

Layers

The technology does so by organizing the available information, in the form of digital geographic data, into categories and then placing each category into a separate map layer. The user can then overlay any combination of layers to produce a composite map showing, for example, the relationship of subway lines to streets, or crime rates to poverty levels, or environmental hazards to a population's health problems (figure 1.8).

One other reason why this layering process that is at the heart of GIS technology works so well in an urban context is that in an urban environment, people are themselves stacked vertically, in tall build-ings and underground. Accordingly, many urban services also occur within this three-dimensional space. Such 3D GIS applications are discussed in more detail in chapter 12.

Using GIS in an urban context

Urban applications of GIS can be divided roughly into three categories: planning, operations, and public information. All of these involve the use of GIS core technology, but with emphasis on differ-ent capabilities.

Planning involves the assessment of trends and development patterns in both space and time, and the formulation of plans to address future problems and opportunities. Planners look at the past as well as present conditions to arrive at a best guess of what will happen, and where it will happen. Since the future is uncertain, planners typically need to look at a range of scenarios before decisions are made. GIS users in the planning field make extensive use of GIS technol-ogy to produce maps based on differing datasets, for *what-if* analyses of alternatives, and to present options and recommendations to the ultimate decision makers, managers, elected officials, and the public.

Municipal and urban operations include delivery of services, regulatory activities, the ongoing maintenance of infrastructure, and response to natural and manmade disasters. Each of these activities requires up-to-date information—in some cases, up-to-the minute information—because of the continuous nature of municipal operations. This information is distributed in various ways to users in city agencies and in the field, depending on the size, sophistication, and management style of the municipality. In some municipalities, each agency has its own GIS. More often, and more efficiently, a citywide GIS is maintained at a central facility. There, geographic data about the city is processed, maintained, updated, and distributed to field units or remote offices when needed. This can be done through printing of up-to-date hard-copy maps or, increasingly, through local and wide area networks (LANs,

WANs), and—for communication with crews in the field—wireless technology. Operational GIS applications rely heavily on the ability of GIS to continually update data and to redraw maps to reflect changes that come from the field and from other sources.

Public information is an increasingly important function of the urban GIS planner because of the increasing use of the Internet and Internet map server technology, which allows for the dissemination of mapped information to a wide audience, twenty-four hours a day. In many cities, the public is very much involved in urban decision making, particularly in dense urban areas. Some cities use their Web sites to provide mapped information about planning proposals, urban amenities, transportation routes, events, and other geographically based information to the public. Similarly, nonprofit and advocacy organizations have become increasingly GIS-savvy, and are also using Internet map server technology to inform the public about issues that concern them and their point of view on issues that legislators are debating.

Another increasingly important function of GIS technology in the urban planning context centers on providing spatial and attribute information for economic development and by extension, job creation. Businesses—particularly large corporations—need site-selection information before they will bring new facilities to an area and tend to gravitate to areas where that information is more easily available. Real estate developers need geographic information about the surroundings of proposed projects, such as availability of transportation, proximity to cultural facilities and commercial services, and socioeconomic information. Increasingly, local economic development agencies are expected to provide this kind of information to the private sector.

Basic GIS functionality

One reason for the rapid adoption and popularity of GIS technology by a wide variety of organizations is its versatility: it can be many things to many users. From the cartographer's point of view, GIS is a dynamic mapping system that adds speed and currency to the mapping process, prompting university geography departments worldwide to adopt GIS as the standard way to make maps. From the perspective of the IT (information technology) manager, GIS is a means of adding the spatial dimension—the *where* of things—to an organization's nonspatial database. For planners, GIS is a way to combine, analyze, and visualize the various kinds of information that describe a geographic area. For operations managers in departments such as sanitation or transportation, GIS is primarily a way to track developments as they occur in space and time, and to manage ongoing operations.

Each GIS user will design a GIS application and produce a GIS project that meets specific needs, but the process always involves these basic steps:

✦ Data acquisition
✦ Data storage and retrieval
✦ Database management

✦ Data processing
✦ Data display and interaction
✦ Data analysis
✦ Data synthesis
✦ Data presentation

As you may have noticed, all of these steps start with the word *data,* indicating the importance of its role in GIS functionality.

Data acquisition means collecting data from various sources. These sources can be primary (you've collected the data yourself through sampling, interviews, inventories, field work, or surveys), or secondary (someone else has already gathered the data that you are going to use, such as census data or existing maps). Data acquisition is among the first steps of a project. Often, an assessment of specific data needs is done first, to determine exactly what data will be necessary in order to carry out the project's objectives. This process is discussed in more detail in chapter 10.

Data storage and retrieval means getting the data into appropriate computer hardware in digital format and in a way that makes it easily accessible and usable. Storing all relevant databases in one system, so that all geographic information can be seen as individual data layers or as overlays of multiple data layers, is a crucial function of a GIS. Data can be entered by scanning, digitizing, geocoding, linking attribute tables to spatial databases, and by other methods. Chapter 7 discusses this function in more detail.

Database management is an important part of any GIS project, particularly those expected to be always available and which require access by many users. Managing data includes functions such as formatting data, creating linkages, and maintaining, updating, and editing—all for the purpose of keeping the databases current and error-free. These activities are discussed in chapter 8.

Data processing means the creation of structured data—data whose spatial nature can be understood and analyzed by the software. Without its spatial component, GIS is just another database storage and retrieval program. Data processing in GIS, then, turns unstructured data into smart data that can recognize spatial relationships among various geographic features. Data processing can include such work as changing map projections, reclassifying data categories, interpolation, triangulation, and conversion from vector to raster. These terms are defined and discussed further in chapters 2 and 9.

Data display and interaction means displaying spatial data (by creating digital maps) and interactively working with it on a monitor. This ability to display data, change it, and look at data in many different ways in turn enables the data-exploration process. Chapters 3, 4, and 5 discuss in detail the topics of data display and exploration.

Spatial data analysis is, for many GIS users, the heart and soul of GIS, and the one through which GIS reaches its full potential as a decision-making tool. The spatial data analysis function sets GIS apart from other data management computer programs. The process refers to various ways of performing complex queries about geographic data, and of creating models to simulate the behavior of geographic processes. These methods are discussed in chapter 9.

Data synthesis and presentation usually refers to the end result of a GIS project. This is where maps are created that effectively visualize and communicate the data to a wider audience, in support of planning and decision-making efforts. Chapter 5 details the issues of map design and visualization.

Although the order of the GIS functions given above is typical of a real-world GIS project, this book's chapters instead reflect a progression of gradually more complex topics—from basic conceptualization of mapping principles, thematic mapping, and map design, to methods of spatial data analysis and GIS project development. By the end of the book, however, readers should possess all the necessary information and techniques for assembling their own projects, from beginning to end.

Typical GIS tasks

Traditional cartography produces static maps that remain fixed in time and, because they are fixed, they begin to become obsolete as soon as they are finished. By contrast, GIS technology is dynamic: it allows you to generate new maps as information changes over time, even minute by minute if necessary.

Maps contain two kinds of information: spatial—a geographic feature such as a road; and nonspatial—information about that road, such as its name, route number, classification, the number of cars on it, even information about when the road was last resurfaced and by which paving company. A single geographic feature can have many nonspatial attributes; the GIS user has control over the selection and display of both geographic features and their attributes. The mapmaking process, because of GIS, has changed to allow not only those with traditional, special cartographic skills to make maps, but also those with access to GIS technology. This change was accomplished through the addition of two specific technologies to traditional cartography: digital visualization and relational database management. Digital visualization, also referred to as computer graphics, uses GIS software to generate map images in digital format. These images can be manipulated and displayed on the user's monitor or printed out. The relational database, discussed further in chapter 8, enables the user to manage attribute data—the data about geographic features, which is stored in multiple, connected tables, and is linked in a way that allows them to be displayed together.

Importing geographic and attribute data

GIS users typically depend on digital geographic data from federal, local, and proprietary sources. Similarly, attribute databases are often generated by organizations outside an urban government. For this reason, GIS software is designed to facilitate the importation of digital mapping and attribute data.

Creating new maps

When digital maps are not available, the GIS user can create them directly, using GIS software, or by converting existing paper maps into digital form. Techniques for this kind of conversion are described in chapter 7. Once a digital map is created, it is stored in the spatial database and can be linked to an attribute database.

Reconciling map scales and projections

A significant benefit of GIS technology is the ability to reconcile the scales and projections of maps that are found in various archives and cover the same geographic areas, but that are drawn at different scales and for different purposes. When these maps are scanned and converted to digital form, the GIS user can display and print them out at the same scale and in the same map projection. Moreover, this reconciliation facilitates the layering of map information, a function basic to the GIS concept. Map scales and projections are discussed in chapter 2.

Organizing mapped information into layers

As previously mentioned, a traditional two-dimensional map on a piece of paper cannot represent all of the elements that interact with one another within the same geographic area. The information about a piece of geography—particularly urban geography—is almost limitless. Infrastructure, transportation, flood plains, topography, demographics, property boundaries, political boundaries, zoning, land uses, and so on, all share the same physical space. GIS technology solves this two-dimensional limitation by separating different kinds of spatial and attribute information into layers that can be overlaid on each other in any combination. Each layer represents a distinct set of spatial and nonspatial data. The nonspatial data that is attached to geographic features remains behind the scenes—ready to be selected, called up, and placed on the map at the discretion of the user.

Layering is a fundamental feature of GIS and predates it. In the late 1960s, Ian McHarg, a visionary landscape architect and a seminal figure in GIS development, introduced a layering system for environmental planners that used tracing paper overlays to reveal the cumulative effect of various environmental conditions on an area. Even before the widespread use of software, architects and civil engineers overlaid their drawings on transparent tracing paper to visualize how diverse elements, such as floor layouts, mechanical equipment, and elevator shafts, would interact with each other in space after construction. Indeed, an architectural plan can be thought of as a map of an interior geography. Architects and engineers still produce layered plans using computer aided design (CAD) software, and most CAD and GIS software can exchange information through import, export, and conversion functions.

Spatial analysis and topology

Spatial analysis is one of the most powerful GIS functions. These functions allow the user to ask questions of a geographic dataset. Using key words, called logical operators, such as *within*, *contains*, or *intersects*, the user can run queries on geographic features and select those that meet certain stated criteria; these can then be displayed as a new map. For example, an urban planner might want to create a 100-foot buffer around a noise-producing highway on a highway layer, then identify the residential neighborhoods (from a land-use layer) and schools (from an education facilities layer) that fall within that buffer. The query language of spatial analysis would allow the planner to select the houses and schools that meet the criteria and create a new map from that information.

Much of this analytical capability depends on a branch of mathematics known as **topology.** Topology describes the location of objects in space relative to each other. It is concerned with whether an object is adjacent to, connected to, contained by, or containing another object, irrespective of their size or shape. These principles of contiguity, connectivity, and containment are the building blocks of topology. Topology makes possible many of the spatial analyses that are discussed in detail in chapter 9 (figure 1.9).

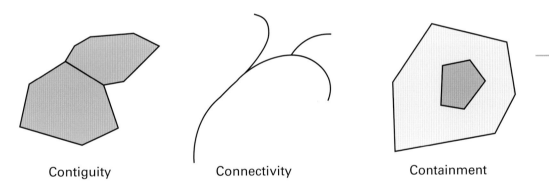

Contiguity   Connectivity   Containment

Figure 1.9 Topological relationships.

Nonspatial attribute database queries

As previously noted, GIS data is composed of both spatial and nonspatial information. Nonspatial data is commonly referred to simply as attribute data. An attribute database can also be queried using the same kind of logical operators used to query spatial datasets. For example, a property tax database might be queried for properties valued at more than $100,000. If there are any properties that meet the criteria, the GIS software will return a map highlighting them. Other standard database management capabilities, such as redefining, reclassifying, and altering attribute data are also available to the GIS user. Queries and reclassification of attribute data are further discussed in chapters 8 and 9.

Linking the map to other images

Although GIS maps, especially when organized in layers, convey a great deal of information, they don't provide the purely visual information that is to be found in photographs and other graphic images. This kind of information is complex and difficult to quantify, but remains of great value to the urban decision maker. Aesthetics, the scale and character of the built environment, street activity, and the current condition of buildings and neighborhoods are some examples of information that must be seen in three dimensions and assessed visually.

GIS software can link photographic and other images to individual features on a digital map. Typically, a symbol on the user interface will indicate that there is a hyperlink to an image; when the user clicks on the symbol, graphics or some other kind of software runs, displaying the photograph. These links are not limited to photographs, but may be other kinds of images, such as graphs, charts or drawings, written reports, or a page from an inspector's log describing site conditions or other information. The use of these kinds of links may become more widespread with the increased use of digital photography and videography.

Connecting GIS and CAD

CAD software is used to create original architectural, mechanical, and engineering drawings. CAD software is optimized for creating and editing detailed drawings, but lacks the topological and cartographic capabilities of GIS. However, CAD drawings can be imported into and exported from most GIS software packages (figure 1.10).

Figure 1.10 CAD drawings can be incorporated into a GIS.

Courtesy of Existing Conditions Surveys, Inc.

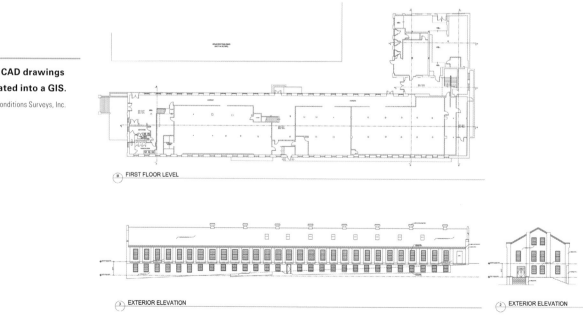

FIRST FLOOR LEVEL

EXTERIOR ELEVATION   EXTERIOR ELEVATION

This capability is important because there are far more CAD users than GIS users worldwide, and drawings of facilities and infrastructure, especially in the urban environment, are commonly produced with CAD software. CAD drawings can be imported as layers in a GIS, or they can be linked to geographic features and referred to through the hyperlink function described above.

Linking objects on the GIS map to CAD drawings provides an efficient way of providing the GIS user with detailed information about specific objects in the urban environment, without imposing engineering standards for accuracy on the entire GIS map.

Three-dimensional GIS applications

GIS can display three-dimensional (3D) models of physical entities such as buildings or terrain, seen in perspective, from any viewpoint. This capability is built in to many GIS software packages and their extensions. Simulated walk-throughs or fly-bys of urban projects and areas can be produced from this data, with the addition of animation software (figure 1.11).

Urban designers and planners may need this capability to assess fully the aesthetics and functionality of urban projects.

GIS is a relatively young discipline: the technology continues to develop and advance, new applications are created, and data is being generated and collected in new ways. As the world continues to urbanize, GIS will become increasingly necessary to urban planners and decision makers, and is expected to have an expanded role in the planning, management, and operation of all urban areas.

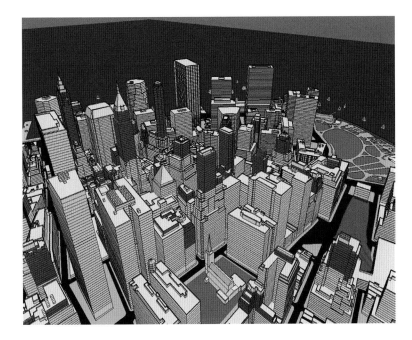

Figure 1.11 3D GIS: building floors colored red meet conditions specified by the GIS user. Reprinted by permission of the Environmental Simulation Center, New York, NY.

Written Exercise 1

GIS on the Internet

Explore some Web sites containing specific GIS projects. Some examples are listed below, and you can do a general Internet search for particular subjects you may be interested in, such as transportation, real estate, or historic preservation.

When you find a project of interest to you that is explained and illustrated fully on the Web site, write a brief report of the project, including the following: the URL, the project name, the researchers' names or institution, the reasons for the study, the time period involved, the data used, where the data came from (if mentioned), the geographic extent of the study area, the software and analytic methodology used (if given), what was done during the study, the results of the research (the findings), and any other pertinent information about the project. Also print out as many of the maps as required to illustrate your report. Your report should be between 500 and 1,500 words.

The Web sites listed below should be considered only as examples of the wealth of GIS information available. Many universities and public agencies, such as the U.S. Environmental Protection Agency (EPA) or the U.S. Census Bureau, are particularly rich sources for GIS information, so spend some time surfing the Web; you can learn a great deal about GIS this way.

Comprehensive GIS Web sites
United States Geologic Survey (USGS)
> *www.usgs.gov*

ESRI
> *www.esri.com*

National Oceanic and Atmospheric Administration (NOAA)
> *www.noaa.gov*

United States Environmental Protection Agency (EPA)
> *www.epa.gov*

Other useful urban GIS Web resources

New York Public Interest Research Group (NYPIRG): Community Mapping Assistance
Project

> *www.cmap.nypirg.org*

World Resources Institute: Where are the poor? Experiences with the development and
use of poverty maps

> *www.wri.org/population*

Relief Web: Emergencies, natural and environmental disasters, armed conflicts, and
humanitarian crises

> *www.reliefweb.int*
>
> *www.reliefweb.int/w/map.nsf/home*

Integrated Data and Workflow Management for Registrars of Voters Using GIS
(Los Angeles County)

> *www.esri.com/industries/localgov/docs/workflowmgt.pdf*

United Nations High Commissioner for Refugees (UNHCR): The Geographic Information
and Mapping Unit

> *www.unhcr.org*

Great Britain Historical GIS Project

> *www.geog.port.ac.uk/gbhgis*

Race Contours (Racial/ethnic changes in the United States)

> *www.usc.edu/schools/sppd/research/census2000/race_census/index.htm*

Urban Data Solutions: 3D urban mapping

> *www.u-data.com*

University of California at Santa Barbara: Project Gigalapolis

> *www.ncgia.ucsb.edu/projects/gig/*

University of Arkansas: CAST (Center for Advanced Spatial Technologies) projects

> *www.cast.uark.edu*

>> Various projects:
>> Mapping agricultural land use in the Mississippi Alluvial Valley
>> National Archaeological Database (then you need to click on one of the projects)
>> Community Asset Development Information Systems (CADIS)
>> Source water assessment and protection
>> National GAP Analysis Program (Biodiversity)

University of Iowa: Department of Public Health Administration projects

www.uiowa.edu/~geog/health/index.html

The Agency for Toxic Substances and Disease Registry (Center for Disease Control): Improving public health through GIS

www.atsdr.cdc.gov/gis/conference98/gisindex.html

U.S. Department of Transportation: Various transportation planning projects

tmip.fhwa.dot.gov/clearinghouse/docs/gis

- Portland Metro, Oregon: GIS database for urban transportation planning
- Orange County (CA) Transportation Authority: GIS for transit planning at OCTA
- Southern California Association of Governments: Access project
- North Carolina DOT: Use of GIS to support environmental analysis during system planning
- Maine DOT: Statewide travel demand model
- San Diego Association of Governments: Multiple species/habitat conservation programs and transportation planning

Using GIS to map crime victim services

www.ojp.usdoj.gov/ovc/publications/infores/geoinfosys2003/welcome.html

References and further reading

Aberley, Doug. 1993. *Boundaries of Home: Mapping for Local Empowerment*. Gabriola Island, B.C.: New Society Publishers.

Allen, John, and Doreen Massey, eds. 1995. *Geographical Worlds: The Shape of the World*. Oxford: The Open University and Oxford University Press.

Berthon, Simon, and A. L. Robinson. 1991. *The Shape of the World*. New York: Rand McNally.

Brody, Hugh. 1998. *Maps and Dreams: Indians and the British Columbia Frontier*. Prospect Heights, Il: Waveland Press.

Chang, Kang-tsung. 2002. *Introduction to Geographic Information Systems*. New York: McGraw–Hill.

Dorling, Daniel, and David Fairbairn. 1997. *Mapping: Ways of Representing the World*. Harlow, UK: Addison Wesley Longman, Ltd.

Dueker, K. J., and D. Kjerne. 1989. *Multipurpose Cadastre: Terms and Definitions*. Falls Church, VA: American Society for Photogrammetry & Remote Sensing and American Congress on Surveying and Mapping.

Hall, Stephen S. 1993. *Mapping the Next Millennium: How Computer-Driven Cartography is Revolutionizing the Face of Science*. New York: Vintage Books.

Harley, J. B. 2001. *The New Nature of Maps: Essays in the History of Cartography*. Baltimore: Johns Hopkins University Press.

Harley, J. B., and D. Woodward, eds. 1988. *History of Cartography, Vol. 1*. Chicago: University of Chicago Press.

Kaiser, Ward, and Denis Wood. 2001. *Seeing Through Maps: The Power of Images to Shape Our World View*. Amherst, MA: ODT, Inc.

McHarg, Ian. 1971. *Design with Nature*. Garden City, NY: Doubleday.

Mitchell, Andy. 1997. *Zeroing In: Geographic Information Systems at Work in the Community*. Redlands, CA: ESRI Press.

Monmonier, Mark. 1993. *Mapping It Out: Expository Cartography for the Humanities and Social Sciences*. Chicago: University of Chicago Press.

Parrott, R., and F. P. Stutz. 1991. "Urban GIS Applications" in *Geographical Information Systems: Principles and Applications,* edited by Maguire, D. J., Goodchild, M. F., and Rhind, D. W. Harlow, UK: Longman.

Plane, D. A., and P. A. Rogerson. 1994. *The Geographical Analysis of Population with Applications to Planning and Business*. New York: John Wiley and Sons.

Schulten, Susan. 2001. *The Geographical Imagination in America, 1880–1950*. Chicago: University of Chicago Press.

Tuan, Yi-fu. 1974. *Topophilia: a study of environmental perceptions, attitudes, and values*. Englewood Cliffs, NJ: Prentice–Hall.

Whitfield, Peter. 1988. *New Found Lands: Maps in the History of Exploration*. New York: Routledge.

Wilford, John Noble. 2000. *The Mapmakers*. New York: Vintage Books.

Wood, Denis. 1992. *The Power of Maps*. New York: The Guilford Press.

Woodward, D., and G. M. Lewis, eds. 1988. *Cartography in the Traditional African, American, Arctic, Australian, and Pacific Societies*. Chicago: University of Chicago Press.

Spatial Data *and* Basic Mapping Concepts

Maps can be described as scale models of reality, and the information about the physical world that is used to construct such models is known as spatial data. Like all models, GIS maps are also simplified representations of reality, limited by the availability of data and defined by design decisions that are made by the GIS user—decisions that require a thorough understanding of spatial data models and mapping principles. In this chapter, we will discuss these concepts and describe the two principal spatial data model types that are used in GIS mapping.

Modeling

Spatial data describes the location and dimension of geographic entities with reference to a map space representing the real world; you need this digital data to create a digital GIS map. Before it can be used, the raw data must be organized, or modeled, into a form that enables it to be stored and displayed as a map or image. The spatial data models that are commonly used in two-dimensional GIS mapping are the **vector model** and the **raster model.** Others, including three-dimensional models, will be discussed in chapter 12.

The vector model

The two-dimensional vector model classifies geographic entities as **points, lines,** and **areas;** areas are also known as **polygons.** All geographic features in vector format are composed of these points, lines, or polygons, and each feature contains information about its geometry, dimensions, and topology—its relationship to other features (figure 2.1).

Point segments are designated in a map space by a pair of x,y coordinates. In a GIS map display, point features may be represented by symbols, such as triangles or squares; by pictograms, such as a flag that represents the location of a government building; or perhaps by a small airplane, to represent the location of an airport.

Line segments are defined by their beginning and end points. The shape of a line is determined by the location (x,y coordinates) of these points, and the intermediate points that connect the straight segments of a line. The more x,y coordinates (or points) describing a line, the smoother that line will appear. Lines are also accorded directionality in a vector GIS representation, in order to assign left or right topological designations. This allows the GIS to determine where the line is in relationship to adjacent polygons.

Polygon features are made up of line segments joined together to form enclosed spaces, such as ZIP Codes or census tracts. The line that encloses a polygon has the same beginning and ending points—the same x,y coordinates.

Note that in the vector model, points located by x,y coordinates in a map space are the basic building blocks of the vector map, because line segments are defined by their beginning and end points, and polygons, in turn, are bounded by lines.

A powerful feature of the vector spatial data model is its flexibility. Much vector mapping data is available for geographic units of various sizes, such as census tracts and counties, with attribute data attached to it. The GIS user can choose the geographic unit that best suits the requirements of the analysis. For example, you may decide to create a map of population density, organizing that information by ZIP Code, by neighborhood or by county, depending on the availability of population data and the intended use of the map. The geographic unit that you select, defined by boundary lines, will become the spatial unit for aggregating and analyzing the data in the GIS map. This unit is assigned a unique identifier for linking to nonspatial database tables.

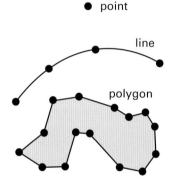

Figure 2.1   In vector formats, points, lines, and polygons represent spatial features.

Another powerful feature of the vector spatial model is the built-in topological structure of the point, line, and polygon features. Topology deals with the relationship of spatial objects to each other and it is this property that allows each feature in the vector spatial model to relate to other features. Each feature in the vector spatial model "knows" where it is in relationship to other features in terms of contiguity, connectivity, and containment. Topology allows you to query the spatial data about these relationships—to determine, for example, a neighborhood's relationship to a park, or a park's relationship to a road.

Vector mapping is basic to nearly all urban GIS applications because features such as roads, buildings, city blocks, and census tracts can be defined as points, lines, or polygons, and related to their nonspatial attributes, such as population or rates of poverty.

The raster model

Other kinds of geographic data will not fit the vector model. Examples include aerial photographs, remotely-sensed data from satellites, or even video. Image-based data sources such as these require another model, called **raster.** The raster model divides a map space into a regular grid of cells, usually square and of identical size. Each cell is assigned a quantitative value that represents the presence (or intensity) of the real-world attribute being measured. This could be temperature, rainfall, concentrations of chemicals in the land, or crop type. In the case of photographs and other remotely sensed images, the value being mapped is reflected light waves and other electromagnetic phenomena. The cell size defines the spatial resolution of the data (figure 2.2).

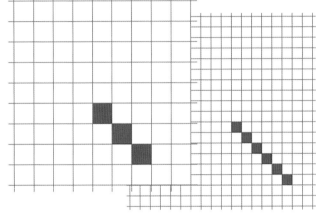

Figure 2.2 Grid cells form the basis of the raster format.

Satellite and other remotely sensed images, for example, are classified by meters (or feet) of resolution. An image with one-meter resolution is composed of grid cells in which each cell represents one square meter on the ground.

In the raster model, the values that are attached to the individual cells are averaged out within each cell to produce a single set of values. In the case of an image with one-meter resolution, each square meter on the ground will be a unit in the spatial data model and will be represented by a single set of values. The basic spatial unit in the raster model is the grid cell. Raster datasets do not use topology as vector datasets do, which can be a limitation in certain types of analyses.

On the other hand, raster images are especially useful when conducting any investigation of physical changes to the landscape over time, since raster data, such as satellite imagery, is collected much more frequently than is vector data.

Raster image formats are identified by their file extensions, and include Tagged Image File Format, or TIFF; bitmap, abbreviated BMP; and Joint Photographic Experts Group, commonly known as JPEG or JPG. Any of these formats can be imported into a GIS software package.

Combining raster and vector

Because raster images often contain visual information that cannot be captured by vector mapping, and because photographs and satellite imagery may be more up-to-date than the available vector data, it is often desirable to add raster layers to a vector map display. When the GIS application is time-sensitive, as in disaster response, it may be imperative to combine the two.

Combining raster and vector images requires that the cells in the raster image be moved and stretched so that the image overlays correctly on a geographically accurate vector map. This process is known as **georeferencing.** Much of the available raster data—such as aerial photography and satellite imagery—is already georeferenced to a latitude/longitude coordinate grid. If the raster data is not already georeferenced, most GIS software contains **rubber-sheeting** capabilities that allow the user to georeference raster images using control points that correspond to points on the vector layer. Rubber-sheeting software stretches the pixels in a raster image to conform to known reference points on a vector map. Once the rubber-sheeting process has been completed, features on the raster layer can be converted to vector format by tracing them onto a vector layer—a process known as **heads-up digitizing.** Alternatively, it may be easier to change the vector spatial data to match up with the raster data so the datasets can align properly. This is accomplished by changing the projection, scale, and other attributes of the vector data so that it registers properly with the raster image. Map projections are discussed later in this chapter.

One common problem when using raster images in GIS mapping is their very large file size. In the raster model, the values for every cell must be stored. In the vector model, only values for the map's point, line, and polygon features are stored, together with geometric and topological information.

Nevertheless, the use of raster imagery is increasingly popular in GIS applications, for the reasons noted above. Also, data compression software is helping to overcome its storage limitation, as is the steady improvement in the power and availability of computer hardware.

When using a vector-based software program, raster files are often incorporated simply as back-drop material, to supplement and add context to the vector data. However, raster data is also useful for spatial analysis, as many processes require grid cell-based spatial data.

Continuous vs. discrete data

Spatial data can also be divided into two other categories: continuous and discrete. **Continuous data** is information that occurs everywhere on the surface of the earth. Examples of continuous data are readings of air quality, temperature, or elevation. Continuous data lends itself to the raster spatial data model, and is usually better at representing spatial and attribute transitions than discrete data. **Discrete data** lends itself better to the vector data model. In this book and in the lab exercises, you will be working mainly with vector data.

Discrete data refers to phenomena that occur only in selected places on the earth's surface and which can occupy only one space at any given time—for example, land use, land ownership, the assessed value of individual property lots, or the median income of urban neighborhoods by census tract. Discrete data is defined by boundaries and sharp lines of demarcation, separating one value from another, often based on the unit of data aggregation, such as ZIP Code or census tract. Discrete data can be point, line, or polygon features, to which attribute data can be attached (figure 2.3).

In an urban setting you are most likely to encounter continuous datasets in representations of the natural environment—air quality, elevation above sea level, precipitation, and the like. Some natural features, however, such as floodplains or wetlands, have discrete boundaries or occur only in selected areas, and so can be adequately described by discrete data also. The built environment, political and administrative districts, and associated attribute data about socioeconomic characteristics or demographics are usually described by discrete data. Some types of population data, however, can also be represented well by continuous (grid cell) data. Population density, for example, does vary continuously and exists everywhere on earth.

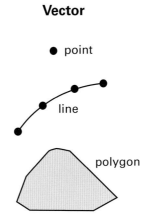

Vector

● point

line

polygon

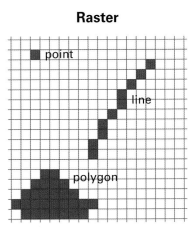

Raster

point

line

polygon

Figure 2.3 **The same data is represented differently in vector and raster formats; the diagram also reflects the corresponding difference between discrete and continuous data.**

Scale

In order to properly interpret information on a map, we almost always need to know something about the scale of the map. Scale refers to the relationship between the size of a feature shown on the map and the size of the feature in the real world. Without knowing the scale of the map, we have no way of being able to estimate distances between places, the size of individual features, or, indeed, how much territory the map covers. If a county is shown as five inches wide on the map, how big is it in real life? Without an indication of scale on the map, we would have to know in advance the size of places of interest in order to make sense of the map.

Map scale is defined as the ratio between map distance and real distance. If our hypothetical county is shown as being five inches wide on a map, and the county is ten miles wide in reality, then the map scale can be stated as "one inch equals two miles." This is an example of a **verbal scale,** or **word statement.** Another way of expressing the scale of this map would be to say the map is at a scale of 1:125,000. This method is called a **representative fraction,** or ratio. Still another way of providing scale information is through a drawing of a scale bar, called a **graphic scale.** These three commonly used methods of communicating information about scale on a map are equivalent. All are correct and are often used together on the same map (figure 2.4).

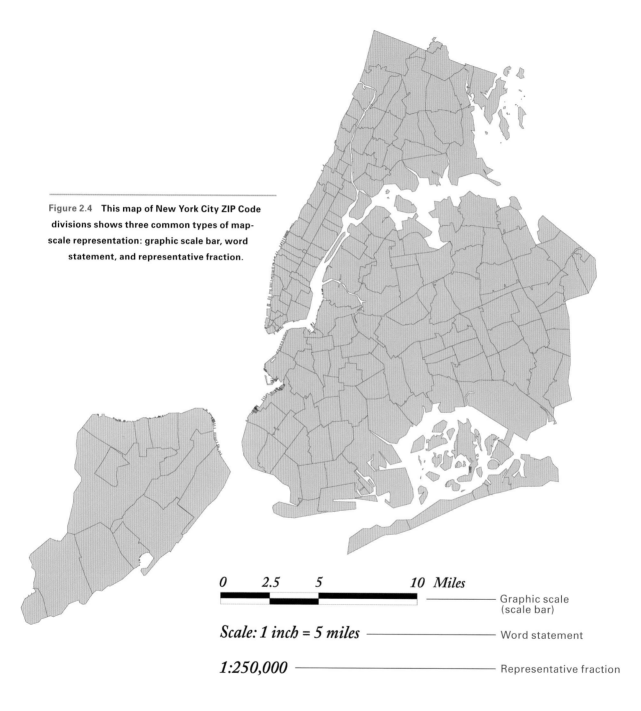

Figure 2.4   This map of New York City ZIP Code divisions shows three common types of map-scale representation: graphic scale bar, word statement, and representative fraction.

0   2.5   5   10 *Miles*
————————————————— Graphic scale
                  (scale bar)

Scale: 1 inch = 5 miles ——————————— Word statement

1:250,000 ————————————————— Representative fraction

Graphic scale is probably the method that is most familiar to the average person, and it is used on most roadmaps and maps in atlases. A graphic scale is a drawing of a bar or line with lengths marked off, starting with zero, and numbers that correspond to each increment, such as 5, 10, and 15, or 100, 200, and 300. Next to the bar or line is generally text indicating the unit of measurement, such as miles or kilometers, in intervals suitable to the map's scale. The lengths of the increments correspond to distances on the map. The bar often continues to the left of the zero, with the major interval unit broken down into smaller intervals. For instance, if the major interval to the right of the zero is ten, the bar to the left of the zero may be divided into units of 5, 2.5, or even 1. The graphic scale bar lets the map user measure an actual distance or feature on the map and compare it to the scale bar, and from that, estimate real-world distance or size. If the map is enlarged or reduced, the scale bar is reduced or enlarged along with it, ensuring that the scale bar is still valid to make accurate measurements.

The word statement or verbal scale method does not use a bar or line, but is simply a sentence fragment expressing the relationship of the distance on the map to distances in the real world, such as, *1 inch equals 100 miles,* or *1" = 100 mi.* This means, of course, that one inch on the map represents 100 miles in the real world. This is very easy for the average person to understand and to translate from the conceptual to the real world. Unfortunately, if the map is enlarged or reduced, the word scale is no longer valid, and this is a major drawback to this method.

The third method is the representative fraction (RF), which is a ratio showing the proportion of one unit of measurement on the map to some number of the same unit of measurement in the real world. The RF *1:100,000* means that one unit of measurement on the map represents 100,000 units of the same measurement on the ground. The representative fraction is usually the preferred choice of cartographers and geographers because it denotes standard scales to people who work with maps. Commonly used scale factors are 1:24,000, 1:100,000, and 1:1,000,000. Most map-literate people will understand immediately from looking at the RF the level of detail that each map will show, regardless of the actual size of the map. A map at 1:24,000 scale converts to a word scale of "1 inch equals 24,000 inches," or "1 inch equals 2,000 feet" (24,000 inches ÷ 12 inches = 2,000 feet). A map at 1:1,000,000 converts to a word scale of "1 inch equals 16 miles," approximately. We reach this figure by calculating that one inch on the map represents one million inches on the ground. Dividing by the number of inches in a foot (12), and then by the number of feet in a mile (5,280) results in about 16 miles.

Another advantage of the representative fraction is that it is unit-free; it is not dependent on any specific unit of measurement. An RF of 1:24,000 could mean 1 inch to 24,000 inches, or 1 centimeter to 24,000 centimeters, or 1 foot to 24,000 feet. Any unit of measurement will be valid as long as the same unit is used for both parts of the equation. Despite these advantages, the representative fraction can still be difficult for many map readers to conceptualize. And, as with the word statement scale, if the map is enlarged or reduced, it is no longer valid.

Most maps contain one or more of these methods. Some thematic maps of widely familiar geographies such as an outline map of the United States may omit scale information altogether, on the assumption that most viewers will understand the scale intuitively; in most cases, scale is not critical to understanding a thematic map. (Thematic mapping is explored more fully in the next chapter.)

Large-scale vs. small-scale

Large-scale and small-scale are terms frequently used in describing maps. These terms are strictly relative, and there are no actual numerical thresholds separating large-scale maps from small-scale maps. Generally speaking, a large-scale map will give the map reader more detail about a smaller area of the earth, and a small-scale map will give the map reader less detail about a relatively larger amount of the earth.

Therefore, a map's scale will give you clues about how much detail is likely to be found in the map. For instance, a map in which one inch represents one mile is obviously going to provide much more detail than one in which one inch represents 1,000 miles. We would compare the scales of these two maps by saying the 1″ = 1 mile map is a large-scale map, and the 1″ = 1,000 miles is a small-scale map. Some people also include the term intermediate or medium scale to denote a scale midway between the two extremes. These terms are strictly relative, however, and do not refer to any absolute thresholds or categories.

The U.S. Geological Survey (USGS) uses the following definitions: large-scale is considered 1:25,000 or larger; intermediate-scale is 1:50,000 to 1:100,000; and small-scale is 1:250,000 or smaller—for instance, 1:1,000,000. We can also add a very large-scale category, for instance, 1:1,000, and a very small-scale category, such as 1:100,000,000.

The terms large and small when describing scale are derived from, and refer to, the denominator in the representative fraction: a small-scale map will have a scale consisting of a fraction representing a small number (1:1,000,000) while a large-scale map (1:10,000) will have a scale consisting of a fraction representing a relatively larger number. The larger the number in the fraction's denominator, the smaller the actual number is, and therefore, the smaller the scale of the map is likely to be. The reverse is also true.

Large-scale maps generally cover a small amount of area in the real world, but show a large amount of detail, whereas a small-scale map generally shows a relatively larger portion of the world, but with much less detail (figure 2.5).

Again, these terms are rather arbitrary, and are not intended to provide much information beyond comparison.

Most planning activities are carried out using large-scale maps, ones showing a relatively small portion of the earth with a lot of detail. That is because most planners' geographic area of interest is a community, county, or urban area. However, some would classify regional scale maps, ones showing multiple states or counties, as intermediate- or small-scale maps. Again, the terms are relative, and on any given project a planner may be dealing with architectural plans of an individual building or a building complex—which are decidedly large-scale—while at the same time working with plans of the entire city, which in this context might be considered small-scale maps.

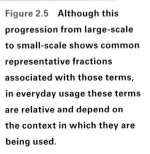

Figure 2.5 **Although this progression from large-scale to small-scale shows common representative fractions associated with those terms, in everyday usage these terms are relative and depend on the context in which they are being used.**

Large scale—1:24,000
Portion of Bronx County, New York

Intermediate scale—1:75,000
Bronx County, New York

Small scale—1:2,500,000
Counties of New York State

Small scale—1:15,000,000
Counties and states of the United States

Temporal scale

When examining spatial data that shows change, trends, movements, or other dynamic phenomena over time—especially with environmental or demographic data—we also must specify the temporal scale. Temporal resolution does not only entail the period of change being measured, but also reflects the time period of data capture. For instance, does the data reflect information about one day, one year, or ten years? How often is the data updated? With temporal data, the level of resolution is equal to the frequency of data collection, with higher frequency of data collection denoting better temporal resolution.

For example, data collected every day, such as remotely-sensed data used to track wildfires, is of a higher temporal resolution than data collected every month, such as satellite images that are used to monitor climatic changes in vegetative cover. Data for many reference maps, such as USGS maps, are collected and updated every few years. Thus, these maps, at a scale of 1:24,000, have high spatial resolution, but low temporal resolution. Conversely, data collected by AVHRR (Advanced Very High Resolution Radiometry) satellite collects data every twelve hours, and thus has a high temporal resolution, but a relatively low spatial resolution of one kilometer per grid cell.

Generalization

When the scale of a map is drastically reduced—from a wall-size map of the world to a world map that will fit in a pocket calendar—detail must be omitted. This requires a process called generalization. There are several ways to generalize geographic data: elimination, reduction, exaggeration, enhancement, collapse, amalgamation, and displacement (Jones 1997) (figure 2.6).

All maps are generalized to some extent, generalization being the process of simplifying and reducing the spatial (and textual) information on a map. By definition, a map is a generalized version of the real world. If it were not, then a map would be at the same scale and size as the real world, which would be impractical and not very useful. Additionally, even if a map were to be created at the same size as the real world that it represents, it would still need to be generalized, since it would be impossible to include every detail of the real world on a map. Mapmaking is by necessity an editing process: the mapmaker must decide what to leave in and what to take out, so that all the desired information fits on the size available for the map. Small-scale maps are much more generalized than large-scale maps (figure 2.7).

When reducing the scale of the map, generalization techniques are necessary in order to reduce the amount of information that will be provided on the map. The main objective is to make the map easily readable. If there are too many unnecessary details, the map will be cluttered, making it difficult for the viewer to discern one feature from another and thus diminishing its overall usefulness.

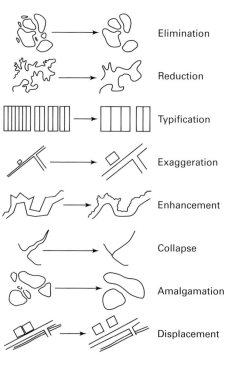

Figure 2.6 Common ways of generalizing map features. Redrawn from Jones, C., *Geographical Information Systems and Computer Cartography*, 1st Edition, © 1997, pp.5, 274. Reprinted by permission of Pearson Education, Inc., Upper Saddle River, NJ.

Techniques of generalization are sometimes referred to as cartographic license, a term derived from the concept of poetic license. Cartographic license gives the cartographer freedom to rearrange, omit, exaggerate, and aggregate spatial features. This is done with an eye towards improving the overall clarity of the map, although the absolute precision of the information may be reduced in the process. In most cases, generalization is a real necessity, as leaving all the information on the map will obscure other features and labels, rendering the map useless.

We can generalize information on a map from a large scale to a small scale; this is a reduction process, in which we subtract information from the map. But we should never work in the opposite direction, from a small-scale to a large-scale map. It is acceptable to edit out what already exists on a map, but not to edit in what we don't have data for. In other words, we should not add information to a small-scale map to give it the appearance and detail of a large-scale map. We can never add detail to a map by a speculative process. This type of embellishment goes beyond cartographic license; it crosses the line over to the more serious act of falsifying spatial information. It is inaccurate and misleading, and should never be done.

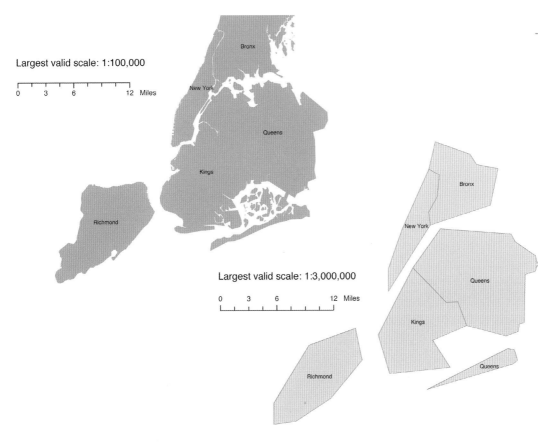

Largest valid scale: 1:100,000

0   3   6   12 Miles

Largest valid scale: 1:3,000,000

0   3   6   12 Miles

Figure 2.7 Two views of New York City, with the same geographic extent and projection, displayed at the same scale and size, but with different degrees of spatial resolution. The difference in the amount of detail available has to do with the original scale at which the data was captured. The bottom map uses data captured at a much coarser resolution (a smaller scale), and will never be able to show as much detail, no matter how big you blow up the map. The top map is able to show much more detail because the original data was captured at a higher level of resolution (a larger scale). The top map is valid at scales of 1:100,000 or smaller, while the bottom map is valid at scales of 1:3,000, 000 or smaller. Therefore, the bottom map is much more generalized.

Geographic extent, geographic unit of analysis, and data aggregation

The portion of the earth shown in a map is called its **geographic extent.** (Geographic extent is another way of specifying relative scale.) The geographic extent of a map specifies the size of the area of analysis and what is contained within it; it is understood that we have not analyzed the data outside of that extent. If I am doing an analysis of all the school districts within New York City, then my geographic extent is New York City.

A map's **geographic unit of analysis,** also referred to as its resolution, is the smallest unit for which data has been gathered or is available. For instance, on a map of the United States that shows median household income for each of the fifty states, the unit of analysis is the state. But if we have a map of one or more states, showing median household income for each county, then the unit of analysis is the county. If, within a county or city map extent, we are showing median household income by census tract, then the census tract is the unit of analysis, and therefore also the map's level of resolution. In my hypothetical analysis of school districts in New York City, my unit of analysis is the school district. This is the same as saying the study's resolution is the school district.

We also say that the data is **aggregated** at the state, county, or tract level, or other geographic or statistical unit (figure 2.8).

This means that the data has been averaged or summarized within the specified geographic unit of analysis. Aggregation is often dependent on the effectiveness of data collection. Planners usually must work with the data they can obtain easily, and are generally not in the position of collecting their own (primary source) data. This limits the unit of analysis to available datasets. Although you can summarize data from smaller units of aggregation to larger units (from county to state, for example), you cannot add detail to a map

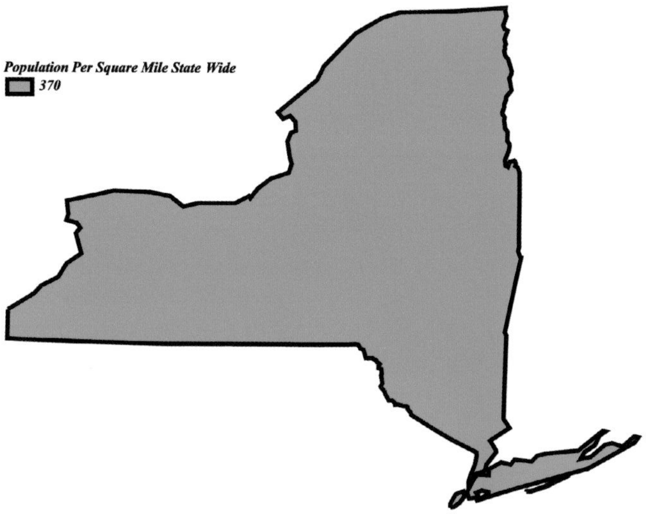

Population Per Square Mile State Wide
370

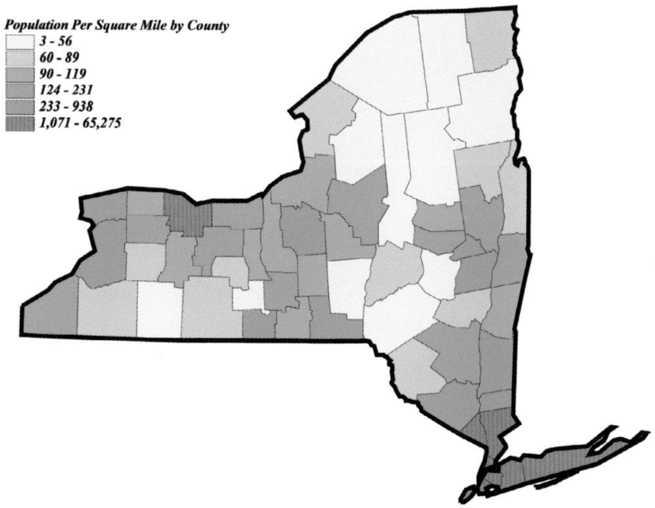

Population Per Square Mile by County
3 - 56
60 - 89
90 - 119
124 - 231
233 - 938
1,071 - 65,275

Figure 2.8   The geographic extent—scale—in both of these maps is the same: New York state. But the unit of data aggregation—resolution—in the top map is the state, and in the bottom map, the county. Higher resolution of units of aggregation will yield a more detailed view of the data.

by disaggregating data from the county level to the tract level. Thus, the aggregation process of attribute data is similar to the generalization process of spatial data: one can combine and average data from smaller units to larger ones, but not the reverse.

When beginning a GIS project, it is important to select a geographic extent and a unit of analysis that will be appropriate to the study. In some cases, these selections are obvious or are predetermined by the problem to be solved or the data at hand, but other times it requires a thorough evaluation of the project design and expected outcomes. For instance, if you are conducting a study to determine where to locate new AIDS prevention and treatment centers within a given metropolitan region, then your geographic extent is probably going to be the municipality, and perhaps some of the surrounding suburban areas, if appropriate. You would need to determine exactly which suburban areas should be included, and the distance from the city center or city limits that the study would cover. These decisions would be based on the availability of data for all possible geographies; the mandate of the partners or collaborators on the project; funding sources and parameters; the physical integration, transportation systems, cultural factors, and social dynamics of the particular metropolitan region; the geography of the existing health care infrastructure, and many other considerations.

Deciding on the unit of analysis can also be challenging. The most important or most pertinent databases for a given study often set the limitations for the unit of analysis to be used. If, for this study, your AIDS case data is aggregated by census tract, then you will most likely need to use the census tract as your unit of analysis. If possible, you would try to obtain all other necessary data, such as demographic and socioeconomic data, at that spatial resolution. But if your AIDS case data was only available at the level of public-health districts, that could present problems because the corresponding demographic data you need may not be available at that level. You may need to use the health district data for the AIDS cases, but use data from ZIP Codes or census tracts for demographic analysis. (In New York census tracts are smaller than health districts.) To resolve this problem, you would need to determine whether the census tract boundaries were within the health district boundaries, and if so, that would let you combine data for the several census tracts that comprise a health district, and then summarize or average the data so that AIDS case data and demographic data could be directly compared. Alternatively, you could choose to analyze the demographic data as it is, at a finer level of resolution than would be possible with the AIDS case data. This would, however, make direct comparison difficult and many statistical tests impossible. Remember, it is not an option to disaggregate the AIDS case data from the health district into a smaller geographic unit of analysis.

Planners must come to terms with the issue of multiple resolutions of datasets—that there can be many different units of data aggregation within one project. Data that you need to use may have been gathered and aggregated by various dissimilar geographic units. In our school district example, in addition to data at the school district level we may need to analyze demographic and socioeconomic data at the census tract level, crime data at the police precinct level, or health data at the health district level. New York City is a good example of this problem. The city is divided into more than two dozen different sets of political, jurisdictional, and service units at the federal, state, and city levels, and most agencies use unique districting plans for administrative and service purposes. In many cases, these administrative boundaries have long historical antecedents and are not easy to change and make consistent. Sometimes administrative boundaries reflect the locations of fixed infrastructure particular to that agency, such as hospitals or firehouses, creating another barrier to congruent boundaries. In addition, there are nongovernmental districts (natural or commercial), such as the area within a certain watershed basin or a marketing catchment area, by which data may be aggregated. These disparate geographic units are not nested hierarchically, for the most part, and none of the boundaries of these areas coincides, creating problems in overlap among districts when analyzing datasets with different geographic units (figure 2.9).

This is a problem common to all municipalities due to the different jurisdictions of city, county, state, and federal governments.

Figure 2.9 There may be dozens of non-coincident units of data aggregation in a given city, reflecting the variety of political and administrative boundaries. Analyzing data from multiple datasets will require overlay analysis. These are just a few of some common units of data aggregation in New York City.

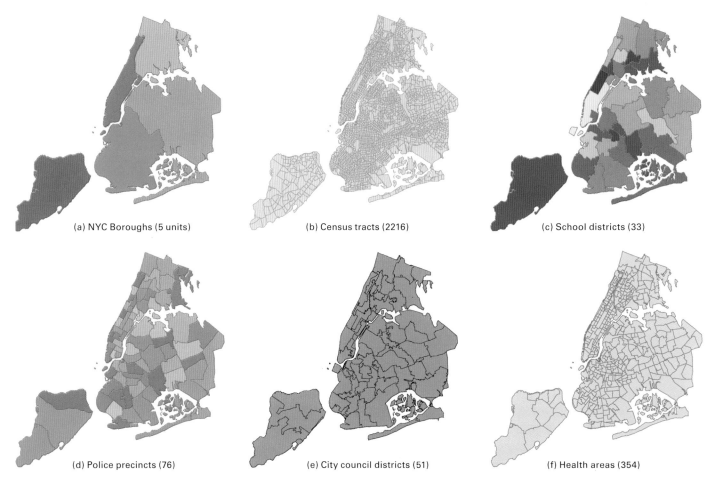

(a) NYC Boroughs (5 units)

(b) Census tracts (2216)

(c) School districts (33)

(d) Police precincts (76)

(e) City council districts (51)

(f) Health areas (354)

Map projections and coordinate systems

In order to make a map, we need to take information from an irregular three-dimensional spherical object—the earth—and transform the information as accurately as possible to a two-dimensional flat surface, the map. The result of this transformative process is called a **projection** (figure 2.10).

Projections use mathematical formulas to transform solid geometry coordinates to plane geometry. Each type of projection method does this in a different way, producing different versions of the same information, with each having a certain amount and type of distortion.

Every flat map of the spherical world is seriously distorted. The flattening of a sphere produces shearing, compression, and tearing of the spherical surface, creating the distortions that every flat map exhibits. The limitations of each kind of projection will dictate what the projection is best used for (figure 2.11).

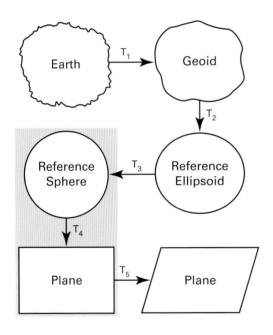

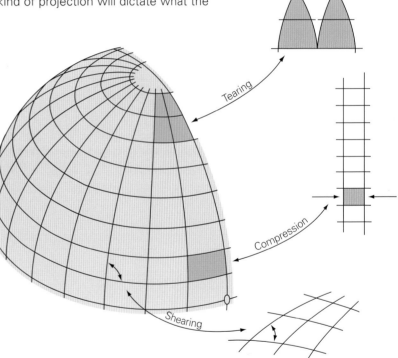

Figure 2.10 **Making a flat map requires several major transformative steps (T): a scale transformation and generalization process are required to convert the actual size and irregular shape of the Earth to a regularized spherical globe (T1 through T3), and then a geometric transformation is required to convert the spherical globe to the flat plane of a map (T4 through T5).** Redrawn, courtesy of JP Publications. From *Map Use: Reading, Analysis, Interpretation (4th Edition)*. By Muehrcke et. al. Reprinted by permission of JP Publication, Madison WI © 2001. JP Publications™.

Figure 2.11 **In transforming the spherical globe to a flat plane, the surface of the globe is torn, compressed, and sheared.** Redrawn from *Cartography: Thematic Map Design*, by Borden Dent. Reproduced with permission of The McGraw–Hill Companies © 1999.

Urban planners generally work with relatively large-scale maps, since they are working with only a small extent of the earth's surface, and the curvature of the earth is not a critical factor. By contrast, on small-scale maps the curvature of the earth can have a significant impact on actual location of features, as well as their accurate graphic representation. Nonetheless, it is important that planners understand the basic principles of projections, especially since planning projects may include spatial data in various projections and coordinate systems. A planner must know how to make all the datasets consistent, and what the implications are for using one projection or coordinate system over another. This remains true even though most GIS software, including ArcGIS, can re-project data on the fly with minimal input from the user.

The need for this knowledge will become obvious the first time you take readings from a GPS device and try to create a spatial database from the resulting coordinates, or when you try to manipulate an aerial orthophoto that is in a different projection than the rest of your data, or when you try to translate decimal degrees to **Universal Transverse Mercator (UTM)** coordinates (discussed in detail on page 49). However, as a planner, you will probably never have to create a new projection or even understand the algorithms and mathematical framework of an existing projection, so the discussion here emphasizes the basics.

Latitude and longitude coordinate system

Projections depend on an accurate and precise coordinate system, defined as a set of numbers that locate a point on the earth's surface. Some coordinate systems are applied worldwide, and some are strictly local. The need for such systems is clear from looking at the shape of the spherical earth.

As this sphere has no natural starting or ending point, people had to develop a system to pinpoint the locations of features, so they could communicate information accurately to each other and be able to locate geographic features reliably in the future. A system that could do this would be an obvious benefit to navigation. The development of a consistent and accurate system of georeferencing was, in many respects, the beginning of what we think of as the modern world.

Although there are a number of georeferencing systems in existence, and these include street addresses, postal codes, and the U.S. military grid reference system, the **latitude and longitude coordinate system** (sometimes shortened to *lat–long*) is the dominant one worldwide. The story of the creation of this system, involving theoretical and practical discoveries in astronomy, physics, meteorology, mathematics, navigation, mechanical engineering, and clock-making, is a fascinating one. The system's development and refinement took at least three thousand years and encompassed many of the world's most significant scientific and technological advances (figure 2.12).

The lat–long coordinate system is premised on the fact that the roughly spherical shape of the earth can, like a circle, be divided into 360 degrees east and west and 360 degrees north and south, with each hemisphere containing 180 degrees. Each degree of latitude or longitude is further divided into 60 equal increments called **minutes,** and each minute is divided into 60 equal increments called **seconds.** The system forms a grid of squares, called a **graticule,** cast like a net over the spherical earth. As long as a starting point has been set for the lines of the grid in both directions, you can pinpoint within feet any location on earth.

How are degrees calculated on the earth's surface? The lat–long system is based on several imaginary lines. The earth turns on its axis of rotation, which is an imaginary line that passes through the center of the earth, connecting the north and south poles. An imaginary plane runs through the earth's center, perpendicular to the axis of rotation, and this is called the **equatorial plane,** corresponding to the place where the sphere is fattest. One can determine the correct latitude of a given location on earth by calculating the angle formed by two lines: the line connecting the point's location on the earth's surface to the hypothetical center of the earth, and the line formed by the equatorial plane that intersects and is perpendicular to the axis of rotation.

Figure 2.12 The latitude–longitude coordinate system is the most common method of georeferencing. *Below:* **The earth is divided into 360°, with lines of latitude and longitude forming an imaginary grid over the earth's surface.** *Above right:* **Any point on the earth's surface can be located using latitude and longitude coordinates, which measure angles formed by the point's intersection with the earth's axis of rotation and the equatorial plane.**

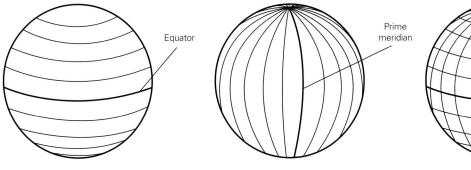

Parallels of latitude

Meridians of longitude

Graticular network

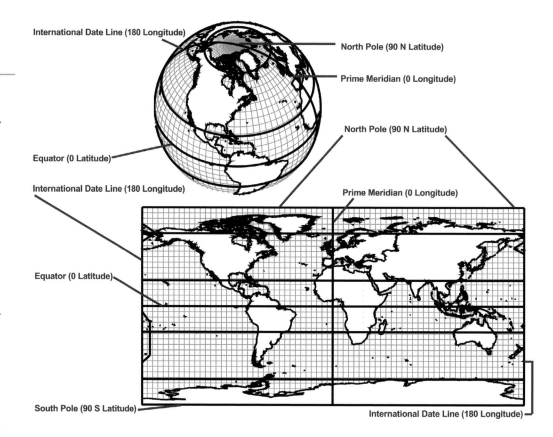

Figure 2.13   The major components of the graticule: Lines of latitude and longitude, the equator, the north and south poles, the prime meridian, and the international date line. It is easy to see here why projections embody so much distortion: The top and bottom lines of the flat map are shown as the same length as the equator, but in reality, as seen on the drawing of the globe, the top and bottom lines represent the poles, and are merely points with no dimension. Therefore, the top and bottom lines have been stretched to fit the rectangular and flat format of the map, resulting in significant distortion in the high latitudes (areas near the poles). ESRI Data & Maps (Juliana Maantay).

The imaginary grid that covers the earth, the graticule, is composed of two sets of lines: Lines of latitude and lines of longitude (figure 2.13).

The imaginary lines going around Earth in an east–west direction are parallel to each other, and are called parallels, or lines of latitude. Lines of latitude run east–west, but measure positions from north and south of the equator. Encircling Earth's thickest middle part is the line of 0° latitude, also known as the equator. North of the equator, lines of latitude extend from 0° to 90° north (the north pole). South of the equator, the lines extend from 0° to 90° south (the south pole). Lines north of the equator are measured in positive numbers, and lines south of the equator are measured in negative numbers. Alternatively, latitude readings are given with an "N" or an "S" after the number to denote whether the line is north or south of the equator. Lines of latitude are unequal in length to each other, with the equator (0° latitude) being the longest line of latitude, and the poles (90° latitude) having no length—the poles are considered to be points. The parallels at 45° are roughly half the length of the equator.

The north–south lines of longitude extend from north pole to south pole and are called meridians. Lines of longitude run north–south, but measure positions from east and west. Meridians are always at right angles to the lines of latitude, and all meridians are of equal length, all extending from pole to pole. The line of longitude that passes through the Royal Observatory in Greenwich, outside London, was established as 0°, and is referred to as the prime meridian. The numbers of the lines of longitude start at 0° and continue both east and west of the prime meridian until they reach 180°, which is at the opposite side of the earth from the prime meridian. The 180° line, which runs through the Pacific Ocean, is called the international date line, and it's where one day becomes the next day. Lines east of the prime meridian are positive, and lines to the west are negative. Alternatively, lines of longitude are designated with an "E" or "W" depending on whether the line is to the east or west of the prime meridian.

The lines of longitude are not parallel to each other as the lines of latitude are, but are farthest apart from each other at the equator and converge at the two poles. Therefore, the amount of the earth's surface that each degree of longitude measures is not consistent; degrees at the equator are larger than degrees of longitude near the poles. The size of a degree of longitude will vary from the longest length measured around the thickest part of the earth at the equator, to the shortest length measured at the poles, where the length of the line of latitude at 90° is literally a point with no dimension, as shown in figures 2.12 and 2.13. Conversely, the size of a degree of latitude remains virtually constant, regardless of its location on the earth's surface.

A commonly used variation on the lat–long coordinate system replaces the minutes and seconds with decimals and is called decimal degrees.

The coordinates of Lehman College in the Bronx are approximately 73°53'40" West and 40°52'23" North . The equivalent decimal degree coordinates for Lehman College are –73.89, 40.87. The negative number for the longitude denotes a location west of the prime meridian, and a positive number for the latitude denotes a location north of the equator.

At this latitude, one degree of longitude is about 49 miles. By comparison, one degree of longitude at the equator is about 69 miles. It is a good idea to be aware of the approximate latitude and longitude of the geographic extent you are working with.

Types of projections

Projection types can be classified according to the method used to produce them. The three main categories of projections are called **azimuthal, conic,** and **cylindrical,** named for the shape from which they are derived. Azimuthal (also known as planar) uses a flat plane; conic uses a cone; and cylindrical, a cylinder (figure 2.14).

The projection process can be envisioned as a flat piece of paper placed on or wrapped around a globe, and the images on the globe transferred to the piece of paper. When the paper is unrolled again into a flat sheet the information has been transferred from the globe (spherical) to the piece of paper (a flat surface). Wherever the paper actually touched the globe as it was placed around it or on it, known as the **points of tangency,** is where the information will have been transferred with the least amount of distortion. No projection is totally distortion free, but the point or line of true scale coincides with the point or line of contact of the projection surface and the globe. Distortion increases with distance from the points of tangency.

Planar or azimuthal

These projections are most commonly used to show the polar regions, since their points of tangency provide less distortion near the poles. Planar projections are less commonly used to depict the equatorial regions or anywhere else on the globe.

Conical

Conic projections have their areas of least distortion in the mid-latitudes, since their points of tangency are along the lines of latitude midway between the equator and the poles. For an area oriented east and west, conic projections are best, since distortion will be minimized along the lines of latitude.

Cylindrical

Cylindrical projections have their areas of least distortion at the equator, since that is where their points of tangency are. Areas towards the poles are increasingly distorted. If we turn the flat paper the opposite way to wrap the globe, so that the points of tangency are along the meridians rather than along the equator, it is called a transverse cylindrical, and has its areas of least distortion along a meridian. Transverse cylindrical projections form the basis of the UTM coordinate system.

Planar/Azimuthal

Conic

Cylindrical

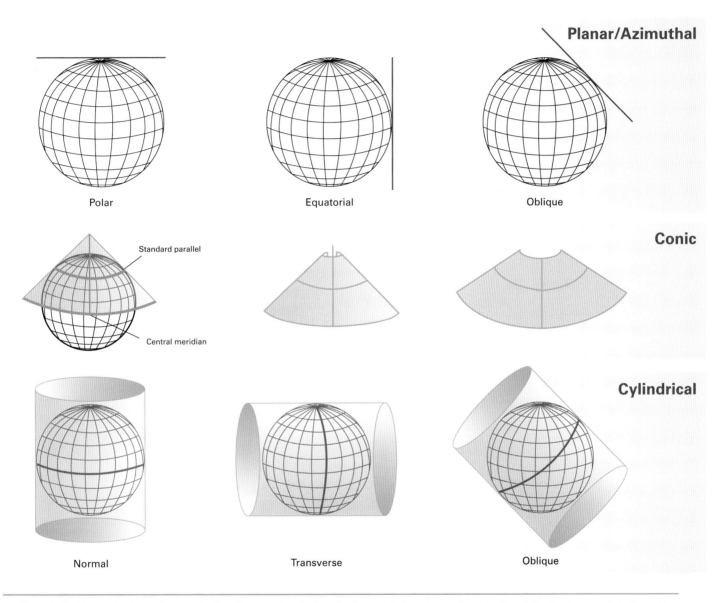

Polar   Equatorial   Oblique

Standard parallel

Central meridian

Normal   Transverse   Oblique

Figure 2.14 **Map projections. Planar projections (top row) can be imagined as a flat piece of paper touching the globe at only one point: the north or south poles, the equator, or any other point on the globe's surface. This is the point of tangency and it represents the part of the projection having the least distortion. Conic projections (middle row) can be imagined as a flat piece of paper wrapped around the globe in a cone shape, with a mid-latitude parallel being the line of tangency, thus representing those parts of the projection having the least distortion. Cylindrical projections (bottom row) can be imagined as a flat piece of paper wrapped around the globe in a cylinder shape. In a normal cylindrical projection, the equator is the line of tangency, while in a transverse cylindrical projection, a meridian is the line of tangency, and in an oblique cylindrical projection, any other Great Circle line is the line of tangency.**

Figure 2.15   The world on one sheet—some common map projections and the spatial properties they each preserve: Mercator (conformal, preserves shape and direction); Gall–Peters (equivalent, preserves area); Miller Cylindrical (compromise, minimizes distortion in several spatial properties, but preserves no spatial property entirely); Plate Carrée (equidistant cylindrical, preserves distance); Robinson (compromise); and Sinusoidal (equivalent).

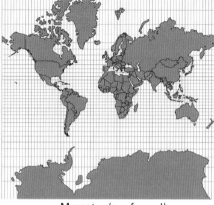

Mercator (conformal)

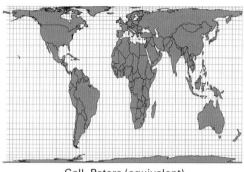

Gall–Peters (equivalent)

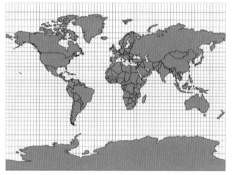

Miller Cylindrical (compromise)

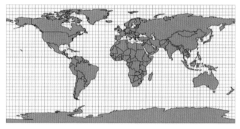

Plate Carrée (equidistant cylindrical)

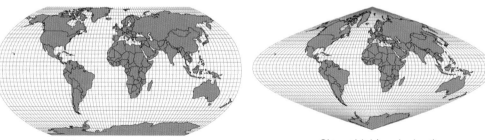

Robinson (compromise)

Sinusoidal (equivalent)

Projections: Conformal, equidistant, equivalent

A three-dimensional model of the spherical earth perfectly portrays the four basic spatial properties of geographical features: area, shape, distance, and direction. All projections contain distortions, since there is bound to be distortion in converting a spherical object to a flat plane, but this distortion is not equal across all four properties. Projections usually retain at least one spatial property that is less distorted than others, so we can say that the projection preserves that particular quality. Some maps depict distances between places more accurately, so we say that projection preserves distance. Other projections preserve cardinal directions better, some preserve shape, and some preserve areal dimensions. No flat projection can portray the four basic properties as well as a globe does.

In addition to classifying projections by the surfaces used to make them (planar, conic, and cylindrical), projections are also categorized by the degree of accuracy that they provide. In this regard, projections called **conformal, equivalent,** and **equidistant** are the principal divisions. Conformal projections have better accuracy with angles, and therefore with shapes and directions; equivalent or equal area projections have truer area measurements, while equidistant projections have better distance measurements.

Other projections

Projections are also named for their creators, some of whom worked as early as the fifteenth century. One of these, the **Mercator projection,** considered a conformal projection, distorts true area, especially in the high latitudes near the poles. It is, however, quite accurate for showing true angles, thus making it most useful for navigation. The **Gall–Peters projection,** considered an equivalent projection, distorts shape very badly, but shows true area, making it suitable for thematic mapping. This is because ratios, densities, rates, and other thematically oriented data require that each square mile be the same size no matter where they are located on the map. The **Plate Carrée** or **cylindrical equidistant projection** maintains the correct distance between every point and the equator, but is not conformal as to shape and direction, and is not equal area. Compromise projections are often used, such as **Miller Cylindrical** and the **Robinson.** These projections do not, strictly speaking, preserve any spatial property, but they do minimize distortion of all properties, hence the compromise label (figure 2.15).

A planner's choice of map projection should be based on the anticipated use of the map. We would probably not use a world map with a Mercator projection as a base for a thematic map showing population per square mile, because a square mile in Africa would not appear the same size as a square mile in Europe. Proper choice of map projection will also be dependent on which part of the world is being mapped. For instance, the **Mollweide projection** is appropriate to use when focusing on areas in the low latitudes (near the equator), while the **Lambert Conformal Conic** (which preserves shape and direction) or the **Albers Equal Area** (which preserves area) are appropriate for areas in the mid-latitudes, and the

Lambert Azimuthal is a good projection to use when focusing on the polar regions, where it minimizes distortion and calculates true distance.

We must be careful when measuring the distance or area of spatial data in a GIS, since measurements will differ depending on the map projection utilized. Figure 2.16 shows how different in size, shape, and area the British Isles can look as plotted out in three common projections.

Projections have been aligned at Land's End, but diverge elsewhere.

☐ Decimal degrees (unprojected)

☐ Gall–Peters (equivalent)

☐ Mercator (conformal)

Figure 2.16 Three different map projections of the British Isles show how dramatically they change the map representation, the distance between points, the shape of the land mass, and the area, and thus, why projection selection is so critical.

Common coordinate systems

In addition to projections, planners need to know about the predominant **coordinate systems.** Sometimes the two terms are used interchangeably, and adding to the confusion is the fact that all coordinate systems depend on a particular projection for their underlying base. Projections are used to transform positions on the earth's curved surface to a flat plane, while coordinate systems are the referencing framework superimposed on top of the projection surface so that position can be measured and estimated.

We have already discussed the most common georeferencing (or coordinate) system, latitude and longitude, but there are several others in active use in the United States: **Universal Transverse Mercator** (UTM); the **State Plane Coordinate** (SPC) system; land-partitioning systems such as the township, range, and section system of the **U.S. Public Land Survey** (USPLS); and other unsystematic methods of establishing property boundaries. In the United States, the UTM and State Plane coordinate systems and their associated projections form the basis of many of the maps produced by the federal, state, and municipal government agencies that planners use.

The UTM coordinate system

The UTM coordinate system was originally developed as part of the military grid referencing system to pinpoint air and surface locations. UTM is organized on a Mercator transverse cylindrical projection, meaning that the line of tangency (the line with the least distortion) is the meridian (as contrasted with an ordinary Mercator cylindrical projection, where the line of least distortion is the equator). In the UTM system, the earth is divided into sixty north–south zones of six degrees of longitude each. Each of these zones has at its center a meridian that represents the line of tangency for that zone, and therefore, distortion is minimal along the central meridian of each zone. Because the zones are only three degrees to either side of the central meridian, the entire zone is relatively distortion-free as well (figure 2.17).

In UTM, the north pole is considered to be 10 million meters north of the equator, and the equator is 10 million meters north of the south pole. Coordinates of a location are given in meters—so many meters north of the equator or the south pole, and so many meters east or west of the zone's central meridian. In order to pinpoint locations in UTM, you must know the zone number as well as the location's coordinates within the zone. UTM is a good coordinate system for urban planning purposes: maps required for most planning activities cover less than a zone or two in width, and maps based on UTM contain minimal distortion. The Bronx is in Zone 18 North, and the UTM coordinates for Lehman College are 593,207.66 meters, 4,525,368.36 meters, which represents the distances east of the zone's central meridian and north of the equator, respectively.

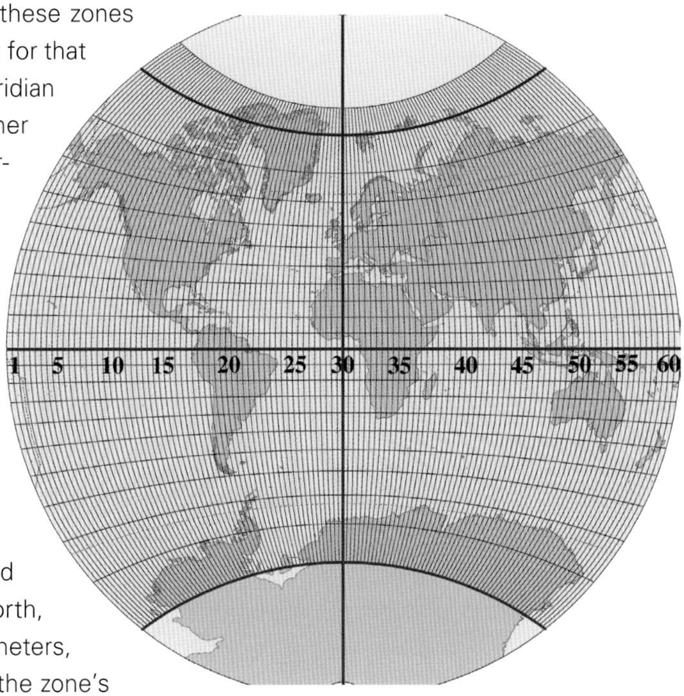

Figure 2.17   **UTM (Universal Transverse Mercator), is a coordinate system that divides the earth into 60 north–south zones of six degrees each. Each zone is centered around a meridian, and distortion is minimized along this line of tangency.**

The State Plane Coordinate system

The State Plane Coordinate (SPC) system is based on the plane geometry of a rectangular coordinate system. This is essentially a Cartesian grid, with a false point of origin established outside the zone, so that the x- and y-axes of the grid have only positive numbers. The false point of origin is located in the lower left corner of the grid, and a given location's position is described by using the coordinates that measure the distance north and east of the false origin. These are called **false eastings** and **false northings** and describe the position of the given location relative to the point of false origin. The unit of measurement in State Plane is generally feet. This system is accurate for the small portions of the earth depicted in a state plane map, and the curvature of the earth is not generally a factor in an area this small.

Each state is determined to be basically north–south or east–west oriented, depending on its major axis. East–west trending states (those that are longest in the east–west direction) generally use a Lambert Conformal projection, and north–south oriented states use a Transverse Mercator. Larger states are divided, if necessary, into smaller components. Some states, like New York State, have two major axes, and thus use two different projections. New York is divided into three north–south oriented zones: the eastern, western, and central zones, each using the Transverse Mercator projection. New York's fourth zone is the east–west oriented Long Island zone, and for that the Lambert conformal projection is used. Some states with panhandles, like Florida, also tend to use both types of projections, but most states, even those divided into multiple zones, use the same projection throughout the state. Each state has its own system (figures 2.18 and 2.19).

As with the latitude–longitude coordinate system, you should have a good idea of the particular State Plane section of your study's geographic extent, as well as its UTM zone.

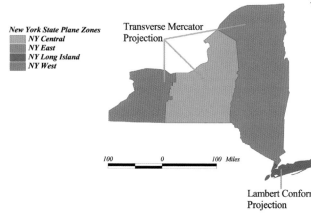

New York State Plane Zones
- NY Central
- NY East
- NY Long Island
- NY West

Transverse Mercator Projection

100   0   100 *Miles*

Lambert Conformal Projection

Florida State Plane Zones
- FL East
- FL North
- FL West

Lambert Conformal Projection

100   0   100 *Miles*

Transverse Mercator Projection

Figure 2.18   State Plane Coordinate (SPC) System Zones for two states, each with east–west and north–south trending sections. Most states with multiple zones have only one directional trend, such as California, Texas, and Illinois, and many states have only one zone, including Montana, New Jersey, Vermont, and North Carolina. The directional trend of the state or SPC zone will determine the type of projection used.

Other land partitioning systems

Other land systems operate on a more local level to identify specific locations. In the United States, the method of recording property boundaries (and therefore location) depends on the history of colonization of the area. Indigenous ways of thinking about the land and location—and for many of these indigenous peoples, the very concept of people "owning" land did not exist—were discarded by colonizing powers in favor of the land ownership concepts and systems already established in their own countries. Each European power also had a different method of granting tracts of land in the New World, which influenced how that land was ultimately partitioned and measured. Spanish, French, and English settlements each exhibited their own distinctive forms of urban design, and so colonized property size and shape reflected these distinctions. All these factors form the basis of present-day cadastral and other local maps in the United States.

In the part of the country that was colonized by northern Europeans first (the Eastern seaboard states), a method of **metes-and-bounds surveying** was used to map property boundaries and to identify locations. Metes-and-bounds surveys establish property lines through physical measurement—the surveyor walking the perimeter of the property and recording the length of each side, the bearing of each line, and any notable landmarks, such as unusually large or different trees.

Figure 2.19 State plane zones of the fifty states show how their orientation depends on whether the state is east–west trending or north–south.

☐ *States*
☐ *State Plane Zones*

In the nineteenth century, the United States had acquired vast amounts of unsurveyed real estate in the wake of the War of Independence and the Louisiana Purchase. In order to more easily measure and market the land, a system to partition it was established by the U.S. Land Partitioning Survey (USLPS). This was accomplished by establishing north–south principal meridians and east–west baselines for the western territories. With each area's principal meridian and baseline as a framework, a grid of six-mile-square land subdivisions

Figure 2.20　Much of the land of the United States is platted using the USPLS system of ranges, townships, and sections. Much of Canada employs a similar system.

Redrawn, courtesy of JP Publications. From *Map Use: Reading, Analysis, Interpretation (4th Edition)*. By Muehrcke et. al. Reprinted by permission of JP Publication, Madison WI © 2001. JP Publications™.

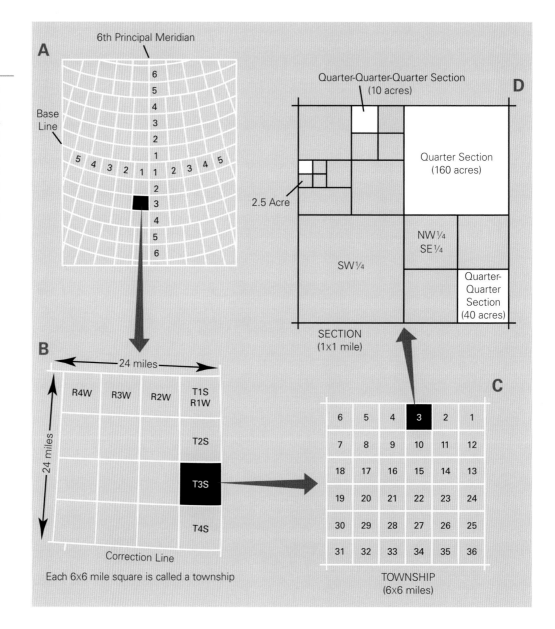

was organized into townships, ranges, and sections. Townships were formed by the east–west lines in the grid, ranges by the north–south rows. Each of these six-mile grid squares was also called a township, and each was further divided into 36 sections of about one mile square each. Each of these contained about 640 acres and was successively subdivided into quarter sections of 160 acres each, then quartered into squares of 40 acres each, and so forth. The grid generally ignored topography, and plains and mountains alike were platted into one-mile squares. This system defines the western part of the United States to this day and gives it the distinctive checkerboard look so clearly apparent when you fly over it (figure 2.20).

Coordinates of a location in this system are given by the numbers of the grid square's township and range in relation to the principal meridian and the baseline, as well as the section number, and the ordinal (north, south, east, or west) direction of the quarter within the section, if applicable. This is a fairly precise method of providing coordinates for a location; in fact, it was actually a very efficient and ingenious way to divide up almost the entire country west of the Mississippi River in a way that allowed any parcel of land, however small, to be accurately located by just a few alphanumeric designators. For example, in figure 2.20, the 10-acre parcel's unique locator identification would be *NW1/4, NE1/4, NW1/4, Sec. 3, T.3S, R.1W, 6th Principal Meridian.* The description starts with the smallest division, and works backwards, up to the principal meridian. Spoken out loud, this description would be "the northwest quarter of the northeast quarter of the northwest quarter of Section 3 of Township 3 South, Range 1 West, Sixth Principal Meridian." A person familiar with the USPLS system would be able to pinpoint that parcel's location without any other information.

Since the USPLS covers so much of the United States, it is worthwhile for American planners to be familiar with it. Understanding how to locate something by the designation of townships, ranges, and sections can be useful in working with current as well as historic maps of much of the United States. The USPLS covers the thirty westernmost states, those that were formed from lands in the public domain. It excludes east coast states, except for Florida, and excludes Kentucky, Tennessee, and Texas; it includes only parts of Ohio and Alaska.

The legacy and implications of property ownership customs and systems are profound for every country in the world. A number of excellent books on this subject, listed at the end of this chapter, make for good background reading for any would-be planner. Planners must have a good familiarity with the cadastral system and its history in the area where they work. They must also have a well-rounded knowledge of the various graphic representations that are available for their geography of interest. In other words, the more maps to which a planner has access, and the more she knows about them, the more complete will be her understanding of her geography.

Written Exercise 2

Maps, scales, and coordinate systems

Find maps of your community in different scales and with different degrees of generalization. At a minimum, look at the following:

+ USGS 7.5-minute quadrangles
+ local cadastral (property block and lot) maps
+ street maps
+ zoning maps
+ public transportation maps
+ tourist maps, with local attractions
+ Sanborn maps, if available. (Sanborn maps were originally intended for fire insurance purposes and covered many urban areas in the United States. They usually contain detailed information at a large scale, such as building footprints, building construction, number of building stories, land use, property lot lines, and a wealth of other information useful to planners. Since these fire insurance maps are often available for many cities starting in the mid-nineteenth century, they are invaluable for tracing historic land use and development patterns, as well as some historic environmental conditions.)

Other specialized maps should be consulted, if you can find them.

Examine all the maps and discuss the level of detail obtainable from each type of map. Compare the scales given on the maps, as well as the amount of generalization required to reduce the map scale.

Determine your own exact latitude and longitude, as well as your State Plane and UTM coordinates and zone. What other cities around the world are on roughly the same latitude as you are? What cities are on roughly the same longitude?

References and further reading

Aronoff, Stan. 1993. *Geographic Information Systems: A Management Perspective*. Ottawa: WDL Publications.

Bernhardsen, Tor. 2002. *Geographic Information Systems*. New York: John Wiley and Sons.

Clarke, Keith. 2003. *Getting Started with Geographic Information Systems*. Upper Saddle River, NJ: Prentice–Hall.

Crouch, Doro P., Daniel J. Garr, and Axel I. Mundingo. 1982. *Spanish City Planning in North America*. Cambridge, MA: MIT Press.

De Mers, M. N. 2000. *Fundamentals of Geographic Information Systems*. New York: John Wiley and Sons.

Johnson, H. B. 1976. *Order Upon the Land: The U.S. Rectangular Land Survey and the Upper Mississippi Country*. London: Oxford University Press.

Jones, Christopher. 1997. *Geographical Information Systems and Computer Cartography*. London: Addison, Wesley, Longman.

Kain, R. and E. Baigert. 1992. *The Cadastral Map in the Service of the State: A History of Property Mapping*. Chicago: University of Chicago Press.

Longley, Paul, M. Goodchild, D. Maguire, and D. Rhind. 2001. *Geographic Information Systems and Science*. New York: John Wiley and Sons.

Morain, S., and S. L. Baros, eds. 1996. *Raster Imagery in Geographic Information Systems*. Santa Fe, NM: OnWord Press.

Price, Edward T. 1992. *Dividing the Land: Early American Beginnings of Our Private Property Mosaic*. Chicago: University of Chicago Press.

Reps, John. 1989. *Town Planning in Frontier America*. Columbia, MO: University of Missouri Press.

Rubinstein, James M. 1999. *The Cultural Landscape*. Upper Saddle River, NJ: Prentice–Hall.

Snyder, John P. 1993. *Flattening the Earth: Two Thousand Years of Map Projections*. Chicago: University of Chicago Press.

Sobel, Dava. 1995. *Longitude: The True Story of a Lone Genius Who Solved the Greatest Scientific Problem of His Time*. New York: Walker Publishing Co.

Thematic Mapping

Planners are both map users and, with GIS, mapmakers. Familiarity with the major categories of maps will help the map user correctly interpret current maps, and help the mapmaker create maps that communicate information effectively and accurately.

One of the most important categories, and one that is used often in planning contexts, is the thematic map. In this chapter, we discuss what a thematic map is, its components, its major subgroups, and its use in planning.

Maps: Tangible vs. mental

In a hierarchy of map types, the top level has two broad categories: tangible maps and intangible, also known as mental maps. All maps can be said to fall into one or the other of these categories—which can be further divided into subcategories—and both are important in planning (figure 3.1).

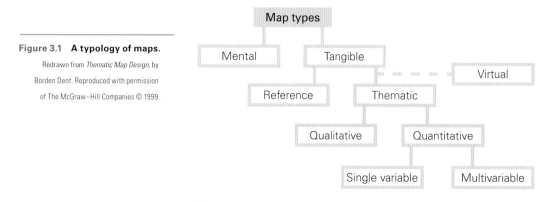

Figure 3.1   A typology of maps.

Redrawn from *Thematic Map Design*, by
Borden Dent. Reproduced with permission
of The McGraw–Hill Companies © 1999.

Mental maps

Mental maps are unique perceptual maps that exist only in our minds. Mental maps help us negotiate our experience of physical space. Many community-led GIS projects and participatory planning processes use mental mapping as a major input to their projects, and it has been shown to be an effective planning tool (Pickles 1995). Mental maps reflect our spatial perceptions and personal knowledge of the geography of interest. For example, we may have in our minds a map of our commuting route from home to work, or a sequential route map of all our neighborhood errands on a day off.

If we were to draw out a mental map, it would probably not reflect geographic reality, but rather the reality of our perception of the geography. For instance, distances that are equal in the real world might exist in our mental maps as longer or shorter than the actual measured distances, depending on how we perceive the distance. Walking a half-mile through a desolate, dangerous, windswept, unshaded, or otherwise uncomfortable terrain may seem longer than the equivalent distance in an aesthetically pleasing, safe, or visually stimulating environment. Distances will also appear to differ depending on the relative friendliness or hostility of the area's inhabitants, the amount of light, the slope of the land, the unevenness of the walking surface, and myriad other factors.

We often have mental maps of places we have never experienced firsthand. We all hold a map of the world in our heads drawn in terms of our own preconceived notions, perhaps based on things we've seen on television, or anecdotal reports we've read or heard from people we know. We have an idea of the geography of these places and this forms the framework of our mental maps. The mental map is, in effect, a collage of all the disparate information we have accumulated about a given geography.

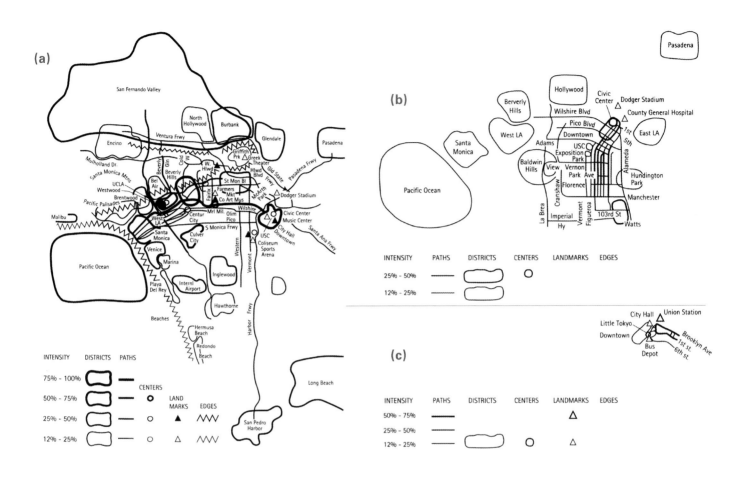

Figure 3.2 How we experience our surroundings depends on who we are: our cultural background, education, income level, gender, age, and physical abilities. Map (a) is a composite of mental maps of Los Angeles compiled by white residents in a predominantly white community, while map (b) is by black residents living in a black community. Map (c) is by Hispanic residents in a community where they predominate.

Dorling, Daniel; Fairbairn, David. *Mapping Ways of Representing the World.* 1st Edition, © 1997 pp. 168–162. Reprinted by permission of Pearson Education, Inc., Upper Saddle River, NJ.

Mental maps tend to reflect and represent the background, life experiences, and knowledge of the mapmaker. The sum total of one's identity definitely affects how one perceives space (figure 3.2).

Many people have a mental map of their city or town, for instance, with sections within the city ranked in terms of safety or risk considerations. Since this is subjective information, everyone's mental maps differ, sometimes significantly. These maps are often intuitive and not considered on a conscious level, but nevertheless form the basis of many decisions that we make throughout our lives. These unconscious perceptions about our local geography also inform many planning decisions.

Mental maps can be productive and practical when used in the planning process. They can be informative, and also help communities achieve consensus and circumvent potential conflicts. For instance, having all stakeholders draw mental maps of the parts of their environment that they deem important, with community or personal landmarks, can help illustrate a consensus of community priorities: what a community considers worthwhile to preserve, improve, or invest in. Compiling and drawing out a composite mental map of problems that

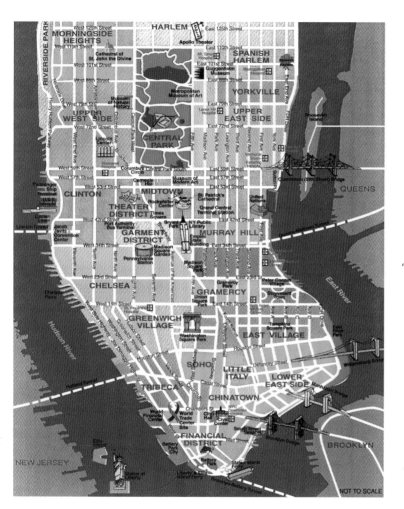

Figure 3.3 There is no political or administrative unit in New York City called the "neighborhood," and therefore there are no official definitions of neighborhood areas or boundaries. Yet everyone (including the Taxi and Limousine Commission) thinks they have a pretty good idea of where these neighborhoods are located. This map represents the TLC's version of Manhattan's neighborhoods, but there might be eight million other versions in New York City, based on the individual's mental map of Manhattan neighborhoods. "The TLC Map" is used with permission of the William H. Brown Company, New York. © William H. Brown Company.

need to be addressed in the community can also be worthwhile. This can aid in resolving locational conflicts, and in helping the community determine where resources are needed or should be directed. Mental mapping is often a good way to jump-start planning for a difficult or contentious process, such as a major re-zoning action or a historic district designation. Graphic portrayal of community priorities by these means can expedite the planning process and help it conform better to the collective vision of the community's future. Often, mental mapping exercises, similar to the one at the end of this chapter, are employed in the public participation process, either as part of a "visioning" session, or of an intensive planning charrette (Aberley 1993). Of course, once a mental map is drawn on paper or on the computer, it ceases to be a mental map and becomes a tangible map, one that can be viewed by more than one person.

Tangible maps

In contrast to mental maps, tangible maps are those that can be viewed by more than one person, and each viewing leads to observation of the same information, or at least the potential for such observation. Tangible maps can be ephemeral and transitory, such as those drawn in the dirt, or they can be physical and permanent, such as hard copies of maps printed on paper, or digital or computer maps, viewed on-screen. Some argue that maps seen only on-screen are not really tangible, since they cannot be held in the hand, and belong more properly to an offshoot category called *virtual*. On the other hand, on-screen maps fit the definition of a tangible map because they can be viewed simultaneously or sequentially by more than one person, and the same information is available to all viewers. And of course, digital maps can usually be printed out, thereby becoming tangible.

Maps: Reference vs. thematic

Virtually all tangible maps belong to one of two categories: reference or thematic. Reference maps are those we use to find our way, such as a road map, or a navigational map for air or sea. Reference maps are also those we use to give us general locational information: a world map showing names and locations of continents, oceans, countries, and capital cities, or a city street map. The majority of maps in atlases are reference maps, showing the locations of geographic entities and features, place names, roads, topography, and so on. These are used specifically for navigational purposes or for general location finding. A cadastral map that shows property boundaries for the purpose of recording property ownership or for tax assessment and collection, might also be considered a reference map (figures 3.4 and 3.5).

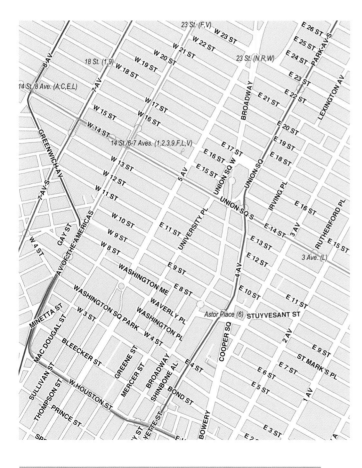

Figure 3.4 **A map showing streets and subway lines is a typical reference map.**

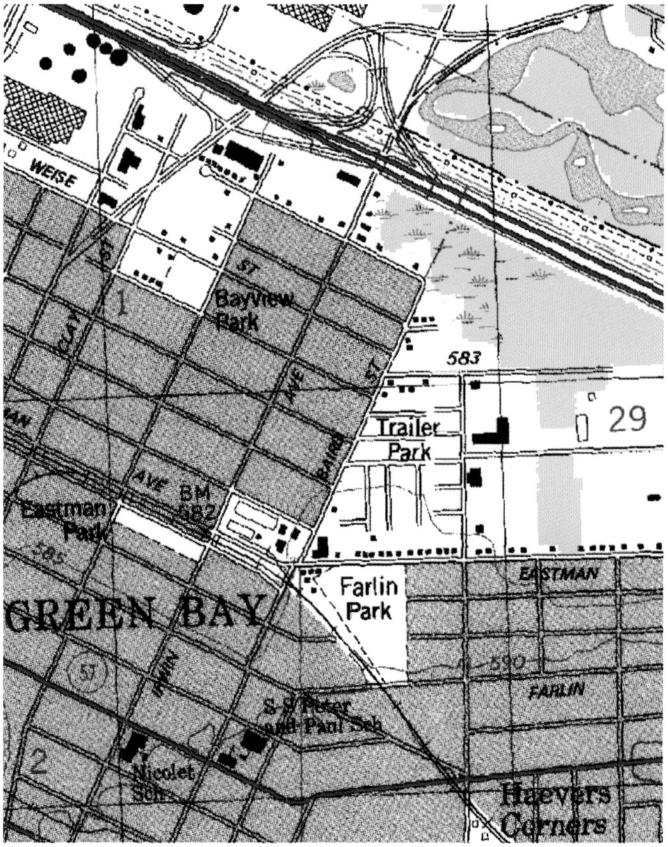

Figure 3.5 **This USGS 7.5-minute quadrangle map is another example of a commonly used reference map.** Courtesy of USGS.

Reference maps are used in planning to provide context to thematic information, and to provide accurate geographic information about political boundaries, highways, streets, railroads, parks, and other landmarks, as well as topographical information. Sources of reference maps and basemap data for thematic mapping are discussed in chapter 6.

A thematic map, by contrast, is not meant to locate anything with any precision. Thematic maps display the distribution of one or more specific variables, such as average household income, crime rate, or incidence of a disease—aggregated by a geographic unit, such as country, state, county, census tract, community district, or ZIP Code. A thematic map may show land use, zoning, or the locations of social service facilities in a city, symbolized by type of service. The name thematic derives from the fact that such a map seeks to portray one *theme* of information usually across some common geographic area, such as students' standardized test score averages in each school district, or median housing price by county, or locations of burglaries within each police precinct (figures 3.6 and 3.7).

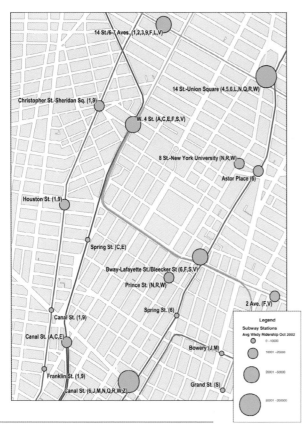

Figure 3.6 **A thematic map showing average weekday ridership at subway stations.**

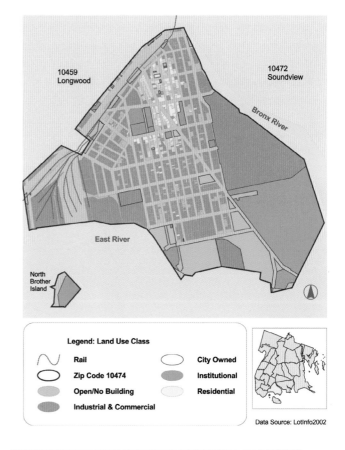

Figure 3.7 **A thematic map of land use in ZIP Code 10474, the Hunts Point area of the Bronx.**

Components of the thematic map

A thematic map is made up of two main components: the geographic basemap, and the overlay of the thematic data. The basemap shows the geography of interest, which must include enough geographic detail so that it is recognizable to the map reader. The thematic information is then placed on top of the basemap. The thematic data is of primary importance in a thematic map, the geographic details of the basemap secondary (figure 3.8).

A thematic map may show more than one theme or variable. For instance, a map may show population density with the locations of hospitals also indicated. Or the map may show locations of homicides as point features, with unemployment rates by census tracts the underlying theme. This enables the map reader to view more than one phenomenon at once, and to note any spatial correspondence between the variables being mapped. This ability to see spatial, or geographic, relationships is one of the great powers of thematic mapping and is discussed in detail in chapter 9.

Because reference maps are designed to show location, they typically show a large amount of geographic detail compared to thematic maps. Since the thematic map is not intended to be used for locational purposes, the small amount of geographic detail available in them is not a significant issue.

Reference maps and thematic maps are not necessarily mutually exclusive, and there is much overlap between the two. Thematic maps showing climatic zones, biomes, or vegetation provinces are sometimes considered reference maps. Conversely, while a topographic map of elevation contours is by definition a reference map, it can also be considered a thematic map, with elevation as the theme. It is not important to classify every map as either reference or thematic, but understanding the basic criteria underlying the identity of these maps and understanding the different purposes served by each type is.

Figure 3.8 Basemap, thematic overlay, and a thematic map combining both.

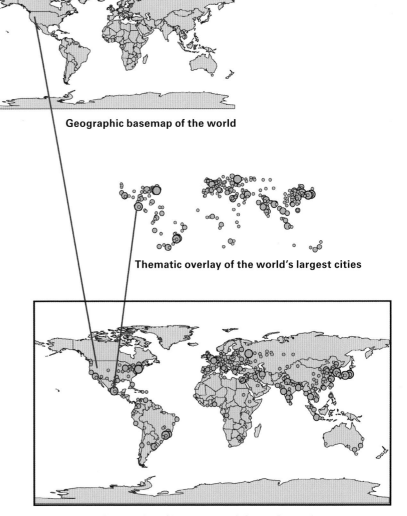

Geographic basemap of the world

Thematic overlay of the world's largest cities

Composite of basemap and thematic overlay

Quantitative vs. qualitative thematic maps

The thematic map category can also be subdivided into qualitative and quantitative, a topic explored further in chapter 4. Qualitative thematic maps have values mapped based on quality (nominal) criteria, not numerical ones. Nominal categories are those having names. Consider a map of the United States with the states shown in two different colors: one color representing those states that voted Democrat in the last presidential election, and the second color representing those states that voted Republican. This would be an example of a nominal or qualitative map, since we are not showing any quantitative data, merely which states had a majority voting for each political party. Another type of qualitative map might show mass transit stations within a metropolitan region, with each station identified and symbolized by the passenger services offered at each station, such as handicap accessibility or transfer facilities. Land-use maps and zoning maps are also qualitative thematic maps (figure 3.9).

Thematic maps may also be quantitative, which means that the values mapped are numerical and comparable. These could include population density per square mile, high school drop-out rates by school district, numbers of historic districts or structures by county, or the amount of tourism-generated dollars per year by state. Other quantitative examples include transit ridership per station, or a map of the central business district with average per-square-foot rental costs of commercial properties (figure 3.10).

Figure 3.9 **This map of zoning districts is a qualitative map because zoning categories are nominal, not numerical.**

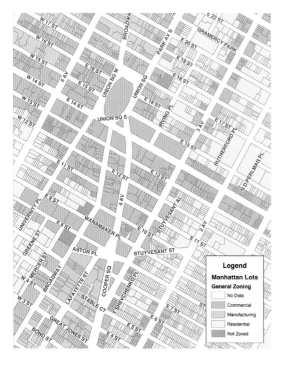

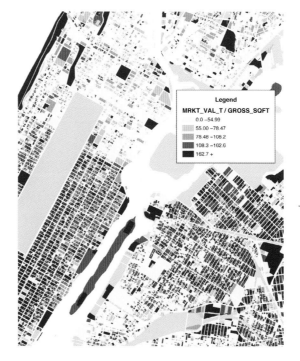

Figure 3.10 **A quantitative map in which property values are numerical and measurable by square foot.**

Types of thematic maps

Creating and working with thematic maps is one of the principal functions of GIS in planning. Other functions—such as basic data exploration, data visualization and analysis, and data display and presentation—also require a comprehensive understanding of which thematic map types are most appropriate for the display and analysis of different kinds of data.

You may never need to produce all the different types of thematic maps, but you should be aware of the range of possibilities. This is important for two reasons: it enables you to better understand and interpret map data produced by others, and it allows you to choose which kind of thematic map will be most appropriate for the data you want to display. The most common types of thematic maps are **dot density, choropleth, proportional symbol, isarithmic, flow,** and **cartogram.** Occasionally, thematic maps are combinations of two or more types of map.

Dot density maps

Dot density maps (also sometimes called dot distribution maps, or areal frequency maps) describe variation in spatial density of some phenomenon. The denser the pattern of dots, the more intensely the phenomenon exists at that location. The dots represent quantities of some variable, such as population, crime, or disease, that have been collected in a defined geographical area, which could be a political or administrative unit such as a census tract, county, state, health district, or police precinct.

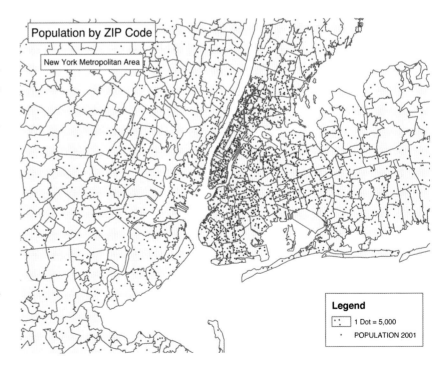

Figure 3.11 A dot density map showing population by ZIP Code.

The ideal dot map would have one dot for each occurrence, thereby preserving and displaying all the data completely, but normally the space available on the map does not permit a one-to-one relationship of dots to mapped phenomena. More typically, a dot density map has a one-to-many relationship with each dot representing several occurrences of the mapped phenomenon. One dot may represent one thousand people, or one dot may represent five hundred occurrences of some event or phenomenon.

Dot density maps are used to plot absolute numbers—counts of things, such as actual people, cases, or events—and should never be used to plot ratios, percentages, or proportions as these cannot be represented by the dot density technique (figure 3.11). Dot density maps are usually intended to give the

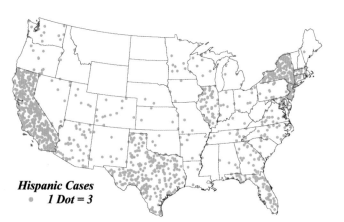

Hispanic Cases
• *1 Dot = 3*

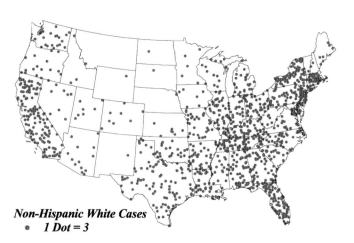

Non-Hispanic White Cases
• *1 Dot = 3*

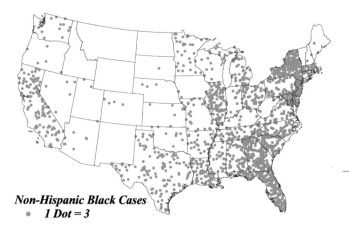

Non-Hispanic Black Cases
• *1 Dot = 3*

map user a visual impression of the distribution and density of the phenomenon, and not to gauge with any great precision the actual numbers of the phenomenon being mapped (figure 3.12).

Because each dot usually represents more than one of the items or occurrences being mapped, dot value and size must be determined when producing a dot density map. In other words, what is the numerical value represented by each dot, and what size dot should be used? This process has become automated in many GIS software programs, and the program will calculate the optimal number to be represented by each dot, based on your data range and the geographic unit used.

Although the software is capable of calculating the most technically correct response, it may not always make the best aesthetic or graphic choice, and the mapmaker will have to recalculate or make manual adjustments to achieve maximum clarity.

Several methods exist for manually calculating how many items or events to assign per dot.

First, examine the range of your data to be mapped. What is the lowest and highest number of your variable per geographical unit? Find the geographical area that has the fewest of these entities to be mapped, and place two or three dots there. Each dot should represent one-half or one-third of the amount of the phenomenon in the geographical area having the lowest value, so that there will be at least two or three dots appearing in the area with the lowest value of the variable being mapped. Each dot, then, is equal to no more than one-half or one-third of the lowest value in your dataset. Furthermore, in the geographic area with the highest value of the phenomenon being mapped, choose dot value and size so that the dots just begin to coalesce (blend together and touch each other) in the highest density of the mapped value. In figure 3.13, all three maps show identical data—the same number of people in each Bronx census tract. However, in each map, the

Figure 3.12 **Dot density map showing cases of tuberculosis by racial and ethnic group in the United States.**

dot represents a different number of people, and this influences how the map looks, as well as how the data is interpreted by the map reader. The map labeled *1 Dot = 500* appears at a glance to be sparsely populated. In reality, each of these census tracts has an average of 4,500 residents, with a total of almost 75,000 people in these 17 tracts. Which dot value gives a better indication of population density? How do you think that might change if population for all the tracts in the entire Bronx were shown? (The empty tract in the center of the area is a shopping center, with no population.)

Changes in dot value and size will produce very different maps. When designing a dot density map, be careful to choose dot value and size that represents the data accurately, without misleading the map user. For example, using an inappropriate dot size or value might result in a geographic area appearing more sparsely or densely populated on the map than it may be in reality (figure 3.14).

Selected Census Tracts

Population
• 1 Dot = 50

Figure 3.13 **For this group of Bronx census tracts, varying the value of the dot produces differences in realistic representation.**

Population
• 1 Dot = 100

Population
• 1 Dot = 500

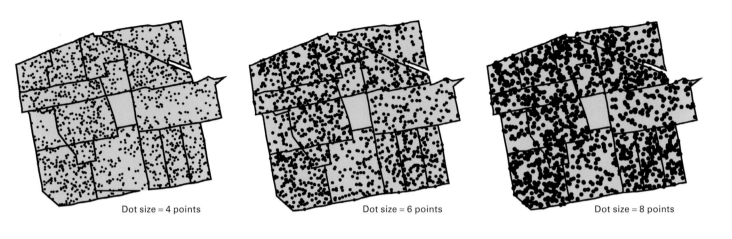

Dot size = 4 points   Dot size = 6 points   Dot size = 8 points

Figure 3.14 Varying the dot size affects the legibility of the map and how accurately it can be interpreted.

Another dot-density design consideration is the placement of the dots within each geographical unit when each dot represents many entities or occurrences. Since the dots represent quantities that have been aggregated by some geographic district or administrative unit, we cannot know from the data exactly where the dots should be placed within the geographic unit to reflect reality. Most GIS software produces a random placement of the dots within each geographical unit, or may distribute the dots more or less evenly throughout each unit, but this probably does not reflect the actual locations of the phenomenon being mapped. For instance, in a map showing population density within a city's census tracts, the dots may be placed evenly throughout each tract, when in fact, parts of some tracts may be industrial areas, parks, or commercial zones where few people live. Few or no dots should appear in those areas.

One way to mitigate this problem is to use data aggregated at the smallest geographic unit possible. The smaller the geographic unit in relation to the overall size of the map, the greater the accuracy of the final dot distribution, because the data is aggregated at a comparatively fine level of resolution (figure 3.15). Another way to address the problem is to apply spatial masks or filters so that parks, water bodies, and other areas where people are unlikely to live are filtered out or masked from the geographical unit. The creation of dot density maps and the application of filters and masks are the subject of lab exercise 3 in part 3.

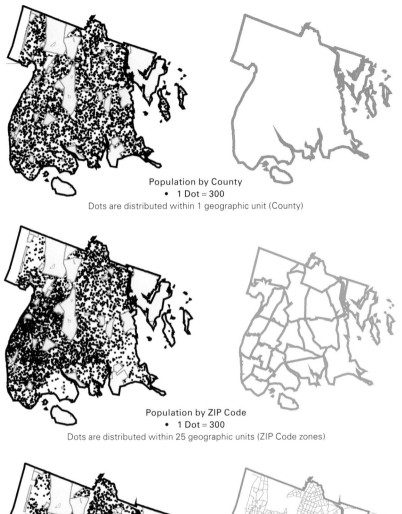

Population by County
- 1 Dot = 300

Dots are distributed within 1 geographic unit (County)

Population by ZIP Code
- 1 Dot = 300

Dots are distributed within 25 geographic units (ZIP Code zones)

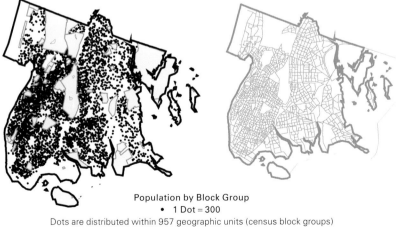

Population by Block Group
- 1 Dot = 300

Dots are distributed within 957 geographic units (census block groups)

Figure 3.15 **All three dot density maps use one dot to represent 300 people, and the dot size and total population are the same in all maps. The maps appear to be different due to the different unit of data aggregation used in each. The smaller unit of aggregation will provide more realistic placement of dots.**

Choropleth maps

Choropleth maps, sometimes also called graduated color maps, are used to show rates, percentages, or ratios of a phenomenon, as aggregated by some geographical or political unit, such as census tract, congressional district, county, state, or country. Choropleth maps should not be used to map absolute numbers—quantities such as the number of schools in a county or the number of people per census tract (figure 3.16).

In a choropleth map, each geographic unit—for example, individual countries in a world map—receives a color, pattern, or tone that designates the value range of the variable. Usually, the colors, patterns, or tones are graduated in hue or intensity to denote the relative magnitude of the variable. For a global map of birth rates by country, for instance, the countries with the highest birth rates would be shown with the darkest shade of a particular color; the countries with a medium high birth rate would be shown with a less intense

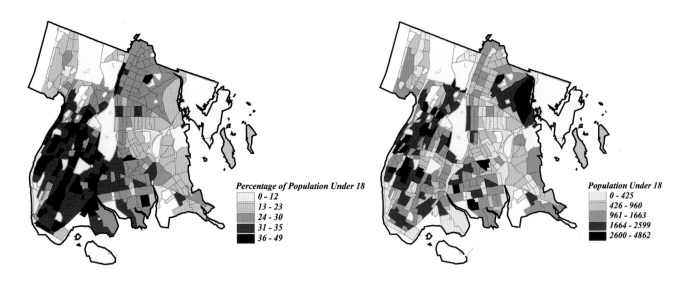

Percentage of Population Under 18
- 0 - 12
- 13 - 23
- 24 - 30
- 31 - 35
- 36 - 49

Population Under 18
- 0 - 425
- 426 - 960
- 961 - 1663
- 1664 - 2599
- 2600 - 4862

Figure 3.16   **The left map is a correct choropleth map, based on the percentage of the population under eighteen years of age, per census tract. The right map is an incorrect choropleth, based on the absolute numbers of children under eighteen, per census tract. The right map gives an erroneous impression of where children are concentrated, because the values are not normalized by the area of the geographic unit (children per square mile), nor are the values shown as a proportion of total population for those units. Therefore, it is impossible to make a meaningful comparison of the values for two or more tracts.**

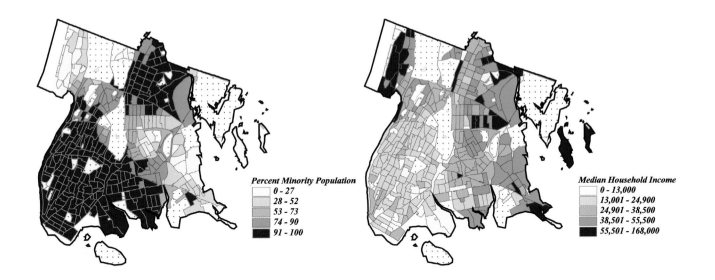

Percent Minority Population
- 0 - 27
- 28 - 52
- 53 - 73
- 74 - 90
- 91 - 100

Median Household Income
- 0 - 13,000
- 13,001 - 24,900
- 24,901 - 38,500
- 38,501 - 55,500
- 55,501 - 168,000

version of the same color; the countries with a low birth rate would be shown with a still lighter shade of that color; the countries with a declining birth rate would be shown with the lightest shade of the color. In this way, the map user can find the value or value range for any given geographic unit, can perceive at a glance where the highest to lowest birth rates are among the units, and can group the units visually and detect any patterns. Intuitively, the progression of shades makes sense: as the rate increases, the color deepens. The map user can also easily compare two or more choropleth maps of different variables to ascertain geographical distributions and any spatial correlation among the mapped variables (figure 3.17).

There are a number of design considerations when using the choropleth technique. The first is to resolve questions about exactly what will be mapped. How large or small an area should the map encompass? Once we decide on the geographic extent of the map, we select the variable or variables to be mapped. If, for example, we are interested in mapping the economic health of a city's commercial districts, we must analyze the available data, as well as the information we are trying to convey, and then select one or more variables to be mapped. What information is most appropriate to the map story we are trying to tell? That is, what variables are adequate numerical proxies for economic health? Dollar per square foot charged for renting commercial space? Dollars generated in retail trade per square foot? Numbers of jobs generated per unit of space? Ratio of occupied versus vacant commercial space? Any of these variables would probably be valid depending on circumstances and purpose, but which one captures the essence of what we are trying to show?

Figure 3.17 **Both maps aggregate data by census tract, and both use the natural breaks classification method with five classes. The method identifies breakpoints by looking for groupings and patterns inherent in the data. It is interesting to juxtapose both of these thematic maps to understand the relationship between the two variables: Ethnicity and income in the Bronx. The maps are virtually mirror images of each other.**

After deciding on variables, the next major decision involves the level of data aggregation to be used. This is often a function of how the data was originally collected: if the data was gathered by ZIP Code or by census block, then that is the smallest unit of geography that we are going to be able to summarize and map, and the map designer does not have a real choice regarding data aggregation unit. When given a choice, however, it is almost always preferable to have the map show a larger number of smaller areal units, rather than a smaller number of larger units. The use of many and smaller areal units allow for a finer resolution of the data and will show greater detail. For instance, a map of average household income by county will show more detail than the same variable aggregated by state (figure 3.18).

Lastly, it is often necessary to convert raw data (which may be in absolute number format) to rates or ratios or percentages. The purpose of the map, the type and range of data, and other factors will influence how the raw data is processed so that it is usable in choropleth mapping. How the mapmaker arrives at the rates, ratios, and percentages will, of course, greatly affect the design of the map. (Raw data processing is taken up in lab exercise 4.) Further discussion of choropleth mapping can also be found in chapter 4.

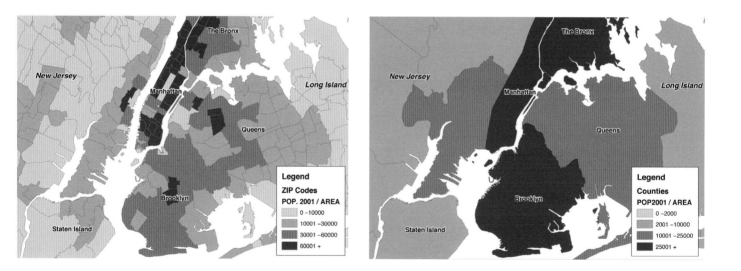

Figure 3.18   Population density at different levels of aggregation. More information can generally be obtained from a map that uses a finer spatial resolution/level of data aggregation. Smaller geographic units of data aggregation will yield more useful information than larger units .

Proportional symbol maps

A proportional symbol map (also called a graduated symbol or graduated point map) shows the relative or absolute size of the entity to be mapped, or the number of the particular phenomenon located at that point, and displays standard symbols at sizes that differ to reflect the relative size of the entity or quantity. The most common symbols used are circles, squares, and triangles, but virtually any symbol can be used effectively, including small pictograms of people to designate population, cars to designate travel volume or traffic congestion, or dollar signs to indicate anything connected with money, whether government expenditures or the daily gross revenues of casinos.

On a national map, for example, cities with the largest populations, such as Los Angeles and New York, are commonly represented by large circles, while medium-sized cities, such as Seattle and Miami, would have smaller circles, and small cities, such as Ann Arbor and Santa Fe, would have the smallest circles. The relative size of the symbol reveals quantitative information about the variable being mapped.

If we are trying to communicate relative amounts, sizes, or degrees of importance of phenomena at specific geographic locations, proportional symbol mapping is one of the most effective techniques to use. But it is appropriate only if the data to be mapped occurs at points, or if the data is aggregated at geographic points within areas (figure 3.19).

An example of data occurring at particular points, and suitable for representation on a proportional symbol map is the location of Toxic Release Inventory (TRI) facilities. A symbol such as a circle or hexagon could represent each facility's location, and the symbol's size would be proportionate to the annual amount of toxic pollutants emitted, measured in pounds. An example of data aggregated by area, mapped at a point, and suitable for representation on a proportional symbol map is tenant housing complaints in a census tract. These could be mapped by a proportional symbol, perhaps an exclamation point, placed at the census tract centroid (the geographical center of the tract's area). Similarly, the rate of childhood asthma hospitalization could be mapped by proportional symbol with data aggregated at ZIP Code centroids (figure 3.20).

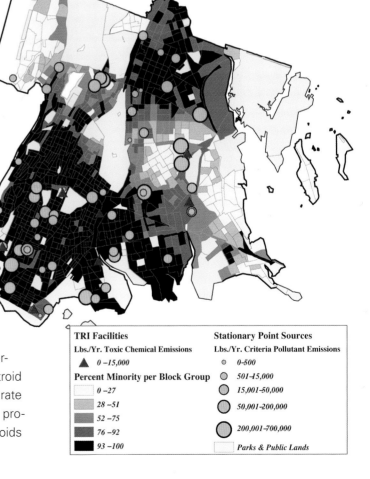

Figure 3.19 The proportional symbols depict the quantity of air pollutants emitted from TRI and Stationary Point Source (SPS) facilities, which are represented as points. These are shown in relationship to the minority population of the Bronx, by census tract.

TRI Facilities		Stationary Point Sources	
Lbs./Yr. Toxic Chemical Emissions		**Lbs./Yr. Criteria Pollutant Emissions**	
▲	*0 –15,000*	◦	*0–500*
Percent Minority per Block Group		◦	*501–15,000*
	0 –27	○	*15,001–50,000*
	28 –51	○	*50,001–200,000*
	52 –75	○	*200,001–700,000*
	76 –92		
	93 –100	☐	*Parks & Public Lands*

Note the difference between these two examples, however: the TRI emissions map entails representing geographical data that occurs at actual point locations, while the tenant complaints or asthma rates use the census centroid (or other proxy) to attach data that has actually been aggregated over the whole census polygon.

Much study has been done on proportional symbol maps to determine how people perceive the relative size of symbols, and how accurately they judge the incremental differences among symbols. Most people do not perceive a circle having twice the area of another circle as being twice as large. Magnitude in area and volume are apparently not as well perceived as magnitude in length, for instance, so area and volume are routinely underestimated. For that reason, absolute scaling (or square root scaling) has been generally abandoned in favor of apparent-magnitude scaling, which exaggerates the size of symbol. In apparent-magnitude scaling, a circle representing an amount of 1,000 would be much larger than ten times the circle representing 100 (figure 3.21).

The size of the symbol is generally calculated automatically by GIS software to conform to standard apparent-magnitude scaling methods. The key goal to remember in symbol size selection is that each symbol should be readily distinguishable from the others. This may be best achieved by manual adjustments in symbol size, based on a standard technique— eyeballing it and estimating.

Proportional symbol maps can utilize data in the form of absolute numbers or counts, or in the form of percentages, ratios, or proportions. They can also depict nonquantitative ordinal (ranked) information, such as *small, medium,* and *large.* The creation of proportional symbol maps is included as part of lab exercise 11.

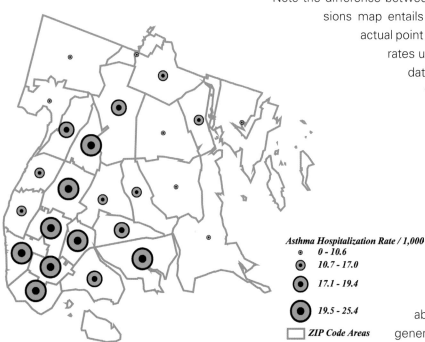

Asthma Hospitalization Rate / 1,000
- *0 - 10.6*
- *10.7 - 17.0*
- *17.1 - 19.4*
- *19.5 - 25.4*

ZIP Code Areas

Figure 3.20 A proportional symbol map, with data aggregated by area, in this case ZIP Codes. Symbols are placed at points, the ZIP Code centroids, that carry the information about the variable—asthma rates by ZIP Code.

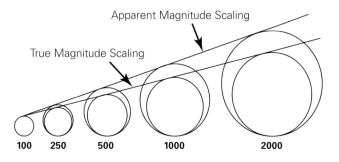

Apparent Magnitude Scaling

True Magnitude Scaling

100   250   500   1000   2000

Figure 3.21 The size differences between proportional symbols must be exaggerated in order for the differences to be perceived accurately by the human eye. Redrawn, courtesy of JP Publications. From *Map Use: Reading, Analysis, Interpretation* (4th Edition). By Muehrcke et. al. Reprinted by permission of JP Publication, Madison WI © 2001. JP Publications™.

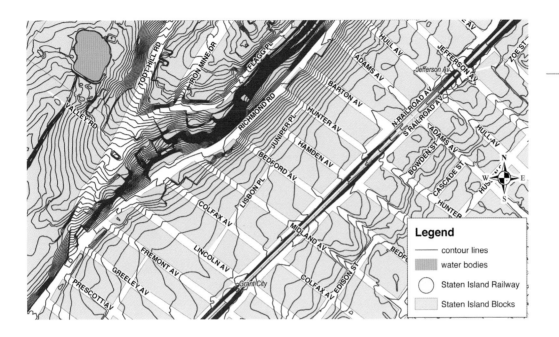

Isarithmic maps

An isarithmic map, also often called a contour map, is one that maps a continuous variable, using lines called isolines to connect places of equal quantity of the variable. The lines demarcate and locate the values. All points along any given line are presumed to have the same value for the particular variable being mapped (figure 3.22).

To create an isarithmic map, the variables to be mapped must be continuous in nature. Discrete occurrences of phenomena, such as the locations of factories within a county, cannot be mapped isarithmically. Most environmental variables, such as air quality, soil moisture, temperature, precipitation, or barometric pressure, are considered to vary continuously and are therefore appropriate for isarithmic mapping. Various human-related variables, such as cost of travel time or median housing cost can also be considered continuous and can thus be plotted with isolines. Variables based directly or indirectly on area, such as population density (population per square mile) or species density, may also be mapped isarithmically, as can disease rates (cases per 1,000 people), percent of households living in poverty, and other variables constructed from ratios or percentages. The most common type of isarithmic map is a topographic map of elevation contours in which the isolines represent locations at the same elevation.

There are many names for different types of isolines, each with a specific purpose. **Isobaths** show depth below a datum (such as mean sea level); **isotherms** show average temperatures; **isobars** show atmospheric pressure; **isohyets** show precipitation; **isospecies** show density of a species; and **isochrones** show travel time from a given point.

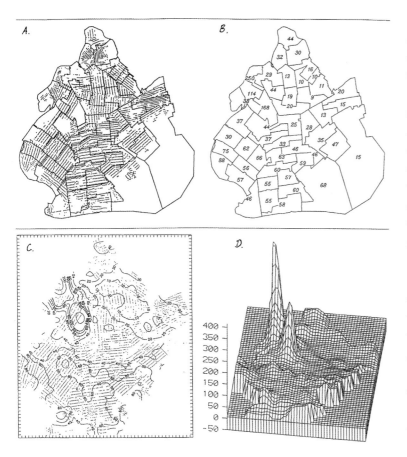

Figure 3.23 Brooklyn, New York, land values expressed as a contour map and as a three-dimensional surface. Drawing A shows the location of block centroids where residential land values were calculated with residual methods; assessor work districts are superimposed. Drawing B shows average block land value ($ per square foot) for each assessor work district. Drawing C shows land values expressed as contours (10, 20, 40, 60, 80, 100, 200, 300 $psf) over a sample of block centroids. Drawing D shows land values as a three dimensional surface. Courtesy of Jack Eichenbaum, PhD.

An isoline map requires using line symbols which are threaded through a series of control points, indicating various quantities of the variable being mapped. Control points, or data points, are locations where measurements were taken or samples were collected, or they are places chosen to represent values within areas (such as ZIP Code centroids). The locations of these points can generally be calculated by their geographical positions (latitude and longitude) or by their x,y coordinate references. The lines link places of equal value. In this way, a value can be determined for every place on the map, even if a sample or measurement was not taken everywhere.

Because most sampling processes are irregular, the points that are ultimately used to determine placement of the isolines are unevenly spaced. The placement of the line of a specific value is dependent on interpolation techniques, which have become fairly standard and are automated in many GIS software programs. Interpolation refers to the process of assigning a data value to a place where no actual data exists for that place, with the assigned value estimated from the values of nearby data points. The assumption fundamental to interpolation is that the value of a point will be closer to the values of points nearer in space than to the values of points farther away. Therefore, the known data points close to the point being interpolated will have a greater effect on calculating its value than points father away will have. There are a number of interpolation methods commonly used, depending on data type, range, sampling technique, purpose of analysis, and other factors. Interpolation is discussed more fully in chapter 9.

The mapmaker must choose the isarithmic interval for the map. Normally, a uniform interval is appropriate, such as an interval of ten, whereby the isolines would represent values of zero, ten, twenty, thirty, and so on. The selection of both the isoline intervals and the lowest isoline value must be done by careful examination of the range of data values. The goal is to show accurately the distribution of values while also displaying sufficient details of the dataset.

In planning, the most frequently used type of isarithmic map is the **isopleth,** in which the isolines connect points representing geographic areas of equal value. These areas are usually political or statistical units having a variable derived from a ratio, percentage, or average attached to each unit. In some ways, these maps are more difficult to construct since the data point used is not the point where the sample or measurement was taken, but is merely the geographic center (the centroid) of the unit by which the data is aggregated. If the geographic unit is large, this presents problems in interpolation, and calls into question the extent of the area that the data value attached to the centroid actually represents (figures 3.23 and 3.24).

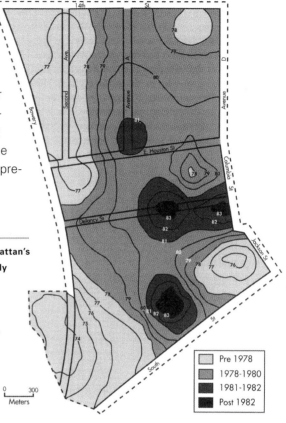

Figure 3.24 **Isarithmic map showing the advance of gentrification. Manhattan's Lower East Side (now called the East Village) in earlier times was primarily a poor immigrant neighborhood consisting of tenement buildings built in the nineteenth century. But in recent years, gentrification has arrived and the area is now trendy and its rents have skyrocketed. Isolines show the date of the onset of gentrification.** Source: Abler, Ronald F., Melvin G. Marcus, and Judy M. Olsen, eds. *Geography's Inner Worlds* © 1992, by Rutgers, the State University. Reprinted by permission of Rutgers University Press. Modified from Smith, Duncan, and Reid, 1989, 'From disinvestments to reinvestment: Tax arrears and turning points in the East Village.' *Housing Studies* 4:238–252. Taylor & Francis Ltd, *http://www.tandf.co.uk/journals.*

Pre 1978
1978-1980
1981-1982
Post 1982

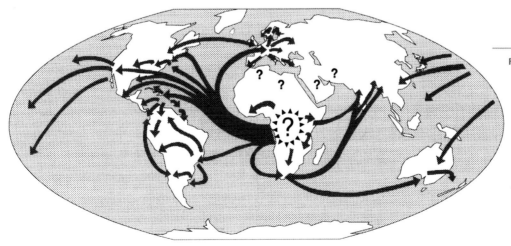

Figure 3.25 **Distributive flow map showing postulated diffusion of AIDS during the 1970s and 1980s.** From *Geography's Inner Worlds* by Ronald F. Abler, Melvin G. Marcus, and Judy M. Olsen, eds. © 1992, by Rutgers, the State University. Modified from "The Origin and Diffusion of AIDS," by Gary Shannon and Gerald Pyle, in the *Annals of the Association of American Geographers* 79:1–24, 1988, with permission of Blackwell Publishing.

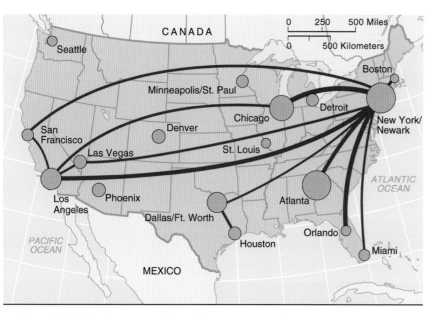

Size of circle indicates number of passenger enplanements, 2002

⊙   equals 20 million enplanements

Width of route line indicates total number of passengers on most heavily trafficked individual routes, 2002.

────   equals 1 million passengers

Figure 3.26   This combination flow map and proportional symbol map shows two different aspects of air passenger travel in the United States. Bergman, Edward F.; Renwick, William H., *Introduction to Geography: People, Places, and Environment*, 2nd Edition. © Reprinted by permission of Pearson Education, Inc. Upper Saddle River, NJ.

Flow maps

A flow map shows volume and direction of the movement of something, such as population or commodities, by means of arrow symbols of different widths. Lines on quantitative flow maps are scaled so that their widths are proportional to the amounts they represent, and the arrowheads point in the direction of the flow movement. Typical variables in a flow map include numbers of people, or the tonnage, volume, or economic value of a commodity. Flow maps are not necessarily accompanied by a legend or a scale, as the amounts depicted by the arrows are often intended to be judged relative to each other by the map user and not by some standard. As with many other types of thematic maps, the information is communicated through a snapshot technique, giving the map user a visual impression of the mapped phenomena, and is not meant to convey exact quantities.

A flow map may show the volume and disposition of a country's exports or imports, or a city's vehicular traffic flow, or the movements of groups in and out of a city or country; flow maps are particularly well-suited to show migration patterns. Distributive flow maps, which don't show quantities but rather direction, origin, and destination of flows, can also display abstract flows such as diffusion of an idea, fashion, or trend (figure 3.25).

Flow maps often are drawn schematically, omitting much of the geographic mapmaker in favor of a block organizational chart. This kind of map generalizes the geography while preserving the data. Flow maps are sometimes also used in combination with other thematic map types, such as proportional symbol maps or choropleth maps (figure 3.26).

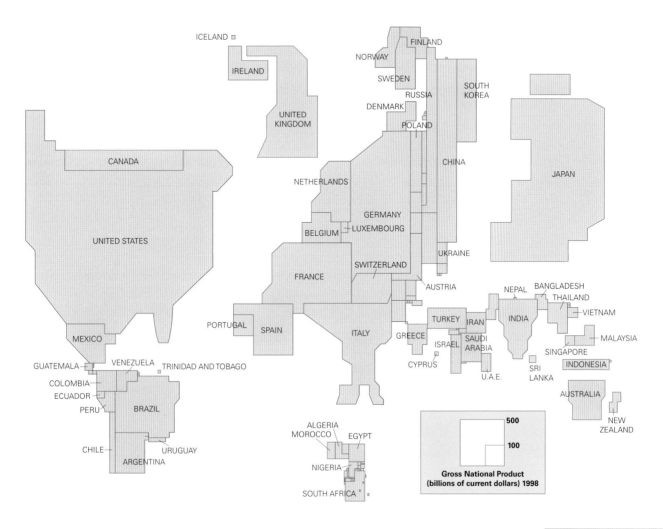

Figure 3.27 This cartogram shows gross national products of countries. Redrawn from Bergman, Edward F.; Renwick, William H., *Introduction to Geography: People, Places, and Environment*, 2nd Edition. © Reprinted by permission of Pearson Education, Inc. Upper Saddle River, NJ.

Cartograms

A cartogram, or a value-by-area map, shows land areas sized to reflect the magnitude of the variable being mapped. Normally, the geographic units shown on maps reflect their real geographic size. Not cartograms—cartograms ignore true geographic size. In other quantitative thematic maps, data is mapped by symbolizing the variable's quantity and placing the symbol in or on the geographic units. In the cartogram, the size of the geographic unit itself is intended to communicate the variable's quantity. For instance, in a cartogram of world population by country, the geographic size of the countries would be drawn in proportion to their population size, not to their geographic size. China, India, the United States, and Indonesia, the world's four most populous countries, would therefore be drawn with the largest geographic extents. The size of the country itself on the map thereby represents the variable of population size (figure 3.27).

The cartogram does not lose data by classification or generalization, and in this way it is similar to the unclassed choropleth map, or the one-to-one dot density map. However, the map user may find it difficult to understand the cartogram, depending on the map user's existing knowledge of and familiarity with the geography being portrayed. Many map-makers choose to include a small inset map with the cartogram. The inset map can be used to remind the map user of the real relationship and physical sizes of the geographic units being mapped, so the information embedded in the altered sizes of the geographic units in the cartogram can be more easily interpreted. The geographic shapes on many cartograms can be highly generalized, often with box-like forms that make no attempt to conform to true shape, resulting in a map that looks more like an organizational chart (figure 3.28).

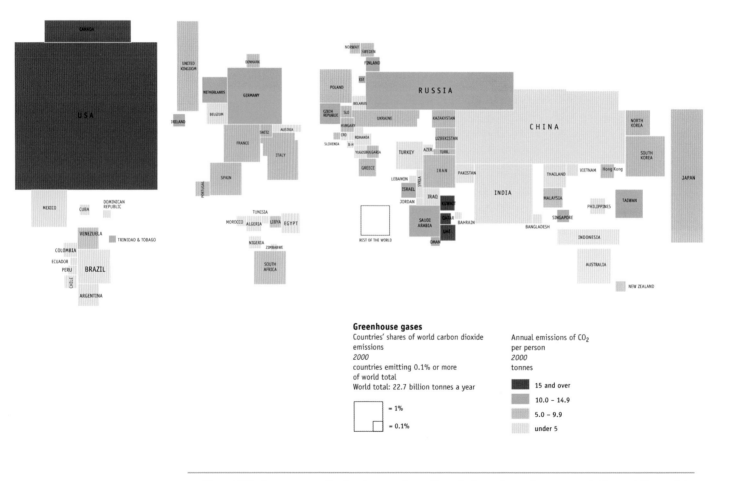

Figure 3.28   A more generalized cartogram shows the emissions of greenhouse gases in the world's countries. Compare to the GNP cartogram. Reproduced from *The State of the World Atlas*, 7th Edition, by Dan Smith (Penguin, 2003). © Myriad Editions Limited/ www.myriadeditions.com.

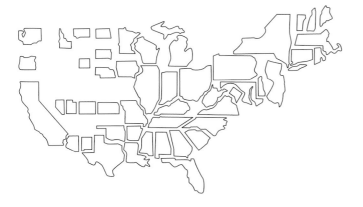
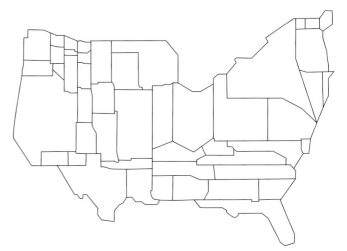

There are two basic forms of cartograms: contiguous and noncontiguous. Contiguous cartograms are mapped to approximate the actual geographic location and spatial relationships of the entities. Noncontiguous cartograms are mapped showing the geographic units as separate pieces, arranged more or less in proper geographic relation to each other, but with no common boundaries. There are advantages and disadvantages to each method. Because of the way cartograms express the theme being mapped by using their size to reflect some variable other than geographic size, contiguous relationships are difficult to maintain. Accurate interpretation of cartograms assumes a high level of knowledge on the part of the map user about the usual appearance of the geographic entities. In the noncontiguous form, correct shape of the geographic entities can be maintained, but the meaning of the map may suffer because countries aren't recognizable, and because true boundary relationships are sacrificed, accuracy suffers too (figure 3.29).

As with flow maps, many cartograms are often used in combination with other thematic mapping methods, such as choropleth or proportional symbol maps (figure 3.30).

Figure 3.29 *(left)* **Noncontiguous cartogram of U.S. population and** *(right)* **contiguous cartogram of U.S. population.** Redrawn, courtesy of JP Publications. From *Map Use: Reading, Analysis, Interpretation* (4th Edition). By Muehrcke et. al. Reprinted by permission of JP Publications, Madison WI © 2001. JP Publications™.

American Indian Population

Distribution of American Indian Population in the Contiguous United States 1990

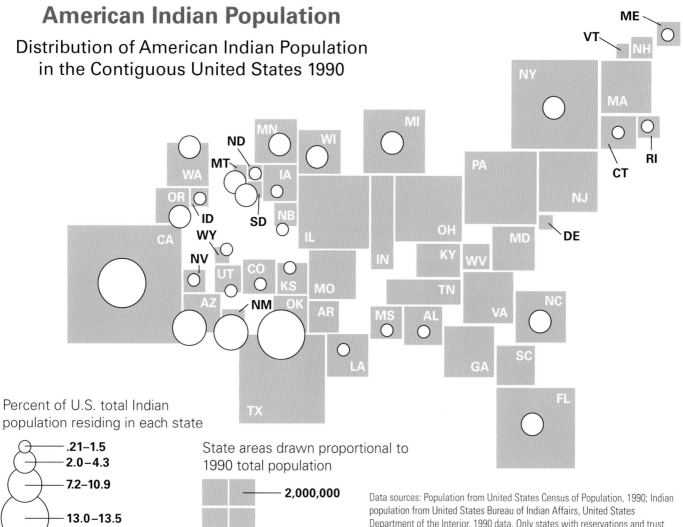

Percent of U.S. total Indian population residing in each state

- .21–1.5
- 2.0–4.3
- 7.2–10.9
- 13.0–13.5

State areas drawn proportional to 1990 total population

2,000,000

© Borden D. Dent, 1992

Data sources: Population from United States Census of Population, 1990; Indian population from United States Bureau of Indian Affairs, United States Department of the Interior, 1990 data. Only states with reservations and trust lands are mapped. States with no such lands had populations accounting for only 11 percent of total Indian populaton in the contiguous United States.

Figure 3.30 A combination map, with both a cartogram and a proportional symbol map. Placing a second variable over a population cartogram may reveal interesting new patterns, or patterns not evident if mapped on geographical space. Here it is clear that the Native American population is concentrated in states having relatively small total population, except for California. Redrawn from *Cartography: Thematic Map Design,* by Borden Dent. Reproduced with permission of The McGraw–Hill Companies. © 1999.

Combination thematic maps

Thematic maps depicting one variable are referred to as univariate maps. However, it is common to plot more than one variable on a map by using two thematic map types, and these are referred to as bivariate maps (figures 3.31 and 3.32). Spatial patterns and distribution can often be more clearly seen by using this technique, explored further in chapter 4.

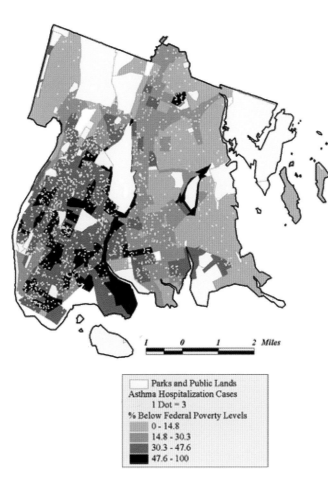

Figure 3.31 **A combination choropleth and dot density map. A dot density theme overlaid on a choropleth theme is an effective way to display two variables on the same map to better visualize their relationship.**

M A P

HATE SPRINGS ETERNAL

Today, over a hundred years after the founding of the Ku Klux Klan, fewer than one in three U.S. hate groups are Klan-affiliated, and most are not based in the South. In other regions, chapters of neo-Nazi and Skinhead organizations far outnumber those of the Klan, as do the ministries of racist religious sects in the West, many empowered by the political rise of the Christian right. Since 1996 the number of U.S. hate groups has jumped 25 percent, solidifying strongholds in rust-belt suburbs and immigrant urban hubs. According to Alabama's Southern Poverty Law Center, which has tracked the phenomenon since 1981, the expansion reflects blue-collar job losses from international trade agreements and burgeoning economic inequity, which, coupled with hundreds of Internet sites, have helped to foment a fresh round of old-fashioned xenophobia. As new census results feed racists' deepest fears and increasing computer access helps spread their voices, the overseas foes of the twentieth century may come to seem tame in comparison to the plugged-in mobs awaiting their turn just down the block.

Figure 3.32 Nominal data displayed over a choropleth map. Location of hate groups in relation to counties with high percentages of minority population, who might become targets of those groups. Copyright © 2000 by Harper's Magazine. All rights reserved. Reproduced from the March issue by special permission.

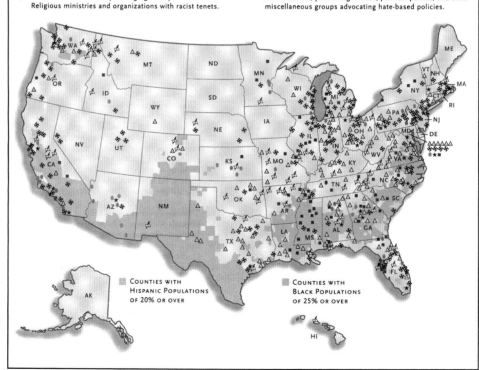

Active U.S. Hate Groups

△ KU KLUX KLAN (**163** chapters)
46 Christian white-supremacist groups, descended from the organization founded in 1865.

⚒ NEO-NAZI (**151** chapters)
24 fascist white-supremacist groups, many promoting atheism or Odinism, based on ancient European polytheism.

⚔ CHRISTIAN IDENTITY (**62** congregations)
Religious ministries and organizations with racist tenets.

⚑ SKINHEAD (**48** chapters)
24 fascist white-supremacist groups, many with Nazi affiliations; most members are under 30 and migratory.

★ BLACK SEPARATIST (**29** chapters)
2 groups advocating race-based hatred.

■ OTHER (**84** organizations)
Media outlets, publishing houses, political parties, and other miscellaneous groups advocating hate-based policies.

COUNTIES WITH HISPANIC POPULATIONS OF 20% OR OVER

COUNTIES WITH BLACK POPULATIONS OF 25% OR OVER

Multivariate maps

A more unusual form of thematic map uses a series of complex symbols to plot out multiple variables. These multivariate maps can be very interesting but are often difficult to interpret, since each symbol contains information about many variables or aspects of variables (figures 3.33–3.36).

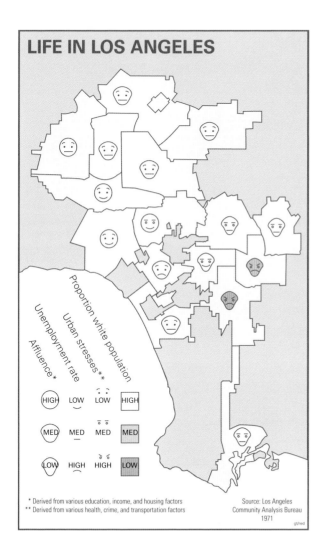

Figure 3.33 **This multivariate map of Los Angeles, circa 1971, shows four sociodemographic factors, each divided into three classes.** Redrawn, courtesy of JP Publications. From *Map Use: Reading, Analysis, Interpretation* (4th Edition). By Muehrcke et. al. Reprinted by permission of JP Publications, Madison WI © 2001. JP Publications™.

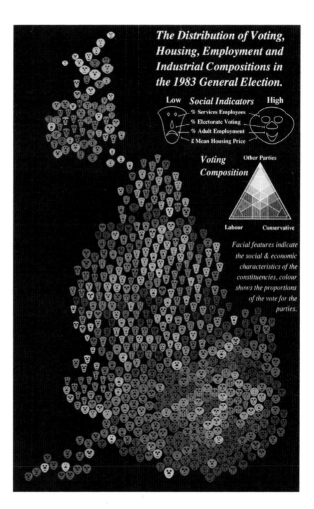

Figure 3.34 **A multivariate cartogram. This map type is particularly effective for identifying exceptions—faces which do not fit in with the crowd.** From *Visualization in Geographical Information Systems*, by H. Hearnshaw and D. Unwin, eds. © 1994 John Wiley & Sons Limited. Reproduced with permission.

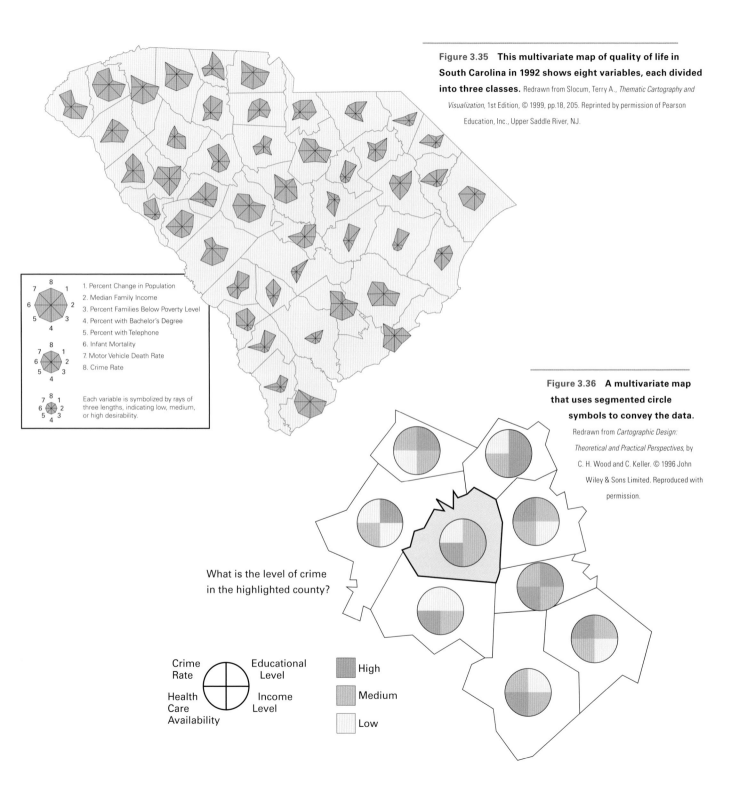

Figure 3.35 **This multivariate map of quality of life in South Carolina in 1992 shows eight variables, each divided into three classes.** Redrawn from Slocum, Terry A., *Thematic Cartography and Visualization*, 1st Edition, © 1999, pp.18, 205. Reprinted by permission of Pearson Education, Inc., Upper Saddle River, NJ.

1. Percent Change in Population
2. Median Family Income
3. Percent Families Below Poverty Level
4. Percent with Bachelor's Degree
5. Percent with Telephone
6. Infant Mortality
7. Motor Vehicle Death Rate
8. Crime Rate

Each variable is symbolized by rays of three lengths, indicating low, medium, or high desirability.

Figure 3.36 **A multivariate map that uses segmented circle symbols to convey the data.** Redrawn from *Cartographic Design: Theoretical and Practical Perspectives,* by C. H. Wood and C. Keller. © 1996 John Wiley & Sons Limited. Reproduced with permission.

What is the level of crime in the highlighted county?

Crime Rate / Educational Level / Health Care Availability / Income Level

High
Medium
Low

Washing 'South' Out of Bronx Mouths; Hoping That 'Downtown Bronx' Will Sound More Uptown

By ALAN FEUER
The New York Times

It has been said that the human spirit is by nature optimistic and that the slightest changes have the power to elevate its mood. When people are feeling mired, they may change hairstyles or brands of beer, but rarely can a neighborhood comfort itself by changing habits all at once.

It can, however, change its name.

Take the South Bronx, for example, where a growing group of residents say they have tired of being thought of as the ugly stepchild of the borough. The neighborhood, they say, has outgrown its reputation as a place of drive-by shootings, poverty and apartment buildings racked by flame.

In an effort to repair the broken window of its image, they have decided to leave the social work to the social workers and the redevelopment to the developers. They have instead have embarked upon a somewhat quixotic mission to improve the image of the neighborhood by changing its name.

"The South Bronx will become the Downtown Bronx, believe you me," said George Rodriguez, who is leading the campaign.

Mr. Rodriguez is the chairman of Community Board 1, which serves Melrose, Mott Haven and Port Morris, and he has lived in the South Bronx for more than 40 years. As befits a man who has placed great faith in the healing power of language, Mr. Rodriguez speaks of his particular tilt at the windmills of prejudice in eloquent, if slightly tortured, terms.

"We are extraordinarily gung-ho in this matter," he said last week. "It has not been undertaken capriciously or contrary to good thinking, but to change the stigma. If you take a helicopter and go upstairs into the sky and look down, you will see that the neighborhood has changed."

He is correct. The South Bronx is no longer the blockswide conflagration that shocked the country during the 1977 World Series when the Goodyear blimp flew above its streets and beamed horrific images of rampant fires and billowing smoke to television viewers across the nation.

Nor is it quite the zone of desolation caught by news cameras that year when President Jimmy Carter tiptoed through the gutted rubble of Charlotte Street.

Nonetheless, the neighborhood has problems not easily fixed by switching a couple of words on its street signs. The South Bronx is part of the poorest Congressional district in the country. Unemployment is nearly 20 percent. More than half the children in the district live in poverty, according to the South Bronx Overall Economic Development Corporation. Half the population relies on public assistance. A third of the households earn less than $10,000 a year.

This onslaught of statistical misery has done little, if anything, to curb the enthusiasm of Mr. Rodriguez and his partners, who say their first order of business is to place new signs on the Bronx side of the Willis Avenue, Madison Avenue and Triborough Bridges saying, "Welcome to the Downtown Bronx."

In the three months since the City section of The New York Times reported on the movement to rename the South Bronx, Mr. Rodriguez has reached out to the boards of several local institutions, from Lincoln Hospital to Hostos College, to generate support. His group is also working with members of the City Council to have the neighborhood's name officially changed.

"I am more than highly optimistic," Mr. Rodriguez said, "that all of this will occur next year."

That said, there is plenty of time to delve into the history of the South Bronx, a term that came into use in the late 1940's, according to Lloyd Ultan, the borough historian.

Mr. Ultan said the term was coined by social workers who had noticed pockets of poverty appearing in the neighborhoods around 138th Street as black and Latino families escaped the crowding of Harlem by flooding northward after World War II. By the 1950's, everything south of 149th Street was considered to be the South Bronx. By the 1960's, it was everything south of the Cross-Bronx Expressway. By the 1970's, Mr. Ultan said, the border had crept north to Fordham Road.

"I remember saying once that if this keeps going, anything south of McLean Avenue would soon be the South Bronx, and McLean Avenue is in Yonkers," he added.

Oddly enough, there was never a section of the Bronx that was thought of as a downtown area, Mr. Ultan said. It is different from Brooklyn, for example, where a clearly defined downtown sits in the neighborhood surrounding Borough Hall. The Bronx had—and still has—transportation and business hubs, but they are far-flung, Mr. Ultan said, from the main government office on 161st Street to the shopping districts of Co-op City and Fordham Road.

To consider the South Bronx as the Downtown Bronx does not, however, offend him.

"Frankly," he said, "it's just as good as any other place."

The South Bronx is hardly the first location in the country that has tried to polish its image with a name change, and it could be instructive to determine whether other towns and cities have benefited by a switch.

"I haven't seen it do a thing for our town, and I've been living here since I was 18," said Shirley Wadding, 58, who is the mayor of Lake Station, Ind., which changed its name from East Gary in 1977. "Basically, we didn't want to be associated with Gary anymore, but all it got us was our own identity."

The residents of Eastpointe, Mich., could be considered experts in the name-change game. The town, on the eastern border of Detroit, has gone through five names.

Founded by Irish Protestants, it was first known as Orange. When Irish Catholics moved in, it became Erin Township. Because it is halfway between Detroit and Mount Clemens, it was later called Halfway. Eventually, it became East Detroit, a name that stuck until 1993, when its residents, perhaps exhausted, settled on Eastpointe.

"I wasn't in favor of that last change," said Harvey Curley, who was mayor then. "I guess I'm the nostalgic, traditional sort."

Some places have found a tremendous boon in changing their names. In fact, one need look no farther than Sleepy Hollow in Westchester County to find an example of improvement by renaming.

Sleepy Hollow, which was known as North Tarrytown as recently as 1996, found inspiration for its new name in "The Legend of Sleepy Hollow" by Washington Irving, who had a large estate there. When the movie "Sleepy Hollow" was released a few years ago, the tourists and investors started coming to town in semi-impressive numbers.

"We get all types of people calling and visiting," said Mayor Philip Zegarelli. "It's a mysterious name with a mystique all its own. It has a type of cachet on the ground."

Unfortunately, the Downtown Bronx lacks a certain cachet and does not quite roll off the tongue with the same sort of spooky allure. Not that Mr. Rodriguez cares a whit.

"I concur somewhat about the name," he said, "but the Downtown Bronx is exactly what it is. You've got the north Bronx, the west Bronx, the east Bronx and the Downtown Bronx.

"Just not the South Bronx, you understand?"

Defining the South Bronx

The unofficial border of the South Bronx moved farther and farther north as the population changed, according to Lloyd Ultan, the borough historian.

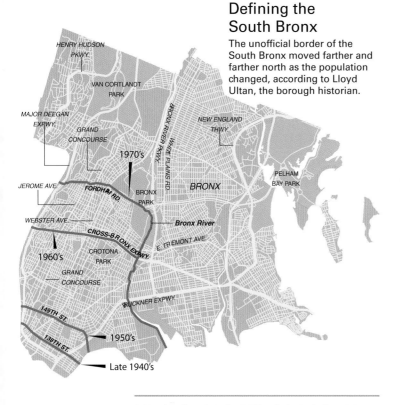

Article and map © 2003 by the New York Times Co. Reprinted with permission.

What's in a name?

A neighborhood or community name has definite connotations for people, conjuring up images based on their own experiences, hearsay, or media reports. These names often refer to poorly-defined geographies, as the story in *The New York Times* describes the area known as the South Bronx.

Over the years, the boundaries of the South Bronx have shifted and expanded, and eventually encompassed much of the Bronx. Since the appellation of South Bronx has come to symbolize everything that went wrong with urban America in the 1960s and 1970s, people now feel it is wiser to dissociate themselves from this name.

An earlier *New York Times* article in 1995 posed the question in its headline: "Mapping the South Bronx: It's South, but South of What?" As these articles point out, the name and associated characteristics of a neighborhood can have profound impacts on public perception, on public and private investment in the community, and on prospects for future improvements.

The collective mental mapping that has occurred, resulting in the expansion of the area falling under the rubric of the South Bronx, is a common phenomena, and one that can be reversed or transformed as conditions (and perceptions) change. There are examples of this in cities all over the country.

In New York's Hell's Kitchen, for example, the area formerly had a reputation as a rough waterfront neighborhood of tenements and industries. Now it is home to a burgeoning theater district, upscale restaurants and shops, and high-rent residences in the former tenement buildings. The neighborhood kept the name Hell's Kitchen, which has altogether different connotations, and indeed, has assumed a certain ironic cachet.

Other communities find it more advantageous to cast off past associations by adopting a completely new name to accompany the new image.

The lesson to be learned from these examples is that place names and their perceived geographic boundaries are important factors in planning, and that often boundaries are imposed on geographies through a process of mental mapping.

Written Exercise 3

Mental mapping

+ Using a blank outline map of your city or town, draw your own mental map of your community: its major roads, landmarks, identifiable neighborhoods, and so on. Highlight those features that are important to community life, as you perceive it. Provide short text descriptors of areas or places.

+ Then, compare your map with an accurate reference map, and make written observations about the differences, but do not change your map.

+ If you are doing this exercise as part of a class or group, circulate or display all the maps so they can be viewed as the individual parts of one imaginary composite map. Note the features that occur on every map, and the features that appear on only one or two maps.

+ What do the mental maps reveal about the perceptions of the individuals in the class regarding their hometown or neighborhood? What are the similarities and differences?

+ How might this information be useful in the planning process?

References and further reading

Aberley, Doug. 1993. *Boundaries of Home: Mapping for Local Empowerment*. Gabriola Island, B.C.: New Society Publishers.

Campbell, John. 2001. *Map Use and Analysis*. New York: McGraw–Hill.

Cuff, David J., and Mark T. Mattson. 1982. *Thematic Maps: Their Design and Production*. New York: Methuen.

Dent, Borden. 1999. *Cartography: Thematic Map Design*. New York: McGraw–Hill.

Dorling, Daniel, and David Fairbairn. 1997. *Mapping: Ways of Representing the World*. Harlow, UK: Addison Wesley Longman, Ltd.

Hartshorn, Truman A. 1992. *Interpreting the City: An Urban Geography*. New York: John Wiley and Sons.

Heywood, Ian, Sarah Cornelius, and Steve Carver. 1998. *An Introduction to Geographical Information Systems*. New York: Prentice–Hall.

Monmonier, Mark. 1991. *How to Lie With Maps*. Chicago: University of Chicago Press.

Muehrcke, Phillip C., Juliana O. Muehrcke, and Jon A. Kimerling. 2001. *Map Use: Reading, Analysis, Interpretation*. Madison, WI: JP Publications.

Pickles, John. 1995. *Ground Truth: The Social Implications of Geographic Information Systems*. New York: The Guilford Press.

Portugali, Juval, ed. 1996. *The Construction of Cognitive Maps*. New York: Springer.

Slocum, Terry, A. 1999. *Thematic Cartography and Visualization*. Upper Saddle River, NJ: Prentice–Hall.

Data Classification Methods *and* Data Exploration

Four

The way data is classified in a thematic map has substantial implications for the map's message, as well as for its aesthetics. In this chapter we discuss what data classification is, why it is important, the different methods of data classification, and how classification can be used in data exploration. We begin with some background on statistical terminology, measurement scales, and mapping quantitative information. While at first glance these numerical concepts may seem unrelated to mapping and GIS, in reality they are crucial components in using and making maps. Planners, therefore, need to be well-versed in these topics in order to benefit from the full potential of GIS as a tool for the analysis and communication of data.

Quantitative vs. qualitative information

Thematic maps may show quantitative or qualitative information. Quantitative maps (figure 4.1) have as their basis the numerical relationships of the variables being mapped— average home value by county is an example of a quantitative map. Qualitative maps (figure 4.2), by contrast, are based on descriptive information, and show location and boundaries of differences of kind or type. Land use shown by property lot is an example of a qualitative map. Qualitative maps make use of nominal measurement scales, while quantitative maps make use of ordinal, interval, or ratio measurement scales. Mapping quantitative data is the main topic of this chapter.

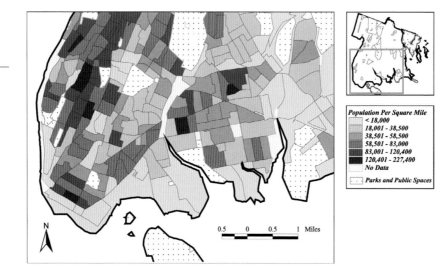

Figure 4.1 Quantitative map, showing numerical categories of data: population density in the South Bronx by census tract.

Figure 4.2 Qualitative map showing named categories of data: land uses in the South Bronx by property lot.

Measurement scales

In thematic mapping, variables are mapped using a measurement scale. There are four basic measurement scales, depending on the mathematical attributes of the variable or phenomena to be mapped—**nominal, ordinal, interval,** and **ratio** scaling.

Nominal scaling

This is the simplest form of measurement—it divides a group into subgroups by grouping similar objects or values into classes with names, hence its *nominal* title. In this measurement scale, there is grouping but not ordering. Therefore, no one group assumes more importance than another group. Examples of nominal scaling maps are land-use classes, such as residential, commercial, or industrial; land-cover classes, such as urban, forested, grasslands, agricultural, water; and land ownership, such as federal land, state land, and private land. Maps showing the location of the major languages spoken in a country, or the principal religions by place, are also using nominal scaling. Classes are distinct, and arithmetic operations between the regions are not possible—it would be like adding apples and oranges. In fact, we cannot perform any mathematical calculations between nominal classes, other than counting the number of occurrences, or frequency, of each class (figure 4.3).

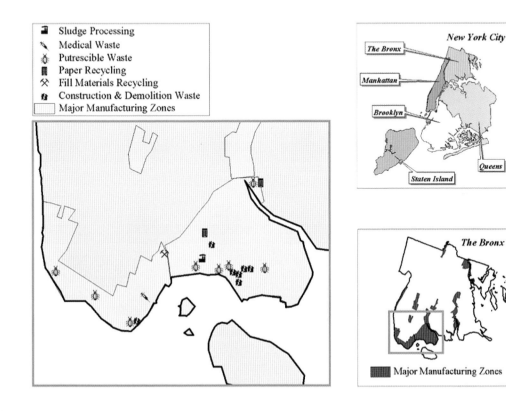

Figure 4.3 These three nominal maps show locations of qualitative features, such as different types of waste-related facilities and major manufacturing zones in the Bronx, and the names of city boroughs.

Ordinal scaling

Ordinal scaling implies a hierarchy of rank—a ranking of classes. Objects are arranged in order of greatest to least number, or as less or greater than some threshold value; these could be the health-based standard for an air pollutant, or the optimal ratio of population to community open space. Ordinal scaling indicates and symbolizes a range from high to low, and thus the relative importance of the classes to each other, but classes are not attached to any particular value. We might have a map of census tracts showing whether their median household income is above or below the federal poverty level, for example, or one showing counties ranked by rates of infant mortality as high, low, or medium. Roads could be classified by type: unpaved, local city street, county road, state highway, federal interstate. There is no named range or intuitive interval between the classes. The highest category could be twice as high as the next category, which could be three times as high as the next lower category. Therefore, we are unable to discern magnitude differences between the classes (figure 4.4).

Figure 4.4 Map using ordinal measurement scaling, showing ranked classes of a variable: high, medium, and low Hispanic proportion of total population by state.

Interval scaling

With this measurement scale, we arrange the classes in ranks, and the intervals between ranks are known. Statistical tests can be performed on data scaled this way. There is no natural origin, and zero is not the starting point, thus the scale is relative and not absolute. A map showing the average annual temperature by county is an example of interval scaling. We cannot say that a county with an average temperature of 80° is twice as hot as one with an average temperature of 40°, because 0° is not the absence of temperature. But we know that with interval scaling, the 1° difference between 39° and 40° is the same as the 1° difference between 79° and 80° because the interval between scale units is consistent.

Ratio scaling

Ratio scaling also involves ordering features with known distances separating the features. The difference is that the ratio–scale magnitudes are absolute, and have a known starting point of zero (the absence of magnitude). An example of ratio scaling is a map showing the percentage of the population that is Hispanic in each state, or the total population per square mile in each county. Most of the thematic maps that we will be creating in the lab exercises will use a ratio measurement scale (figure 4.5).

Moving from nominal to ratio scaling allows increasing amounts of information to be gleaned from the data in terms of statistical strength and predictive capabilities.

Statistics and mapping

The goal of statistics in general, and for mapping in particular, is to summarize values—to calculate the best numerical descriptor for a dataset or geographic unit, to express associations and connections between two variables, and to make deductions about the data in terms of estimates, predictions, and degrees of significance. The measurement scale used to make a thematic map will govern what type of statistical operations (if any) can be performed on the data. In turn, the statistical techniques available will govern the kind of map that can be produced. Statistical mapping is another term for quantitative thematic mapping.

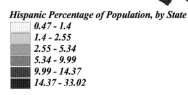

Hispanic Percentage of Population, by State

	0.47 - 1.4
	1.4 - 2.55
	2.55 - 5.34
	5.34 - 9.99
	9.99 - 14.37
	14.37 - 33.02

Figure 4.5 Map using ratio measurement scaling: Hispanic population as a percentage of total population by state.

Ratio, proportion, and percentage

The three terms ratio, proportion, and percentage are related, but the differences among them are important to understand. Planners often use derived variables in mapping—so called because they are derived from applying ratios, proportions, or percentages to values.

A **ratio** expresses the relationship between data. The number of items in one class is divided by the number of items in another class. Gender ratio is a good example—if there are twenty males in the room and ten females, the room's gender ratio is 2:1. Population density is another ratio, defined as the number of people within a certain area, divided by the area's extent. One thousand people within two square miles equals five hundred people per square mile.

A **proportion** is the number of items in one class compared to the total of all items. If the number of males in the room (20) is divided by the total population of both males and females (30), the proportion of males to the total population would be 0.66. Another example might be the number of non-Hispanic blacks in a census tract (100) as compared to the total population of that tract (1,000). This would work out to 0.1 (100÷1,000).

A **percent** is the proportion multiplied by one hundred. Using the example above, non-Hispanic blacks make up ten percent of the population of that census tract.

Variables, values, and arrays

Variables are the subjects of the data values, such as elevation, income, and population density, that we attach to a location. They are the terms that describe the location. If the geographic coordinates of a place are designated by a series of x,y coordinates, then the descriptive variable could constitute the z-coordinate. (The z-coordinate can be used for virtually any variable; elevation is a commonly used z-coordinate.) Variables normally contain the information we are interested in analyzing.

In tables and spreadsheet terminology, variables are often called **fields.** These fields form the **columns** in a table or spreadsheet, which are the vertical lines of information. Information about individual occurrences, called **records,** appear as **rows,** or horizontal lines of information in tables or spreadsheets. Individual observations are called **values.** Values appear each in their own cell of the spreadsheet. Values arranged in ascending or descending order are called an **array** (figure 4.6).

Population and Population Density of Canadian Provinces, 1991

Variable, or Field (Column)

Array (field of values organized in ascending or descending order)

Derived Field (Pop1991/Area)

Name	Area	Pop1991	Pop91_sqmi
Prince Edward Island	2207.134	129765	59.389020
Nova Scotia	21199.186	899942	44.110480
New Brunswick	27996.778	723900	26.007760
Newfoundland & Labrador	152523.521	568474	3.961215
Yukon Territory	186112.055	27797	0.150310
Manitoba	251261.199	1091942	5.157409
Saskatchewan	251738.905	988928	4.488028
Alberta	256873.619	2545553	10.231320
British Columbia	364449.495	3282051	9.142941
Ontario	380275.566	10084885	29.308860
Quebec	585860.072	6895963	13.163780
Northwest Territories	1320278.815	57649	0.045341

Record (Row)

Value (Individual Cell)

Figure 4.6 **The components of a table.**

Frequency distribution and histograms

A **frequency distribution** is an ordered array of numbers that shows the frequency of the occurrence of each value. The **relative frequency** is that proportion of the observations belonging to a particular class. Let's say we have a community with 1,000 property lots, and 750 of them are zoned for residential land use, 200 for commercial, and 50 for industrial. We want to generate a relative frequency table of land use by property lot. The relative frequency of the residential lots is 750÷1,000—or the residential lots divided by the total number of lots, which equals 0.75. The relative frequency of the commercial lots is 0.20, and of the industrial lots is 0.05. All relative frequencies for a dataset always add up to 1.0 (table 4.1).

Histograms are special graphs that use information from the relative frequency table (figure 4.7). These are described in more detail in chapter 5.

Land-use category	Frequency	Relative frequency
Residential	750	0.75
Commercial	200	0.20
Industrial	50	0.05
TOTAL Community A	**1,000**	**1.00**

Table 4.1   **Relative frequencies of land use by property lots in two communities.**

Land-use category	Frequency	Relative frequency
Residential	500	0.50
Commercial	300	0.30
Industrial	200	0.20
TOTAL Community B	**1,000**	**1.00**

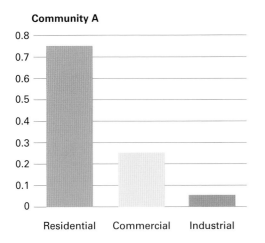
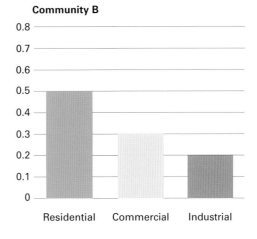

Figure 4.7   **Histogram of land-use classes for two communities.**

Summarizing data with statistics

One of the ways to summarize a dataset is to develop a number, called a **central tendency,** that describes some central characteristic about the dataset. The **mode,** the **mean,** and the **median** are types of central tendency measurements. One of the most commonly used central tendencies is the **arithmetic mean,** or **average.** For example, if we wanted to characterize the age of all the students in a class to obtain a typical age, we would add the students' ages and divide by the number of students. This gives us an average. This average could then describe the age of students in a class, without the need to consider the entire dataset of ages. This is the point of using central tendency: to provide an accurate representation, characterization, or summary of a dataset—in this case, the students' ages.

In conjunction with the central tendency, it is often helpful to calculate an indicator of how much to trust that central tendency number or how accurately the number characterizes the dataset. One way to determine this is to examine whether the numbers in a dataset are clustered closely around the central tendency number or whether they are spread out. This scattering, or concentration, is called a **measure of dispersion,** and it is a useful guide to showing how the data range is distributed.

The central tendency selected depends on the measurement scale of the dataset. Nominal, ordinal, interval, and ratio measurement scales each have an appropriate central tendency method, as well as a method for calculating the effectiveness of the summarizing descriptor.

Nominal scale: The mode and the variation ratio

With the nominal scale, we are dealing with data that is not quantified, and therefore the statistical analysis of data is limited. The mode is the central tendency measurement used with data in a nominal scale. The mode is simply the name or number of the class having the greatest frequency in the dataset. For instance, suppose our dataset of land uses within Community A contains 1,000 property lots, with 750 in residential use, 200 in commercial use, and 50 in industrial use. The mode, in this case, is the land-use category with the greatest frequency of property lots: the residential, with 750 lots out of the 1,000 total.

As we did with the central tendency, we want to estimate how accurately the mode reflects the dataset, and so we calculate a number between 0 and 1 called the **variation ratio.** This is the proportion of features or events not in the modal class. The smaller the variation ratio, the more accurate the mode is considered. If all features of a dataset are in the same class (in other words, all are in the modal class), the variation ratio would be 0.

The variation ratio (V) is calculated by dividing the frequency of the modal class by the total number of features or events in all the classes of the dataset, and then subtracting that number from 1. In our example of land use in Community A,

$$V = 1 - (750 / 1,000)$$
$$V = 1 - 0.75$$
$$V = 0.25$$

In this example, the variation ratio is a good summarizer of the data, since most of the records (the property lots) are in the modal class, (750 lots out of 1,000 are residential) and the variation ratio is correspondingly small. With the land-use data for Community B, however, the mode of 500 does not describe the data as well as the mode does in Community A, and the larger variation ratio of 0.50 for Community B supports that assessment.

$$V = 1 - (500 / 1,000)$$
$$V = 1 - 0.50$$
$$V = 0.50$$

The mode and variation ratio can also be determined for interval and ratio data.

Ordinal scale: Median and percentiles

Data in the ordinal scale is ranked data, and therefore the number that describes central tendency is the **median,** which is calculated by finding the midpoint in a ranked array of the data.

If we put the values in order by means of an ascending or descending array, we simply count from the top or bottom of the list until we reach the halfway point. The center value is the median—the midway point in the values, where fifty percent of the values are above and fifty percent of the values are below the median. If there is an even number of records, the middle two values are averaged.

The median does not necessarily summarize the data very well. We could have two very different datasets, yet exactly the same median for both, so the median is not necessarily a reliable way to compare datasets.

Suppose we have a list of unemployment rates by ZIP Code for two communities, and we want to summarize and compare the data for each community (table 4.2).

Community #1	Community #2	
1	4	
2	4	
3	4	
3	5	
5	**5**	**Median unemployment rate**
5	12	
5	18	
6	20	
6	25	
6	30	

Community #1	Community #2	
1–6	**4–30**	**Data range**

Community #1	Community #2	
3	5	**25% quartile**
6	20	**75% quartile**
3	15	**Interquartile range**
1.5	7.5	**Quartile deviation**

In the graphic, the median for each of the communities is identical, yet the datasets tell a very different story about each community. Although the median of 5-percent unemployment is a relatively low rate and occurs in both communities, note that each ZIP Code in Community #1 has a similar level of unemployment and that they tend to be close to the median 5-percent value; Community #1 is more homogeneous with respect to unemployment rates. But in Community #2, although the median unemployment rate is also 5 percent, several of the Community #2 ZIP Code areas have rates far higher than the median value. This community has much greater variation in unemployment rates. It would not be possible to assess which community has a more serious unemployment problem by looking at the median alone. In fact, it would not be accurate to say that they had the same levels of unemployment, and thus the datasets portray two very different communities with respect to unemployment rates.

Quartiles

To differentiate between two datasets with identical medians and discover which one indicates a community having the more serious unemployment problem, we can examine the dispersion of the values around the median by computing **quartile** values. The first quartile, also called the twenty-fifth percentile, is the number that separates the lowest 25 percent of the observations from the upper 75 percent. The second quartile is the median, and the third quartile is the number that separates the upper 25 percent of the observations from the lower 75 percent; this is called the seventy-fifth percentile. The upper quartile minus the lower quartile is called the **interquartile range,** and it identifies the half of the values that are closest to the median value. The interquartile range divided by two gives the **quartile deviation,** which is a measure of how the values are distributed around the median. A low quartile deviation number indicates values more closely packed around the median and therefore a median that is a better summarizer of the dataset.

In our unemployment example, we can see that the 5 percent median is a better summarizer of the ZIP Codes' unemployment rates in Community #1, because it has a quartile deviation of 1.5. The higher quartile deviation of 7.5 in Community #2 indicates that the values are not packed as tightly around the median as in Community #1. Therefore, the median represents the values in Community #1 better than it does in Community #2. By looking at the quartile deviation, you can also see that Community #2 might have some values that are extremes, while the quartile deviation value in Community #1 indicates that most of the unemployment rates are likely to be relatively low and similar to the median value.

As a point of interest, if we calculate the modes for the two datasets, Community #1 has a mode of 5 percent unemployment, while Community #2 has a mode of 3 percent unemployment. In this case, the modes do not adequately describe the datasets and, in fact, point us in the wrong direction. The median can also be used with interval and ratio data.

Interval and ratio scales: The arithmetic mean and the standard deviation

Data in interval and ratio scales is more useful in geographic analysis because more statistical tests can be applied to them. The statistic most often used as a one-number index or descriptor to describe the array of values is the arithmetic mean, more commonly known as the average. The mean is derived by adding up all the values in the dataset and dividing by the total number of individual values in the dataset. For instance, if a town contained three ZIP Codes with populations of 6,000, 11,000, and 13,000 respectively, we calculate the average population of the ZIP Codes as the sum of all three ZIP Codes (30,000) divided by 3 (the number of ZIP Codes). The average (or mean) population per ZIP Code in this town is 10,000.

The statistic used to describe dispersion around the mean is the **standard deviation.** Similar to the method described above, the standard deviation allows us to evaluate how closely the numbers in the dataset are packed around the mean—in other words, how well the average (or arithmetic mean) describes or summarizes the set of numbers.

The standard deviation is calculated by obtaining the square root of the variance, which is defined as the difference between each value in the dataset and the average; the difference is squared, those squared differences are added, the result divided by the number of values. The smaller of two standard deviations indicate values more closely packed around the mean, and therefore the mean is shown to be a better summarizer of the dataset.

As with the median, you could have two very different datasets and still have the same mean, which is why the standard deviation is so useful. For example, table 4.3 lists the annual income of ten households in two different city blocks. Both have the same mean, or arithmetic average. Despite having an identical mean, the datasets actually portray two very different blocks. In Block #1, the data range is from $6,000 to $70,000, meaning the annual incomes of the residents are quite heterogeneous. It is doubtful that the $25,000 mean value adequately represents the typical household income of that block. On the other hand, Block #2, with the same mean income, has a range of $20,000 to $30,000, so the values are packed quite closely around the mean. The annual incomes of the residents of Block #2 are more homogeneous, and therefore the mean does a better job of representing the typical household income in that block.

Table 4.3   Two datasets with the same mean values but different standard deviations: income for ten households in each of two blocks.

Block #1	Block #2
$6,000	$20,000
$8,000	$21,000
$10,000	$22,000
$13,000	$23,000
$15,000	$24,000
$18,000	$25,000
$20,000	$26,000
$22,000	$27,000
$33,000	$28,000
$60,000	$29,000
$70,000	$30,000

$275,000	$275,000	Aggregate household income by block
$25,000	$25,000	Mean household income
$6,000–70,000	$20,000–30,000	Data range
21,250	3,316	Standard deviation

The standard deviation helps us better understand the nature of the datasets, and how well the mean represents the values. Block #1 has a high standard deviation, letting us know that the values are not tightly clustered around the mean, and so less representative. Block #2 has a much lower standard deviation, letting us know that the mean value of $25,000 is more representative of the typical household income in that block.

Methods of classification

Maps displaying rates, percentages, or ratios—such as in the interval or ratio measurement scales with choropleth maps, or absolute numbers in ranges—such as in proportional symbol or dot density mapping, must generally classify the numerical information they are displaying so that map viewers can more easily see the differences and similarities in a geographic region. Classifying the data means creating ranges or classes out of the dataset and assigning each geographical unit to a class.

For instance, let's say we have rates for a disease by U.S. state. This means that, theoretically, we could have as many as fifty different values for the disease rate, assuming that each of the fifty states had a different and unique rate. In most cases, it would be ineffective to display each state in a different color or pattern to correspond to each unique rate, because it would be difficult at best to see any spatial pattern: there would be too much information, and each state would essentially be in its own class. It is much more effective to classify data, to put it into similar groups.

Classification is most often used with quantitative data, but qualitative (nominal) data can be classified as well. Classification involves organizing values in an array along an imaginary number line, then grouping the numbers according to some criteria of similarity. The mapmaker usually attempts to form groups so that the numerical differences *within* the groups are less than the differences *between* the groups.

Normally, thematic maps use at least three and as many as seven groupings, or classes, of data; five is the standard number. Studies have shown that the human eye cannot differentiate among more than eleven classes, and even maps with seven classes can lead to confusion. Conversely, a map with fewer than three classes is generally inadequate to convey the variations in data that drive the creation of the map. The four most common techniques for data classification are known as **equal interval, quantile, natural breaks, and standard deviation.**

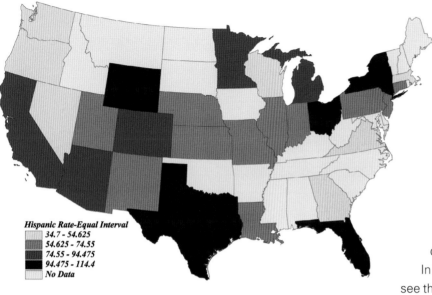

Hispanic Rate-Equal Interval
- 34.7 - 54.625
- 54.625 - 74.55
- 74.55 - 94.475
- 94.475 - 114.4
- No Data

Figure 4.8 Hispanic rates of death from heart disease, equal interval classification. Rates are deaths per 100,000, age adjusted to 1970 population. "No data" indicates too few numbers to calculate a stable estimate. States shown are those with a Hispanic rate calculated.

Equal interval

The equal interval, or equal step method of creating classes, is perhaps the most straightforward. It divides the range of values in a dataset so that each subset has an equal range. In a dataset with values of 0 to 100, to be divided into five classes evenly, data ranges for the subsets would be 1–20, 21–40, 41–60, 61–80, and 81–100. Each class is composed of the same interval: 20. Class boundaries are defined by dividing the range of values (the maximum value minus the minimum value) by the number of classes desired.

In looking at the legend in the figure 4.8 map we see that the data range for the rate of heart disease in the Hispanic population is 34.7 to 114.4 per 100,000. The data range is therefore 79.7, and if we want four classes, we divide the range by 4 and end up with an interval of 19.92. Each class therefore contains a range of 19.92. This method is useful if the data values are fairly evenly spread throughout the range, and are not skewed to one extreme or the other. If they were, we could end up with a map in which the data values fall mainly in one class; most of the geographic units would receive the same color or symbol and thus prevent us from discerning any spatial pattern in the data.

Quantile

In the quantile method of classifying data, the data range of our example remains the same, but the class intervals do not. The class intervals in the quantile method are formed by arranging the data values in ascending order, and developing classes based around the quantile ranges, as described below. The quantiles can be any number of classes, and in the example shown, figure 4.9, there are four classes. (Classes divided into four are called quartiles; five-class divisions are called quintiles.) Quantile ranges depend on the number of actual values in the dataset, as opposed to arbitrarily defined intervals of the equal interval method. Using this methodology, to find the quartile (four-class) ranges in our example, the dataset is split in half to find the median and then halved again.

In figure 4.9, the median value is 65.5 because 65.5 is the average of the two values in the middle of the data array. The data classes are formed by putting a quarter of the data values (in this example, the first group of seven data values) in the first class, the second quarter of data values (the second group of seven data values) in the second class, the third quarter in the third class, and the fourth quarter in the fourth class. The actual data values therefore determine the class boundaries, and each class should have the same number of data values. In our example, there are twenty-eight states with values, so each class of the four classes in the quartile method will have seven states. This provides a very differentiated map since each class is equally represented. Although it looks symmetrical because it results in the numerical equality of geographic units in each class, in fact the quantile method may not be the best way to represent certain datasets, especially if many of the values fall into one extreme. This, for instance, would lead to similar data values being broken up into separate classes and represented by different class symbols, which may be misleading since the data values may in reality be very similar.

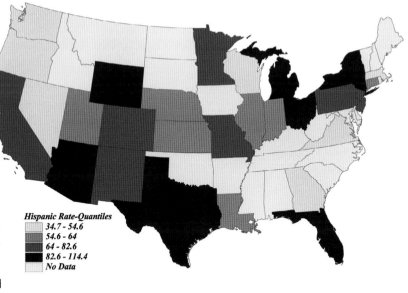

Hispanic Rate-Quantiles
- *34.7 - 54.6*
- *54.6 - 64*
- *64 - 82.6*
- *82.6 - 114.4*
- *No Data*

Figure 4.9 **Hispanic rates of death from heart disease, quantile classification.**

Natural breaks

In the natural breaks method, rather than dividing the dataset into classes based on the quantile ranges, class breaks (or class boundaries) are placed where they seem to occur naturally. The object is to create classes so that each grouping is internally consistent and alike (homogeneous), while the differences between the groupings is distinct and noticeable (heterogeneous) (figure 4.10).

For a small dataset, a scatterplot can be drawn and class breaks determined visually. For instance, in figure 4.11, assuming we want to divide the data into four classes, the array of data values in ascending order reveals natural groupings of values between 34.7 and 44.9, then a break in the data values; then the next class is formed from 48.2 to 64.0, followed by another break; then 67 to 86, then another break; and then the values in the last class are 101.8 to 114.4.

For large datasets, mathematical algorithms are generally used to develop natural breaks class boundaries. One such algorithm is called Jenks Optimization; it seeks to form groups that are internally homogeneous while also maximizing heterogeneity among classes. Jenks Optimization is the basis for the natural breaks classification process in ArcGIS. The natural breaks method is most useful as a good, all-around classification method for many purposes. For example, it can be employed when arbitrary class boundaries, used in equal interval method or classification based on mean (standard deviation) or median (quantile), are unsuitable. Since it is tailored to each dataset, the natural breaks method often brings out the nuances of data missed by other methods.

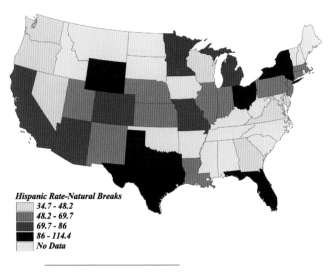

Hispanic Rate-Natural Breaks
- 34.7 - 48.2
- 48.2 - 69.7
- 69.7 - 86
- 86 - 114.4
- No Data

Figure 4.10 Hispanic rates of death from heart disease, natural breaks classification.

Figure 4.12 Hispanic rates of death from heart disease, standard deviation classification.

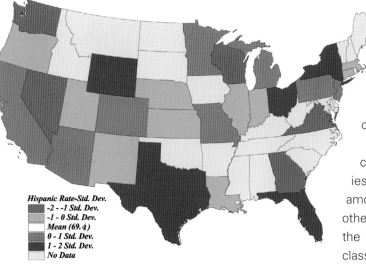

Hispanic Rate-Std. Dev.
- -2 - -1 Std. Dev.
- -1 - 0 Std. Dev.
- Mean (69.4)
- 0 - 1 Std. Dev.
- 1 - 2 Std. Dev.
- No Data

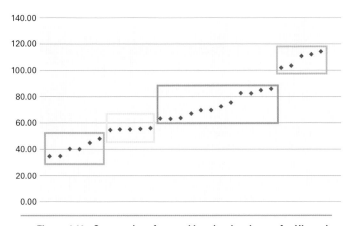

Figure 4.11 Scatterplot of natural breaks visual array for Hispanic rates of death from heart disease.

Standard deviation

The standard deviation method of classification groups the data values into classes based on the mean and the standard deviation. Each class represents one (or one half) standard deviation above or below the mean (arithmetic average) for the dataset. In our example, the mean of the Hispanic rate for deaths due to heart disease is 69.4. The standard deviation for this dataset is 23, and so each class interval is approximately 23. Note that standard deviation classification can result in classes containing values outside the actual range of the data, due to the way standard deviations are constructed. For instance, in this example, if we subtract the standard deviation from the mean of 69.4, we get 46.4 as our first class break below the mean and 23.4 for our second class break below the mean. The minimum value in this dataset is 34.7, so the second class break includes values that are outside the data range. Likewise, if we add 23 to the mean of 69.4, we get 92.4 as our first class break above the mean and 115.4 for the second class break above the mean. The maximum value is 114.4, so again the standard deviation class breaks are outside the limits of the actual data range.

The standard deviation classification method is especially useful when performing longitudinal studies (studies comparing different time periods) or for comparisons among datasets that vary widely in their mean, median, or other central tendency. In standard deviation classification, the mean and the standard deviation are the basis of the class boundary formation, and so the mean and standard deviation of different datasets can be compared with the mean and standard deviation of each other.

When using the standard deviation classification method, it is customary to use two different color ramps, indicating classes of standard deviation above and below the mean. This is called a dichromatic color ramp. In figure 4.12, orange indicates standard deviation above the mean, and the more intense orange shows increasing deviation. Purple shows standard deviation below the mean, and the more intense purple shows increasing deviation in the opposite direction.

Different maps from the same data

An alteration in the number of classes or a change in the class limits can produce dramatic differences in maps. As the previous maps show, the data can be presented quite differently depending upon which classification method is used. Notice that the states of Arizona and Michigan, for instance, are in the highest class in the quantile method, but only in the third-highest class in the other methods—even though all the maps used four classes to divide the dataset. Other states, like New York, Texas, and Florida, are in the highest class in all classification methods. These differences are highlighted in figure 4.13.

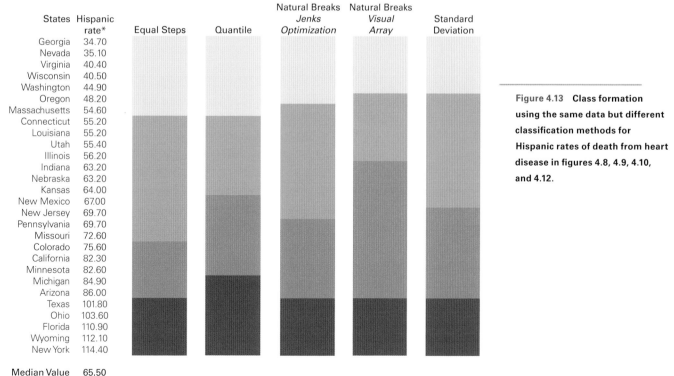

States	Hispanic rate*
Georgia	34.70
Nevada	35.10
Virginia	40.40
Wisconsin	40.50
Washington	44.90
Oregon	48.20
Massachusetts	54.60
Connecticut	55.20
Louisiana	55.20
Utah	55.40
Illinois	56.20
Indiana	63.20
Nebraska	63.20
Kansas	64.00
New Mexico	67.00
New Jersey	69.70
Pennsylvania	69.70
Missouri	72.60
Colorado	75.60
California	82.30
Minnesota	82.60
Michigan	84.90
Arizona	86.00
Texas	101.80
Ohio	103.60
Florida	110.90
Wyoming	112.10
New York	114.40
Median Value	65.50
Mean Value	69.40

* Deaths per 100,000, age adjusted to 1970 total U.S. population.

Figure 4.13 Class formation using the same data but different classification methods for Hispanic rates of death from heart disease in figures 4.8, 4.9, 4.10, and 4.12.

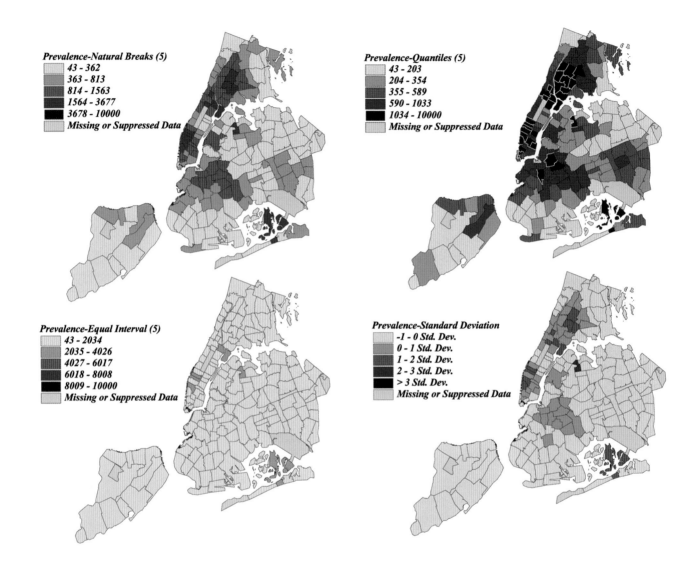

The maps of Hispanic death rates from heart disease show only mild differences, but in many cases, different classification systems will produce wildly different presentations of the same data. For an example, compare the Hispanic death rate maps with maps showing AIDS prevalence (figure 4.14).

Issues in classification

Choice of classification method can have profound impact on the message presented in the map and may influence public policy decisions made on the basis of the mapped information. Therefore it is imperative to examine several ways of classifying the data and selecting the most honest one (figure 4.15).

Some cartographers believe that thematic maps should always be presented with more than one classification scheme, to let map users know that there are multiple ways of looking at the same data. Alternatively, the thematic map could be accompanied by an explanatory note stating that the data classification depicted on the map is only one of many possible ways of looking at the data. We will be experimenting with classification methods in lab exercise 4.

A mapmaker can create classes using virtually any interval, but the classes must be chosen carefully to create meaningful presentation of the information.

Generalization and simplification are intrinsic to the process of classifying mapped data because values are grouped with others that are similar and the exact data values are not shown on the map. Because of this, specific facts may be absent from the map, even though a better general sense of the data may be communicated. Classification allows the mapmaker flexibility in presentation and considerable control over the map's meaning.

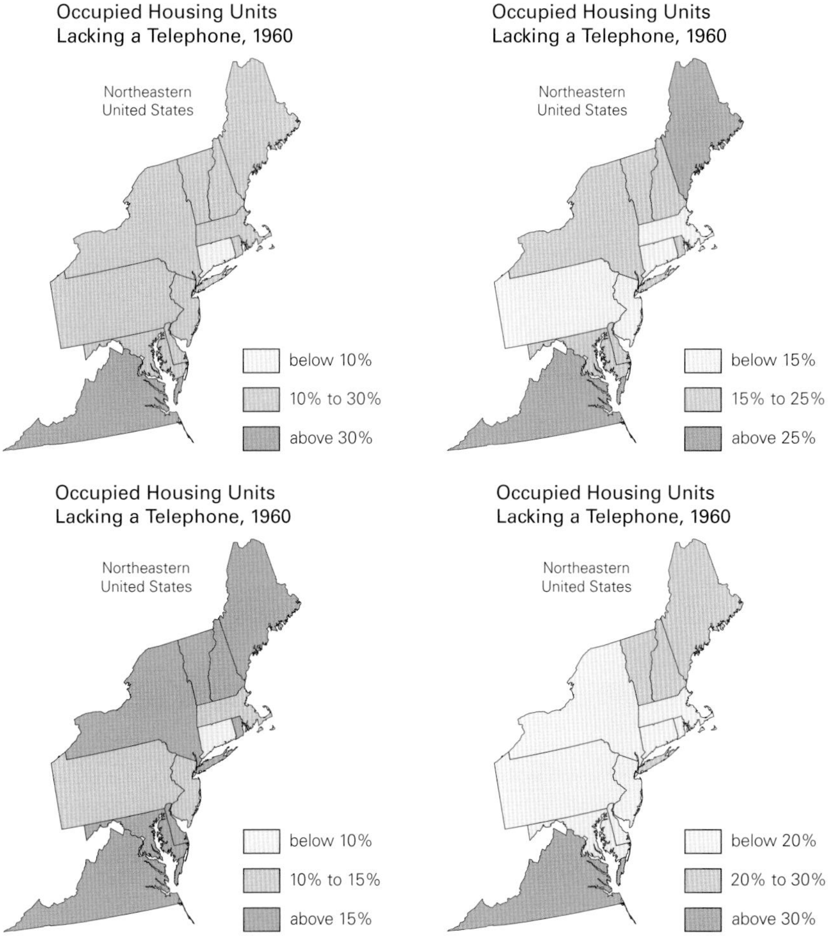

Figure 4.15 **These maps by Mark Monmonier show that when different sets of class breaks are applied to the same data, different-looking choropleth maps will result. Class breaks can be manipulated to yield choropleth maps supporting politically divergent interpretations, and class formation can have profound impacts on a map's message. This, in turn, can affect public policy, budgeting, and decision making; mapmakers must be vigilant in presenting as complete and unbiased a view of the data as possible.** Redrawn from *How to Lie with Maps*, 2nd Edition, by Mark Monmonier. Reprinted by permission of the University of Chicago Press, © 1996.

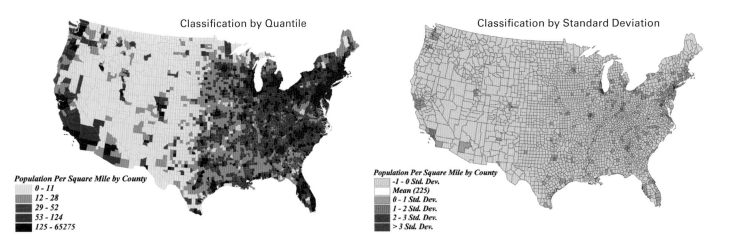

Classification by Quantile

Classification by Standard Deviation

Population Per Square Mile by County
- 0 - 11
- 12 - 28
- 29 - 52
- 53 - 124
- 125 - 65275

Population Per Square Mile by County
- -1 - 0 Std. Dev.
- Mean (225)
- 0 - 1 Std. Dev.
- 1 - 2 Std. Dev.
- 2 - 3 Std. Dev.
- > 3 Std. Dev.

Figure 4.16 These two maps present identical data using different classification methods. Which way yields a better perspective of the data? Can you learn more about population density in the United States by using both versions of the data?

The nature of the dataset helps in selecting a classification method. There isn't any single optimal way of classifying data for maps. You have to consider the purpose of the map, the particular statistical profile of the dataset to be mapped, and how best to simplify the data—whether the method is to be based on graphic principles or on more formal mathematical rules. Developing class intervals should be done on a trial-and-error basis to develop the most suitable approach. With GIS, this is much easier to do. And sometimes, as mentioned above, it is wisest to present the data using two different classification methods, as is done in figure 4.16.

Unclassed choropleth maps plot data using a unique color or pattern for each differ-ent value by geographic unit. Values are not grouped into classes. The map will have as many different colors or patterns as there are different values in the dataset. Unclassed choropleth maps are more accurate than classified ones, because no information is lost in the classification–aggregation process. However, the burden shifts to the map reader to generalize, simplify, and interpret the map. Map interpretation may be more difficult, because there may be more colors or patterns to differentiate, and distribution and pat-terns of the values may not be obvious. With unclassed data, the mapmaker relinquishes control over the map's message (figure 4.17).

Data exploration

Data exploration means looking for overall trends in the data, and relationships among data-sets; it can be thought of as playing with the data to reveal unknowns and to point the way toward avenues for further productive analysis. This is a very different process from making maps for presentation. In data exploration, the goal is not to communicate information, but to examine data to find any patterns or unexpected trends, as well as to be able to characterize accurately the data's structure and content. Since each dataset is unique, data exploration can be a trial-and-error process in which the planner is looking at the dataset in a multitude

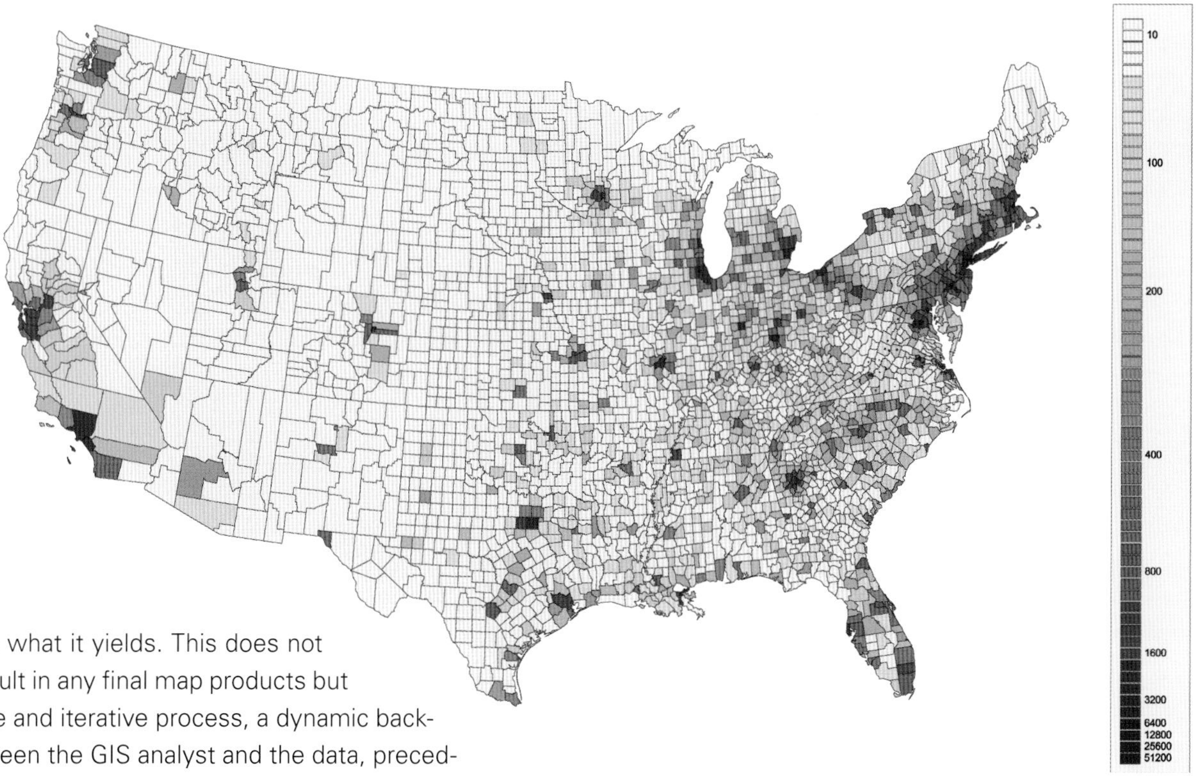

Figure 4.17 **Unclassified map of population density: every county is represented by a unique graduated color symbol, eliminating loss of information due to classification. Theoretically, it is possible to ascertain the actual population density for any county in the country using an unclassified map. This is not the case for a map whose data has been classified, which yields only approximations of actual population densities. However, unclassified maps are often more difficult to interpret and to discern spatial patterns because of the numerous and varied colors used.**

of ways to see what it yields. This does not necessarily result in any final map products but is an interactive and iterative process, a dynamic back-and-forth between the GIS analyst and the data, preceding a more formal data analysis.

Data exploration is not a totally unstructured free-for-all, however. With experience, an analyst will have an idea of which methods and techniques of data exploration will work better on a particular dataset, thus shortening the exploration process and reducing the trial-and-error component. However, since you, the analyst, don't know what you will find, it is impossible to fully plan out the research design in advance, and data exploration remains an adventure for even the most experienced GIS analysts. The ability to explore data graphically is one of the most important benefits of GIS and is often the most exciting part of an analysis. Most GIS software programs allow simultaneous viewing and manipulation of associated maps, tables, and charts, and the hows and whys of doing so will be the subject of the next chapter.

One simple method of data exploration is to generate descriptive statistics about the dataset. The first step is to know the minimum, maximum, range, mean, median, standard deviation, and other statistical measures of the dataset. Line graphs, bar charts, scatterplots, frequency histograms, and other methods of graphing the data are also usually helpful in exploration, especially if the graphing function is dynamically linked to the mapping software, so that features highlighted and updated on the map are highlighted and updated on the corresponding graphs, and vice versa.

Classification itself can also be an effective method of exploration. By looking at multiple classification methods, changing the number of classes, and changing the map type, the analyst or planner can see patterns and relationships that can inform the future direction of the research, suggest additional data requirements, and guide more complex statistical analysis. Unclassified maps can also aid in data exploration and are an excellent starting point for visual analysis.

Spatial aggregation, which is a kind of reclassification or generalization, can also be a constructive way to explore data. Spatial aggregation consists of grouping geographic units into larger entities, to help discern broad patterns in the data that might not be visible in the smaller geographical units. It is defined as "the process of combining smaller spatial units, and the data they contain, into larger spatial units by dissolving common boundaries and lumping the data together" (Heywood 1998). For example, if we mapped a city's census tracts for percent of population below the federal poverty level, we may get a checkerboard appearance. If we aggregate the tracts into larger units, such as sections or districts of the city, thereby collapsing the data for many tracts into one polygon, we might be able to see the broader spatial pattern.

Another exploratory technique is to map the same variable in two or more ways on the same map. For instance, mapping a proportional symbol map of absolute numbers of disease cases, then overlaying it on a choropleth map of disease rates can be instructive in allowing us to view absolute numbers of cases in juxtaposition with rates, which are standardized for population numbers. We can also compare maps based on different ways of looking at the same data. For example, one could plot the economic characteristics of a city's residents by census tract in a variety of ways, such as the percentage of population below the federal poverty line on one map, and income plotted by the mean and standard deviation on another map. By plotting a variable in several ways, we can often begin to see patterns or gain insights about the information that may not have been apparent when looking at the variable in only one way. Figure 4.18 provides an example of this.

Mapping **bivariate** (two variables) or **multivariate** (more than two variables) data can also be a fruitful way to examine spatial patterns, dispersion, and distribution of phenomena. Plotting a dot density map of absolute numbers of elderly residents, and putting that layer over a choropleth map showing the percent of the population of the young—those less than 18 years old—could provide an interesting conjunction of the young and old in a community. Or, a choropleth map of income by census tract could be overlaid onto a proportional symbol map showing the percentage of the adult population without a high school diploma. The result would presumably tell us something about the relationship between these two factors.

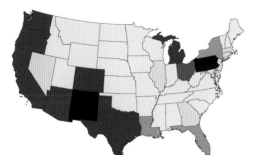

Hispanic Deaths Due to Stroke (Rate per 100,000)

18 - 22.26
22.26 - 26.52
26.52 - 30.78
30.78 - 35.04
35.04 - 39.3
No Data

Hispanic Deaths Due to Stroke
• *1 Dot = 2 Deaths*

Hispanic Deaths Due to Stroke
(Rate per 100,000)
Hispanic Rate

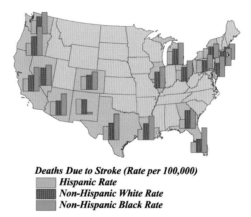

Deaths Due to Stroke (Rate per 100,000)
Hispanic Rate
Non-Hispanic White Rate
Non-Hispanic Black Rate

Figure 4.18 Several ways to visually explore data on deaths due to strokes: *(clockwise from upper left)* choropleth map, dot density map, bar graph symbol, proportional symbol map. Each map contributes something different to our understanding of the overall picture of Hispanic death rates for stroke. Rates are deaths per 100,000, age adjusted to total population, 1970.

Exploration can also be done by examining change over time; indeed, this is sometimes necessary. A sequence of maps can be generated showing a given variable plotted as a dot density, proportional symbol, or choropleth map for various time periods. Amounts or percentages of change can also be plotted to show another view of change over time. Flow maps and cartograms are also useful exploration techniques for spatial and temporal data. In addition, specialized software, software extensions, and scripts can aid in data exploration, enabling the creation of animation sequences, graphing, geostatistical analysis, and three-dimensional transformations. Data exploration techniques can also be used to map the uncertainty factor inherent in a dataset.

Written Exercise 4

Thematic map interpretation

Part 1

(A) Thematic map description exercise: Answer the following questions about each of the thematic maps from chapter 3 that are listed. The first one has been completed.

1) Figure 3.20 (map of asthma hospitalization cases)

Q *What is the geographic extent of the map (the geographic area covered by the map)?*
A Bronx County, New York City

Q *What is the unit of analysis (the geographic unit at which data is aggregated)?*
A ZIP Code

Q *Is it a qualitative or quantitative thematic map?*
A Quantitative

Q *What is the map's theme, and what type of thematic map is it?*
A Theme: the numbers of asthma hospitalization cases.
Type: proportional symbol map

Q *What type of measurement scale is used (nominal, ordinal, interval, or ratio)?*
A Ratio (scale with a known interval between ranks, and a starting point of zero—the absence of value)

Q *What spatial generalizations can you make by interpreting the thematic data on this map?*
A Asthma hospitalization cases are most numerous in the southern and central areas of the Bronx.

Answer the same questions for the following maps:
Figure 3.7
Figure 3.12
Figure 3.17
Figure 3.33

(B) Descriptive statistics exercise: *In the chart below, calculate the following statistics for the data ranges given: minimum, maximum, range, mode, median, and mean, and enter them where indicated. Several values have already been calculated for you. Use your answers in the next part of the exercise.*

Community districts	% of adults without a high school diploma	Ranked array	Min.	Max.	Range	Mode	Med.	Mean	Std. Dev.
Bronx 1	57	22	22	57	35	29	43	41	11.78
Bronx 2	56	29							
Bronx 3	49	29							
Bronx 4	47	29							
Bronx 5	49	37							
Bronx 6	50	38							
Bronx 7	38	47							
Bronx 8	22	49							
Bronx 9	37	49							
Bronx 10	29	50							
Bronx 11	29	56							
Bronx 12	29	57							
Manhattan 1	9								16.84
Manhattan 2	10								
Manhattan 3	42								
Manhattan 4	13								
Manhattan 5	5								
Manhattan 6	5								
Manhattan 7	10								
Manhattan 8	5								
Manhattan 9	35								
Manhattan 10	36								
Manhattan 11	44								
Manhattan 12	44								

(C) Data classification exercise: Using the descriptive statistics and values of the datasets from part B on the previous page, calculate class boundaries (class breaks) for each of the Bronx and Manhattan datasets, using the following classification methods: equal interval, quantile, standard deviation, and natural breaks. Use the chart below.

Equal Interval (Equal Steps): Use 3 Class Groupings		
	Bronx	Manhattan
1st Class		
2nd Class		
3rd Class		

Quantile: Use 3 Class Groupings		
	Bronx	Manhattan
1st Class		
2nd Class		
3rd Class		

Standard Deviation: Use Intervals of 1 Std. Deviation		
	Bronx	Manhattan
−1 to −2 Std. Dev.		
0 to −1 Std. Dev.		
Mean		
0 to +1 Std. Dev.		
+1 to +2 Std.Dev.		

Natural Breaks: Use 3 Class Groupings		
	Bronx	Manhattan
1st Class		
2nd Class		
3rd Class		

Answer these questions: Which classification method would yield the most interesting and varied map? Which would show the data most accurately?

Figure 4.19 (on page 122) shows an example of the final classified maps for the Bronx.

Part 2

Thematic Map Interpretation: *Find an interesting example of a thematic map in a recent newspaper or magazine. Cut out or copy the entire article with the map, and write a few paragraphs explaining the map. Be sure to answer questions such as: What is the purpose of the map? What is the map's story? How is the information graphically portrayed? What spatial information does the map contain? What is happening where? What is the geographic unit used to aggregate the data? How is the data classified? Can you think of any factors that may account for the distribution or pattern shown on the map?*

Carefully examine the map's title, legend, and content, and comment fully on the information and meaning contained in those elements. Use the example below as a guide.

Poor Workers Finding Modest Housing Unaffordable, Study Says

By LYNETTE CLEMETSON
The New York Times

With the rise in housing costs outpacing that of wages, there is no state where a low-income worker can reasonably afford a modest one- or two-bedroom rental unit, according to a study issued today by the National Low Income Housing Coalition, an advocacy group based in Washington.

The study, "Out of Reach 2003," compared wages in each state, large metropolitan area and county in the nation against fair market rents as determined by the Department of Housing and Urban Development.

In 40 states, the study found, renters need to earn more than two times the prevailing minimum wage to afford basic housing. In the most expensive states—Massachusetts, California, New Jersey, New York, Maryland and Connecticut—they must earn more than three times the minimum wage.

The federal minimum wage, last increased in 1997, is $5.15 an hour, though some states put their minimums higher, topped by Alaska, at $7.15. The nation's median hourly wage last year was roughly $12 an hour, according to the Economic Policy Institute, a research organization that advocates on behalf of low- and middle-income workers.

The gap between affordability and housing costs has been increasing, the study found, and researchers and advocates for low-income workers say the implications cut across a variety of social issues.

"When low-income families are paying so much of their income on housing, they are left to skimp on other necessities like food, medicine, clothes and time spent with children," said Sheila Crowley, president of the National Low Income Housing Coalition.

For each state and locality, the study determined a "housing wage": the income a full-time worker must earn to afford a modest rental unit. "Affordable" rent was defined as no more than 30 percent of the household income, the standard calculation used by HUD. A "fair market rent" was also based on a HUD standard, typically the 40th percentile in a given housing market (the 50th percentile in the markets that are tightest and most expensive).

The national median housing wage necessary to afford a standard two-bedroom rental unit, the data indicated, is $15.21 an hour, up 37 percent from 1999, when the organization began collecting detailed statistics.

In Massachusetts, the most expensive state for housing, the housing wage for a modest two-bedroom apartment is $22.40 an hour. In the District of Columbia, whose metropolitan area is one of the most expensive, it is $23.42 an hour. But even in West Virginia, the most affordable state, a worker would need to earn $8.78 an hour to afford basic housing, considerably more than the state's $5.15-an-hour minimum wage.

Renting at Minimum Wage

Workers who make minimum wage — $5.15 to $7.15 an hour — cannot afford a two-bedroom rental unit at fair market rents, according to a report released yesterday by the National Low Income Housing Coalition.

Hourly wage a worker must earn to afford a two-bedroom rental unit

$8.78   10.00   12.50   15.00   17.50   23.42

Ms. Crowley said more subsidized housing was needed for low-income families. Not all housing experts agree.

"We're not, in this time of rising deficits, going to be able to subsidize our way out of any housing gap," said Howard Husock, director of case studies at the John F. Kennedy School of Government, at Harvard. "We need to find ways to increase wages of low-income families and to create more two-wage-earner households, through programs like those that encourage marriage."

Title of map: Renting at minimum wage

This map, based on a study prepared by the National Low Income Housing Coalition, shows the hourly wage a worker must earn in order to afford a two-bedroom rental unit in each state of the United States. The data has been aggregated by state, with the average fair market rental rates based on the federal Housing and Urban Development figures reflecting the fortieth percentile rental rate in that housing market. This figure is then used to calculate the minimum hourly wage required to be earned by a worker in each state, assuming that workers should spend no more than 30 percent of their household income on housing costs, and using that criteria to define affordable housing.

In the map, each state is colored with one of five gray values as a solid fill. Graduated shades from white to black represent the range of hourly wages, from a low of $8.78 (in West Virginia) to a high of $23.42 (in Washington, D.C.). The hourly wages have been divided into five classes. The first (lowest) class range is $8.78–$10.00. This is shown by a white fill (the lightest color). Workers living in states within this class must earn between $8.78 and $10.00 per hour in order to afford to rent a two-bedroom unit costing thirty percent of their income. The second class is $10.00–$12.50, and is shown by a light gray fill. The third class is $12.50–$15.00, shown by a medium gray value; the fourth class is

$15.00–$17.50 and is shown by a dark gray fill, and the highest class, $17.50–$23.42, is shown by black (the darkest value in the map's graduated color ramp). Workers living in states within this class must earn between $17.50 and $23.42 per hour in order to afford to rent a two-bedroom unit costing 30 percent of their income. The darker the state's gray value on the map, the higher the hourly wage must be for a worker to afford a two-bedroom rental unit in that state. The classification system shown in the map's legend is a modified equal interval method, using an interval of $2.50, adjusted to accommodate the actual high and low values in the data range.

The states falling into the first (lowest) class range are mainly in the Deep South and parts of Appalachia. The second class is comprised of various states in the Midwest, Southeast, and Northern Plains states. The third and fourth classes contain states somewhat scattered geographically. The fifth and highest class is made up of California and several highly urbanized states and locations in the Northeast, such as New York, New Jersey, Massachusetts, Connecticut, Maryland, and Washington, D.C. In these places, it requires more than three times the federal minimum wage ($5.15 per hour) to afford a modest two-bedroom rental unit.

It is likely that the class divisions of hourly wages required for affordable rental units correspond to differences among the states in average wages and concomitant average rental prices. For instance, the states in the lowest hourly wage class on the map are also most likely the states with the lowest average incomes and/or the lowest average housing prices. Conversely, the states in the highest hourly wage class on the map probably tend to be the states with the highest average incomes and/or the highest average housing prices.

Clearly, in those states in the highest classes, many workers earn far more than the minimum wage, but for those workers at the low end of the pay scale, it is very difficult to find decent affordable housing appropriate to their earning power in those states. In fact, there is no state where the federal minimum hourly wage of $5.15 would provide a worker with enough income to afford the average two-bedroom rental rate. Possible solutions might be to increase the amount of subsidized affordable housing, to raise the minimum wage (as some states such as Alaska have done), to increase the earning power of workers through training and education, to encourage the formation of multiple-earner households, or to provide tax breaks to businesses in hopes of spurring the creation of higher-paying jobs.

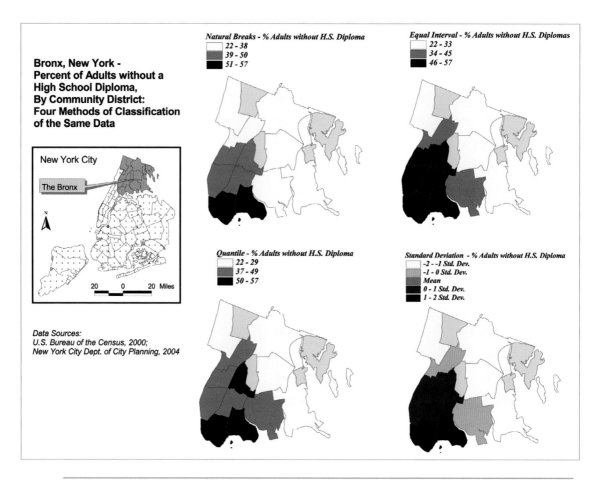

Bronx, New York - Percent of Adults without a High School Diploma, By Community District: Four Methods of Classification of the Same Data

Natural Breaks - % Adults without H.S. Diploma
- 22 - 38
- 39 - 50
- 51 - 57

Equal Interval - % Adults without H.S. Diplomas
- 22 - 33
- 34 - 45
- 46 - 57

Quantile - % Adults without H.S. Diploma
- 22 - 29
- 37 - 49
- 50 - 57

Standard Deviation - % Adults without H.S. Diploma
- -2 - -1 Std. Dev.
- -1 - 0 Std. Dev.
- Mean
- 0 - 1 Std. Dev.
- 1 - 2 Std. Dev.

New York City

The Bronx

N

20   0   20 Miles

Data Sources:
U.S. Bureau of the Census, 2000;
New York City Dept. of City Planning, 2004

Figure 4.19   As you develop the class breaks for the community districts in the Bronx and Manhattan, compare the classes to the maps, and try to determine which classification method would yield the most interesting and varied map, and which would show the data most accurately. This may differ for each borough. Also, compare the data ranges and the class breaks of the Bronx to those of Manhattan, and think about the particular problems of mapping posed by each dataset. The green areas on the map represent parks, golf courses, and other public spaces.

References and further reading

Hearnshaw, H., and D. Unwin, eds. 1994. *Visualisation in Geographical Information Systems*. Chichester, UK: John Wiley and Sons.

Heywood, Ian, Sarah Cornelius, and Steve Carver. 1998. *An Introduction to Geographical Information Systems*. New York: Prentice–Hall.

MacEachren, A. M., and R. D. F. Taylor. 1994. *Visualization in Modern Cartography*. New York: Pergamon Press.

Rogerson, Peter, A. 2001. *Statistical Methods for Geography*. London: Sage Publications.

Thiessen, Heiner. 1997. *Measuring the Real World: A Textbook of Applied Statistical Methods*. New York: John Wiley and Sons.

Tukey, John W. 1977. *Exploratory Data Analysis*. Reading, MA: Addison–Wesley.

Data Visualization
and Map Design

One of the primary purposes of a map is the presentation and communication of data to the map reader. Use of effective presentation techniques will enhance the clarity and accuracy of the map's message. In this chapter, map design guidelines are outlined to aid the mapmaker in creating maps that successfully communicate the thematic data, while also helping the map reader to explore the map and gain a deeper understanding of the information in it. Charts and graphs are often integrated within map layouts to emphasize or expand upon aspects of the data, and this requires a solid understanding of design and data presentation techniques.

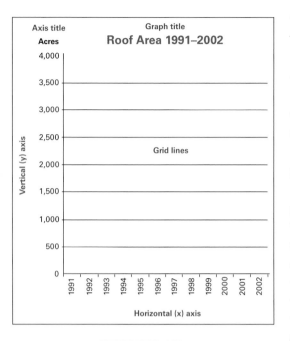

Figure 5.1   **The components of a graph.**

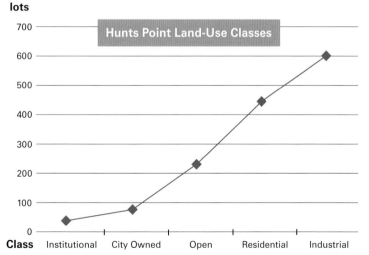

Types of charts and graphs

The terms *graph* and *chart* are sometimes used interchangeably, but there is a difference: Graphs depend on mathematical values for their construction, while charts—whether drawings, diagrams, or pictograms—do not. Many graphs make use of two axes, a horizontal x-axis and a vertical y-axis, forming a grid to display information about one or more variables (figure 5.1).

Some of the common types of graphs are **line, bar, surface, pie, histogram,** and **scatterplot.**

Although some data clearly works better with one or another type of graph, often the type of graph used is a matter of personal preference. With most graph-making software, including ArcGIS, it is very easy to experiment with different kinds of graphs using the same data, and to explore the data in multiple ways before deciding how to display the data to best advantage. The same variables can be shown using many different types of graphs; there is usually not one correct or incorrect type of graph for each dataset. One can show the same data using many different types of thematic maps.

Line graphs

A line graph consists of a line connecting points plotted on a grid. The points represent the values of the variable being graphed. The line's path shows the value of one variable in relation to another (figure 5.2).

Time periods, for example, might be shown on an x-axis, and the variable whose relationship to time we wish to examine—population, birthrate, tax revenues, or homicides—is placed in the y-axis.

Figure 5.2   **Line graph.**

Dependent and independent variables: y-axis and x-axis

One of the variables on a line graph may be thought of as dependent on the other. The independent variable is usually plotted along the x-axis, and the dependent variable along the y-axis. Some examples of paired variables include (the independent variable is listed first) temperature and disease rates, average household income and rental rates, and alcohol consumption rates and accident rates. The terms *dependent* and *independent* do not necessarily imply that a cause-and-effect relationship exists between the variables—only that there may be some association between the two, and that the independent variable may have some bearing on the dependent variable (figure 5.3).

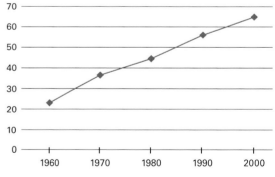

Figure 5.3 **Line graph with dependent and independent variables.**

Graph scales

Graph scales are arbitrary. In other words, there are no rules governing how much horizontal or vertical space a certain value should take along an axis. Because of this, graphs are often prone to misinterpretation on the part of the reader. Sometimes, deliberate distortion is introduced into a graph when a scale is used that tends to emphasize or de-emphasize some aspect of the data. Another common error can occur when one axis is longer than the other; if so, a change in value of the variable will appear very different in the graph than if both axes were the same length. To avoid unintended exaggeration of the data, both axes should be approximately the same length (figure 5.4).

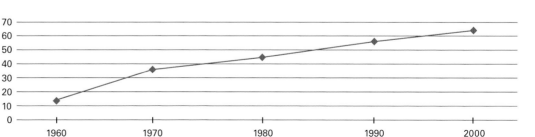

Figure 5.4 **To avoid unintended exaggeration of the data, both axes should be approximately the same length.**

Both axes in a line graph should start with a zero baseline. This means that the lower left corner of the graph (where the numerical values for both axes begin) will have the value 0. A common graphing pitfall is an axis that starts without a zero baseline. Using a starting number other than zero makes it difficult to judge the relative importance of the value changes and can be used to mislead. You can correct this problem by including a zero baseline, or, alternatively, by indicating a break in the axis. (However, keep in mind that a zero baseline is generally not the standard starting point for an x-axis that represents time.)

Line graphs have either **arithmetic** or **logarithmic scales,** and each type has unique characteristics.

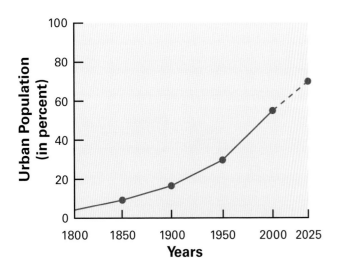

Arithmetic line graphs

On an arithmetic line graph, the spacing between data values is uniform. Each unit of vertical or horizontal distance indicates equal changes in value, anywhere on the graph. In other words, a timeframe of five years should be shown as five times as long as a timeframe of one year (figure 5.5). A graph using nonuniform spacing is either drawn carelessly or is intended to mislead the reader.

Logarithmic line graphs

Other kinds of line graphs employ a logarithmic, or ratio, scale. Rather than starting at zero, a logarithmic scale starts at an appropriate number, such as one, and increases until it reaches ten times the original value, such as ten. Additional exponents of ten are then used if needed. The next tick along the axis would be 100, then 1,000, then 10,000, and so forth. Log scales are useful when the variable to be plotted has a very large range, and it would not be possible to see differences at one end of the scale using conventional scaling. The variable spacing of log scales prevents equal distances on the graph from representing absolute differences in value; instead, equal measurements on the graph correspond to constant percentage change in value.

Graphs with a log scale on only one axis (usually the y-axis) while the other axis remains an arithmetic scale are called semilogarithmic or semilog graphs. With a semilog graph, the slope, shape, and direction of the lines delineate trends in the value changes (figure 5.6).

Figure 5.5   In an arithmetic line graph, each axis is divided into units that represent equal amounts of time (x-axis) and percentages (y-axis) so that five units, for instance, are shown as five times longer than one unit. Redrawn from *Urban Geography,* by David Kaplan, James O. Wheeler and Steven R. Holloway. Reprinted by permission. John Wiley and Sons © 2003.

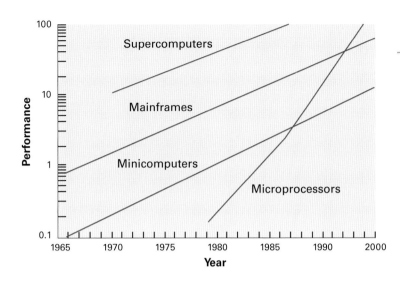

Figure 5.6 **In a semilog line graph, the x-axis (years) is arithmetic, but the y-axis (performance) is logarithmic; in other words, units are measured on an exponential scale.** Redrawn from "Computer Technology and Architecture: An Evolving Interaction," by John L. Hennessy and Norman P. Jouppi, in *Computer*, September 1991 (Vol. 24 No. 9) © 1991 IEEE.

Multiple line graphs

A graph with several lines plotted on it is known as a multiple line graph. Because each line represents a different variable, these graphs can be confusing. There are fewer interpretation problems if all the lines are plotted using the same scale, but in some cases the various lines indicate different types of quantity and each has its own scale. Be careful interpreting multiple line graphs, as you cannot reach any conclusions about the data by comparing the different lines, except for the determination that there is or is not a trend relationship. And even if such a relationship exists, it does not imply a causal relationship. We cannot tell from a graph, for instance, which variable is the cause and which is the effect, or if indeed another variable is responsible for the pattern (figure 5.7).

Figure 5.7 **Multiple line graph.**

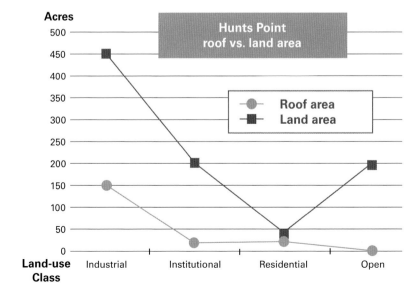

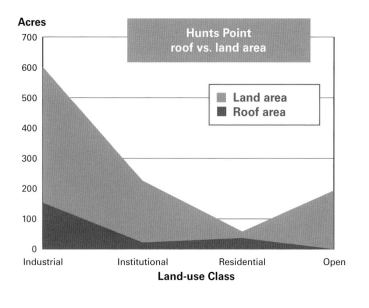

Figure 5.8   **Surface graph.**

Surface graphs

Although these graphs look similar to line graphs, they are plotted and interpreted very differently. Values on a surface graph are presented as colored or patterned areas on the grid. This results in a cumulative view of the data; the graph's top band or line reflects a total value—the accumulation of all the values below it. On a surface graph, values are calculated as the difference between the bands or areas. A line graph's values, by contrast, are plotted as individual data points, and are calculated by measuring from the baseline (figure 5.8).

Either absolute numbers or percentages can be shown on a surface graph. If percentages are displayed, the total of all the bands on the grid must equal 100 percent.

Bar graphs

A bar graph, also called a column graph, consists of a series of bars or columns, each bar representing a particular data range, such as a population subgroup, income range, or a time period. The bars may be oriented on the grid using either a vertical or horizontal baseline. The bars in any one graph should always be the same width so that the reader can accurately judge the area of each bar, and hence, its proportion of the variable being graphed.

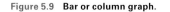

Figure 5.9   **Bar or column graph.**

Bar graphs can illustrate either absolute or percentage values. The individual bars or columns can be subdivided horizontally, using a stacked bar graph, to show the values of a single category for different time periods or the values of different categories for a single time period (figure 5.9).

As with line graphs, bar or column graphs should have a zero baseline. The two bar charts in figure 5.10 use the same data, but one has a zero baseline and the other does not. Visual bias is introduced in the graph without a zero baseline. Note that the difference between the bars in the left graph is much more dramatic than between those of the right; this discrepancy can mislead graph readers by overemphasizing the difference between the values represented by the bars.

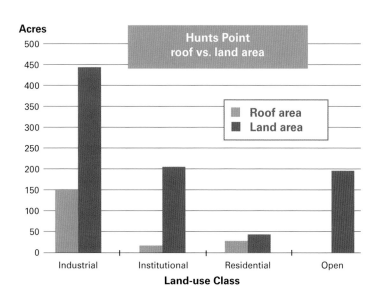

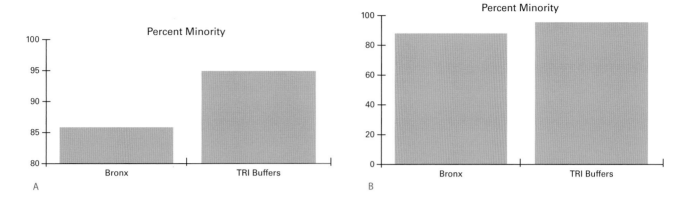

A

B

Figure 5.10 **Bar graphs should always use a zero baseline on the y-axis so as not to mislead the graph reader. Compare graphs A and B, both showing the same data: percent minority population in the Bronx and in the TRI (Toxic Release Inventory) buffer. The y-axis of graph A begins at 80 percent, and therefore the differences between the two bars are emphasized in an exaggerated way. The y-axis of graph B begins at zero, and the bars are shown in a more realistic proportion to each other. They do not misrepresent the data or lead the graph reader to perceive a larger difference between the bars than there really is.**

Pie graphs

A pie graph is a circle divided proportionally into segments, or wedges, with each wedge representing a proportion of the total. For example, a pie graph could represent the zoning categories of land in a city. Each wedge represents the proportion of land in each different zoning category, with all the wedges of the pie graph totaling 100 percent of the land. When making a pie chart, one must be careful to ensure that the wedges represent nonoverlapping parts of the whole, that they add up to 100, and that the total is not included as one of the wedges within the pie. Generally speaking, the use of pie graphs drawn at an angle, or three-dimensional pie graphs, is discouraged due to difficulties in accurate interpretation of the data (figure 5.11).

Figure 5.11 **Pie graph.**

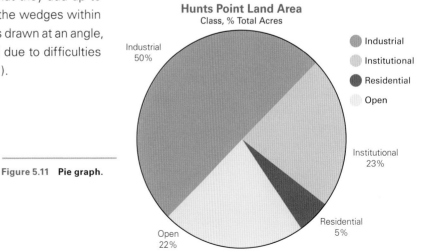

Multiple and proportional symbols

Numerical data can be represented on a map or in a graph using an appropriate repeated symbol, with each symbol representing a certain quantity. For instance, we might use a pictogram of a child to represent 1,000 children per school district. By counting the number of symbols in each school district and multiplying by 1,000, we can tell the total number of children in each district area of the map. This technique can also be used in graph form.

As discussed in chapter 3, we can also use proportional symbols in graphs, such as graduated circles, other geometric figures, or pictograms. Graphs using multiple or proportional symbols can be effective in presenting simple data in a visually interesting way. Pictograms can be sized according to the **absolute** or **square root** method of scaling, or the **apparent-magnitude** or **perceptual** method of scaling. With perceptual scaling, symbols are exaggerated in size rather than drawn in absolute proportion as described in chapter 3's section on proportional symbol maps (page 73). The symbol sizes selected in perceptual scaling are meant to correspond to the differences that can be easily differentiated by the human eye and mind (figure 5.12.)

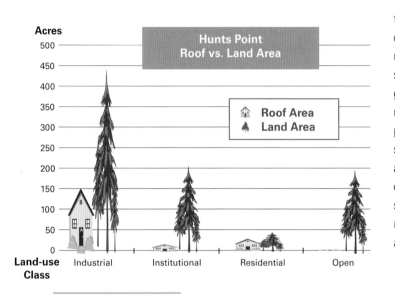

Figure 5.12 Proportional symbol graph.

Figure 5.13 Scatterplot, or scatter diagram, showing a strong negative relationship between the percentage of adults without a high school diploma and recycling rates in each community. As the percentage of adults without a high school diploma rises, the percentage of solid waste that is recycled in each community declines.

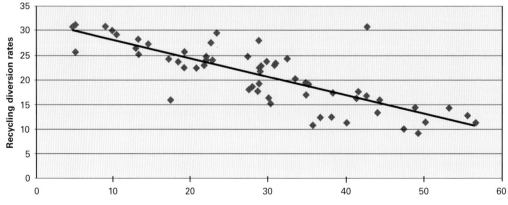

Scatter diagrams

Scatter diagrams, also called scatterplots, are used to display the relationship between two variables, such as the annual number of building violations in a neighborhood and the mean household income of that neighborhood; the percentage of residential land in a community district and its crime rates; or an area's distance from the city center and its population density. A scatter diagram is a group of points placed on the graph's grid according to the values of the two variables. Each dot represents a pair of values. As with other graphs, the value of the independent variable is measured on the x-axis, and the dependent variable is measured on the y-axis.

A trend line can be drawn through the cloud of dots on a scatter diagram. This line characterizes the relationship between the variables. An upward sloping trend line, from lower left to upper right, denotes that the relationship is positive: as the value of the independent variable increases, the value of the dependent variable also increases. A downward sloping trend line, from upper left to lower right, indicates a negative relationship, where the value of the dependent variable decreases as the value of the independent variable increases. How well the trend line fits the concentration of dots is an indication of how strong the relationship is between the variables. If the dots are clustered close to the line, the relationship is strong. If the dots are widely dispersed away from the line, the relationship between the variables is weak (figure 5.13).

Histograms

Histograms are a special type of bar graph: The x-axis contains the data values or ranges of values, and the y-axis reflects the frequencies of each value or range. The bars are scaled so their heights reflect the number of values in each class. Histograms provide immediate information about the distribution of the data. The viewer can see at a quick glance how many values are in each class, and what the most and the least prevalent values are.

Relative-frequency histograms, in which the y-axis is scaled to the relative frequency rather than the absolute frequency, are considered more useful than simple histograms. In the relative frequency histogram, the heights of all the bars add up to 1, making it easier to judge how much of the total each class represents.

It is often helpful to display a histogram next to a map. The viewer will get a more thorough view of the data, both spatially and nonspatially (figure 5.14).

Figure 5.14 A relative frequency histogram. Approximately one fourth of New York City's community districts have between 4.9 and 13.05 percent of their populations living below the federal poverty level. Nearly one-third of the districts have between 13.06 and 21.21 percent below poverty, and about 14 percent of the districts have more than 37.53 percent of their population below the poverty level.

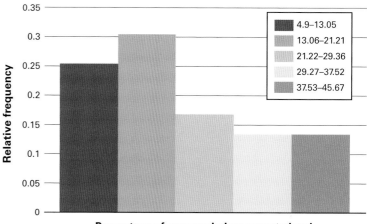

Percentage of persons below poverty level, NYC community districts

Legend:
- 4.9–13.05
- 13.06–21.21
- 21.22–29.36
- 29.27–37.52
- 37.53–45.67

(y-axis: Relative frequency; x-axis: Percentage of persons below poverty level)

Integrating graphs and maps for data visualization

Graphs are often integrated into maps, with the charts or graphs essentially acting as the thematic overlay on top of a geographic basemap. Using graphs on a map can be an effective way of displaying data in a spatial context and can be especially useful in exploring comparative rates or ratios (figures 5.15 and 5.16).

Graphs can be used in a map in place of symbols. Occasionally, they can supplement other types of symbols, such as choropleth fill patterns or colors.

Figure 5.15   An integrated map and graph—comparative mortality rates by race or ethnicity. The bar graphs are used as the thematic symbol on the map. Rates are deaths per 100,000, age adjusted for 1970 total U.S. population.

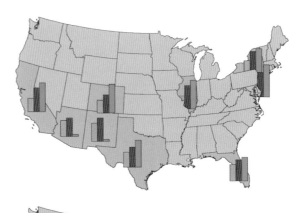

Breast Cancer Mortality Rates
Hispanic Rate
Non-Hispanic White Rate
Non-Hispanic Black Rate

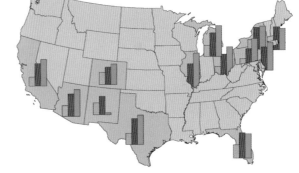

Lung Cancer Mortality Rates
Hispanic Rate
Non-Hispanic White Rate
Non-Hispanic Black Rate

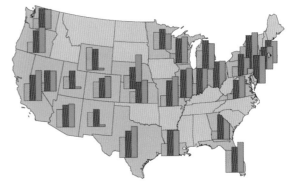

Heart Disease Mortality Rates
Hispanic Rate
Non-Hispanic White Rate
Non-Hispanic Black Rate

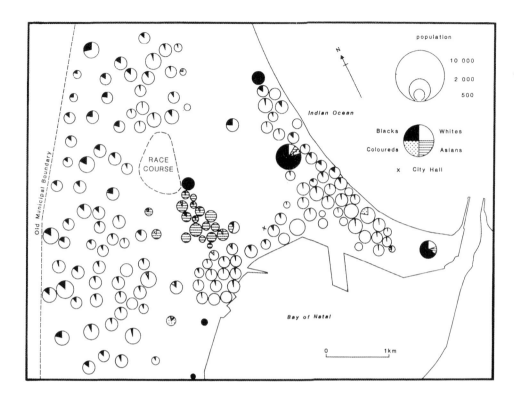

Figure 5.16 A proportional symbol/pie graph is used as the thematic overlay on this map.

From *The Atlas of Apartheid* by A.J. Christopher. Reprinted by permission. Taylor & Francis/ Routledge © 1994.

Data visualization

Data visualization, sometimes referred to as Visualization in Scientific Computing, or ViSC, is defined as "the use of computer technology for exploring data in visual form . . . and the use of computer graphics for acquiring a deeper understanding of data" (Visvalingham 1994). Visualization can also be thought of as "the interplay between technology and the human mind" (Davies and Medyckyj-Scott 1994). Maps are, and always have been, rich sources of data. GIS increases exponentially the richness of data, and many of the functions of GIS make it possible to look at the data in many different ways and from various viewpoints. We can manipulate the data, examine its statistics, plot graphs of it, classify and reclassify it with different schemes of class breaks and classification methods, and look at multiple views of the data at the same time. We can also explore and display movement in the data in two major ways: by looking at changes in data values over time using static maps of each time period, or by using an animated sequence of maps. We can also look at the movement of the spatial object by changing the viewpoint, rotating spatial features, viewing them in three dimensions, or by creating a virtual "fly-through" of the data. All of these techniques attempt to provide us with insights into the data through what is essentially a dynamic creative process of brainstorming.

Data visualization experiment using GIS

In an experiment conducted in Northern Ireland by Carol McGuinness and her colleagues, the strategies for using information between a group of experienced GIS users and a group of GIS novices was compared. "Eighteen subjects were presented with a database of nine variables relevant to machine peat cutting [in the mountains of Northern Ireland]. . . . Variables were available as layers in the ArcPlot application and included incidence of cutting, type of peat, contours, roads, rivers, rainline, rain stations, areas of natural beauty/scientific interest, and density of wading birds. Subjects were asked to use the database to consider how environmental and social variables were related to the incidence of machine cutting, giving particular attention to potential conflicts between economic and environmental issues. . . [Subjects] were free to plot and replot any individual or combinations of the nine variables during the session. Experts made use of most of the available information, at least at a cursory level, with all experts looking at either eight or all nine data layers. Experts, however, considered fewer layers at the same time (an average of only 2.2) than did novices (with a mean of 2.6 and one novice who overlaid all nine variables at the same time). While experts looked at fewer variables simultaneously, they were much more likely to go back and re-examine combinations (an average of 30 vs. 9 re-examinations for experts vs. novices, respectively). These findings seem to be in keeping with the idea that a successful procedural schema for spatial problem solving involves an interactive process of setting subgoals and narrowing in on a solution through a process of seeing potential solutions and then applying a self-assessment procedure (engaging in reasoning why) to determine whether they are viable. From the task performance data and the summaries subjects were asked to write, there is clear evidence that experts obtained more information and were able to provide more adequate assessments of the situation. Experts made more frequent reference to spatial relationships, and protocol analysis suggests that they approached the task in a more systematic and structured way. . . . [T]he experts imposed domain-specific schemata that allowed them to categorize what they saw and that provided a structure for organizing the process of data exploration. . . . [T]he novices tended to engage in more local and superficial hypothesis testing. . . . [U]nless an analyst can rely on an appropriate event schema for logically structuring analysis, the GVIS [geographic visualization] session may become a purposeless random walk through the data that results in feature identification only by chance. We have an opportunity to facilitate visualization tasks (particularly for nonexperts) by making systematic analysis the default option" (MacEachren 1995:397–398).

Research originally published as: McGuinness et al. (see reference list for complete citation)

Researchers of data visualization have noted that experienced map readers are more likely to find patterns than novices. An experienced map reader has "a whole dictionary of names for grouped entities which he or she carries around in her head, allowing her to read and understand maps more quickly and effectively" (Edwards 1991).

The Ireland study cited in the sidebar on the previous page is not intended to discourage novice GIS users, but rather to demonstrate that expertise and eventual success in data visualization is acquired through practice. Although this seems obvious, many novice GIS users expect to get expert results without putting in the time required to actually become an expert. However, it is also true that advances in the default analytical capabilities of GIS software may help put experts and novices on a more equal footing.

Map design and composition

As we've discussed previously, an important use of maps and GIS is data exploration and visualization. This is mainly an interactive and iterative process, resulting in working maps that are not necessarily suitable for presentation or even understandable to other viewers. Nonetheless, presentation remains one of the primary purposes of maps and GIS: to communicate information about a place or topic to the person reading the map. Therefore, a map for presentation must make that information as clear and concise as possible.

Throughout history, mapmakers have developed certain cartographic conventions to aid the map user in interpreting the information contained in the map. Principles of design and composition play a crucial role in organizing and displaying the information to be conveyed by the map. Knowledge of the main elements of a map, graphic design principles, theories of color, and basic concepts of typography help the mapmaker to effectively communicate the map's intended message.

Map readers can most easily interpret maps that have the least amount of abstraction, so one of the goals of the map designer should be to minimize it.

Cartographic conventions

As casual map readers, most of us are used to seeing certain things presented in certain ways on maps. There is generally a title, a legend, and a scale bar, in addition to the map image itself. North is usually at the top of the map, and land masses are typically bounded by a heavy outline. In many maps, certain colors mean specific things: water is blue, forests are green, and deserts are tan. Features that are colored differently on a map generally differ in some important respect in reality, and are intended to be interpreted differently. Place names on the map are understood to represent the names of the areas they are written over or near. Names with large letters are usually to be considered more important for the specific purpose of the map than names with small letters.

We understand all these things and they seem obvious to most of us because we have seen them used many times on many different maps. Orientation, subject content, symbology, scale, labeling, colors, and text, as well as relative size and position of the map elements, are essentially clues to the map user as to how the map is to be interpreted. Maps that are successful in communicating data are well designed. A well-designed map often makes use of these cartographic conventions in order to enhance clarity and promote comprehension of the map data. It is assumed that most maps designed for planning purposes will be thematic maps, as opposed to reference maps, although many of the same design principles apply to both.

Main map elements

When we look at a map—especially most professionally produced maps, but even our own hand-drawn maps—we notice certain standard elements and certain conventions appearing in almost every map. These elements, and their orderly and logical arrangement, assist the map reader in obtaining the fullest set of information possible from the map (figure 5.17).

Figure 5.17   The main elements of a map.

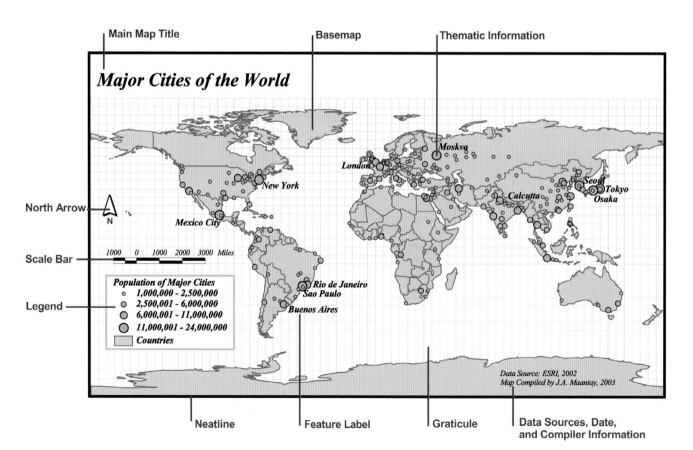

Maps generally contain the following elements:

Title

The title or titles help the map user by stating the subject of the map, the geographic region, perhaps the time period to which it applies, and other important aspects of its content.

Map image

This is the map itself, consisting of a geographic basemap, usually with an overlay of some thematic information. This is obviously the main focus of the map layout or composition. In a thematic map, the geographic basemap is generalized, to emphasize the thematic information.

Legend

The legend shows map symbols and their meanings, enabling the map user to interpret the map information. The legend might show colors or symbols, some of which may be conventional while others are unique to a particular map. Conventional symbols include railroad tracks represented as a solid line with small perpendicular cross-lines, a river represented as a wavy blue line, or a city shown as a circle. On the other hand, the meanings of some symbols may be unique to a map, such as a red triangle representing a toxic release facility or a green square representing a public school.

Scale

This is a special feature common to many maps but virtually unique to maps as opposed to any other graphics. Most maps contain a scale bar or some other indication of how to measure distances on the map, as compared to the distance in the real world.

Orientation

This refers to information in the map that tells us how to look at the map relative to the earth's surface. Most maps conventionally show north to be on the top of the map page, but not always. Even when north is on top, it is a cartographic convention to include a north arrow in the map page to guide the map reader and orient her as to direction.

Data sources and map author information

It is customary to provide the map user with information about the sources of the data used in the map, as well as the map author's or compiler's name and affiliation, and the date when the map was compiled. It is especially important to give credit to the data sources so the map user understands the nature of the data used in compiling the map, and can verify data values, accuracy, method of data collection and aggregation, and any other aspects of the data, if required.

Basemap feature labels

Some thematic maps use labels to identify important geographic features in the basemap, such as major cities, rivers, roads, or political/administrative areas. These may prove helpful to the map user, especially if the geography is unfamiliar, or if the geographic features have some relevance to the thematic information being presented.

Graticule

Graticules are the lines of latitude and longitude or other coordinate system forming a grid to give the map user a reference point. These are often omitted in thematic maps.

Neatlines

These are frames around the map elements, including borders around the map image itself, the legend box, the insets, or the title. They do not serve any technical purpose, but simply help the map reader in visually organizing the information contained in the map.

Insets

Sometimes additional small maps are included on the map page to provide the map reader with contextual information about where in a larger geography the map area lies. Insets can show a larger area than the main map, thereby indicating where the main map is in relation to a continent or the world—acting as a "You Are Here" sign. Conversely, an inset can show a detailed version of an area from the main map, such as a street map of the capital city on a country map. In other cases, insets show different or more detailed information about the area in the main map, such as population density or the location of important landmarks (figure 5.18).

Graphs, histograms, scatterplots

It is often very helpful to the map user's understanding of the data to provide graphs as an accompaniment to the map image. The graph might be a bar or pie graph showing proportional comparisons of an aspect of the thematic variable, or it might be a frequency histogram or a scatterplot showing the dispersion of values of a variable, and the relationship of one variable to another, respectively. These graphs or charts can appear as independent elements in the map composition and can be used to display information that is not shown spatially on the map. Graphs and charts can appear on the map as symbols, as part of the thematic information on the map image itself, as discussed earlier.

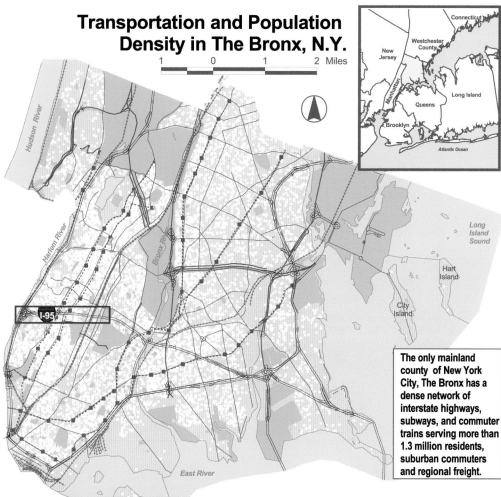

Transportation and Population Density in The Bronx, N.Y.

The only mainland county of New York City, The Bronx has a dense network of interstate highways, subways, and commuter trains serving more than 1.3 million residents, suburban commuters and regional freight.

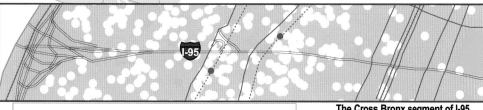

The Cross Bronx segment of I-95, highlighted in above inset, has the highest daily volume of vehicular traffic in the United States.

Data Source: LotInfo 2002;
U.S. Bureau of the Census TIGER

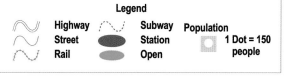

Legend

	Highway		Subway	Population
	Street		Station	1 Dot = 150 people
	Rail		Open	

Figure 5.18 Inset maps: the large map in the center of the layout is the main map, showing Bronx County, which has the same boundaries as The Bronx borough. The map on the upper right is an inset map that helps situate for the map reader the location of the main map (the county) within the context of the entire metropolitan area. The inset map at the bottom of the main map shows a detail of an area within the county. In addition to such context and blow-up maps, other types of inset maps may be included to show a particular theme or themes pertaining to the geography of the main map.

Hierarchy of map elements

The title, map image (basemap with thematic information), and legend are the most important elements in a map composition. Scale and orientation are also desirable elements to include, but are of secondary importance, and are often omitted from thematic maps. This is because most thematic maps use a geographic basemap of areas familiar to the map user, so scale and orientation are assumed to be known, or are not critical in conveying the thematic information to the map user. The other elements, such as feature labels, neatlines, charts, graticules, and inset maps are optional.

When designing a map layout, the mapmaker must decide what elements are necessary for the map user's correct understanding of the information, or if a particular element leads to visual clutter and so detracts from clarity of communication. The mapmaker must always keep in mind that the main purpose of the presentation map is to communicate the thematic information, and that all design decisions should be made to enhance map understanding. Maps that appear very complex or intricate may actually hinder map communication and comprehension.

Graphic design

Creating an effective map layout requires arranging the map elements into a meaningful and attractive design because that design has a great deal of influence on the ultimate usefulness of the map. The map can succeed or fail in its goal of adequately and honestly conveying its own information based on the decisions made in the graphic design stage. Because of innate or culturally induced perceptions, and our prior experience with reading maps, certain mapping conventions and ways of representing data will help make the map a more successful communication tool. These conventions are not the same as universal laws; the principles of good graphic design are quite subjective.

A good graphic design strikes a balance between conventions that have proven effective and subjective aesthetic judgment. There are few absolutes in map design and composition, and many decisions will be based on aesthetic opinion and personal judgment. While many conventions are valid, and it is difficult to produce a bad map if they are taken into account, it should be noted that there are many examples of excellent and effective maps that break one or more of the conventions. As you become more experienced in designing map layouts (especially if you incorporate user feedback into improving your work), you will be better able to discern when you should follow the conventions and when it is best to push the envelope of cartographic design. Cartographic license, noted in the chapter 2 discussion of map generalization, also plays a factor. Effective communication through maps often entails taking some cartographic license in the graphic design decisions as well.

Perhaps the best way of learning how to design good maps is to study the maps made by others, critically analyzing them as to what works well and what doesn't, and visualizing various improvements.

Before preparing a new layout of your own, it is helpful to imagine the layout in your mind's eye. A mental or sketch layout can save a lot of time in the long run and lead to better-designed maps because it encourages you to think through some of the common design issues in advance and work out any problems. Sketching a rough draft of the layout on paper, boxing in the main map elements, and trying a few different placement options will help you see what works graphically, and how much material can fit on the page while also maintaining legibility and clarity. Experimentation is indispensable in the creative process of map design, and computerized mapping with GIS enhances our ability to easily visualize and display alternative solutions. The map design process is iterative, and at each step you may need to go back to a prior step and make adjustments or refinements.

Some of the factors that should be taken into consideration when designing a map layout are visual balance, hierarchy, placement of map elements, visual contrast, line weight, and figure-ground relationships, discussed below, as are color and typography, two additional design tools.

Visual balance

Visual balance results in proper harmony and proportion of map elements. When we design a map layout on a page, we must place the map elements in such a way that our eye is led through the page in a natural sequence. This is both aesthetically pleasant and results in effective communication, since the hierarchy of information can be made obvious through good visual balance.

Paradoxically, visual balance is more often achieved with an asymmetrical layout rather than with a page divided into two or more equal sections. This is because visual balance requires visual tension, which is more readily achieved with asymmetry. It is essential to be aware of the visual weight of objects in a map layout. The various map elements, whether contained within boxes or not, should be placed so that heavier appearing elements are towards the bottom of the page, with lighter elements on top. In experiments with people's perceptions about visual weight, it was found that regularly shaped entities appear heavier than irregularly shaped ones, and bright colors appear heavier than dark ones to most people (Dent 1999).

Visual balance focuses on the optical center of a page, which is generally a point above the actual geometric center of the page. The eye will focus first on the element at optical center, and then be led in a certain direction by the placement of the other elements. The effective placement of elements within the layout gives the eye direction. A layout with good visual balance tends to be more interesting and alive than one with poor visual balance. Although concepts of good visual balance are very subjective, the manner in which a layout space is proportioned will be perceived by map readers as either dynamic, and therefore having good visual balance, or static, with poor visual balance.

An important but often forgotten map element is the so-called leftover blank space on the page—the areas between major map elements. This white space should be judiciously proportioned so that map elements do not appear to be scattered or floating arbitrarily on the page. A well-designed map will use the white space as the glue that bonds the elements together into a coherent and visually pleasing layout (figure 5.19).

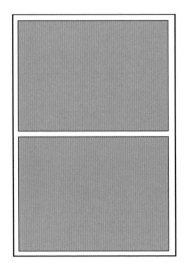

**Symmetrical layout:
no visual excitement.**

**Asymmetrical layout:
enhanced visual interest.**

**Symmetrical layout:
static design.**

**Asymmetrical layout:
dynamic design.**

Figure 5.19   Visual balance.

Hierarchy

Not all map elements are of equal significance, and the hierarchy of elements becomes a critical design component. If the main map elements are considered to be the title, legend, and map (basemap with thematic overlay), these items will necessarily be larger and more prominently placed than secondary elements such as the scale bar and north arrow. Font size may also vary hierarchically. For instance, the main title should be larger and more prominent than subtitles, data source, or scale bar notations. Likewise, within the map itself, the font sizes of place names will vary hierarchically. For instance, country names will appear larger than state names, which in turn will be larger than county or city names. Names of larger cities will be larger than names of smaller cities.

Placement of map elements

The proper placement of map elements on the page is necessary for the creation of a harmonious and informative layout. Depending on where we place map elements, we not only convey information about the relative importance of these elements to the map user, but we also can use placement to enhance clarity, promote a balanced and attractive page, and generate interest and excitement about the map among map users. It is usual to place the main title, for example, near the top of the page and the credits for the data sources and map author information near the lower left or right hand corner. This is because we normally read a page from top to bottom, so the more important information should be near the top of the page, where our eye starts reading.

Visual contrast

Contrast is a central characteristic of maps, since it is through contrast that we perceive differences. Visual contrast can be achieved in a map layout by varying the texture, color, or line weight of elements. The objective is to make it easier for the map reader to differentiate among features or elements (figure 5.20).

Line weight

Outlines of geographic features should be displayed legibly enough to be seen, neither so thin that they can't be seen, nor so thick that they interfere with other lines or features. These are among the considerations of line weight. Line weight can also be used as an indication of the hierarchy of the map features. For instance, local streets are usually portrayed with a thinner line than that used for a major highway. The line weight is not meant to accurately reflect relative road widths, but is rather a shorthand way of establishing hierarchy of symbols. Varying the line weights helps the map user to quickly interpret information and adds visual interest to the map (figure 5.21).

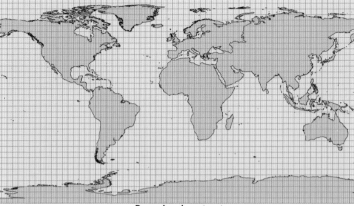

Poor visual contrast

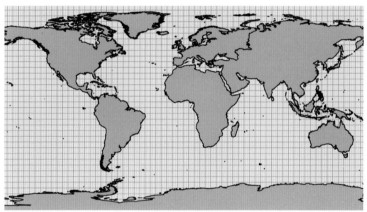

Better visual contrast

Figure 5.20 **By altering line weights, colors, and figure-ground juxtapositions, better visual contrast can be achieved, heightening map readability.**

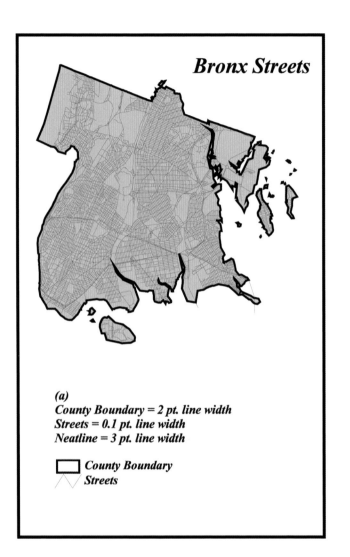

(a)
County Boundary = 2 pt. line width
Streets = 0.1 pt. line width
Neatline = 3 pt. line width

☐ *County Boundary*
⋀ *Streets*

(b)
County Boundary = 0.1 pt. line width
Streets = 1.5 pt. line width
Neatline = 0.5 pt. line width

☐ *County Boundary*
⋀ *Streets*

Figure 5.21 In these maps of streets in the Bronx, line weights give the map reader clues about the relative importance of map elements or geographic features.

Figure-ground relationships

Figure-ground relationships are part of a perceptual issue—how our eyes and mind make sense of two-dimensional figures and transform them into a three-dimensional under-standing of space—that is, which parts of a map we see in the foreground and which in the background. It is crucial for the map reader to easily understand, for example, which part of the map represents water and which part land. In order to reduce visual confusion, the map reader must be able to see at a glance which boundaries create the figure (usu-ally assumed to be the land), and which create the background (usually assumed to be the water). The appropriate use of color, pattern, texture, and line weight enable the map user to differentiate between the figure and the ground. For instance, water is typically a darker tone, while land masses appear white. The structure of the figure-ground relationships can dramatically enhance the readability of the map.

Figure-ground design decisions also come into play with map symbolization and the use of geographic framework grid. In the case of symbols placed on top of a basemap, map users understand that the basemap continues underneath the symbols, and the symbols are not actually interrupting the basemap—in effect, the symbols are in front of the basemap (figure 5.22).

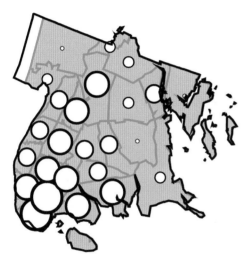

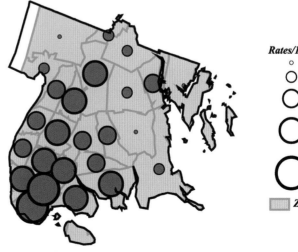

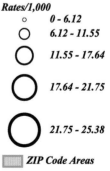

Figure 5.22 The colors or tones used for the symbols and background elements can influence how figure-ground relationships are perceived by the map reader. These maps show asthma hospitalization rates for children.

Rates/1,000

○ *0 - 6.12*

○ *6.12 - 11.55*

○ *11.55 - 17.64*

○ *17.64 - 21.75*

○ *21.75 - 25.38*

▦ *ZIP Code Areas*

In the case of a graticule placed over the map, it is less confusing for the map reader if the grid is placed over the water or the land, but not both. This in turn depends on which part of the map is considered the background. The interrupted graticule, perceived as being "behind" the map, helps define the figure-ground relationship (figure 5.23).

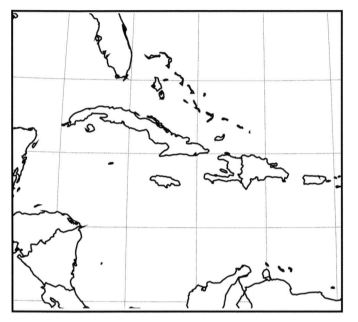

Line weights varied:
Graticule = 0.1 point
Country Boundaries = 1 point
Neatline = 2 point

Line weights the same:
Graticule = 0.5 point
Country Boundaries = 0.5 point.
Neatline = 0.5 point

Figure 5.23 Varied line weights as well as position of elements relative to each other reinforce correct figure-ground relationships.

Color

There are no hard and fast rules for designing with color, and each person sees colors somewhat differently. Color is therefore one of the most subjective of the design elements in mapmaking. As individuals, we all have our unique color preferences, and the meanings of colors also may have strong associations, depending on cultural traditions. For instance, in contemporary American culture, green is often associated with nature, the environment, and health, while red is almost universally perceived as a warning to stop, or of danger. In this way, color can have a strong undercurrent of meaning and can impart a subconscious message to the map reader.

Color in map design

Colors are used to simplify mapped data and to unify and organize mapped information. Colors aid in the communication of data and often help make data more legible, enabling sharper differentiation of nuance. Color also can provide emphasis and draw the map reader's attention to selected elements. In addition to these functional uses, color is also used to make the map attractive and interesting, and to create a certain impression.

Color vs. grayscale

Maps can be created in grayscale (white, gray, black) or in color. Each has its benefits and drawbacks. Grayscale is often preferred for many mapmaking purposes because it is easier and less expensive to reproduce than color. Reproduction of black and white and grayscale maps generally does not result in loss of information, whereas reproduction of color maps is fraught with difficulties in accurately reproducing the colors that the mapmaker originally intended, and this can lead to reduced transmission of information. Full color, however, is generally more attractive and visually stimulating, in addition to being less limiting in a design sense. Color is usually preferred when there is complex information to be presented, because it allows far more design choices to the mapmaker.

Color conventions

With color, as with other graphic design building blocks, there are certain conventions that are used for maps. Conventional color symbolization includes blue for water, tan for desert, green for forests, and so on. There are also conventional color schemes or color ramps for things like elevation, temperature, or depth below water. In urban planning, land uses follow a conventional color scheme—purple for industrial, brown for institutional, red for commercial, yellow for residential, and so on.

When colors are used in thematic mapping for portraying a variable's numeric range (such as percentage of households per tract owning automobiles), it is usual to use a monochromatic color ramp of graduated shades, with the lowest numbers represented by the lightest value in the ramp, and the highest numbers represented by the darkest value. Intuitively, the map reader can understand that the highest percentage or concentration of a variable is found at the darkest spot; the strength of the color corresponds to the strength of the variable. Although this makes intuitive sense, there are exceptions to this convention, and occasionally it may work better graphically to have the highest concentration of the variable be represented by the lightest value instead—for example, if a choropleth map also displayed point symbols or extensive labeling on the darkest part of the map, making the symbols or text difficult to read. This reversal is usually acceptable as long as the legend clearly states the corresponding colors and values.

In most cases there is wide latitude in color choice for symbols, basemap outlines, typography, and other map elements. Bear in mind that colors are influenced by the colors that surround them, so care must be taken by the mapmaker in the juxtaposition of colors. There are also considerations of legibility of labels in color selection.

Value, hue, and saturation

These are terms that describe aspects of colors. **Hue** is what we think of as the name of a specific color—red, blue, green—and each such hue has its own wavelength in the visible spectrum. A **value** is the quality of lightness or darkness of a color—the amount of white or black that is added to a hue. Perception of a color's value is influenced by the background color. **Saturation** is an indication of color purity, as compared to a neutral gray. A fully saturated color will be the most brilliant version of that color, containing no gray. These terms are used in GIS when developing custom colors or color ramps, and it is worthwhile to experiment with customizing colors by altering the hue, saturation, and value settings.

Color contrast

The use of contrasting colors can support the recognition of figure-ground relationships. Warm colors such as reds and oranges appear to advance, while cool colors such as blues and greens appear to retreat. Likewise, high brightness and deeply saturated colors advance, while darker and less saturated colors recede. Colors that appear to advance should be used for figures, and the colors that appear to recede should be used for the ground. Contrast in value (lightness or darkness of a color) is an effective way to create clear figure-ground distinctions.

Legibility

The choice of colors can increase or reduce legibility on a map layout. In a full color map, symbols or letters in color must be placed on background colors that will enhance their legibility and clarity. For instance, red letters or symbols on a green ground are barely legible in most cases, while black letters or symbols on a yellow ground, or green on white, are highly legible. Experimenting with colors is very easy and straightforward in ArcGIS, so any necessary changes can be made rapidly and easily.

Typography

The primary function of lettering on maps is to communicate information. For reference maps, type is used to identify place names and to label other elements, such as scale bars and graticules. For thematic maps, the emphasis is less on identifying place names and more on providing explanatory text, legends, and titles. Typefaces can also lend a certain personality to the map, and the proper selection of type style, size, and color can improve the map's appearance and add interest.

Type fonts and size

A typographic font is a complete set of all the upper and lower case letters of one size and style of type face, including numbers and punctuation marks. Normally, a specific type style is available in various sizes that are called points, which are the standard units of measurement for type. Seventy-two points equal one inch of letter height (figure 5.24).

a (8 points)

a (12 points)

a (24 points)

a (72 points)

Figure 5.24 Font sizes.

A letter of 4 to 5 points is about the smallest size that can be read without straining. The type on this page is 9.6 points. On an 8½×11-inch map page, legends and other explanatory text are often in the 10- to 14-point range, while titles tend to be larger, perhaps in the 18- to 24-point range, or even larger if the map layout is on a larger page. Ancillary text, such as data sources and the mapmaker's name, tend to be smaller—8 or 9 points. Labels of place names on the map are often slightly smaller, from 6 to 10 points. In addition to different sizes, typefaces are usually available in several variants, such as boldface and italic. Boldface is a blacker, thicker version of the normal weight type, and italic is the slanted version of the upright normal form (figure 5.25).

GIS for the Urban Environment   (Normal)

GIS for the Urban Environment   (Boldface)

GIS for the Urban Environment   (Italics)

Figure 5.25 Normal, bold, and italic typefaces.

Typeface styles

There are hundreds of typeface styles to choose from within most GIS and computerized mapping programs. Most typefaces fall into one of two major categories: serif and sans serif. Serifs are the small strokes that look like brackets at the ends of individual letters; serif typefaces have those strokes on each character. Popular serif typefaces include Times Roman, Garamond, and Courier. Sans serif typefaces do not have the brackets on each letter, and are therefore more simple in form, giving the impression of a cleaner, more modern appearance. Sans serif typefaces include Arial, Helvetica, and Univers. Typefaces with serifs are considered easier to read than sans serif typefaces because serif letters are more distinct, making them easier to recognize letters even when the whole word or letter is not visible (figure 5.26).

Figure 5.26 Serif and sans serif typefaces.

GIS for the Urban Environment   (Garamond, 12 point, serif typeface)

GIS for the Urban Environment   (Arial, 12 point, sans serif typeface)

The typeface selected for the various elements of the map will have a significant effect on the overall design and composition of the layout. GIS software makes it easy to experiment, however. It is generally advisable to use the same typeface throughout the map, creating variation through the use of boldface or italics, or italics of different sizes. If you must use more than one typeface, select those that are distinct from each other, yet harmonious. Using too many different typefaces will add clutter and confusion to the map. Decorative typefaces, such as Old English Blackface, script, three-dimensional, or shaded should be avoided altogether for maps. The basics are best.

Legibility of type

Communication of information, the main purpose of putting type on maps, will not be achieved unless the text is legible. Legibility depends on correct selection of typeface style, size, color, placement, and spacing. As mentioned above, serif typefaces are usually easier to read than sans serif. Whether a serif or sans serif typeface is selected, using upper and lower case letters will be more effective than all capital letters. Words formed with all capital letters are usually reserved for labels of large areas, such as countries, (as distinguished from labels for point features). Italics are conventionally used to label water features. Typeface sizes should be selected for optimum legibility, but also for their ability to communicate information about the relative importance of elements: more important elements or features should have larger type as labels or text.

In maps, letters are often placed on different color backgrounds. Therefore, the mapmaker must ensure the text is legible when placed against all the various colors on the map. The color of the letters themselves will obviously be an important consideration in how legible the text is. Strive for maximum contrast between letter color and background color, or the letters will blend into the background and be hard to read. You can also change the color of the font as necessary to be seen against different color backgrounds.

The placement of lettering on a map also has implications for legibility. Letters used as labeling for points, lines, and areas should not interfere with or be obscured by any other feature. This can sometimes severely limit your ability to place names on the map, depending on the density of the map features. Standard orders of preference for label placement for points have been developed: the label at the upper right of the point is the most preferable placement, then upper left, then lower right, and so forth. However, there is little consensus, and some software (including ArcGIS) has an automatic labeling function as well as a software extension that offer a number of possibilities for optimal placement of individual labels.

A linear feature should be labeled along as much of its length as possible, while avoiding conflict with other labels. This is usually accomplished by spacing the letters generously along the feature. It is possible in ArcGIS software to fit a label to a curved linear feature, such as a river or road. Large areas on the map can be labeled within the area itself, provided the entire label fits within the boundaries of the area. The label should be stretched, if necessary, to take up the entire width of the area, from left boundary to right boundary. This will serve to reinforce a sense of the actual extent of the area. If all capital letters are used for large area labels, large amounts of space between letters does not hinder readability, since the letters will be perceived to belong together. If an area is too small for the label to fit within it, the same rules as for label placement for points apply.

Labels should always be placed on the page in the normal left-right reading direction and should not be placed with vertical lettering or upside-down.

As with all the graphic design guidelines, these recommendations about typography are just that—recommendations. There are many exceptions to the rule, and occasionally what works best breaks some (or all) of the rules. However, following the above guidelines will likely result in clear and effective maps. Once the basic skills are learned, bolder and more imaginative maps are possible.

Written Exercise 5

Map design critique

Based on the above design guidelines, examine the map below and give a detailed critique of it. Take into consideration aesthetics, composition, balance, use of color and typography, and map content analysis and presentation. Discuss as many of its shortcomings as you can find, and give suggestions for improving it.

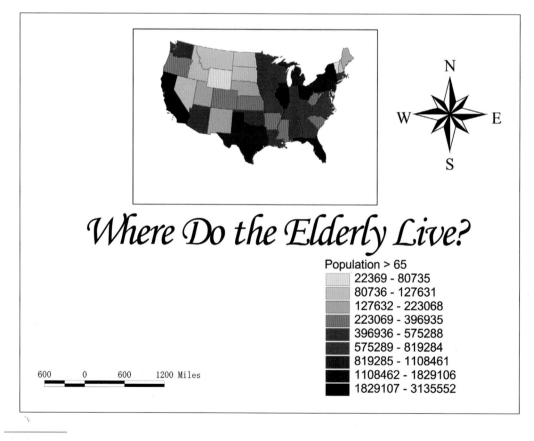

Figure 5.27

References and further reading

Brail, Richard K., and Richard E. Klosterman, eds. 2001. *Planning Support Systems: Integrating Geographic Information Systems, Models, and Visualization Tools.* Redlands, CA: ESRI Press.

Brown, Judith R., Rae Earnshaw, Mikael Jern, and John Vince. 1995. *Visualization: Using Computer Graphics to Explore Data and Present Information.* New York: John Wiley and Sons.

Davies, C., and D. Medyckyj–Scott. 1994. The Importance of Human Factors in Visualization, in *Visualization in Geographic Information Systems,* H. M. Hearnshaw and D. J. Unwin, eds. Chichester, UK: John Wiley and Sons.

Dent, Borden. 1999. *Cartography: Thematic Map Design.* New York: McGraw–Hill.

Edwards, G. 1991."Spatial Knowledge for Image Understanding," in *Cognitive and Linguistic Aspects of Geographic Space.* D. M. Mark and A. U. Frank, eds. Amsterdam: Kluwer Academic.

Kraak, M. J., and F. J. Orneling. 1996. *Cartography: Visualization of Spatial Data.* Harlow, UK: Pearson Education Ltd.

MacEachren, A. M. 1995. *How Maps Work: Representation, Visualization, and Design.* New York: The Guilford Press.

McGuinness, C., A. van Wersch, and P. Stringer. "User Differences in a GIS Environment: Protocol Study." Proceedings, 16th Conference of the International Cartographic Association. Cologne. May 3–9, 1993. pages 478–485.

Robinson, A. H., J. L. Morrison, P. C. Muehrecke, A. J. Kimerling, and S. C. Guptill. 1995. *Elements of Cartography.* New York: John Wiley and Sons.

Tufte, Edward. 1983. *The Visual Display of Quantitative Information.* Cheshire, CT.: Graphics Press.

Visvalingham, M. 1994. Visualization in GIS, Cartography, and ViSC, in *Visualization in Geographic Information Systems,* H. M. Hearnshaw and D. J. Unwin, eds. Chichester, UK: John Wiley and Sons.

Wood, C. H., and C. P. Keller, eds. 1996. *Cartographic Design: Theoretical and Practical Perspectives.* Chichester, UK: John Wiley and Sons.

Sources *of* Urban Data

In previous chapters we have shown how you can use GIS to integrate and visualize information about phenomena that share the same geography. This chapter will be devoted to the information itself—what data sources are available, where to find them, and how to evaluate them.

A wealth of data

In recent years there has been a dramatic increase in the availability of digital geographic data, driven by increased investments in digital mapping projects and developments in Global Positioning Systems (GPS) and remote sensing technology. Governments have become increasingly aware of the need for GIS data, and the private sector is recognizing the market potential of mapped information for business use. Data conversion—the digitizing of hard copy maps for use with GIS—is proceeding at an accelerated pace in urban areas around the world. Meanwhile, the private sector continues to develop its own mapped data products. Commercially available GIS products often prove to be excellent data sources for urban GIS users. The following are a few examples of this increased availability:

+ Digital street mapping has been produced by the Census Bureau for the entire country and is updated quarterly by commercial data providers. In addition, a number of commercial data providers offer street mapping that is more accurate than the census maps.
+ Nationwide digital mapping of the physical environment, natural hazards, and man-made hazards is under way by the federal government.
+ Business locations, traffic volumes, property data, telecommunications data, updated demographics, and other urban datasets have been mapped by the private sector and are commercially available.
+ Tax lots are now being digitized in most urban areas in the United States. As a result, fine-grained digital mapping of urban areas and lot level property data is increasingly available.

Meanwhile, the GPS and remote-sensing industries have added imagery, three-dimensional vector data, increased positional accuracy for field-generated data, and environmental information to the growing list of GIS data resources. These technologies have made it possible to conduct GIS analysis in real time, or with information that may be no more than a few hours old. Available and relatively inexpensive remote sensing datasets now include:

+ Photogrammetry at one-meter resolution, available for most urban areas in the United States.
+ Three-dimensional vector mapping of the natural and built environments produced by lidar technology, as discussed in chapter 12.
+ Satellite imagery at one-meter resolution, available worldwide.

In the past, much of this information was simply unavailable in digital form. Now, the challenge is to discover its existence, find it, access it, assess its quality, and evaluate its relevance to your task.

Choosing the appropriate data

Before starting your search for available digital mapping and data, you must make some decisions. Since each layer in your GIS application will differ according to its attributes, you will be making these decisions layer by layer. The general criteria for choosing data, however, will be the same. These can be divided into three categories: relevance, appropriate level of detail, and quality (meaning its accuracy and currency). As will be discussed later, **metadata**—data about the data—is available for most GIS datasets. Metadata allows you to evaluate the suitability of a data source for your analysis.

Relevance

Maps can be thought of as graphic depictions of selected attributes of objects, seen from a particular point of view and projected onto a flat surface. The number of attributes that define any object is infinite, and determining which ones are relevant to your analysis is a crucial first step in setting up your GIS database.

At the beginning, it is advisable to cast a wide net.

For example, if you were to do an analysis explaining the location of the two central business districts in Manhattan, Midtown, and the Wall Street area, you might look for property data, demographic data, economic data, and infrastructure data. You might understandably ignore Manhattan geologic data. But in doing so, you would overlook an important fact about the business districts—their location is primarily due to the fact that bedrock is close to the surface in both areas. That means it was easier to construct tall buildings in those locations than elsewhere on the island.

Because the choice of appropriate data is crucial to the success of a GIS application, the data collection process should be preceded by a thorough study of the client organization's information needs and the area you are studying.

Data collection by others

Your search may include databases that have been collected by others for their own purposes and are not relevant to your analysis in general, but which contain one or more fields of relevant information. As discussed in chapter 8, the relational database management system that is part of a GIS enables you to extract this information and incorporate it into your own database. Assume, for example, that you are doing a study of daytime population in an urban business district. Mapped census data will show you the residential population (where people live) but census estimates of working population (where people work) are less reliable. One better way to estimate the daytime population is to use property tax data. Tax assessors' databases typically contain fields showing land use and building square footage. Commercial, manufacturing, and public building square footages, divided by an estimate of square feet per worker, can be used to estimate total daytime population. Business location databases, available from several commercial firms, can also be used. These databases locate businesses, classify them by type, and contain a field showing number of employees per location. You might want to consider using both the tax and business data sources and comparing them to arrive at the best estimate of daytime population.

As discussed further in chapter 10, the collection of attribute data from various sources is an important part of the GIS development process, and a thorough search may contribute significantly to the quality of the final product. It is a good idea to allow ample time for searching government and commercial databases for relevant information, for extracting that information, and for integrating it into your GIS application.

Level of detail

In chapter 3 we discussed the classification (ranging) of attribute data for thematic mapping, and how the classification of attribute data is under the control of the user. Mapped data, on the other hand, comes in geographic units that have been predetermined by the data provider—data such as census blocks, census tracts, ZIP Codes, counties, states, streets, and highways, to name a few. Although you can combine geographic units later in your analysis, you must first decide on the basic unit that is appropriate to your analysis before you order the data. The choice of a geographic unit for a GIS layer is analogous to the choices you make when you classify attribute data to create a thematic map. There is a similar trade-off between clarity and level of detail in the choice of base geographic unit to

Figure 6.1 Population density by ZIP Code in New York City.

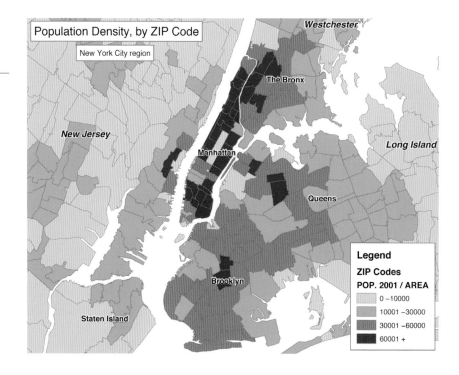

map, such as between ZIP Code and county. As you increase the size of your geographic unit, your attribute data is averaged out over a larger area, resulting in a more generalized picture of the data.

Figures 6.1 and 6.2 show population density in New York City by ZIP Code and by borough, respectively; boroughs are counties in New York City.

The borough geographic unit might be suited to the needs of a metropolitan transportation planner looking for an overview of population density in the region, but a city planner looking for a more detailed picture of density distribution within the city would choose the ZIP Code geography. Your choice of geographic unit will depend on the purpose of your analysis, the availability of relevant attribute data and the size of the geographic area under consideration.

Also, effective presentations can be made by overlaying GIS layers at differing levels of detail. For example, community district boundaries can be overlaid on property values mapped at lot level. The eye first takes in the larger picture, but the detail is there in the background for closer examination.

Commonly available geographic units from the U.S. Census and other sources are discussed on the following pages.

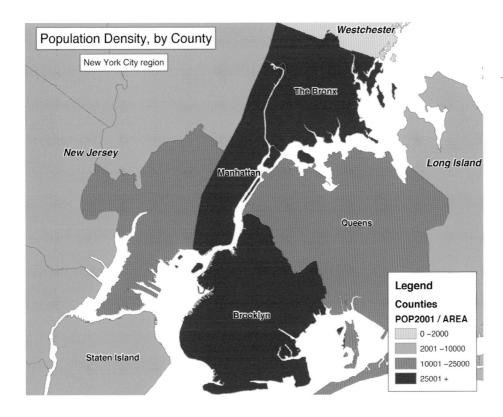

Figure 6.2 New York City population density by county. Each of the five boroughs of New York City is a county.

Quality

Having decided on what data to look for and the appropriate level of detail required, you will need to determine how accurate and how up-to-date this data should be for your analysis. Since decision making in the real world is based on available information that often leaves something to be desired, GIS presentations should include a quality assessment of both the spatial and attribute data that is used in the analysis. It is easy to be fooled by the good-looking graphics produced by GIS software. Your assessment of data quality should address—again, with respect to the purpose of your analysis—the reliability of your attribute data, the accuracy of your map data, and the currency of both. A discussion of data quality should be a part of any GIS presentation. If your presentation consists entirely of maps, footnotes would be appropriate.

Map accuracy

Many of the currently available digital mapping products were originally created from existing hard-copy maps and images that were drawn at specified map scales. Digital mapping is constructed of points, lines, and polygons. Hard-copy maps are converted into digital maps by locating selected points on the hard-copy map and identifying them either as point entities or as the nodes of lines or polygons. Points, defined by their x-, y-, and z-coordinates, are the basic building blocks of digital mapping. The question then arises: how accurate are these points relative to each other, to known landmarks, and to their location on the earth's surface? The answer depends on the scale of the original hard-copy mapping and imagery. If you use a map of the world to digitize points in New York City, you may be off by several miles. If you are using a detailed map of Manhattan, your margin of error might be measured in feet. Metadata and product literature for digital mapping products should cite the scale of the hard copy map from which it was digitized, expressed in standard notation—for example, 1″=2000′: one inch equals 2000 feet (architectural/engineering notation), or 1:24,000: one unit equals 24,000 units (mapping notation). GIS product literature normally uses the latter.

The Federal Geographic Data Committee (FGDC) is a nineteen-member interagency committee composed of representatives from the Executive Office of the President, Cabinet-level and independent agencies. The FGDC is developing the National Spatial Data Infrastructure (NSDI) in cooperation with organizations from state, local, and tribal governments; the academic community; and the private sector. The NSDI encompasses policies, standards, and procedures for organizations to cooperatively produce and share geographic data.

To cite the original hard-copy map scale to indicate the accuracy of the digital map may seem counter-intuitive and confusing to beginning GIS users, because GIS software does spatial calculations in spatial units—such as feet or meters or miles—not in map scales.

Once an existing hard-copy map has been digitized and entered into the spatial database of a GIS, a computer can print out new hard-copy maps at any desired scale.

National map accuracy standards

The metadata file that accompanies a geospatial dataset may indicate that it meets a national standard for positional accuracy. The Federal Geographic Data Committee has published national map accuracy standards, but there are others, including the *National Map Accuracy Standards of 1947,* and *Accuracy Standards for Large Scale Maps,* published by the American Society for Photogrammetry and Remote Sensing (ASPRS) in 1990. If national map accuracy standards are used, they will be referenced in the metadata file that accompanies the data, together with information on how to access the documentation.

It should be noted that geospatial accuracy standards are statistical rather than absolute. These standards refer to confidence levels or indicate a percentage of points on the map that will meet a given standard for positional accuracy. If your analysis requires a high degree of geographic accuracy for specific points, it is best to check their position in the field using GPS (as discussed in chapters 7 and 12) or traditional survey techniques.

Attribute data accuracy and currency

The nonspatial attributes of mapped geographic entities are subject to error: street names can be wrong, or a county road can be misclassified as an interstate highway, or any number of other problems can be present. Commercially available map databases that are updated on a continual basis have an advantage over government data that is updated less often, because mistakes are found and corrected as a part of the updating process. It follows that the more often this is done, the more quickly errors will be corrected. If you are looking at a large metropolitan area, you will probably find that the quality of both geospatial and attribute data varies widely among localities within the same region.

The effect of time on map accuracy depends on the geographic unit. County boundaries, for example, are unlikely to change. The same boundaries that were in place twenty years ago will probably not differ from the current boundaries. But other geographic units such as ZIP Codes and tax parcels are constantly changing. Each layer in your GIS will have its own time sensitivity. For example, you can overlay recently updated parcel mapping on an older map layer showing school district boundaries that have remained constant for thirty years.

In addition, individual GIS applications will have their own requirements for data currency. Data that is a few hours old may be useless in an emergency management context, whereas a demographic analysis that uses year-old data may be acceptable.

The importance of currency for attribute data also depends on its intended use. Socioeconomic analyses rely heavily on the decennial U.S. Census, but as a consequence, by the end of the following decade, the data may be too old to be useful. To fill this gap, several commercial sources and the Census Bureau itself provide estimates and projections for census data to keep it current across the succeeding years. Demographers use a variety of techniques to produce these estimates and projections. These include cohort survival models based on population by age, utility reports on the number of residential meters installed, and even personal interviews, formal and informal, with local planners. Updating techniques vary by data provider, and it is difficult to quantify improvements in accuracy over the decennial census data. In general, it is best to assume that the most recent update contains the most accurate data. However, it is important to read the metadata file that should accompany the dataset, to evaluate sources and projection methods for updated data. An Internet search for U.S. Census data will produce a number of sources for projected census data. It is, of course, important to identify these estimates and projections as such, and to distinguish them from current data that can be verified. As is the case with mapping accuracy, each analysis will call for a different level of currency.

Finally, the overall quality requirements of your GIS data—the reliability, accuracy, and currency of spatial and attribute data—need to be assessed with respect to the purpose of your analysis. This is particularly true of map accuracy, because of the high costs and file sizes associated with detailed and highly accurate mapping. In the 1980s, when GIS was first coming into common use by local government, a number of jurisdictions in metropolitan areas made the same mistake. They required engineering levels of map accuracy for their GIS applications without first implementing a GIS using standard—and inexpensive—digital mapping that was sufficiently accurate for planning purposes. The result was unnecessarily high costs and time delays—sometimes of several years—before GIS products could be made available to users.

When specifying data to be collected for a GIS project, you want to be sure that you are not paying for data that you don't need. In chapter 7 we will discuss the linking of GIS mapping to CAD drawings to provide detailed spatial data in areas where an engineering level of accuracy is required. A commonly heard rule-of-thumb has it that "the best technology is the lowest level of technology that will do the job." This would not be a bad approach to take when setting standards for data quality.

Metadata

Metadata describes geospatial and attribute data, providing the information that you will need to evaluate its suitability to your analysis and to transfer it to your GIS database.

Metadata files should always accompany GIS data. Without metadata, there is no way to judge its accuracy or its lineage objectively.

The Federal Geographic Data Committee (FGDC) has set standards for the content of metadata files at the federal level. These standards are mandated for GIS data produced by federal agencies. Executive Order 12906, *Coordinating Geographic Data Acquisition and Access: The National Spatial Data Infrastructure,* signed in 1994 by President Clinton, states that "each agency shall document all new geospatial data it collects or produces, either directly or indirectly, using the standard under development by the FGDC, and make that standardized documentation electronically accessible . . ." The metadata content standard has been published by the FDGC as FGDC-STD-001-1998: *Content Standard for Digital Geospatial Metadata.*

In addition to the federal government and most GIS data clearing houses, many state and local agencies also adhere to the FGDC standard for metadata content. Other data sources may not meet this standard, and it may be necessary to contact the data provider for further information.

Organizing metadata

The FGDC divides metadata information into the following ten categories:

Identification

Includes descriptions of the data source with contact information, key words in the dataset, the time period covered, the purpose for which the data was collected, the frequency of updates, the spatial area covered by the data, access limitations, and security classifications.

Data quality

Includes reports on attribute accuracy, positional accuracy of the spatial data, and data lineage.

Spatial data organization

Provides information on vector and raster objects, including vector and raster object types, vector topology level, vector and raster object counts, and the spatial reference methods used.

Spatial reference information

Includes map projections and horizontal and vertical coordinate system definitions.

Entity and attribute information

Provides an overview and detailed descriptions of entities and attributes.

Distribution
Includes information on distributors, and technical requirements for accessing the data.

Metadata reference
Gives information about data providers and use limitations.

Citation information
Includes title, details of sources and contacts, series names, and online links (if any).

Time period
Shows the time period(s) covered by the data.

Contact information
Details of contact information.

Graphical representation of the Federal Geographic Data Committee's
Content Standard for Digital Geospatial Metadata

FGDC-STD-001, June 1998

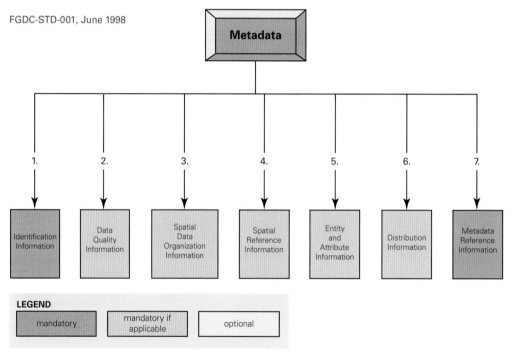

Figure 6.3 The Federal Geographic Data Committee's metadata content standard for general categories of information to be included in metadata files. By Susan Stitt, courtesy of U.S. Geological Survey, Denver, CO.

The U.S. Decennial Census

The primary source of mapped demographic information in the United States is the Census Bureau's decennial census. Mapped data from the 2000 census is available directly from the census Web site, and also from a number of commercial data providers. The value added by private data providers includes demographic projections—typically five years into the future—and current-year updates for the intervening years between decennial census dates. These estimates, like the census data itself, become more reliable as the data is averaged out over larger geographic areas. That is, the chances of a population projection being off the mark are greater at census block level than at census tract level.

The census geographic hierarchy is as follows:

Census region
The United States is divided into four census regions—Northeast, Midwest, South, and West. Each of the four census regions is divided into two or more census divisions.

Census division
Groupings of two or more states within a census region.

State
The fifty states and District of Columbia.

County (or statistically equivalent entity)
There are 3,141 counties in the United States; in Louisiana counties are called parishes. (Alaska has no counties.) When looking at large geographic areas, the county is a convenient and often used geographic unit. There is a wealth of socioeconomic data available at county level from the Census Bureau, the Department of Labor, other government agencies, and the private sector.

County subdivisions and places
Urbanized areas in the United States are identified by the Census either as county subdivisions or as places—depending on their legal and administrative relationships with counties and states. The Census defines an urban area or urban cluster as "a densely settled territory, which generally consists of a cluster of one or more block groups or census blocks, each of which has a population density of at least 1,000 people per square mile at the time; and surrounding block groups and census blocks, each of which has a population density of at least 500 people per square mile at the time; and less densely settled blocks that form enclaves or indentations, or are used to connect discontinuous areas with qualifying densities." Minor civil divisions (MCDs) are also classified as county subdivisions. Most cities, towns, and villages in the United States are classified either as minor civil divisions or as incorporated places.

The Census Bureau uses the following designations to identify other urban areas or clusters within the larger metropolitan area:

Census designated place (CDP)

CDPs are the statistical counterparts of incorporated places. They are defined as "settled concentrations of population that are identifiable by name but are not legally incorporated under the laws of the state in which they are located." There are no population size requirements for CDPs.

Consolidated city

Defined as "a unit of local government for which the functions of an incorporated place and its county or minor civil division (MCD) have merged." The combination of county subdivision and census designated place serves to identify the urbanized areas within a county or metropolitan area, and data is available from a number of commercial providers, along with associated demographic data.

Census tract

In the 2000 Census, the entire landmass of the United States, Puerto Rico, and the Virgin Islands was covered by census tracts. Census tracts are sized primarily by population, although geography is also taken into consideration. The optimum population size of a census tract is 4,000 people, with a usual range of 1,500 to 8,000 people. The geographic area covered by a census tract varies widely according to population density. A center city tract may cover a few city blocks, whereas census tracts in outlying suburban areas often cover several square miles. You can get a good sense of population distribution in an urban area from a census tract map without looking at the associated attribute data. Densely populated areas will show up as clusters of relatively small census tract polygons.

Census block group

The Census Bureau, in cooperation with local governments, delineates block groups within census tracts. Block groups are clusters of census blocks that have some affinity such as land use or population density. They often are small urban agglomerations or areas of higher than average density within the census tract. Block group population ranges from 600 to 3,000 people, with an optimum population of 1,500. Block groups provide the best level of detail for many urban GIS applications and are often the smallest geographic unit available from private sector data providers.

Census block

The census block is the smallest unit in the census hierarchy of geographic units, and consists of "areas bounded on all sides by visible features, such as streets, roads, streams, and railroad tracks, and by invisible boundaries, such as city, town, township, and county limits, property lines, and short, imaginary extensions of streets and roads." Although the census does not define an optimum population size for the census block, a typical urban census block might contain 200 people. Geographically, census block areas vary widely between urban and rural areas. In the center of a city, a census block may be no larger than a city block. All of the territory in the United States, Puerto Rico, and the island areas of the United States is mapped at census block level. It should be noted that the availability of census block data is subject to federal privacy rules, and census block data is less likely to be reliable than data averaged out over larger geographic areas such as block groups or tracts, because of the uneven number of people who actually fill out and return the census forms on which the data is based. Larger geographic areas tend to average out anomalies that may be found at block level.

The Census Bureau also maps geographic areas without reference to the hierarchical relationships listed above. These are known as inventory units. They include:

Congressional, state, and local electoral districts

Political districts follow the boundaries of census blocks or tracts. Not all local voting district boundaries are available.

School district

The Census Bureau obtains the boundaries and names for school districts from state officials. For Census 2000, the bureau tabulated data for three types of school districts: elementary, secondary, and unified.

Traffic analysis zone

A traffic analysis zone (TAZ) is a statistical entity delineated by state or local transportation officials for tabulating traffic-related data such as journey-to-work and place-of-work statistics. A TAZ usually consists of one or more census blocks, block groups, or census tracts.

ZIP Code Tabulation Area (ZCTA)

The ZCTA is new to the 2000 census. This geographic unit approximates the geographic coverages of ZIP Codes, but follows the boundaries of census blocks. (See U.S. Postal Zone Mapping, page 172.)

Metropolitan areas

At the urban level, the Census Bureau produces statistics for counties or groups of counties that surround a core city or urban area, and are considered to have strong social, economic, and commuting ties with the core. Prior to June 2003 these were known as **metropolitan statistical areas** (MSAs), **prime metropolitan statistical areas** (PMSAs), and **consolidated metropolitan statistical areas** (CMSAs). PMSAs represented areas with populations of 1,000,000 or over. CMSAs were agglomerations of PMSAs in major urban areas such as New York and Los Angeles. In the six New England states, metropolitan areas were defined as **New England County metropolitan areas** (NECMAs).

In June 2003, the Census Bureau redefined metropolitan statistical units, designating them as **core based statistical areas** (CBSAs) and dividing CBSAs into two categories: **metropolitan statistical areas** and **micropolitan statistical areas.** CBSAs with urban core populations of 50,000 or more are defined as Metropolitan Statistical Areas, and CBSAs with urban core populations of 10,000 to 50,000 as Micropolitan Statistical Areas. **Major metropolitan statistical areas** with core populations of 2.5 million or more may be divided into **metropolitan divisions,** representing distinct social and economic centers within the larger metropolitan area, and CBSAs that have a high degree of commuting interaction are grouped into **combined CBSAs,** although statistics continue to be available for individual CBSAs within the combined areas. The six New England states continue to have their own metropolitan statistical units, now known as **New England city and town areas** (NECTAs). The basic building blocks of NECTAs are minor civil divisions (cities and towns), rather than counties, a reflection of New England's town- and village-based development history. As of December 2003, 93 percent of the U.S. population lived in core based statistical areas: 83 percent in metropolitan areas and 10 percent in micropolitan areas.

Maps and detailed information about core based statistical areas and the core based statistical area concept can be found on the U.S. Census Web site.

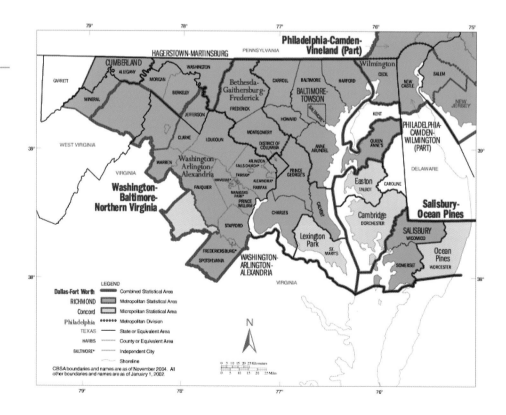

Figure 6.4 Census geographic hierarchy for the area of Maryland and the District of Columbia, including statistical areas, counties, and independent cities.

U.S. Department of Commerce Economics and Statistics Administration. U.S. Census Bureau.

U.S. Census TIGER/Line files

In the 1970s, the Census Bureau's DIME (Dual Independent Map Encoding) project introduced address-matchable digital street mapping for selected regions in the United States. Address matching lets the GIS user geocode street addresses contained in an attribute database (for example, a customer list) and mark them with points or symbols on the map. Often the terms "address-matching" and "geocoding" are used synonymously, although there are differences in the two procedures.

The successor to the DIME project is the U.S. Census TIGER (Topologically Integrated Geographic Encoding and Referencing) mapping system, first introduced on a nationwide basis with the 1990 census. TIGER/Line files include street-level mapping with address ranges for each street segment in the United States. (The address-matching capabilities of TIGER/Line files are described in chapter 7.) TIGER/Line files are an important source of urban digital mapping. In addition to their address matching capabilities, they are often used as base-mapping for GIS presentations. TIGER/Line includes street and highway classifications, waterways, political boundaries, and the boundaries of census tracts that do not conform to streets. Some local governments maintain their own TIGER/Line-based mapping, with updates and corrections based on surveys and local transportation agency files. TIGER/Line files in ASCII format may be downloaded free of charge from the Census Web site, or ordered on CD-ROM for a small fee. Commercial data providers update the TIGER/Line address data on a quarterly basis and add graphic quality by separating streets, roads, and other features. These value-added TIGER/Line products are considered by many urban GIS users to be well worth the expense because of their superior graphic quality and because the mapping is regularly updated.

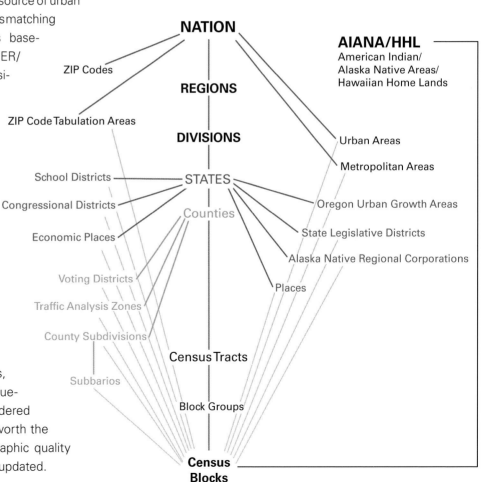

Figure 6.5 **The standard hierarchy of U.S. Census geographic entities.** U.S. Census Bureau 2000.

U.S. Postal zone mapping

There are some 50,000 postal ZIP Codes in the United States, and this geographic unit is often used by business in lieu of the census tract for marketing studies and territory management. Most of the demographic data that is available for census tracts is also available commercially for ZIP Codes. ZIP Codes are generally thought of as areas (polygons), but an increasing number are points on the map representing individual buildings or facilities within ZIP Code areas. If you use ZIP Code areas as your geographic unit for a GIS analysis, you will need to include the data from the point ZIP Codes within these areas. The advantage of using the ZIP Code as the geographic unit for a GIS presentation is its familiarity: people know their ZIP Codes, but very few people know their census tract number. ZIP Codes can also work well as basemapping for GIS presentations. Like census tracts, they have population data associated with them. As population becomes denser, the ZIP Code area becomes smaller, and ZIP Code mapping of a metropolitan area gives a good graphic sense of the distribution of densely populated areas.

ZIP+4 point mapping is also commercially available, and is used by business for direct mail campaigns and micro marketing. There are several million ZIP+4 points in the United States, and it is not unusual in dense urban business districts to find several ZIP+4 Codes in a single building. At present, very little attribute data is available for ZIP+4 point mapping.

U.S. Geological Survey (USGS) mapping

Digital mapping products that are available from USGS and its business partners include:

7.5-minute quadrangles

The USGS has divided the United States into 7.5-minute quadrangles, and provides both hard-copy and digital mapping of these quadrangles—usually referred to simply as "quads"—through business partners. Since one degree of latitude or longitude is equivalent to 60 minutes, 7.5 minutes represents one eighth of a degree. The area covered by one degree of latitude and longitude varies according to geographic location: in the Northeast, this area is approximately 40 miles (east–west) by 60 miles (north–south), or about 2400 square miles. Most USGS 7.5-minute quads are at 1:24000 scale, a relatively large-scale format that permits a good deal of detail. (Some quads maps, and some layers in the 1:24,000-scale maps were digitized at a scale of 1:100,000.) The 1:24,000-scale permits road widths to be delineated (unlike TIGER's single-line street mapping), and most of the features shown on hard-copy USGS quadrangle maps have been digitized for GIS use and separated into vector layers.

These 7.5-minute quadrangles may prove to be the most detailed mapping available for an urban area, depending on which mapped features are relevant to the GIS application. They may be particularly useful for basemapping. However, in many areas this mapping is ten or more years old and may require updating.

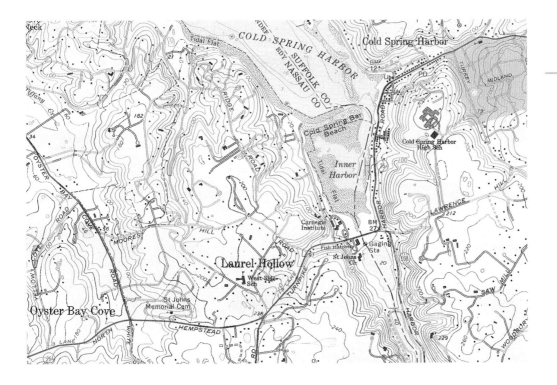

Figure 6.6　**Raster image of a paper USGS 7.5-minute quadrangle map.** USGS.

Digital orthophoto quadrangles (DOQQ)

The USGS produces aerial photography in digital form, usually at one-meter resolution, for selected areas in the United States. This map series covers one quarter of a 7.5-minute quadrangle and is available for a fee from USGS partners, who are listed on the USGS Web site.

Digital Elevation Models (DEM)

Digital elevation models are arrays of points that include a z-coordinate, representing the height above sea level of the terrain at each point. DEMs can be used to produce contour maps and three-dimensional views of terrain. Several software packages and GIS utilities are available to translate DEMs into contour maps and 3D images of terrain, using standard inter-polation methods. See chapter 12 for additional details.

Figure 6.7　**Detail of USGS digital orthophoto quadrangle (DOQ). Although this image has the fuzzy look associated with one-meter resolution, it reveals the layout, site design, and character of housing developments in the area.** USGS.

Other federal digital mapping

Other federal agencies that provide digital mapping include:

- The Department of Agriculture, Soil Conservation Service: soil mapping in selected areas
- The National Oceanic and Atmospheric Administration (NOAA): coastal and bathymetric (underwater) mapping
- The Environmental Protection Agency (EPA): hazardous site mapping
- The Federal Emergency Management Agency (FEMA): floodplain mapping of riverine and coastal areas

GIS data produced by state and local government

Some state, county, and municipal governments produce maps that include:

- Cadastral mapping: tax blocks and lots
- Infrastructure
- Public transportation
- Local administrative districts
- USGS mapping updated and corrected at local level
- TIGER/Line files updated and corrected at local level
- Zoning and other regulatory districts
- Community districts and neighborhoods
- Enterprise zones and business improvement districts

Commercial sources

GIS data that is available from commercial sources include:

- Updated and corrected government mapping and attribute data
- Cadastral imagery (scanned images of tax maps)
- Georegistered satellite coverages and orthophotos
- Telecommunications franchises and service areas
- Landmarks
- Natural and manmade hazards
- Business point locations, with detailed SIC (Standard Industrial Classification) code, sales volume, and number of employees
- Vehicular traffic volumes by street segment
- Rail and subway lines
- Bank deposits and other microeconomic data

✦ Detailed, hand-drawn, or (more recently) CAD-produced block-and-lot maps with building footprints and annotated building information. These are available for most cities in the country from the Sanborn Map Company. Sanborn maps date from 1867 and are updated annually. They are useful for several GIS related purposes, including the identification and mapping of hazardous waste sites from historical maps. Raster images of Sanborn maps are available, and the maps are updated annually (figure 6.8).

✦ Photogrammetry, satellite imagery, scanned paper mapping, and lidar three-dimensional GIS data. These are detailed further in chapter 12.

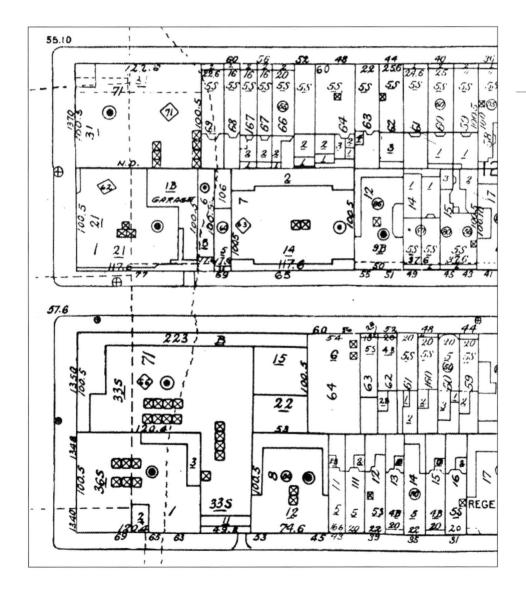

Figure 6.8 Detail of Sanborn map sheet (Manhattan Land Book).

Sanborn Map is provided by the Sanborn Map Company, Inc.

The Internet

Many GIS data sources are listed on the Web sites of federal, state, and local agencies and of commercial data providers. Much GIS data can be downloaded directly from these Web sites, and indeed, downloadable GIS data from the Internet is becoming an important source. A wealth of census data, including attribute data, can also be downloaded from the Census Bureau Web site.

Data clearinghouses

GIS data clearinghouses either provide links to geographic data or let you download it directly. Two good nationwide clearinghouses that contain or reference extensive GIS data are:

+ The NSDI Geospatial Data Clearinghouse: *www.fgdc.gov/nsdi/nsdi.html*
+ Geography Network: *www.geographynetwork.com*

The NSDI (National Spatial Data Infrastructure) includes the Web-based National Geospatial Data Clearinghouse, which documents all data with metadata as required by executive order. This clearinghouse describes itself as "a collection of over 250 spatial data servers that have digital geographic data primarily for use in GIS, image processing systems, and other modeling software. These data collections can be searched through a single interface based on their descriptions, or metadata."

In addition to nationwide clearinghouses, GIS data can be found in clearinghouses that are maintained by state and regional government, and by universities. Municipal clearinghouses are mostly in the discussion stage as of this date.

An important private sector resource for clearinghouse information is Geography Network, hosted by ESRI, which provides maps, data, and data links to a voluminous number of geographic sources worldwide, both government and commercial. Geographic data files can be downloaded and saved, or imported directly into ArcGIS applications. Live online mapping services are also accessible, and a GIS Web browser, ArcExplorer Web Services Edition, lets you create GIS maps while online, from any location.

Figure 6.9 Census tract boundaries in the five boroughs of New York City. Smaller tract sizes indicate greater population density (approximately the same number of people in a smaller area)—a census tract map provides a rough overview of relative population densities within an urban area.

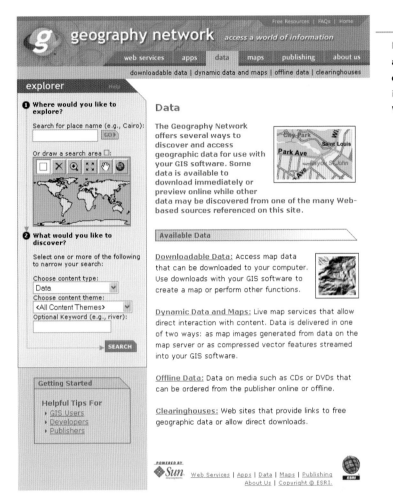

Access to urban GIS data

Since September 11, 2001, security concerns have resulted in restrictions on the availability of some urban GIS data to the public. These restrictions encompass urban infrastructure, detailed mapping of buildings and other facilities, aerial photography, and satellite imagery. However, a consistent national security policy on access to GIS data has yet to be developed. Aerial photography and satellite imagery, for example, are often restricted by public agencies, but are available from commercial sources. Some data GIS is sold to the public—often with license restrictions—by local government agencies. Other GIS datasets can be obtained from public agencies, but only through provisions of the Freedom of Information Act. Data availability among local jurisdictions, as we have emphasized previously, varies widely in both depth and consistency.

Written Exercise 6

Evaluation of an urban dataset

Evaluate the usefulness of the data that you retrieved from the Internet for any of the lab exercises. Discuss data quality issues and limitations. Give an assessment of the metadata, where available. If metadata has not been provided for the dataset that you select, create simple metadata documentation.

References and further reading

Decker, Drew. 2001. *GIS Data Sources*. New York: John Wiley and Sons.

Huxhold, William E. 1991. *An Introduction to Urban Geographic Information Systems*. Oxford: Oxford University Press.

National Research Council. 1999. *Distributed Geolibraries: Spatial Information Resources*. Washington, D.C.: National Academy Press.

Oswald, Diane. 1997. *Fire Insurance Maps: Their History and Applications*. College Station, TX: Lacewing Press.

Shekhar, Shashi, and Sanjay Chawla. 2003. *Spatial Databases: A Tour*. Upper Saddle River, NJ: Pearson Education.

Thompson, Morris M. 1987. *Maps for America: Cartographic products of the U.S. Geological Survey and others*. Reston, VA: U.S. Geologic Survey.

Mapping Databases

Much of the data that you will want to use for GIS analysis will not yet be spatially enabled, but it probably contains one or more geographic references, such as ZIP Code, street address, or latitude and longitude. Fortunately, attribute data with spatial references is available for a wide variety of geographic entities in the world. Where digital mapping is unavailable, you can create your own vector maps with GIS editing tools. You can also scan hard-copy maps, then move and stretch them to match a geographically accurate vector layer in your GIS. In a process known as georeferencing, you can then create vector map layers linked to your attribute tables. This chapter covers the three principal ways to map data: point encoding, linking existing mapping of lines and polygons to your attribute tables, and vector mapping to link to your attribute tables.

Point geocoding

Most attribute data describes objects or conditions—customers, incidents, physical objects, street addresses, and the like—that can be represented by points on a map. These point objects are normally geocoded over a base GIS layer that orients the viewer and places the points in a geographic context. **Point geocoding** is the most common way to map an attribute database that contains a field referring to geographic location.

Geocoding by coordinates

Unmapped tables can be point geocoded, provided that they contain columns that locate each object (or row) in the table geographically by latitude and longitude, by state plane coordinates, or by some other two-dimensional (x,y) grid system. Tables describing objects such as electric utility poles, for example, often contain latitude and longitude columns. The GIS user interface will prompt you to identify these columns, and will then create a map layer showing the poles as points or as symbols chosen by the user. The process of geocoding entire tables is known as **batch geocoding.** Most GIS software packages include a user interface that automates the batch geocoding process and produces reports on the number of objects that have been successfully geocoded. The ability to batch geocode large attribute databases is a powerful GIS feature and is at the heart of many urban GIS applications such as crime reporting, traffic accident data, and infrastructure management (figure 7.1).

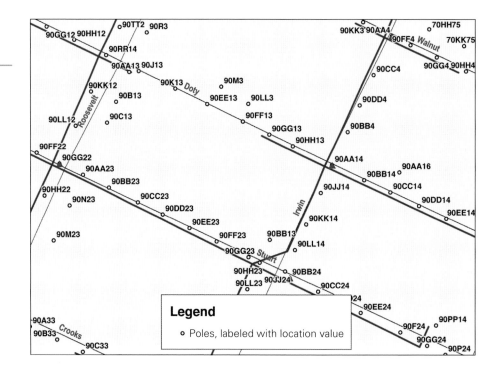

Figure 7.1 Electric utility poles geocoded from a table containing lat–long fields, which can be generated in the field using GPS equipment and mobile GIS software for PDAs, such as ArcPad. Wisconsin Public Service Corporation.

Geocoding to the centroids of polygons

Points can also be matched to the centroid of a geographic unit such as a county, a municipality, an election district, a ZIP Code, or a census tract. GIS software allows you to point geocode your attribute database to these center points.

Geocoding by street address

Point locations in a table may contain a column with a street address. The process of point geocoding using the street address is also referred to as address matching. Address-matched Census TIGER/Line files, as noted in chapter 6, have been refined and updated by local government in some urban areas. This process is finding particular value in conjunction with 911 emergency assistance projects now under way in many parts of the country. These projects update addresses in municipal GIS databases so that emergency responders can locate a caller in distress using an address-matching program. This 911 updating should significantly improve the quality of address-matching data in the future, especially in the outer reaches of metropolitan areas. In addition, a number of private companies are producing address-matched maps for the United States and for other countries. Address-matched TIGER/Line mapping is available at no cost, but fee-based commercial mapping is generally more accurate and is updated as often as quarterly.

In address matching, each street segment is mapped as a line object. The segment is linked to an attribute table that contains address ranges for the left and right sides of the street. Let's assume, for example, that the street addresses for a block on Jones Street start at number 100 and end at number 200. If the right side of the street contains the even numbers, addresses on the right side will start at 100 and end at 200. The left-side numbers will start at 101 and end at 199. The street segment will be mapped as a line, and the associated attribute table will look like this:

Street Name	Segment	City	County	State	From right	To right	From left	To left
Jones Street	15	Anytown	Oneida	NY	100	200	101	199

The *From right, To right,* and the *From left, To left* columns contain the address ranges.

Address-matching capability, built into the GIS software or included in a commercial address-matching package, will interpolate addresses between 100 Jones Street and 200 Jones Street. If your street number is 150, the software will map your address halfway down the street.

The ability to map—in theory—every address in America, and in other parts of the world where the data is available, significantly extends the reach of GIS. In practice, there are problems when address matching an unmapped database. First, there is the question of geographic accuracy. Commercial address-matched street mapping in the United States is available at 1:24,000 scale, but TIGER mapping is only available at 1:100,000 scale or better. The positional accuracy of TIGER/Line street mapping varies according to location.

Furthermore, 150 Jones Street may not actually be located halfway between numbers 100 and 200, particularly on an urban street. The geographic accuracy of address matched points may be an issue when these points are overlaid on detailed urban mapping.

More importantly, there is the question of the *hit rate:* the percentage of addresses in an unmapped database that are found and mapped by your address matching software. Even with the best data, it is rare to achieve a hit rate of 100 percent. First, the address data must be in a format recognized by the address-matching software, and often it is not. If, for example, the software calls for the street number and the street name to be in the same column, your database must also follow this convention. If it does not, the address will not be matched. Secondly, the software must recognize alternative abbreviations, such as "Ave." and "Av.," and must also be able to recognize alternate names for the same street—such as "59th Street" and "Central Park South."

Even more problematic is the quality of the address data in the unmapped attribute database. As previously mentioned, 911 emergency response programs across the country are assigning street numbers to properties throughout the United States, but at the present time, hit-rate percentages in undeveloped areas can be in the low 70s. In developed urban areas, by contrast, most address-matching databases can support hit rates of 90 percent or better.

High-quality, address-matched data can significantly improve hit rates, and a commercial high-end address-matching program may be well worth the cost. Commercially available programs and some GIS software include functions such as automatically cleaning the unmapped database. This involves correcting format and spelling errors, providing look-up tables for alternate spellings and street names, and assigning latitude/longitude and nine-digit ZIP Codes to addresses. In addition, these technologies allow the user to map unmatched addresses at fallback locations. If, for example, a street address cannot be found, the user has the option to geocode the address at the ZIP Code center point, or to map the point manually. The mapped points can be color coded to indicate level of geographic accuracy.

Figure 7.2 A street address located using an address-matchable TIGER/Line file. This is an example of the basic, government-issue address matchable files, available nationwide at no cost. Commercial data providers sell TIGER-based street mapping, with more basemap layers and better graphics; they also offer frequent updating of street geography and addresses.

Unmapped attribute tables

Suppose that you want to produce a thematic map that shows the number of abandoned buildings in your study area, and that classifies them by ZIP Code, as shown in figure 7.3. Suppose further that you have an unmapped attribute table showing the number of abandoned buildings within the five-digit ZIP Code, and a mapped ZIP Code boundary table that contains no attribute data except for the ZIP Code number. With only these tables, you cannot produce this map using a geocoding routine, because geocoding produces only points on a map, and the ZIP Code boundaries are polygons.

So you will need to merge the unmapped attribute table with the mapped ZIP Code table. As we will discuss in more detail in chapter 8, the relational database design of a GIS allows you to merge tables that share a common (key) column, provided that the key column contains a unique value for each row. In this case the key column—present in both tables—is the five-digit ZIP Code number.

The following unmapped table is extracted from the 2001 New York City Department of Finance property tax database. The selected fields show the number of abandoned properties, by ZIP Code.

This table was merged with a mapped ZIP Code table, using the ZIP_CODE column as the key column, to produce the map shown in figure 7.3. Unmapped tables that describe linear objects such as roadways can also be merged with mapped tables, to show, for example, the maintenance history of a road segment.

ZIP_CODE	total_abandoned
10002	33
10014	2
10018	3
10011	2
10010	1
10036	1
10019	6
10022	1
10029	56
10035	81
10026	131
10027	248
10037	38
10030	162
10039	94
10031	70
10032	12
10033	12
10040	4
10034	5
10463	1

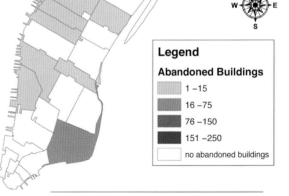

Legend

Abandoned Buildings

	1 –15
	16 –75
	76 –150
	151 –250
	no abandoned buildings

Figure 7.3 **A map of abandoned buildings in Manhattan, produced by merging an attribute table showing total abandoned buildings by ZIP Code with a mapped ZIP Code table.**

Creating vector mapping

When mapping does not exist for objects in your unmapped database, you will need to create new vector mapping—preferably on a new GIS map layer. There are many ways to do this:

Table digitizing

Table digitizing is the traditional means of data conversion. This is the process of translating hard-copy maps into digital format for use in a GIS. A typical urban digitizing project would be the mapping of tax parcels using existing surveyors' maps. The hard-copy map is fixed to a digitizing table. Using a specialized mouse, known as a puck, which is equipped with crosshairs, the digitizer scans over a hard-copy map and clicks on points on the map. These points are then recorded in the spatial database and identified as point entities or as the nodes of lines and polygons. Grid coordinate values describing the geographic location of the digitized points are entered into the spatial database by the GIS software, along with the geometry and topology of the resulting geographic entities. Large table digitizing projects are often outsourced to firms that specialize in this work, located both in the United States and overseas.

On-screen editing of spatial databases

GIS software includes map editing tools that allow a user to create new vector map layers. For example, you may have a database of field reports describing the condition of fire hydrants that have yet to be mapped. The fire hydrants can be located from descriptions of their distance from street intersections. They can then be placed on the map as points or symbols, and overlaid on a street map layer. The field report table can be linked to the map table by fire hydrant ID, present as a key field in both tables.

On-screen data conversion

On-screen data conversion is the process of converting raster images to vector format by tracing over scanned images on the computer screen. On-screen digitizing software that includes features such as automatic snapping to raster lines has been available in the CAD world for some years, and more recently, GIS software providers have added on-screen digitizing features to their programs. Georeferenced photogrammetry and satellite imagery are now readily available from both from government and from commercial sources at resolutions ranging from one to ten meters, and it is now relatively easy to create vector mapping by tracing over these images on the computer screen (figure 7.4).

Georeferencing

Georeferencing is the process of moving and stretching scanned raster images to conform to known points on a geographically referenced grid; this process is also known as **rubber-sheeting.** Once an image—perhaps a hand-drawn paper map from an emergency response team—has been scanned in, software routines included in most GIS packages allow the user to georeference the map image to existing geographically accurate vector maps. The user identifies points on the scanned image and matches them to points on the vector map, and the software does the rest. The degree to which the scanned image conforms to the vector map depends on the number of points that are identified; the more points, the more accuracy. Once a scanned image has been georeferenced, on-screen digitizing can be used to create vector mapping as described above. Several of the case studies in part 2 underscore the significance of georeferencing for urban GIS applications.

Scanned images and digital photographs

Hard-copy images such as photographs, documents, and paper maps can provide important information about GIS map features, and these images can be scanned in raster format, using the relatively inexpensive image scanners that are available today. Photographs taken with a digital camera can be transferred directly to computer in raster format. Photographs and other images can convey information about features on a GIS map that cannot be described in attribute data tables. Photographs of buildings or neighborhoods, for example, make it possible to assess conditions in the neighborhood, and to give a visual impression of how buildings and their environment are actually being used. Scanned architectural and engineering drawings provide detailed information about individual map features such as buildings or infrastructure, and scanned documents can also provide commentary and other information that is best communicated verbally.

Using functionality within the GIS software, or by using programming languages such as Microsoft Visual Basic, these images can be linked to the vector map. When so linked, the user can click on a symbol placed at the appropriate location on the map and bring up a photograph, document, or other raster image.

GPS

Finally, GPS, the Global Positioning System, is increasingly used to provide in-the-field, point-level spatial data for GIS applications. GPS equipment records the latitude and longitude of points on the earth and provides an accurate, spatially referenced database that can be point geocoded as previously described.

Figure 7.4 Combined vector and raster map. Vector lines overlaid on a USGS DOQQ image, showing part of a New York City neighborhood. The DOQQ aerial photograph (a raster image) is georeferenced and can be used as a guide for creating a vector layer. USGS DOQQ provided by IntraSearch, Inc./MapMart.com.

Written Exercise 7

Data acquisition

Assume that you are planning the route for a light rail transit line through the downtown area of your city. The data sources available to you include satellite imagery at one-meter resolution, Census 2000 tract mapping and data, Census TIGER address-matchable street mapping, and an unmapped, lot-level property tax database containing land-use data, total square footage of commercial buildings, number of housing units per lot, and the street address of each lot.

Evaluate and map the demand for the light rail service along various routes in the downtown area, based on the number of people living and working at locations along these routes.

Discuss how you would use these data sources, together with the mapping techniques discussed in this chapter, to produce your GIS presentation.

References and further reading

De Mers, M. N. 2000. *Fundamentals of Geographic Information Systems*. New York: John Wiley and Sons.

Geocoding in ArcGIS. 2004. Redlands, CA: ESRI Press.

Harmon, John E., and Steven J. Anderson. 2003. *The Design and Implementation of Geographic Information Systems*. New York: John Wiley and Sons.

Hohl, Pat. 1997. *GIS Data Conversion Strategies, Techniques and Management*. Santa Fe, NM: OnWord Press.

Montgomery, Glenn E., and Harold C. Schuch. 1998. *The GIS Data Conversion Handbook*. Chichester, UK: John Wiley and Sons.

Silver, Mike, and Diana Balmori. 2003. *Mapping in the Age of Digital Media: The Yale Symposium*. New York: John Wiley and Sons.

Attribute Data *and* Relational Database Management Systems

Eight

A GIS can generate a road map from its spatial database. But unlike traditional mapping, it also contains an alphanumeric attribute database that will alter the images as attribute data is altered; if a route number is changed, that change can be recorded in the attribute database and the map will display the new route number. The GIS links the attribute database to the spatial database that produces the map, using relational database design principles. Traditional database management techniques can manage both the attribute data and the images that display this data graphically. But first you must filter, combine, and update your data to suit the requirements of your GIS application, and to do this effectively, you need to understand the principles of database management. This chapter discusses these principles, and describes operations most commonly used to assemble, manage, and map GIS databases.

Attribute database tables

Attribute data, unlike spatial data, can be simply organized into a matrix of rows and columns, with each row reserved for an object and each column reserved for an attribute of that object. Let's assume, for example, that you have data about individual lots in a city block. As shown in table 8.1, known attributes of the lots are separated into columns, with column headings such as lot ID, lot area, and so on. The individual lots are listed as rows, with the appropriate attribute value entered under each column heading. Note that an attribute value can be a number, a dollar value, a date, or an alphanumeric description (in database terminology, known as a **character string**).

The database table is the basic building block of all attribute **database management systems** (DBMS). Later we will discuss how this simple matrix of objects and their attributes enables **relational database management systems** (RDBMS) to work with multiple tables and to link together the spatial and attribute databases of a GIS.

Importing attribute tables into GIS

As is the case with mapping, GIS users work primarily with imported attribute data. Typically, the data that you will want to import into your system is either in spreadsheet or text format, or exists in tables from popular database management technology such as Microsoft Access or SQL Server, or from companies such as Oracle and IBM.

Spreadsheets

Spreadsheets, such as Microsoft Excel, are organized as two-dimensional matrices of columns and rows, and are easily imported into GIS software and converted into a database table.

A spreadsheet can be thought of as a very limited kind of DBMS that is optimized for calculations.

Text files

Text files are also known as **ASCII** files, because of their nearly universal adherence to the American Standard Code for Information Interchange (ASCII) format. A text file can be converted to a GIS database table, provided that the text is organized into rows and columns, and the positioning, or delimiting, of columns and data values is indicated, so that data values will appear under the correct column headings in the GIS database table. Commas, spaces, or tab keys are used to produce text files known as **comma delimited, space delimited,** or **tab delimited,** and you will often encounter text files in one of these formats. You will also encounter undelimited text files accompanied by documentation that specifies the number of spaces to be reserved for each column. These are known as **fixed-length files.**

BLOCK	LOT	LANDUSE_GEN	LANDUSE_DETAIL	LOT_AREA_SF	NUM_FLOORS	RES_UNITS
458	2	Commercial	STORE BUILDINGS	1919	4	0
458	3	Industrial	LOFT BUILDINGS	1334	4	0
458	4	Commercial	HOTELS	5593	5	31
458	6	Commercial	GARAGES & GAS STATIONS	12204	1	0
458	11	Public/ Cultural Services	ASYLUMS & HOMES	15705	7	0
458	22	Residential	WALK-UP APARTMENTS	1357	3	4
458	23	Residential	WALK-UP APARTMENTS	1311	3	3
458	24	Residential	TWO-FAMILY DWELLINGS	1311	3	2
458	25	Residential	TWO-FAMILY DWELLINGS	1311	3	2
458	26	Residential	WALK-UP APARTMENTS	1288	3	6
458	27	Residential	WALK-UP APARTMENTS	4400	4	8
458	29	Commercial	STORE BUILDINGS	2200	7	4
458	30	Open Space/ Recreation	MISCELLANEOUS	21800	0	0
458	31	Commercial	OFFICE BUILDINGS	1635	4	3
458	32	residential	WALK-UP APARTMENTS	3263	6	28
458	34	residential	TWO-FAMILY DWELLINGS	1638	4	2
458	35	residential	TWO-FAMILY DWELLINGS	1526	4	2
458	36	residential	WALK-UP APARTMENTS	2830	3	3
458	38	Commercial	GARAGES & GAS STATIONS	2519	1	0
458	40	Vacant Land	VACANT LAND	1407	0	0
458	41	Vacant Land	VACANT LAND	1675	0	0
458	42	Residential	WALK-UP APARTMENTS	1650	3	3
458	43	Residential	WALK-UP APARTMENTS	1536	2	3
458	44	Residential	ONE-FAMILY DWELLINGS	1600	2.5	1
458	45	Residential	TWO-FAMILY DWELLINGS	1600	4	2
458	46	Residential	WALK-UP APARTMENTS	1602	4	4
458	47	Residential	PRIMARILY RESIDENTIAL, MIXED USE	1280	3	2
458	1	Commercial	STORE BUILDINGS	1787	4	2

Mainframe file formats

A common issue for planners in city agencies is geographic data that is needed for a GIS application, but which exists only on the city's legacy mainframe system, and in an uncommon file format such as EBCDIC (Extended Binary Coded Decimal Interchange Code). In such cases, a professional with mainframe experience will need to convert these files into an ASCII or other Windows-compatible format before you can import them into the DBMS.

Table 8.1

Creating database tables

When geographic data is generated in-house, GIS software lets the user create new tables, and define their structure. Before entering your own data into an attribute database table, you create a new table and define the table's structure by entering a column-heading name, column width, and data type for each field. The three commonly used data types for database fields are **character**—also often called text or string—**number,** and **date.** In addition, most DBMS break down the number category into subgroups that specify whether the number is an integer, decimal, or other common number type. (Some fields with number values are defined as character fields because the data contains leading zeros—for example, the ZIP Code 02135. If defined as a number field, the system will remove leading zeros where it finds them and compromise your database.) GIS and other DBMS software typically include user-friendly dialog box interfaces for creating and modifying tables.

Working directly with external database management systems

You can also work directly with external DBMS, using ODBC connections, or middleware such as the ESRI Spatial Database Engine (SDE). ODBC translates requests made by your GIS into language that is understood by the external DBMS, and provides direct access to attribute data files that reside there, letting you query the data as though it resided in your own GIS. Most GIS software comes with ODBC drivers for the major database applications.

SDE and other spatial data middleware programs let you work directly with external database systems, and allow you to control access to the database.

Managing database tables

Managing the information contained in a single table will be made easier through the use of wizards within software platforms such as ArcGIS. The syntax for entering these commands will vary according to platform, but all will let you select fields for display, query the database, and perform other basic database management functions.

Table 8.2

BLOCK	LOT	LANDUSE_GEN
458	2	Commercial
458	3	Industrial
458	4	Commercial
458	6	Commercial
458	11	Public/Cultural Services
458	22	Residential
458	23	Residential
458	24	Residential
458	25	Residential
458	26	Residential
458	27	Residential
458	29	Commercial
458	30	Open Space/Recreation
458	31	Commercial
458	32	Residential
458	34	Residential
458	35	Residential
458	36	Residential
458	38	Commercial
458	40	Vacant Land
458	41	Vacant Land
458	42	Residential
458	43	Residential
458	44	Residential
458	45	Residential
458	46	Residential
458	47	Residential
458	1	Commercial

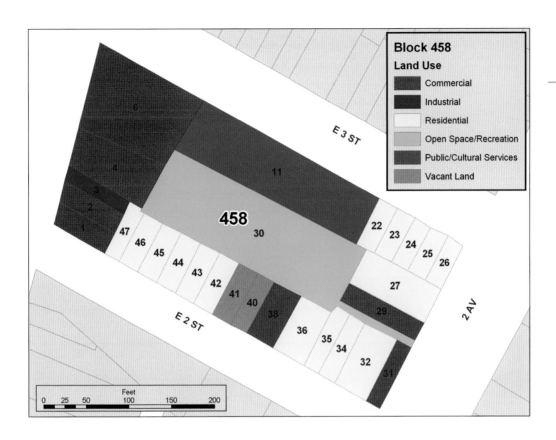

Figure 8.1 **Lots color coded by land use, using attribute data from table 8.3.**

Selecting fields from a database table

Assume that you want to show only the block ID, lot ID, and general land-use fields from table 8.1. Doing so gives you the result shown in table 8.2. As we will discuss later in this chapter, the attribute data shown in table 8.2 can be linked to the lot-level map in figure 8.1, showing lots in block 458 color coded by land-use category.

Queries

Perhaps the most well-used database management tool is the **query** function. Queries enable you to filter the data in your tables to meet your specifications. You can select a subset of rows that meet your conditions, or a group of fields from the larger table. The industry standard language for managing databases is Structured Query Language (SQL), and database queries are often referred to as SQL queries. SQL, first introduced by IBM in 1982 and adapted as a standard in 1986 by the American National Standards Institute (ANSI), allows you to manage databases using English language commands and **Boolean algebraic operators.** Queries and Boolean operators are discussed in more detail in chapter 9.

BLOCK	LOT	LANDUSE_GEN	LANDUSE_DETAIL	LOT_AREA_SF	NUM_FLOORS	RES_UNITS
458	23	Residential	WALK-UP APARTMENTS	1311	3	3
458	24	Residential	TWO-FAMILY DWELLINGS	1311	3	2
458	25	Residential	TWO-FAMILY DWELLINGS	1311	3	2
458	26	Residential	WALK-UP APARTMENTS	1288	3	6
458	27	Residential	WALK-UP APARTMENTS	4400	4	8
458	32	Residential	WALK-UP APARTMENTS	3263	6	28
458	34	Residential	TWO-FAMILY DWELLINGS	1638	4	2
458	35	Residential	TWO-FAMILY DWELLINGS	1526	4	2
458	36	Residential	WALK-UP APARTMENTS	2830	3	3
458	42	Residential	WALK-UP APARTMENTS	1650	3	3
458	43	Residential	WALK-UP APARTMENTS	1536	2	3
458	44	Residential	ONE-FAMILY DWELLINGS	1600	2.5	1
458	45	Residential	TWO-FAMILY DWELLINGS	1600	4	2
458	46	Residential	WALK-UP APARTMENTS	1602	4	4
458	47	Residential	PRIMARILY RESIDENTIAL, MIXED USE	1280	3	2
458	22	Residential	WALK-UP APARTMENTS	1357	3	4

Table 8.3

Now let's assume that you are interested only in those lots in table 8.1 that have residential land uses. Using a query dialog box like the one shown in figure 8.2, you can tell the system to display only those rows in table 8.1, where *LANDUSE-GEN* equals *Residential*. Table 8.3 shows the results of this query.

Tables and views

Note that table 8.3 is not considered an actual table. It is merely a view of a table that shows only those rows that met your conditions. It will disappear when you exit your program, unless you choose to convert it into a new database table and save it.

Filtering, sorting, and other operations can be performed on views, as well as on base tables, and you can update the base table by changing the values in a view.

Using views to restrict data access

Access to the database can be restricted to views and controlled by password. For example, a view of a personnel department database that includes salary information will be restricted to a small group of users, but a view of the same database, based on the query to show names and phone numbers, can be made available to a larger group.

Summary statistics

Most database management systems will calculate summary statistics for number fields, such as minimum value, maximum value, range, average, sum, and standard deviation. This is particularly useful for summaries and reports, where averaging household income for census blocks in an urban area, or totaling the number of housing units in a neighborhood is necessary.

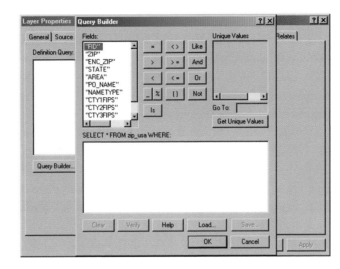

Figure 8.2 The ArcGIS 9 Query Builder screen.

Grouping data

Suppose that you want to break down the total number of lots in block 458 by general land-use category, and display the average lot area for each land use type. You can do this by grouping rows that have the same general land use, using the LANDUSE_GEN field. Because this is a character field, the rows that have the same character values will be grouped together and the COUNT field will be added. The results are shown in table 8.4.

BLOCK	LANDUSE_GEN	COUNT	AVG_LOT_AREA_SF
458	Commercial	7	3900
458	Industrial	1	1334
458	Public/Cultural Services	1	15705
458	Residential	16	1844
458	Open Space/Recreation	1	21800
458	Vacant Land	2	1541

Table 8.4

Derived fields

Now let's assume that you have another table in your database that tracks assessed land values. Fields in this table include lot area in square feet (LOT_AREA_SF) and land value (LAND_VAL). If you are interested in assessed land value per square foot, you can create a new field derived from the others and call it LAND_VAL_PSF. Initially the values in LAND_VAL_PSF will be 0. You can specify that the values in the new LAND_VAL_PSF field will equal the values stored in the LAND_VAL field, divided by the values stored in the LOT_AREA field. The software will populate the LAND_VAL_PSF field with these derived values, as shown in table 8.5.

Table 8.5

BLOCK	LOT	LOT_AREA_SF	LAND_VAL	LAND_VAL_PSF
458	2	1919	210000	109.43
458	3	1334	200000	149.93
458	4	5593	900000	160.92
458	6	12204	800000	65.55
458	11	15705	2000000	127.35
458	22	1357	200000	147.38
458	23	1311	400000	305.11
458	24	1311	400000	305.11
458	25	1311	400000	305.11
458	26	1288	150000	116.46
458	27	4400	350000	79.55
458	29	2200	400000	181.82
458	30	21800	700000	32.11
458	31	1635	300000	183.49
458	32	3263	500000	153.23
458	34	1638	500000	305.25
458	35	1526	450000	294.89
458	36	2830	500000	176.68
458	38	2519	138000	54.78
458	40	1407	150000	106.61
458	41	1675	170000	101.49
458	42	1650	400000	242.42
458	43	1536	380000	247.4
458	44	1600	400000	250
458	45	1600	400000	250
458	46	1602	250000	156.05
458	47	1280	400000	312.5
458	1	1787	250000	139.9

Joining columns to produce a unique ID field

The lots in block 458 are uniquely identified by the value in the LOT field. However, these values may be duplicated in the surrounding blocks. Furthermore, block numbers in Manhattan may be duplicated in the other four boroughs of New York City. If you want to zoom out and look at your lot level data citywide, you will you need to add a BOROUGH field (the borough number for Manhattan is 1), and then create a UNIQUE ID field that combines the values in BOROUGH, BLOCK, and LOT into a single number. The new field uniquely identifies lots citywide because no two lots have the same BOROUGH, BLOCK, and LOT number. This new combined field is known as a **concatenated** field. As discussed below, a unique ID field is required for every table that is part of a relational database, and it is frequently necessary to concatenate fields in order to arrive at a unique identifier for each row in your table, as shown in tables 8.6 and 8.7.

UNIQUE_ID	BORO	BLOCK	LOT	LANDUSE_GEN
1004580002	1	458	2	Commercial
1004580003	1	458	3	Industrial
1004580004	1	458	4	Commercial
1004580006	1	458	6	Commercial
1004580011	1	458	11	Public/Cultural Services
1004580022	1	458	22	Residential
1004580023	1	458	23	Residential
1004580024	1	458	24	Residential
1004580025	1	458	25	Residential
Etc.				

Table 8.6

UNIQUE_ID	BORO	BLOCK	LOT	LOT_AREA_SF	LAND_VAL	LAND_VAL_PSF
1004580002	1	458	2	1919	210000	109.43
1004580003	1	458	3	1334	200000	149.93
1004580004	1	458	4	5593	900000	160.92
1004580006	1	458	6	12204	800000	65.55
1004580011	1	458	11	15705	2000000	127.35
1004580022	1	458	22	1357	200000	147.38
1004580023	1	458	23	1311	400000	305.11
1004580024	1	458	24	1311	400000	305.11
1004580025	1	458	25	1311	400000	305.11
Etc.						

Table 8.7

Exporting tables and views

Once you have completed an analysis of your attribute data, you can export your data tables and views in ASCII format, for use in reports or for insertion into other DBMS. Several GIS software packages, including ArcGIS, allow you to export your data in formats that can be read directly by other DBMS.

Relational databases

So far we have looked at some options for managing individual database tables. If properly designed, these tables can be incorporated into a relational database that is capable of combining, managing, and updating information that is stored in several tables. As previously noted, the relational database is an indispensable element of GIS, first because it enables the system to relate attribute data to spatial data, and secondly because it can manage data stored in more than one database table. The use of GIS to plan and manage a complex urban environment involves the integration of data from multiple sources, continually updated by multiple users, and this cannot be done efficiently without the use of an RDBMS.

Good database design

Many agencies and organizations in recent years have centralized their data into a data warehouse that is managed by an RDBMS. The design of very large databases that serve several departments should be done by IT professionals who have a thorough knowledge of the functioning of the organization. Planners who keep in mind the following general database design principles can enhance the efficiency of their own agencies' operations:

+ As described on the next page, all tables must contain key fields—which contain a unique value for each row—that enable them to be joined on a row-for-row basis.
+ The database should be expandable. Experience has shown that only a small percentage of the ultimate information needs of organizations are envisioned at first by users. Expandability is one of the benefits of a relational database, as new tables can be added and related to existing tables via the key field.
+ Access privileges and updating responsibilities should be defined from the beginning. Access and updating privileges can be granted to tables, to fields within tables, or to views of the database that include data fields from several tables.
+ The relational database structure should be documented in a metadata table that describes the individual tables, their fields, domains, and constraints.
+ Tables that are part of a relational database should be limited, if possible, to a single category of information.

Key fields

A properly designed relational database is composed of related tables, each of which describes different attributes of the same objects. Tables in a relational database are related to each other by a key field, also known as a linking field, that is present in each table, and which contains a unique value for each row. For example, tables 8.6 and 8.7 can be matched row for row because they each contain a field, UNIQUE_ID, with the same values in both tables. Database tables that include key fields are the building blocks of a relational database. Key fields, unlike the other fields, are duplicated for matching purposes in all of the tables that make up the database.

Combining data from two or more tables

Relational database management systems give you the option to join tables, or to select fields from several tables and combine them into a new view of your database that contains only the information of interest. If, for example, you want to look at the relationship between land use type and assessed land values, you can select LANDUSE_GEN from table 8.6 and LAND_VAL_PSF from table 8.7 to produce table 8.8.

UNIQUE_ID	BORO	BLOCK	LOT	LAND_USE_GEN	LAND_VAL_PSF
1004580002	1	458	2	Commercial	109.43
1004580003	1	458	3	Industrial	149.93
1004580004	1	458	4	Commercial	160.92
1004580006	1	458	6	Commercial	65.55
1004580011	1	458	11	Public/Cultural Services	127.35
1004580022	1	458	22	Residential	147.38
1004580023	1	458	23	Residential	305.11
1004580024	1	458	24	Residential	305.11
1004580025	1	458	25	Residential	305.11
Etc.					

Table 8.8

Joining tables with an unequal number of rows

Since the tables that make up a relational database are often created and updated by different individuals or departments, your database may include tables of unequal length. For example, rows may have been added by the mapping department to table X as a result of an update to a housing subdivision, but not to table Y, which is maintained by the tax assessor. When one table contains more rows than another, you have the option to join them into a new table or view that is either restricted to the smaller (matching) set of rows, or which includes the unmatched set of rows from the larger table. These are known respectively as **inner** and **outer joins.**

Importing data from an external database table

But tables do not always have to be merged in their entirety in order for them to share data. You may want to import data values from an external database table (not part of your relational database) into one of your own database tables. You can do this if the external table includes a matching key field. Suppose that you want to import zoning information from a planning department table that lists zoning by lot in a field labeled LOT_ZONING, and which also contains a LOT_ID field that can be matched to the UNIQUE_ID field—the key field—in table 8.7. The RDBMS will allow you to add an empty ZONING field to table 8.7, and populate this field with the data values from the LOT_ZONING field in the planning department table—as shown in table 8.9.

Table 8.9

UNIQUE_ID	BORO	BLOCK	LOT	LAND_USE_GEN	LAND_VAL_PSF	ZONING
1004580002	1	458	2	Commercial	109.43	C6-1
1004580003	1	458	3	Industrial	149.93	C6-1
1004580004	1	458	4	Commercial	160.92	C6-1
1004580006	1	458	6	Commercial	65.55	C6-1
1004580011	1	458	11	Public/Cultural Services	127.35	R7-2
1004580022	1	458	22	Residential	147.38	R7-2
1004580023	1	458	23	Residential	305.11	C6-1
1004580024	1	458	24	Residential	305.11	C6-1
Etc.						

This feature is frequently used to integrate data into a single database from a variety of external sources.

Adding the spatial dimension

A simple way to query your database spatially is to create buffers around points, lines, and polygons. GIS software enables the user to select geographic features and to create buffers around them at specified distances. Features that lie within the buffers will be selected. Figure 8.3 shows properties that lie within 100 meters of a subway line. These types of spatial queries are discussed in more detail in chapter 9.

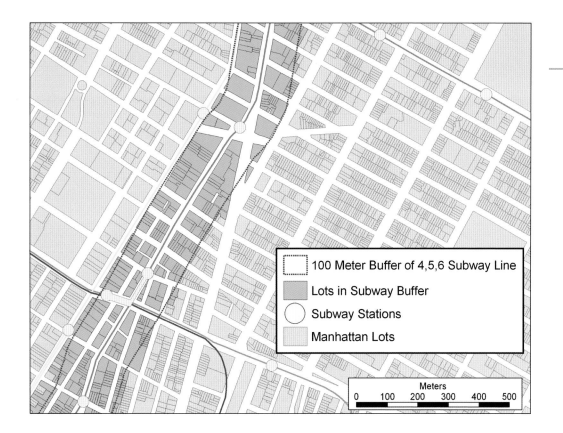

Figure 8.3 **The results of a spatial query: a buffer shows properties that lie within 100 meters of a subway line.**

A more targeted spatial selection is also easily accomplished. For example, table 8.2 (city block 458) lists two lots, numbers 40 and 41, as vacant land in the LANDUSE_GEN field. Perhaps you are considering these two lots as sites for a mid-rise housing project, and you want to know how many stores are to be found within walking distance, or one-half mile, of this block. You can produce a new map that shows the locations of stores within this radius by combining a spatial query with an attribute query.

If you have an area-wide version of table 8.1, you can search the LANDUSE_DETAIL column for the character string *store buildings,* and add the spatial condition that the store buildings must fall within a circle of one-half mile radius, drawn around the center point of block 458. This will return a view that meets both conditions, as shown in figure 8.4.

Figure 8.4   The results of a combined attribute and spatial query, showing store buildings within walking distance of proposed development sites.

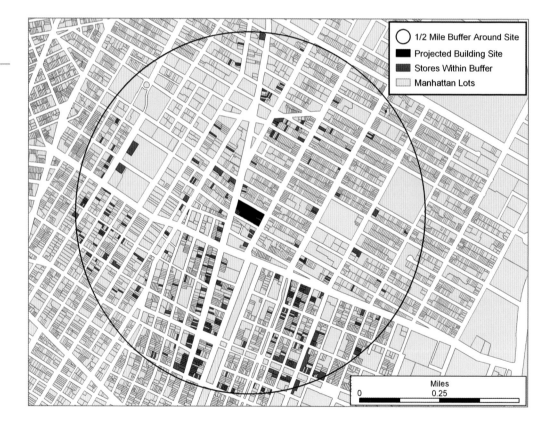

RDBMS and data integration

The proper integration of mapping and attribute data from various sources is key to the successful implementation of a GIS. This is particularly true in an urban environment because urban density concentrates events geographically, and therefore more of those events have the potential to conflict with one another. Relational database technology makes it possible to organize disparate datasets into a single system, but the assembly of data from multiple sources presents challenges to the designer of a GIS application.

In addition to determining what data will be needed, standards will have to be set for data quality. Is Census 2000 data good enough? Will tract-level geography be acceptable? These are questions that require an in-depth knowledge of the client's information needs. Data from external sources is likely to be uneven in quality, and this calls for careful documentation in the metadata files, together with a realistic assessment of the overall quality of the assembled data. As we will discuss in chapter 10, integration of the many data sources that are to be found in an urban environment is one of the major challenges to be met when designing a database for an urban GIS application.

Written Exercise 8

Integrating data using an RDBMS

Using the Internet, locate and download one or more tabular datasets pertaining to your geographic location and containing variables that you would be interested in mapping. Discuss alternative ways that the tabular dataset(s) could be developed into a spatial dataset, and what would be required to integrate the tabular dataset with a spatial dataset you acquired in the written exercise in chapter 6. How could an RDBMS be used to integrate these datasets?

References and further reading

Adam, Nabil R., and Aryya Gangopadhyay. 1997. Database Issues in *Geographic Information Systems*. Kluwer International Series on Advances in Database Systems, Vol. 6. New York: Springer.

Barker, F. Scott. 2004. *Access 2003*. New York: John Wiley and Sons.

Date, C. J. 1999. *An Introduction to Database Systems*. Boston: Addison–Wesley.

Chappell, David and J. Harvey Trimble. 2001. *A Visual Introduction to SQL*. New York: John Wiley and Sons.

Chang, Kang-tsun. 2002. *Introduction to Geographic Information Systems*. Boston: McGraw–Hill.

Hernandez, Michael J. 2003. *Database Design for Mere Mortals: A Hands-On Guide to Relational Database Design*. Boston: Addison–Wesley.

Rich, David. 2002. *Relational Management and Display of Site Environmental Data*. London: Taylor and Francis/CRC Press.

Stanczyk, Stefan, Bob Champion, and Richard Leyton. 2001. *Theory and Practice of Relational Databases*. London: Taylor and Francis/CRC Press.

Svenonius, Elaine. 2001. *The Intellectual Foundation of Information Organization*. Boston: MIT Press.

Williams, Brian K., and Stacey C. Sawyer. 2003. *Using Information Technology: A Practical Introduction to Computers & Communications*. Boston: McGraw–Hill/Irwin.

Methods *of* Spatial Data Analysis

Spatial analysis is the heart of GIS. Without spatial analysis capabilities, GIS would be merely a computerized mapping and spatial database storage utility. All GIS software includes at least some spatial analysis functions, which range from simple queries to complex modeling and geostatistical operations. Planners should have at least a passing familiarity with the different types of analyses, although the simple analytical methods will often suffice for many planning purposes.

Types of analysis

The most widely used types of spatial analysis are:

- ✦ Query, also known as a phenomenon-based search or a containment search within a spatial region
- ✦ Reclassification
- ✦ Buffer analysis, sometimes called proximity analysis or proximal search
- ✦ Overlay, including operations such as union, intersect, and spatial join
- ✦ Interpolation
- ✦ Modeling
- ✦ Geostatistical analysis
- ✦ Expert systems, also known as rule-based reasoning

A considerable amount of overlap among all of these types can be expected. Functions such as queries, reclassifications, buffering, dissolving, and interpolation are generally performed on one data layer at a time, while overlay functions involving processes such as clip, merge, union, intersect, and spatial join necessarily involve more than one layer.

Figure 9.1 Selection by attribute, or phenomenon-based search. Countries with populations greater than 150,000,000 are highlighted in blue.

Simple query: Data search and analysis

The simple query is one of the most basic of spatial analytical methods. It involves searching for spatial features that meet certain criteria, or which are located in specified, constrained areas. A query is answered when the software selects and displays a set of features (or records) that meets the criteria. The selected features will be highlighted on the map as well as in the attribute table. This kind of data search is also known as a **phenomenon-based search,** because we are using one particular attribute, or phenomenon, to search a data layer for features that belong to a certain category, or which are found within specified constraints.

For instance, if we are working with a spatial database of the world's major cities, we might query to find all the cities with a population of five million or greater. The resulting map would highlight the cities that meet the criteria, and the underlying attribute table would

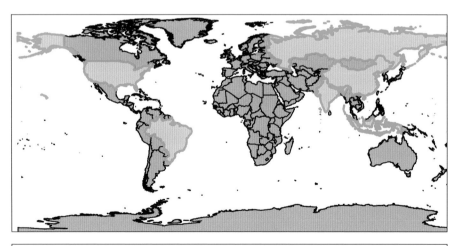

Shape*	FIPS_CNTRY	CNTRY_NAM	POP_CNTRY	SQKM_CNTRY	CURR_TYPE
Polygon	BR	Brazil	151525400	8507128	Cruzeiro Real
Polygon	CH	China	1281008318	9338902	Renminbi Yuan
Polygon	ID	Indonesia	189331200	1910842	Rupiah
Polygon	IN	India	894608700	3089282	Rupee
Polygon	RS	Russia	151827600	16851940	Ruble
Polygon	US	United States	258833000	9450720	Dollar

highlight the corresponding records. Obviously, this query will work only if the database contains a field for population size. We could use the same database of the world's major cities to select the names and locations of all the cities whose physical dimensions exceed 500 square miles, or which are national capitals; in both cases, the table would presumably contain a field for physical size or for capital city status. In ArcGIS, a phenomenon-based query is known as a *selection by attribute* operation (figure 9.1).

Selection by location

We might also create a query to find all the major cities within a certain geographical unit such as a country or a continent, or within a certain arbitrary boundary such as a rectangle or circle placed over a portion of our map. In ArcGIS this type of query is called a **selection by location** operation. This kind of search can also be called a **containment search within a spatial region;** the spatial region can be a predefined window, such as a circle or rectangle, or an existing spatial feature that acts as a kind of cookie-cutter of the spatial database (figure 9.2).

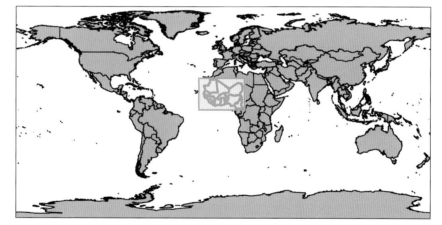

Figure 9.2 Selection by location, or containment search within a spatial region. All the countries within the blue rectangle are selected.

FID	Shape*	FIPS_	GMI_CNTRY	CNTRY_NAME	POP_CNTRY	SQKM_	CURR_TYPE
26	Polygon	BN	BEN	Benin	5175394	116514.7	CFA Franc
45	Polygon	CM	CMR	Cameroon	13218480	466306.6	CFA Franc
79	Polygon	GA	GMB	Gambia, The	936026	10677.6	Dalasi
82	Polygon	GH	GHA	Ghana	16698090	239980.9	Cedi
93	Polygon	GV	GIN	Guinea	62420070	246076.5	Franc
110	Polygon	IV	CIV	Ivory Coast	13498860	322215.5	CFA Franc
131	Polygon	LI	LBR	Liberia	2902441	96296.03	US Dollar
146	Polygon	ML	MLI	Mali	9744733	1256747	CFA Franc
151	Polygon	MR	MRT	Mauritania	2204077	1041570	Ouguiya
162	Polygon	NG	NER	Niger	8797739	1186021	CFA Franc
164	Polygon	NI	NGA	Nigeria	97228750	912038.6	Naira
184	Polygon	PU	GNB	Guinea-Bissau	1085777	33635.39	Peso
198	Polygon	SG	SEN	Senegal	8116554	196910.7	CFA Franc
201	Polygon	SL	SLE	Sierra Leone	4551746	72531.43	Leone
221	Polygon	TO	TGO	Togo	4048365	57299.69	CFA Franc
233	Polygon	UV	BFA	Burkina Faso	10164690	273719.1	CFA Franc

Reclassification

Reclassification can be performed on both attribute and spatial data, and in an urban planning context, both forms of reclassification are commonly necessary. Reclassification entails transforming or combining data in new ways, resulting in new information not currently in the map or table. For example, the attribute values in a field called *Median Household Income* could be recalculated from U.S. dollars to euros; a field for *Income Data for the Year 2000* could be recalculated using a reverse-inflation index to show the equivalent in 1990 dollars; a field for a physical area could be recalculated from square miles to square kilometers. These examples involve arithmetic operations such as addition, subtraction, division, and multiplication, but exponential, square root, and trigonometric functions can also be used.

In all these examples, only the attribute values are being reclassified; there is no change in spatial data—no polygon boundaries modified, or relocation of lines or points. Reclassification of attribute data often results in **derived data fields,** where new fields are created with values that have been manipulated or recalculated based on data from existing fields. The result is a new set of attributes that wasn't given in the existing database, but which is nonetheless useful to planners. For example, from two census tract data fields called *Total Population* and *Age Cohort 65 and Over,* a planner could create a new derived field labeled *Percent Elderly.* In lab exercise 4, the derived data field *Percent Minority* is a reclassified one, created from the sum of the various minority categories divided by the total population for each census tract, multiplied by 100.

Spatial data can also be reclassified. Reclassification of spatial data is typically done to aggregate similar or related spatial data categories into larger ones. When doing this kind of reclassification, categories should be nested hierarchically, and there should be no overlaps or gaps. For example, a zoning map and associated database may contain thirty different land-use categories, including a variety of residential, commercial, and industrial zones. For a generalized zoning map of the entire city, you may want to show only the three major categories (residential, commercial, and industrial), requiring you to combine all the residential subcategories into one category, all the commercial subcategories into another, and all the industrial subcategories into a third. This is called a spatial reclassification because the polygons that represent individual zoning subcategories are merged together by group, the boundaries between like polygons are being dissolved, and new, larger polygons are being formed from the merged zones. Each of the original thirty zone types can be made to fit into one of the three new categories, and none of the original thirty zone types can occur in more than one of the new categories—this is what is meant by no gaps and overlaps (figure 9.3).

Likewise, a zoning map may contain twelve different subcategories of residential zones, each delineating the density and type of housing permitted for each lot in the city. To create a more generalized map of the entire city, you need to collapse the twelve categories into three: low-density one- and two-family dwellings, medium-density dwellings, and high-density multiple dwelling buildings. This will require reclassifying the original category assignment of each property lot in the database. Rather than twelve residential zone categories, the map and attribute database will reflect only three.

These processes could be accomplished in two ways: either by reclassifying the attributes or by dissolving the boundaries between like polygons. In the first method, you would create a new field that would contain the values of the simplified zoning categories. (The new values would be calculated by a logical expression, such as *zoning categories 1, 2, 3, 4 = 1; 5, 6, 7, 8 = 2; and 9, 10, 11, 12 = 3.*) Once the original twelve zoning categories are collapsed into three new categories in a new field without any of the original data having been altered, the variable can now be mapped using the new field with the three new categories. In the second method, the boundaries between the polygons that share the same new zoning category would be removed, thus changing the spatial data. This method is discussed in more detail below.

Spatial reclassification is often done to simplify a map, by merging several similar categories into fewer classes. This makes the map more understandable. Reclassification can result in the loss of some information, so the data in the original categories should whenever possible be left in the database in case it is needed in the future. Caution should be exercised when the reclassification of data results in the creation of more categories than were present in the original data. These more finely grained data categories necessarily involve some speculation on the part of the mapmaker or analyst, thus potentially reducing the credibility of the data. Spatial reclassification should always result in fewer categories than the original, rather than more.

Geoprocessing

Geoprocessing is the general term given to a variety of operations on spatial data in which data layers are combined in different ways to yield new spatial or attribute information. Typical geoprocessing operations include overlay analysis and buffering, and functions usually known by the actual action that is performed on the spatial data, including **dissolve, merge, clip, union, intersect,** and **spatial join,** also known as **point-in-polygon analysis.**

Figure 9.3 Reclassification of many zoning categories into three major categories.

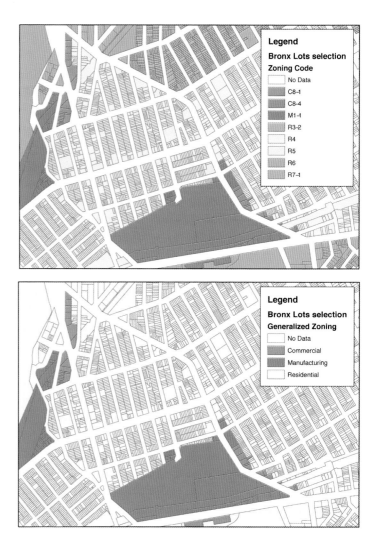

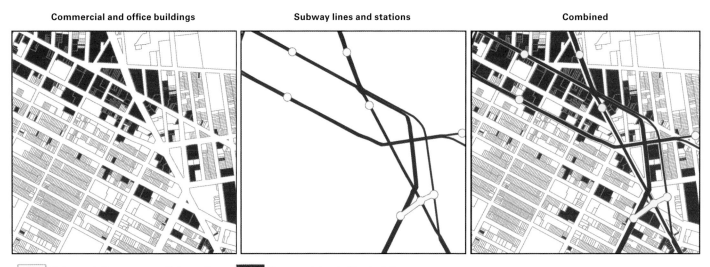

Commercial and office buildings   Subway lines and stations   Combined

☐ Subway stations ——— Subway lines ■ Commercial and office buildings

Figure 9.4 **Overlaying two or more layers of data often yields additional information not available or readily apparent in individual data layers.**

Overlay operations

Overlay operations, which allow you to derive new information that doesn't exist in any single layer, constitute one of the most useful functions of GIS. To understand overlay operations, imagine taking multiple layers of data, each layer drawn on a separate transparent sheet of film, lining up the sheets on a light table or overhead projector so that they can be viewed all together, and then observing all the new information that appears from the overlapping polygons, lines, and points.

For instance, suppose we had several layers of information about a community: a polygon layer showing median household income by census tract, a point layer representing the locations of property crimes, and a polygon layer showing property crime rates by police precinct boundaries. If we overlay these three layers, we can obtain a composite perspective that lets us make inferences about the relationships among crime rates, location of specific types of crime, and household income. Or suppose we had a data layer of a community's commercial property lots, with information about market value per lot, and a data layer of subway lines. By overlaying the two layers, we can examine the relationship between proximity to major public transportation routes and market value of commercial properties—and see whether such properties are worth more if they are close to a subway line (figure 9.4).

In these examples, the overlay function allows us to make visual comparisons and to combine data.

Buffers-proximity analysis or proximal search

Buffers can be considered a variation on overlay analysis. They are created to examine attribute or spatial data within a certain distance of a particular feature. Also called proximity analysis or proximal search, buffering involves creating a new polygon around an existing feature. The feature being buffered can be a point, line, or polygon (figure 9.5).

Buffering is demonstrated in the lab exercise for this chapter, in which the demographic characteristics of the population living within one-half mile of Toxic Release Inventory (TRI) facilities, represented as points, are located. The circular buffer results in a new polygon layer. Buffering lets the map analyst examine other features and data that fall within the buffers, or perform an overlay analysis to determine exactly what features (such as schools, or characteristics of the general population) lie within the buffer polygons.

Lines can also be buffered. If, for instance, we wanted to find out if children attending schools near major highways suffered higher rates of absenteeism than other children, we would buffer each highway (a line feature) to a distance of 500 feet, creating a buffer polygon 500 feet to either side of the highway's centerline. Schools (point features) found within the highway buffer polygons could be selected for analysis of absenteeism rates and then be compared to schools outside the buffered areas, or to all schools.

Polygons can also be buffered. For example, we might need to find all the property lots near other lots that are being proposed for re-zoning, so that those property owners can receive legally required notification. We could select the polygons that are being proposed for the zoning change and create buffers of the required notification distance from the outside boundaries of these lots. This would result in buffer polygons around the selected property lots, allowing us to generate a list of all the affected property owners within distance of the lots proposed for re-zoning.

Figure 9.5 Point, line, and polygon features can be buffered at one or multiple distances.

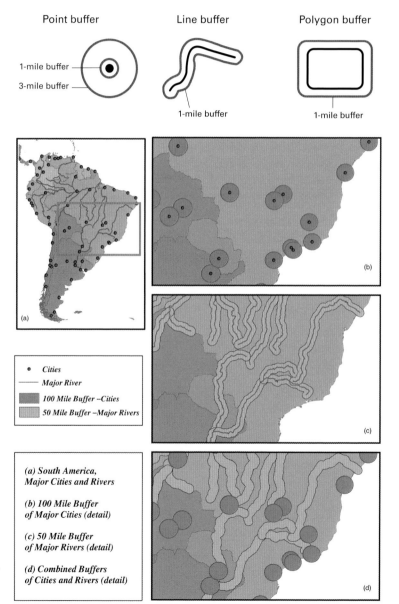

Figure 9.6 Buffers can be used individually around one class of features or in combination with multiple features.

Figure 9.7

Dissolve

As we noted earlier, the **dissolve** function is a type of reclassification. It is used to eliminate the boundaries of polygons by aggregating spatial data features based on an attribute—meaning that all the boundaries between polygons that share a certain attribute will be dissolved. For instance, if your database of census tract boundaries was dissolved by city council districts, your result would be a layer of tracts whose individual boundaries have been eliminated, and the tract polygons merged together based on council district. The result: All the tracts located within a particular council district become a single polygon. By dissolving tract boundaries, we can easily obtain summary statistics on the tracts within the jurisdiction of each council district—not only a count of the number of tracts, but also sums, minimums, maximums, averages, standard deviations, and other calculations of any field that was part of the original attribute table for tracts.

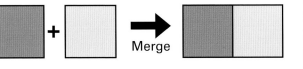

Figure 9.8

Merge

Merge is used when you have two or more geographically adjacent layers, each containing the same type of features, and you want one layer that contains all their features. Each layer to be merged must have at least some of the same fields and be represented by the same type of geographic entity (point, line, or polygon). If one layer has additional fields or different fields from the initial datasets, these fields will not appear in the new layer. You might use the merge operation if you had a street dataset for a city, and you wanted to create a seamless database of adjacent suburban street layers by appending the suburban street layers to the urban street layer, thus creating a metropolitan area street database.

input theme   clip theme

Figure 9.9

Clip

Clipping is used when you need to reduce the spatial extent of one or more datasets to the geographical extent of another dataset—for example, a dataset of ZIP Codes for an entire city, of which only one portion is relevant. We could take the spatial data showing the boundaries of the community of interest, and use it as the basis to clip the citywide ZIP Code spatial data. We would end up with just the ZIP Codes in the community of interest. The clip function acts as a kind of cookie-cutter—with the community boundary data as the cutter, and the city-wide ZIP Code data base as the cookie dough. Clipping is often used when some of the datasets have a much larger extent than the area of interest, and the analyst needs to minimize processing time and storage space; it is particularly useful for creating manageable study areas that are representative of the larger dataset.

Clipping is also often used in buffer analysis to extract information in other layers that pertain only to the area within the buffers. The buffer polygons act as the cookie-cutters in such a case.

Union

The **union** function is one of the most powerful of the geoprocessing tools. It will overlay two datasets, producing a new dataset that has the combined attributes of both, whether or not they overlap in extent. For example, suppose we had a dataset of watershed or drainage basin boundaries within a municipality, and another dataset of property lots.

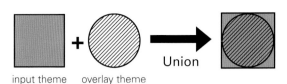

input theme overlay theme

Figure 9.10

We want to be able to see which properties are contained within each watershed, including those properties that could be split between more than one watershed; we also want to see if any watersheds are split by any property lots. The union function will preserve the attributes of both datasets, even those property lots that do not overlap the watershed boundaries or those watershed boundaries that do not overlap any property lots. Each resulting polygon will contain the attributes (data fields) of both the property lot and the watershed it belongs to. Our new data layer will show the total extent of both input data layers; it will contain areas meeting either criteria: belonging to the watershed dataset or the property lot dataset.

Intersect

The **intersect** function is similar to the union function, but less inclusive. Intersect is used to overlay two layers to create a layer that has the combined data of the two input datasets, but contains only features that fall within the spatial extent of both.

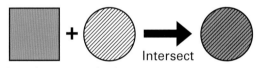

input theme overlay theme

Figure 9.11

Consider a layer showing all the flood-prone areas of a city, and a second layer showing historic preservation districts. If we use the intersect function, we can create a new layer showing areas common to both—historic districts that are flood-prone. Any features meeting both criteria (being both flood-prone and being located within a historic preservation district) will appear in the new dataset; features not meeting both criteria will not be included. (As it happens, the seemingly inconsequential words *and* and *both* in this context will play an important role in more complex operations, as will be described below in the section on Boolean logic.)

Spatial join (point-in-polygon analysis)

There are two main types of join processes in GIS: table joins and spatial joins. They are completely different, and would never be performed on the same data. In a table join, only attribute data from two different datasets is joined; this process is not strictly considered a type of spatial analysis, while spatial joins are. A table join connects two attribute tables, linked on a common field, such as a ZIP Code or census tract number. Thus, the underlying spatial boundaries remain the same. Indeed, a table join is not possible if the underlying spatial data is not the same—it must be performed on two layers composed of the same points, lines, or polygons, and that have a matching field in the attribute table.

The spatial join, however, joins two layers that have different spatial boundaries or properties, such as a point layer and a polygon layer, or a line layer and a polygon layer. This spatial join function joins two sets of spatial data together to assign the attributes of one to the other.

Suppose we had a point shapefile locating all automobile thefts in a city, and wanted to find out the neighborhood characteristics where the crimes were most prevalent. We could perform an analysis of demographic information and other indicators of the census tracts where the thefts occurred, but the auto theft database does not contain data giving the census tract number. To do the demographic analysis, we would first have to know the census tract number for each of the auto theft locations. This could be accomplished with a spatial join. By spatially joining the two databases and using an overlay process, the software would calculate where each auto theft is located by census tract, and append the tract number and all other attributes of the census tract data to the auto theft database. In this case, the software finds the relationship between the points and the polygons—thus, the spatial join is also known as a point-in-polygon analysis.

Furthermore, in a spatial join, the software actually performs a type of overlay analysis with the point and the polygon shapefiles, and determines within which polygon each point lies. When the spatial join was completed in this example, the data could be summarized and you could determine how many auto thefts occurred in each census tract. From there you could develop rates per census tract, and examine those rates in light of demographic, socioeconomic, or other characteristics of the tract (figure 9.12).

Spatial joins are possible not only with point and polygon layers, but also using line and polygon layers, such as streets (lines) and counties (polygons), or rivers (lines) and watersheds (polygons).

Shape	TRI_ID	ZIP_Code
Point	1	10461
Point	2	10457
Point	3	10474
Point	4	10461
Point	5	10451
Point	6	10460
Point	7	10454

Figure 9.12   Point-in-polygon analysis (spatial join). In this example, TRI facilities have been geocoded to street address location, but the TRI database lacked a ZIP Code field. To locate each TRI facility in a ZIP Code, the software analyzes the topological relationship of each point and the polygon boundaries, and determines the ZIP Code polygon designation for each point. The fields in the ZIP Code database are then appended to the TRI database, allowing summarizing of relevant fields.

Optimal location/suitability analysis

Location analyses, also known as suitability analyses, use overlay and Boolean logic to perform queries whose results are locations that meet very specific sets of criteria. These types of analyses are also referred to as multicriteria evaluation, or MCE. The criteria generally apply to more than one layer of information.

A realistic scenario for such an analysis: a city planner is assigned to determine the best location for a lead poisoning education, prevention, and treatment facility. Based on expert knowledge about lead poisoning and the particular city in question, criteria would be developed to pinpoint locations that would best address the requirements. The planner would need several layers of data to solve this problem, based on the criteria and thresholds. These might include demographic and socioeconomic data from the census, locations of major highways, housing data on the age of buildings, and the locations of past lead poisoning cases. Depending on the complexity of the analysis, many other datasets could also be required. Criteria in this case could be based on possible risk (high numbers of people potentially affected), potential for exposure (whether lead poisoning is a likely occurrence in that location), and vulnerability (a large proportion of people without adequate access to preventive health care, health education, and other resources). The analysis would be based on the assumption that the optimal location for a lead poisoning center would be at the intersection of places of high risk, high exposure, and high vulnerability. These terms would all have to be defined and based on defensible thresholds.

An optimal area for a lead poisoning center might have large numbers or a large percentage of children per census tract, since children are most vulnerable to lead poisoning. The planner might establish a certain numerical threshold as a cutoff point—greater than 25 percent of the population under age 15 or a minimum of 500 children. This data would address the risk component of our optimal location analysis.

1) Risk submodel

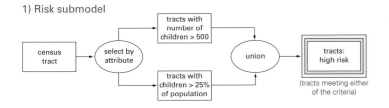

Figure 9.13   **This figure and those on pages 220 and 221 show a possible flow chart for an optimal location analysis of a proposed lead poisoning prevention, education, and treatment center. This submodel selects those tracts containing substantial populations at risk from lead exposure.**

The planner might also obtain a point dataset showing locations of past cases of childhood lead poisoning, as well as the locations of areas known to have a high potential for lead exposure, such as old housing stock (which has a high probability of having lead-based paint or lead water pipes) or areas close to major highways (which have the potential for nearby soil contamination from vehicular emissions of leaded fuels). This data would address the potential exposure component of the optimal location analysis.

2) Exposure submodel

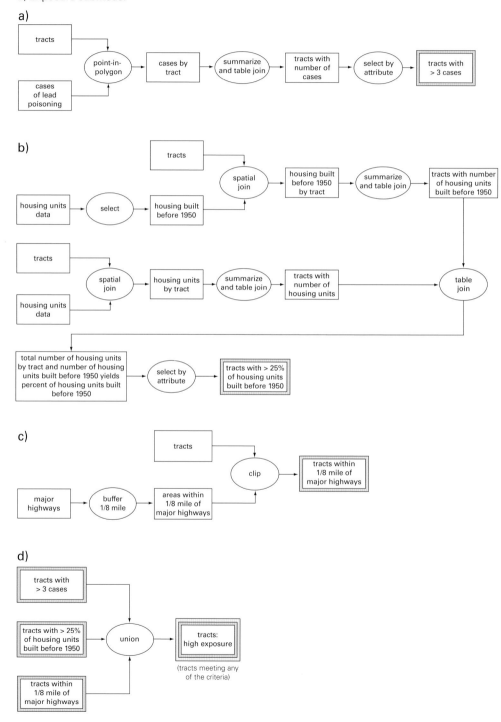

a)

tracts

point-in-polygon

cases of lead poisoning

cases by tract

summarize and table join

tracts with number of cases

select by attribute

tracts with > 3 cases

b)

tracts

spatial join

housing units data

select

housing built before 1950

housing built before 1950 by tract

summarize and table join

tracts with number of housing units built before 1950

tracts

spatial join

housing units data

housing units by tract

summarize and table join

tracts with number of housing units

table join

total number of housing units by tract and number of housing units built before 1950 yields percent of housing units built before 1950

select by attribute

tracts with > 25% of housing units built before 1950

c)

tracts

clip

tracts within 1/8 mile of major highways

major highways

buffer 1/8 mile

areas within 1/8 mile of major highways

d)

tracts with > 3 cases

tracts with > 25% of housing units built before 1950

union

tracts: high exposure

(tracts meeting any of the criteria)

tracts within 1/8 mile of major highways

The planner could then look at areas where mean or median household income is low, where adult educational attainment is low, and where health care access is low. This would be an indicator of vulnerability. Cutoff points for these criteria would have to be established. Expert judgment would determine whether a specific household income should define low income or whether federal poverty levels should. Similarly, a threshold for low educational attainment and low access to health care—such as a lack of health-care facilities within a certain distance of the home—would have to be established. These thresholds might be set by health experts, consultants hired by the funders of the project, the governmental agencies sponsoring the project, or other experts, depending on the project and the technical or statistical nature of the criteria. Often, however, the planners conducting the preliminary GIS analyses set the thresholds as a first screening level.

3) Vulnerability submodel

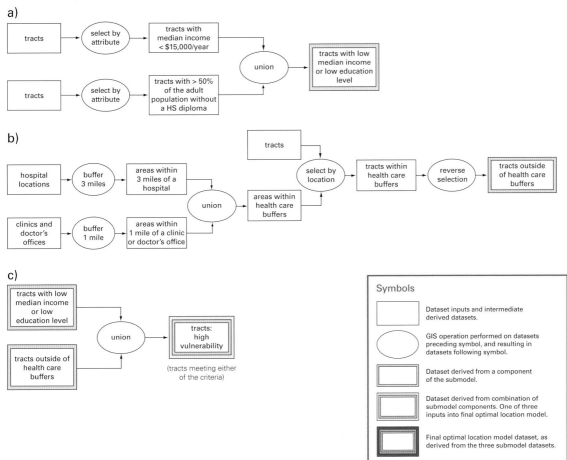

Once the criteria are finalized, it is a relatively simple matter to find the optimal location: one that meets all of the criteria. In a more complex type of analysis, discussed below, the criteria can also be weighted or ranked. By overlaying the different data layers pertaining to the criteria, the planner can locate the areas that meet all the criteria.

4) Optimal location for a lead poisoning center

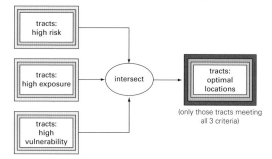

NOTE The flow chart represents a very simplified version of an optimal location model. In the real world, the above might constitute a relatively quick and easy preliminary analysis, perhaps designed to identify likely areas to focus on more closely. An actual model would incorporate some element of ranking or weighting the importance of each criteria, as well as potentially many more criteria and more complex analysis involving, for instance, public transportation routing analysis and the delineation of the catchment areas of existing health care facilities.

Boolean operators

Boolean logic, or Boolean algebra, is used in most GIS overlay operations. The main Boolean operators are *and, or, not,* and *xor.* The *and* operator finds all those areas where all of two or more criteria or conditions are met and where these areas coincide spatially. The *or* operator finds all those areas meeting at least one of two or more criteria or conditions, whether or not they coincide spatially. The *not* operator finds the areas not meeting one or more of the criteria or conditions. The *xor* (exclusive or) finds those areas meeting one of two or more criteria or conditions, but not both. **Venn diagrams** are schematic representations often used to graphically portray Boolean operators. The circles represent different sets of criteria or conditions, and the shaded areas of the circles represent the areas meeting those criteria, according to the particular Boolean operator in effect (figure 9.14).

Using our optimal location analysis of a lead poisoning center as an example, the Boolean algebra might be employed as outlined below. Keeping our three categories or criteria of risk, exposure, and vulnerability, the analysis would require three separate unions, creating three separate selections. Each selection would represent the tracts meeting any of the conditions, that is, the risk selection would include all the tracts that are either composed of 25 percent or more children, or which have more than 500 children total. After the three unions are performed, the three sets of selected tracts are intersected, using the *and* operator. Our final selection of tracts has to meet the conditions of high risk, *and* high exposure, *and* high vulnerability (as defined for us or by experts). The obvious optimal choice for locating our lead poisoning prevention center is where all three conditions are met (figure 9.15).

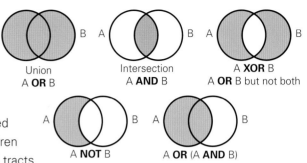

Figure 9.14 Venn diagrams of Boolean operators. As an example, the circle A might represent industrial zones in an urban area, and the circle B might represent neighborhoods of high poverty in the same urban area. The first diagram shows areas meeting either condition—either an industrial zone OR an area of high poverty. The second diagram shows areas meeting both conditions—an industrial zone AND an area of high poverty. The third diagram shows areas meeting either condition, but not both conditions (XOR)—an industrial zone or a high poverty area, but not both together. The fourth diagram shows areas meeting only one condition—an industrial zone, but *not* an area of high poverty, which excludes those industrial areas that are also high poverty areas. The fifth diagram shows a more complex use of Boolean operators: areas meeting one condition or the intersection of that condition with the other—an industrial area, whether or not it is also an area of high poverty.

1) Tracts—High risk

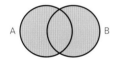

A Tracts with number of children > 500
B Tracts with children > 25% of population

2) Tracts—High exposure

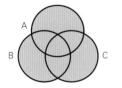

A Tracts with > 3 lead poisoning cases
B Tracts with > 25% of housing units built before 1950
C Tracts within 1/8 mile of a major highway

3) Tracts—High vulnerability

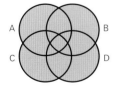

A Tracts where median household income < $15,000/yr
B Tracts where > 50% of the adult population has no high school diploma
C Tracts more than 3 miles from a hospital
D Tracts more than 1 mile from a clinic or doctor's office

4) Tracts—Optimal location for a lead poisoning center

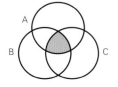

A Tracts—High risk
B Tracts—High exposure
C Tracts—High vulnerability

Figure 9.15 Venn diagrams of the optimal location analysis of a lead poisoning center.

<table>
<tr><td>

1 **RISK UNION**
SELECT TRACTS WHERE
Children 0–15 = 25% or more of total population
OR
Children 0–15 => 500.

2 **EXPOSURE UNION**
SELECT TRACTS WHERE
Childhood lead poisoning cases => 3
OR
Housing built before 1950 => 25% of housing stock
OR
Within 1/8 mile of major truck route or
limited access highway.

3 **VULNERABILITY UNION**
SELECT TRACTS WHERE
Mean household income is < $15,000/year
OR
No high school diploma => 50% of adult population
OR
More than 3 miles from a hospital
OR
More than one mile from a doctor's office or
out-patient clinic.

4 **OPTIMAL LOCATION INTERSECTION**
SELECT TRACTS WHERE
Tracts are *high risk*
AND
Tracts are *high exposure*
AND
Tracts are *high vulnerability.*

</td></tr>
</table>

The Boolean queries in the box on this page show the kinds of expressions that you would enter to select the census tracts meeting all the conditions. Each of the first three selections would be saved as a new polygon layer selection. For the fourth query, the three sets of selected tracts would be intersected to find only those tracts meeting all three sets of conditions.

Modeling

The suitability analysis/optimal location analysis above also represents an example of a simple model. Models simulate some aspect of the real world to provide an estimate of existing conditions where data is incomplete, unmeasured, or difficult to obtain, to predict future conditions, to explore and evaluate alternative scenarios, or to solve for an optimal location.

Models are divided into three main types: analog, conceptual, and mathematical; any of the three can be used in GIS. Analog models are also called physical models, since they are tangible or scale models where one unit of measurement on the model represents "x" units of measurement in the real world. A scale model of a house, an airplane, or a car is an analog model. Maps are analog models as well, since there is a direct and logical relationship between the parts of the model and the object in the real world. Maps are, in effect, scale or analog models of geographic reality.

Conceptual models simulate a real world process or entity by diagrammatically expressing the relationships that make the process work. Conceptual models are often depicted as flow charts showing sequential and interrelated steps to be taken to effect a certain result, such as a diagram outlining the steps necessary for a property owner to obtain a zoning variance or amendment from the city planning department (figure 9.16).

Mathematical models express a process by means of numerical formulas. Many physical processes, such as soil erosion, storm water runoff, stream flow, air pollutant dispersion, or the movement of contaminants through soil and groundwater, can be expressed as mathematical models. A mathematical model requires that the process involved is generally well understood, and can be simulated and predicted accurately and consistently, with known equations. However, many processes in planning cannot be reduced to a series of numbers. Many planning processes and decision-making exercises are more qualitative, and depend on more subjective and expert judgment.

Some spatial models are also called cartographic models, since they involve the analysis and integration of the information on map layers, either manually using overlay transparencies, or with the help of a GIS. Cartographic modeling depends on Boolean algebra, and so the technique has also been called map algebra (or sometimes "mapematics"), defined as the application of manual or automated map-based calculations to solving a problem—

as differentiated from applying mathematical calculations (Berry 1995). Cartographic models may be considered a type of conceptual model, but since they use map layers, they also incorporate elements of analog models. Cartographic models can take several forms: binary, ranked, rating, and weighted. These forms are discussed below in the order of increasing complexity. Process models, also discussed below, are usually considered mathematical models, although there is occasionally some overlap between cartographic and process models.

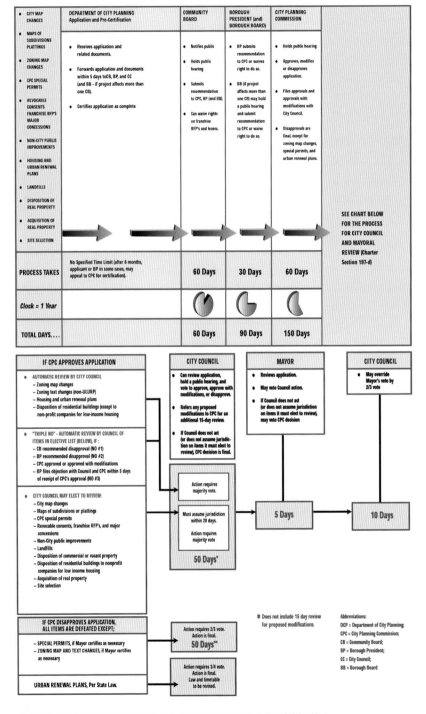

Figure 9.16 Flow chart of the land-use decision-making process in New York City, known as the Uniform Land Use Review Procedure.

Source: Bytes of the Big Apple™ Tax Block & Lot Files © 1995, 1997, 1998, 2002, 2004, 2005 and Pluto™ © 2003, 2004, 2005. NYC Department of City Planning. All rights reserved.

Binary model

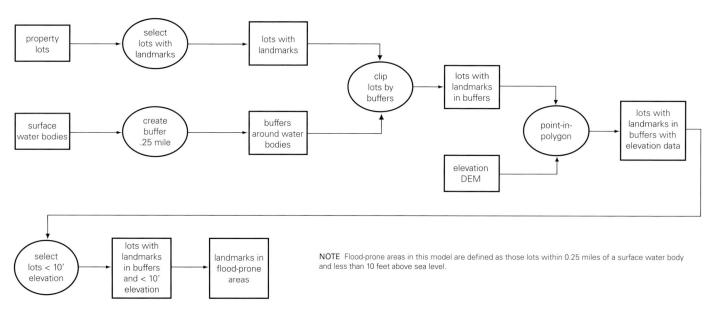

NOTE Flood-prone areas in this model are defined as those lots within 0.25 miles of a surface water body and less than 10 feet above sea level.

Figure 9.17 Flow chart of a binary model used to find flood-prone landmark structures. These criteria are simplified, and a real model would take more factors into account to identify flood-prone areas.

Binary models

Binary models are among the simplest. This type of model takes various parameters that are relevant to the problem at hand, then searches the geographical extent to find areas that either meet or do not meet the selected criteria. They are called binary models since each stage of the model yields a yes or no response to the question of whether a given area meets the criteria. The output of a binary model is in binary format: 1 (true) for map features that meet the selection criteria and 0 (false) for map features that do not (figure 9.17).

Suppose that we need the locations of structures or districts in a city that carry a landmark designation and which are located within a flood plain, or are otherwise in danger of flood damage. We could start with a layer of all property blocks and lots within the city, and simply query the database as to whether the properties contain landmark structures or whether they are within a designated landmark district. Property lots are thereby divided into yes (1) or no (0) (thus, binary) categories: either they do or do not contain a landmarked structure or they are or are not within a landmark district. This would be the first of several binary operations that we would apply to answer the question.

Next would be the determination of parcels considered flood-prone. This would likely be at least a two-step process. Using a layer of the city's surface water bodies, such as neighboring rivers or lakes, we would develop a new layer of coastal land areas by buffering those parcels within a certain specified distance of the water body. We would then overlay the buffered water-body layer with a property parcel layer. Land would either fall within that buffer distance of a water body, or not—again, a binary question. Of the parcels determined to be on coastal land, we would then determine which parcels lie below a certain elevation threshold measured by mean height above sea level. This query would yield parcels within coastal areas and below the threshold elevation. These, then, would be considered the flood-prone parcels. Riverfront properties on a high cliff would not be included in the data layer of selected parcels because they would exceed the elevation threshold. Low-lying properties far from a water body would also not be considered flood-prone.

Onto this layer we would then overlay the set of landmark-status properties. Landmark properties would either fall within the flood-prone areas or not—yet again, a basic binary question. Parcels either meet the criteria absolutely, or do not, and are therefore not considered flood-prone. The output of a binary model requires multiplying the yes (1) or no (0) results of the queries to yield a yes or no final answer. If any of the initial query results are 0, the final answer will also be zero, since any number multiplied by 0 is always 0.

The *yes* or *no* nature of the binary model limits its use for nuanced analyses, and may yield faulty or misleading results. As an example of this problem, suppose that our thresholds are the following: the parcel must be within one-quarter mile of a water body, and also be less than 10 feet in elevation to be considered flood-prone. This binary method would eliminate parcels from the selected set that are .251 miles from a water body, and 10.1 feet in elevation, which is a very near miss for both criteria—in reality, the parcel would likely be flood-prone. However, the binary model also eliminates parcels that are .0001 miles from a water body, but are 10.1 feet in elevation (a parcel virtually in the water, but a couple of inches above the elevation threshold), also potentially a flood-prone parcel. Additionally, it eliminates parcels that are 0.251 miles from a water body but are 0 feet in elevation, or even below sea level. Obviously, all these three parcels should logically be considered flood-prone, but the binary model would eliminate these parcels from consideration, and they would be deemed safe from flooding.

Note the limitations of this kind of analysis: It does not rank likely candidate parcels in danger of flooding in order of seriousness; it yields only a *yes* or *no* for each parcel; only parcels meeting all the criteria will be labeled flood-prone or at risk. If we want a more sophisticated assessment, we can move to the next level—the ranking model and the rating model.

Ranking models and rating models

Ranking models also use a binary yes (1) or no (0) for their outputs, but in an additive fashion, rather than multiplying the criteria results as in the binary model. In the three-criteria example of flood-prone landmark parcels (landmark status, proximity to water body, elevation), the results could therefore be a

> 0: *no flood risk to landmark parcels* (0 + 0 + 0 = 0);
> 1: *low risk* (1 + 0 + 0 = 1);
> 2: *medium risk* (1 + 1 + 0 = 2);
> or 3: *highest risk* (1 + 1 + 1 = 3).

By adding the different values, we arrive at a composite ranking of risk.

In a ranking model, the risk classes established per criteria are not based on the same standard, but rather are ordinal. In other words, a cumulative risk ranking of 2 is higher than a risk value of 1, but not necessarily twice as risky. The rating model, on the other hand, employs a consistent scale among the criteria. Since a given number always represents the same degree of risk, regardless of which layer it is from, the range of criteria values can thus be added and averaged, resulting in a more nuanced assessment of risk (table 9.1).

Rating models use criteria having a range of values. Using a rating model for our flood-prone parcels could yield not only *yes* or *no* values, but also values ranging from low-risk to high-risk along a 1-to-9 scale. By adding values for different criteria and simply averaging by the number of criteria, you can arrive at a composite evaluation of risk. A value of 9 would be the highest level in this three-criteria example: (9 + 9 + 9) divided by 3 (the number of variables or criteria). A value of 1 would be the lowest: (1 + 1 + 1) ÷ 3. At that point, the determination of what risk level would be designated as *flood-prone* would need to be made. The use of a ranking or rating model alleviates the problem of parcels being flood-prone in real life but not being designated as such, as they would be in a binary model.

Rating Model	
Rating buffer: Distance to surface water body in miles	
9	0–.25
8	.26–.50
7	.51–.75
6	.76–1.00
5	1.01–1.25
4	1.26–1.50
3	1.51–1.75
2	1.76–2.00
1	> 2.00
Rating elevation above sea level in feet	
9	0 or less
8	1–3
7	4–10
6	11–15
5	16–20
4	21–25
3	26–30
2	31–40
1	> 40

Table 9.1 **Priority factors in a rating model. After the various operations are carried out, the attribute table for the landmark parcels would also contain a field for elevation rating, as well as a field for the rating of the distance to a water body. A new field could be created, with values derived by adding the elevation rating and the distance-to-water body rating. This would yield a risk from flooding rating. The map could then be plotted using the column Flood Risk Rating.**

Weighted models

The weighted model lets you designate one criterion to be more important than another so you can answer a spatial question in a more meaningful way. Suppose in our simple model of flood-prone parcels we decided to include a fourth parameter, soil type. Our city engineers have assured us, however, that soil type is less of a predictor of flooding potential than distance to water body and elevation would be. We can estimate risk values for each criterion, as we did in the rating model, and then add in the extra step of assigning a weight to each criterion. This weight factor is also often referred to as an importance ratio. Each criterion's rating is then multiplied by its weight, and then added to the other criteria and averaged by the sum of its weights.

Let's say we have a parcel with a landmark value of 9 (high risk), a coastal proximity value of 8 (high risk), an elevation value of 6 (moderate risk), and a soil value of 9 (high risk). The landmark and coastal proximity layers are weighted at 3, elevation is weighted at 2, and soil type is weighted at 1.

$$[(9 \times 3) + (8 \times 3) + (6 \times 2) + (9 \times 1)] \div 9 = 8$$

That means we have a total weighted average of 8 for that particular parcel. Our highest possible weighted average in this example is 9, so 8 represents a high-risk parcel in terms of flooding potential. Altering the weights assigned to each layer will drastically impact the results. The weights are usually established somewhat arbitrarily, since it is not necessarily known that elevation is twice as important in calculating risk potential of flooding than soil types (table 9.2).

Rather than simply using a scale of 1 to 9 to designate the risk rating of the variables in the model, you can also use the actual data values of the variables. This removes one of the subjective elements in the model by eliminating the need for the GIS analyst or other expert to decide on thresholds corresponding to the various ranks. However, using actual data values will require that these values are transformed to a common scale. In our example of flood-prone parcel determination, the two variables are distance to the coastline, and elevation above mean sea level. Although the values of both variables may be given in feet, the scale of the first variable (distance to the coast) could be 0–100,000 feet, while

Table 9.2 Importance weights in a weighted model. In this model, the values are rated, and the variables receive weights according to magnitude of risk. The values' ratings are multiplied by the weights, the resulting values are then added and averaged, to yield a weighted average risk measurement for each landmarked lot. Each of the three rated variables would have a corresponding new field created and calculated by multiplying the rating and the weight. A fourth new field would be created to calculate the average weighted risk per landmark lot, by adding and averaging the three rated and weighted variables.

Weighted Model	
Rating soil types (weight = 1)	
9	Very flood prone
8	
7	
6	
5	
4	
3	
2	
1	Not flood prone
Rating elevation above sea level in feet (weight = 2)	
9	0 or less
8	1–3
7	4–10
6	11–15
5	16–20
4	21–25
3	26–30
2	31–40
1	> 40
Rating buffer: distance to surface water body in miles (weight = 3)	
9	0–.25
8	.26–.50
7	.51–.75
6	.76–1.00
5	1.01–1.25
4	1.26–1.50
3	1.51–1.75
2	1.76–2.00
1	< 2.00

the scale of the second variable (elevation) is likely to be only 0–1,000 feet. Therefore, to account for this vast difference in data range, the two sets of values must be transformed to comparable scales, so that the arithmetic operations of a weighted model can be performed on them. The linear transformation equation is very simple. It entails taking each value to be transformed, and subtracting the minimum value in the dataset from it, then dividing the result by the range of the data. (The range of the data, as you will remember from chapter 4, is the difference between the maximum and the minimum values in the dataset.) This standardization process transforms each value to a number between 0.0 and 1.0, making comparisons among variables valid. Models using linearly transformed values are referred to as weighted linear index models.

Process models

Many process models are mathematical, although some can be conceptual. Process models, as the name implies, are used to determine the expected outcomes of various processes, based on known relationships, sequences, and occurrences of events that comprise the process. These process models are often executed outside the GIS environment, using independent programs. The spatial results obtained through the model can then be brought back into the GIS for data exploration and visualization, but the GIS software is not actually running the model. Software developers have been incorporating more complex modeling capabilities within some GIS software, enabling many hydrological and natural hazards models, for instance, to be integrated more closely with GIS functions, and allowing the importation of many external models to be incorporated and run within the GIS.

Process models are useful for comparing different future scenarios—what might occur to specific locations or entities under one set of conditions as opposed to another. For example, in a study of soil erosion in agricultural areas, a model could compare how much soil loss from erosion could be expected after twenty years of crop production using terracing of land, versus the same twenty years of crop production without terracing. For this type of model, the user would need to provide several inputs, including elevation, slope, slope length, soil series, soil depth, the degree of rockiness in the soil, the type of crop or land cover, and precipitation rates. From these primary inputs the model could derive secondary inputs such as the soil's water availability, oxygen availability, nutrient availability, and potential erosion hazard, ultimately yielding suitability classes of soil for crops, and thus, a comparison of the erosion potential with and without terracing (Burrough 1993) (figure 9.13).

The inputs for many models, such as air pollution dispersion or transport of contaminants in ground water, need to be fairly precise, and numerous inputs are required to fill in all the blanks in an equation. For instance, a model designed to estimate air pollutant dispersion from a point source requires inputs such as the height of the smokestack, the prevailing wind direction and speed, the exit velocity of the pollutants from the stack, the type and quantities of the pollutants emitted, the facility's elevation, and the configuration of nearby structures, among other factors.

Universal Soil Loss Equation

$A = R \times K \times L \times S \times C \times P$

Where, A = Soil Loss (tons/ac)
R = Rainfall
K = Slope Erodibility
L = Slope Length
S = Slope Gradient
C = vegCover & Mgt.
P = Control Practices

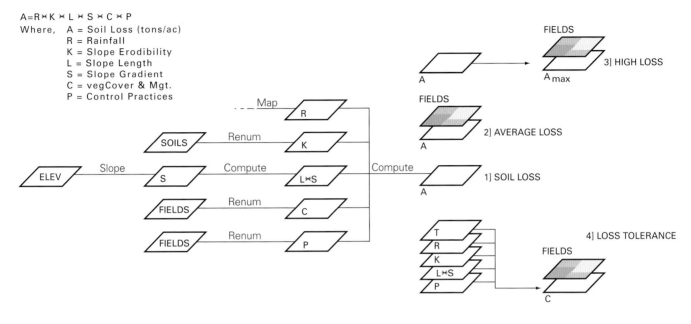

Obtaining all the necessary inputs may be difficult, expensive, and time-consuming. In many cases, the required information may have to be estimated.

Figure 9.18   Flow chart of a process model: The Universal Soil Loss Equation (USLE). Redrawn from *Spatial Reasoning for an Effective GIS*, by Joseph K. Berry. Reprinted by permission of John Wiley and Sons © 1996.

Expert systems and rule-based reasoning

Expert systems use human expertise and knowledge to create model rules that guide the software in problem solving. Rule-based reasoning uses a decision tree to direct the computer's actions. The decision tree generally comprises an **IF > THEN > ELSE** sequence of rules, with alternative scenarios given, depending on the answer to the *if* condition for each step of the tree. *Yes* answers permit the process to continue unimpeded to the next step, the **THEN** branch of the decision tree. *No* answers direct the process to the *else* branch of the decision tree.

Rule-based reasoning is similar to artificial intelligence (AI) and is based on the idea that certain decision-making processes can be codified into rules which, if followed automatically, will yield results similar to those made by a knowledgeable person. An expert system requires not only the rules, but also a knowledge base, and an inference engine to interpret the rules. Expert systems can incorporate fuzzy logic into the analysis, thereby introducing some amount of subjectivity and flexibility in how the rules are applied.

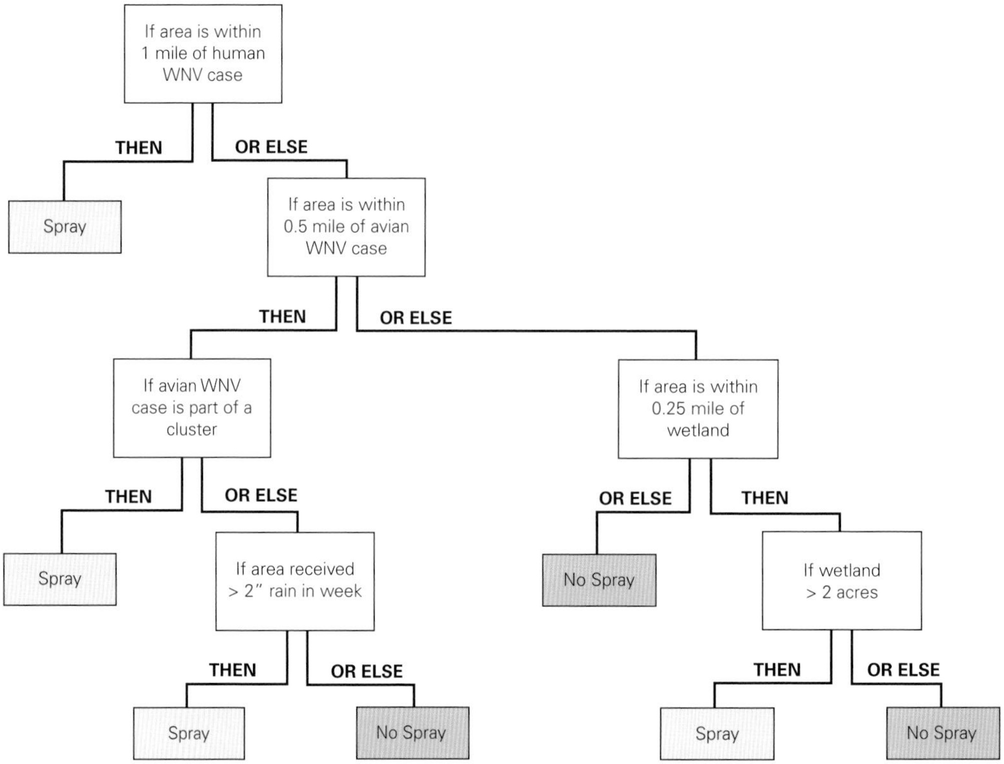

Figure 9.19 **Flow chart of an expert system, showing IF > THEN > ELSE rules. This represents a hypothetical decision tree for determining which areas should receive pesticide spraying in response to the threat of West Nile virus (WNV).**

Expert systems linked to GIS are becoming more frequent, especially in assessing risk from environmental impacts and natural hazard events. Expert systems can also help evaluate the consequences of various management strategies, which in turn can enhance certainty in decision making (figure 9.19).

Interpolation

Interpolation is a process of intelligent guesswork, used to make estimations of values for areas without data. If data is not available for every area to be modeled, interpolation is required to provide us with values in locations where data is missing. Interpolation requires some number of sampled points—the more points, the better the interpolation. The estimated values for the unsampled locations are based on the values in the locations that were sampled.

Suppose we are studying soil contamination in an urban watershed. Soil sampling and analysis is an expensive undertaking, so we have only been able to retrieve about three dozen soil samples, each of which has been analyzed for lead and zinc content. In order to estimate the cost of remediating the soil—that is, to clean it up and make it suitable for new housing development—we need to be able to estimate the levels of contamination in all the areas, including those areas that have not been sampled. Interpolation will give us a lead and zinc value for every location in the watershed, based on the lead and zinc values in the sample locations.

To perform most types of interpolation in most GIS programs, you must start with a point layer of sample data, or a point layer of data aggregated at points, such as ZIP Code centroids.

Interpolation estimates values at unsampled places on the basis of the "first law of geography," first promulgated by Waldo Tobler, and also known as the law of spatial autocorrelation: everything is related to everything else, but near things are more related than distant things (Tobler 1979). Spatial autocorrelation quantifies the degree to which nearer and more distant things are interrelated.

Sample selection

The sample need not be a physical sample of soil, water, or air, but could also be a count of an item of interest at specific sample locations, such as a species count. We might count the number of trees present on certain selected streets to interpolate the number of trees on unsampled streets. This would give us a reliable picture of tree density of the whole city, while requiring only that we count trees on a limited number of streets. Likewise, we might count vehicles traveling on certain major thoroughfares of the city, arrive at an hourly average, and interpolate the sampled vehicular traffic to the unsampled major highways in the remainder of the city.

In addition to measured magnitudes and counts of things in absolute numbers, sampled values can also be based on ratios, rates, and percentages. Population density, disease rates, crime rates, and median and average values, where the data is attached to points, can all be interpolated.

The interpolation will be more reliable if the sample locations have been systematically chosen with the intention of obtaining representative sites, both in terms of geography and content. Sample sites can be selected based on a random process, a predetermined, uniform grid of some sort, a transect of a certain size, pattern, and extent that cuts a swath through the area of interest, or a combination of random and uniform. Systematic sampling with local random allocation often is a good solution and is possible with a uniform grid and with samples taken at random locations within each grid cell.

Thiessen polygons (Voronoi diagram)

One of the simplest forms of interpolation, and therefore one of the easiest to understand, is the **Thiessen polygon** method. This method creates polygons from irregularly spaced point data, based on the idea that a point of unknown value can be assigned the values of the point with known values that lies closest to it. The closest point of known values to another point with unknown values is determined by creating polygon boundaries around all points with known values (the sample points). This is done by bisecting the distance between points of known values, and connecting the lines so that they form a continuous net of contiguous polygons. It is then assumed that any point of unknown value within a given polygon has the same values as that of the known point that lies at the centroid of the polygon. All points within the polygon thus take on the attributes of the centroid (figure 9.20).

The obvious problem with this method of interpolation is that it creates dramatic and somewhat arbitrary boundaries between values. Thiessen polygons are not the best way to represent continuous data, since in real life continuous data rarely has such definite lines of demarcation between values. This method is not capable of taking in to account transitional values. It assumes that everything within a given polygon is homogeneous (another erroneous assumption, in most cases, especially with natural data such as rainfall) and that values change sharply midway between sample points.

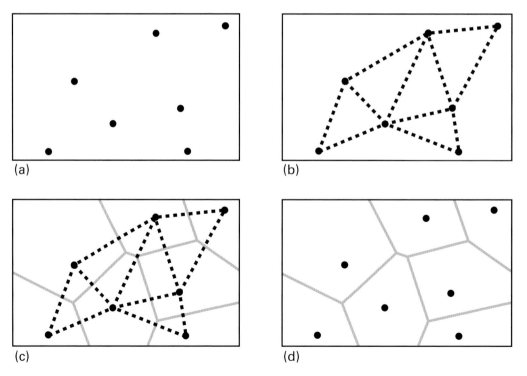

Figure 9.20   Theissen polygons, a simple interpolation method, are formed in the following way: (a) Sample point locations are plotted; (b) the points are joined by lines forming triangles; (c) perpendicular bisectors of the original lines are drawn; (d) the original lines are removed, resulting in a set of contiguous polygons. All points within any given polygon are assumed to have the same value as the polygon centroid (the original sample point). Redrawn from *Geographic Information Systems in Ecology*, by Carol Johnston. Copyright © 1995. Reprinted with permission from Blackwell Science Publishing.

(a)   (b)

(c)   (d)

Nevertheless, Thiessen polygons are very useful for certain applications. Suppose for a watershed study we needed to categorize a county by land-use type. The best spatial and attribute data available to us comes from the county tax assessment office. This data consists of property lot centroids, but not the actual property lot lines. The centroids have a database of information attached to them, including the land use in each lot. But we have no way of knowing where the boundaries of the lots are, and consequently do not know the boundaries of the land uses, so we cannot map land use as polygons, only as points. This will not suffice for the purposes we need land-use data for, which is to model the impacts of different land uses on the watershed in terms of water quality and quantity, for which we need areal data. Thus, we need to estimate the boundaries of the various property lots in order to assign land-use categories to each lot. Using the Thiessen polygon method, we can approximate the amount of land use in each category by estimating where the boundaries are between land-use types. Although the Thiessen polygons created around the lot centroids may not precisely reflect the actual property lot boundaries, this type of data is appropriate for this method of interpolation, since land use in reality often does change at the boundaries of property lines, usually without any gradation or transitional areas.

Thiessen polygons are appropriate for qualitative or nominal data that numerically-based interpolation methods would not address in a meaningful way. It is also a useful method for data that represents a phenomenon with rapid, abrupt variation. Using the Thiessen polygon method, the GIS analyst starts with a data layer of sample points and ends with a data layer of irregular polygons. Thiessen polygons are less appropriate for interpolating data that varies continuously through space, such as rainfall or temperature data, as opposed to data aggregated by census tracts or ZIP Code areas, which have discrete boundaries.

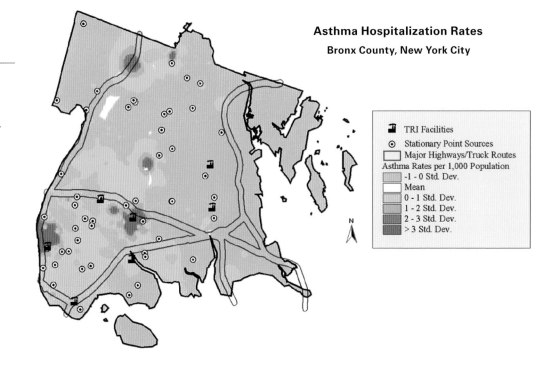

**Figure 9.21 Interpolation by
inverse distance weighting
(IDW), using census tract
centroids as the sample points.**

Inverse distance weighted

The inverse distance weighted (IDW) method of interpolation takes sample points of known values and interpolates a continuous surface from the values of those samples. It assigns values to unknown points based on the values of the nearest sample points of known value. The inverse distance weighted method employs two alternative ways of selecting the sample points that are used in calculating the unknown values. IDW uses either a specified number of nearest neighbors (for instance, the six or eight known sample points that are nearest to the unknown point), or it uses a fixed radius of a specified distance from the unknown point. The fixed radius method, in its calculations of value of the unknown point, includes the number of sample points that are within that specified radius. This number will vary depending on the density and regularity of the sample points. The values of sample points used in the estimate are weighted according to distance from the point of unknown value, with closer points accorded more weight in determining the unknown values. The influence of the known values, therefore, diminishes with distance. Estimated values can never be less than or greater than the range of the sample point values. In IDW interpolation, the GIS analyst begins with a layer of points of known values (discrete data), and ends up with a continuous data layer consisting either of isolines (contour lines) or a raster (regularized grid) of cell values (figure 9.21).

Spline

The spline method of interpolation also uses sample points of known values to estimate the value of unknown points, but rather than creating a distance weighted average, as IDW does, it fits an imaginary surface through the sample points of known value, similar to a rubber sheet that has been draped over the tops of the sample points, assuming that the sample points have a vertical dimension (a z-value) in proportion to the attribute being estimated. Unknown points are estimated from the selected number of sample points nearby. The spline method generally produces a smoother interpolation, and the estimated points may be given values that are outside the range of the known values of the sample points. As with the IDW method, spline interpolation requires a data layer of sample points, and returns an isoline or a grid layer of continuous data (figure 9.22).

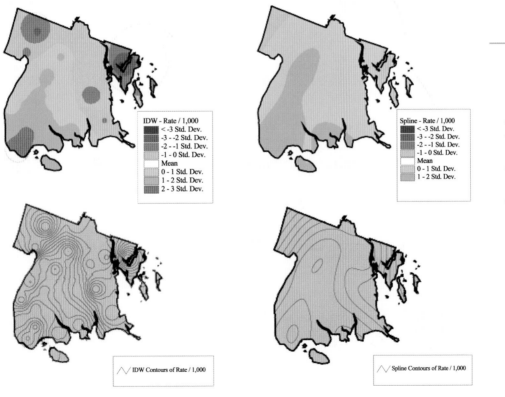

Figure 9.22 Comparison of interpolation methods: Inverse Distance Weighting versus Spline, using ZIP Code centroids as the sample points. Compare the difference between IDW interpolation using census tract centroids in figure 9.21 and IDW interpolation using ZIP Code centroids in this figure.

Kriging

Kriging is an interpolation method that is often preferred by analysts because it accounts for random variation in its estimating process. The method takes its name from that of a South African mining engineer who developed a geostatistical method of estimating mineral ore bodies. Kriging is more complex mathematically than the others outlined above and, many would argue, is a truer representation of reality, since it takes into account not only the proximity of the sample points, but also their directions.

Kriging uses the values of the sample points in several ways: First it calculates a general trend surface, or constant mean value, from all the sample points, referred to as the structural component. Then it looks at localized variation, called the spatially correlated component. Then it takes into account a random noise or residual error component, and combines the three components into a continuous surface of estimated values. Kriging uses a moving window approach to selecting the sample points, but sample values along the direction of the trend have more influence than values going against the trend. This method is best with regularly sampled data showing a strong trend. As with the spline method, kriging can estimate values beyond the range of the samples (figure 9.23).

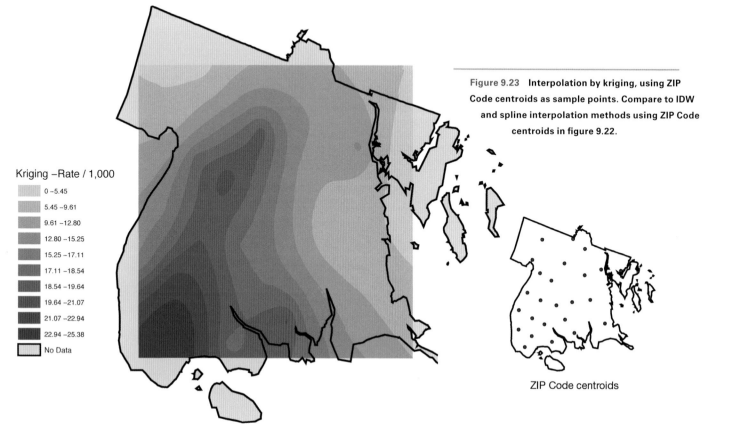

Figure 9.23 Interpolation by kriging, using ZIP Code centroids as sample points. Compare to IDW and spline interpolation methods using ZIP Code centroids in figure 9.22.

Kriging –Rate / 1,000

- 0 –5.45
- 5.45 –9.61
- 9.61 –12.80
- 12.80 –15.25
- 15.25 –17.11
- 17.11 –18.54
- 18.54 –19.64
- 19.64 –21.07
- 21.07 –22.94
- 22.94 –25.38
- No Data

ZIP Code centroids

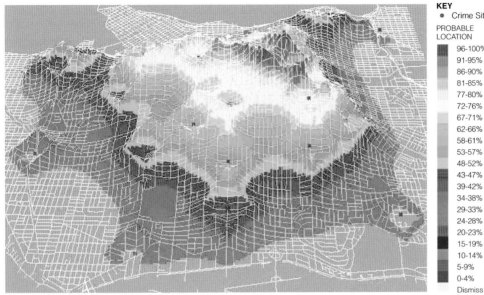

KEY
● Crime Site
PROBABLE LOCATION
96-100%
91-95%
86-90%
81-85%
77-80%
72-76%
67-71%
62-66%
58-61%
53-57%
48-52%
43-47%
39-42%
34-38%
29-33%
24-28%
20-23%
15-19%
10-14%
5-9%
0-4%
Dismiss

The above map was generated by geographic profilers working on a 1995 case involving a series of insurance-agent robberies in Vancouver, British Columbia. The red dots represent the crimes and the red shading shows the region where the offender will most likely base his operations. This is usually referred to as his anchor point.

Geographic Profiling: One Example in Canada

Investigators searching for the Washington area serial shooter are using geographic profiling — the science of predicting where criminals live based on where they commit crimes.

The theory behind geographic profiling is that most criminals will operate relatively close to home in a comfort zone defined by where they live, work, play or commute. Psychologists call this the least-effort principle.

Geographic profilers use specially designed software that takes the locations of a related series of crimes along with the known and predictable pattern of a serial criminal to produce a map of the probable location of the criminal's anchor point (where the criminal leaves from or returns to after the crime). This can reduce an investigators' prime search area by more than 90 percent.

Source: Environmental Criminology Research Inc.

Figure 9.24   GIS is increasingly being used for law enforcement applications. The technique of geographic profiling uses a predictive model that interpolates from points of crime occurrences.

Copyright © 2002 by The New York Times Co.

Reprinted by permission.

Regardless of the method of interpolation used, the good analyst makes sure to review the results very carefully and to note any anomalies. The software may estimate values admirably based on complex mathematical formulas, but the results may not necessarily make sense. In other words, always understand your data and investigate results that don't make sense. At this point in technology, software is not usually able to pick up counter-intuitive results—that is still the province of humans.

Written Exercise 9

Using GIS for problem solving

Find an article in a current newspaper or news magazine that reports an issue or problem that could be solved using GIS. In no more than three pages, summarize the issues, describe how GIS could address the problem, and briefly outline a methodology for carrying out the project and what spatial analytical techniques might be used. Discuss the datasets that would be necessary to conduct the study analysis. Attach the article to your paper. You may also find it helpful to create maps that illustrate the issues or problems in your project proposal, or locate maps on the Internet that present the issues graphically.

Use the newspaper article (next page) and project proposal (page 242) as an example.

McGreevey Seeks Purity Standards for 2 Bergen Reservoirs

By ROBERT HANLEY

The New York Times

RIVER VALE, N.J., March 11

Gov. James E. McGreevey expanded his campaign to clean up New Jersey's drinking water supplies today by proposing that the state's highest purity standards for untreated water be applied to two Bergen County reservoirs.

"We need to take the most stringent measures to protect our drinking water supplies," Mr. McGreevey said during a news conference on the banks of one of the reservoirs, Lake Tappan. "This fundamentally must be our commitment to families across New Jersey."

He said protection of drinking water is among the state's most important quality-of-life issues, along with preserving open space and curbing sprawl.

He also said that he was the first New Jersey governor to protect reservoirs with the water-purity standards known as Category 1, which have traditionally been used for pristine trout streams.

Mr. McGreevey proposed Category 1 status today for Lake Tappan, a 3.5-billion-gallon reservoir created in 1967; for Woodcliff Lake, a 100-year-old, one-billion-gallon reservoir about 10 miles west of here; and for about 30 streams that flow into the two reservoirs. About 750,000 people in Bergen and Hudson Counties get water from them. The reservoirs are owned by a private utility, United Water New Jersey, which said it supported the governor's proposal.

The governor's announcement opened a process of hearings, public comment and analysis that the state's environmental commissioner, Bradley M. Campbell, estimated would take about a year to complete.

Legally, Category 1 protection, which applies to water before it reaches purification plants, bars a measurable degradation of the water quality in a stream, river or other body of water. The regulations intended to achieve this goal include creating 300-foot buffer zones along designated streams, rivers, lakes and reservoirs; tightening treatment levels in sewage plants discharging wastewater into them; and increasing the amount of runoff water that must be transferred into the ground instead of into streams.

Governor McGreevey started his water-quality campaign last April when he proposed nine reservoirs in the state for Category 1 designation. Mr. Campbell, who accompanied the governor here, said today that he expected the regulations for those reservoirs to be approved by his department next month.

Developers have fought the proposals, and Mr. Campbell's aides said today that state officials had proceeded cautiously in writing the regulations in hopes the rules would be solid enough to withstand expected court challenges.

Mr. Campbell said it was unlikely that the regulations for today's nominations would be prepared more quickly than last April's; he noted that the mayors of the 22 towns in the watershed of the two reservoirs would help him choose which portions of the 30 feeder streams would be classified as Category 1.

Last April, Mr. McGreevey nominated the Oradell Reservoir, a sister reservoir south of Lake Tappan and Woodcliff Lake, for Category 1. Since then, environmentalists have urged him to add Lake Tappan and Woodcliff Lake, as well as their feeder streams, to his list because all the water in them eventually flows into Oradell Reservoir.

Environmentalists in River Vale said the governor's choice of Lake Tappan had a certain urgency because a developer has proposed about 100 town houses on 26 acres on the reservoir's western shore.

Burton Hall, chairman of the town's open space advisory committee, said he was thinking about asking Mr. McGreevey for a building moratorium if the developer's plans were approved before Lake Tappan got tighter protection.

Aides to the governor and Mr. Campbell said they doubted either official had the legal authority to order such a moratorium.

Using GIS to apply and implement category 1 water purity standards to Lake Tappan and Woodcliff Lake in Bergen County, New Jersey

After decades of relentless growth, New Jersey is the most densely populated state in the United States and Bergen County, in the densely populated northeastern portion of the state, one of the most crowded counties. In *The New York Times,* Governor James McGreevey claimed that curbing sprawl, preserving open space, and protecting drinking water are the most important quality-of-life problems facing the state. To help address these problems, the governor proposed that the state's highest water purity standards (category I) be applied to two of Bergen County's reservoirs. The project proposed in this paper would use GIS to implement these new standards in the affected watershed towns.

United Water New Jersey (UWNJ), a private utility, supplies water to more than 750,000 customers in sixty towns in densely populated Bergen and Hudson counties. UWNJ owns four reservoirs in the upper part of the Hackensack River watershed, which occupy 113 square miles and are interconnected with adjacent water systems. Three of the reservoirs—Oradell, Woodcliff Lake, and Lake Tappan—are in Bergen County. Oradell Reservoir is already designated to meet category 1 purity standards for untreated water "traditionally . . . used for pristine trout streams," according to *The Times.* Because all of this "water eventually flows into Oradell Reservoir," Woodcliff Lake, Lake Tappan, and segments of thirty streams will also have to meet category 1 standards. Category 1 protection includes "300-foot buffer zones along designated streams, rivers, lakes, and reservoirs [and requires] tightening treatment levels in sewage plants discharging wastewater into them and increasing the amount of runoff water that must be transferred into the ground instead of into the streams."

There are twenty-two municipalities in the watershed of these reservoirs, making jurisdictional issues a key part of any plan, especially when local input is needed to define applicable segments of the thirty feeder streams. The initial objective (Phase I) of the GIS project is to identify all land uses and activities within the watershed, and especially within or near the buffer areas that are likely to adversely impact water quality. Pollution prevention practices can then be put in place, and in Phase II of the project, GIS could be used to inventory and model water quality in the watershed. By assessing existing conditions and predicting future conditions, the GIS analysis will help determine what changes in land use, development density, and other activities would be required to maintain category I standards.

Figure 9.25

Figure 9.26

Project steps, phase I

1 Develop a timetable for project implementation, and define project goals.
2 Obtain the exact wording and intent of category 1 designation and standards, including restrictions placed on land uses and activities within 300-foot buffer areas.
3 Organize key stakeholders, define responsibilities, gain long-term commitment. Understand and map the multilevel jurisdictional boundaries included in the watershed, and learn which public and elected officials are responsible for each area.
4 Assemble the following data layers to incorporate into the GIS:
    a Boundaries of municipalities in UWNJ watershed/water supply area
    b State assembly districts in the water supply/watershed area
    c Boundaries of the entire UWNJ supply system
    d UWNJ watershed boundaries
    e Surface water bodies (reservoirs)
    f Surface water bodies (rivers and streams)
    g Tax lots (land-use and owner data)
    h Tax lots (zoning)
    i Roads within watershed
5 Clip rivers and streams to study area. Gather input from all twenty-two municipalities to define feeder stream segments.
6 Buffer the surface water bodies to 300 feet.
7 Intersect buffers with road layer, and determine roads to be designated "no salt" roads in winter snow conditions.
8 Intersect buffers with tax lot/land-use data.
9 Using the intersected layer, identify all golf courses, farms, and other sources of nonpoint water pollution within and adjacent to buffers.
10 Identify all point sources of water pollution within and adjacent to buffers.
11 Identify property owners of lots within and adjacent to buffers, and contact them for pollution prevention planning assistance.
12 Review zoning designations for all lots in or adjacent to buffers, and determine those areas where zoning/town master plan should be considered for revision, based on potential water quality impacts.

See the New Jersey Department of Environmental Protection's Web site Maps for Mayors. *www.state.nj.us/dep/gis/maps4mayors.htm*

References and further reading

Berry, Joseph. 1995. *Spatial Reasoning for Effective GIS*. New York: John Wiley and Sons.

Burrough, P. A. 1993. *Principles of Geographical Information Systems for Land Resources Assessment*. Oxford: Oxford University Press.

Chrisman, N. R. 2001. *Exploring Geographic Information Systems*. New York: John Wiley and Sons.

Clarke, Keith C., Bradley O. Parks, and Michael P. Crane. 2002. *Geographic Information Systems and Environmental Modeling*. Upper Saddle River, NJ: Prentice–Hall.

Fotheringham, A. Stewart, Chris Brunsdon, and Martin Charlton. 2000. *Quantitative Geography: Perspectives on Spatial Data Analysis*. London: Sage Publications.

Goodchild, Michael, Bradley O. Parks, and Louis T. Steyaert, eds. 1993. *Environmental Modeling with GIS*. Oxford: Oxford University Press.

Isaaks, Edward, and R. Mohan Srivastava. 1989. *An Introduction to Applied Geostatistics*. New York: Oxford University Press.

Johnston, Carol A. 1998. *Geographic Information Systems in Ecology*. Malden, MA: Blackwell Science.

Longley, Paul, Michael Goodchild, David Maguire, and David Rhind. 2001. *Geographic Information Systems and Science*. New York: John Wiley and Sons.

Mitchell, Andy. 1999. *The ESRI Guide GIS Analysis, Volume 1: Geographic Patterns and Relationships*. Redlands, CA: ESRI Press.

Tobler, Waldo. 1979. "Cellular Geography" in *Philosophy in Geography,* edited by Steven Gale and Gunnar Olsson. Boston: D. Reidel Publishing Co.

Voogd, H. 1983. *Multicriteria Evaluation for Urban and Regional Planning*. London: Pion.

Walford, Nigel. 1997. *Geographical Data Analysis*. Chichester, UK: John Wiley and Sons.

GIS Project Development and Institutional Issues

Ten

There is little doubt the use of GIS in organizations is expanding. Most large municipal agencies and utilities use GIS in at least one department for planning and operations, while smaller government agencies are catching up in terms of usage. Manufacturers and retail chains increasingly use GIS for such functions as site selection, strategic marketing, and sales. The real estate industry is using GIS for site selection, property management, property appraisal, and brokerage applications. Non-profit organizations and the health industry are using GIS in urban areas to determine and manage services at the neighborhood level. Each of these organizations will have its own structure and management style, and the process of setting up and maintaining a GIS will vary accordingly. This chapter is devoted to a discussion of principles and guidelines for managing the process.

Driving forces

Whether increased demand results from new applications or the expansion of existing ones, the GIS development process is driven by institutional issues: the size of the organization, who will have access to data, who will benefit from the GIS product (such as management, the public, planners, regulators, and operations staff), and who has the budget to implement a GIS project and maintain the GIS database. And—as will become quickly apparent—development is driven by management style and process: by how departments interact and their traditional ways of accessing and sharing data, whether in a centralized manner or on a department-by-department basis.

Proposing a GIS

GIS projects usually begin when someone identifies a simple, well-defined need for mapped information—for example, paper zoning maps for planners and the public. From that point, project requirements have a tendency to expand as the benefits of GIS become apparent to others. Once you have introduced the technology—new for most people—to potential users, the challenge is to keep control of the development process as additional users and new uses are identified. The initial GIS application, designed to meet a specific need, may be thought of—and communicated to potential users—as the pilot project for a wider application that will have the capacity to serve multiple GIS users and additional clients. The development of a GIS project is an iterative process that involves a good deal of back-and-forth between designers and potential users as costs and benefits are estimated and re-estimated, choices are made, and the project is defined and redefined.

Individuals and groups may play more than one role in the development of a GIS project, and for purposes of clarity we will categorize these roles as follows:

- Proposer: managers or professional staff who initiate the project.
- Designer: professional staff or consultants responsible for design and application development of the GIS.
- Database administrator: responsible for administering the GIS database.
- Database maintenance staff: responsible for updating and correcting the database.
- Users: staff with GIS training who will produce GIS product for themselves or others.
- Clients: nontechnical consumers of the GIS product. These could include planning and operations staff, management, elected officials, and the public. The GIS product might be paper maps and data, screen shots, images delivered over the Internet, or GIS mapping and data inserted into presentation media such as Microsoft PowerPoint.
- GIS coordinating committee: a steering committee that meets regularly throughout the process and includes representatives of all interested parties.

Initiating a project

Regardless of the size of the organization, the proposal process should include the following initial steps:

✦ Establish a GIS coordinating committee that includes representatives of all interested groups, and designate a lead member with GIS training and experience. This should be a standing committee that initiates the project and continues to meet throughout the design, development, and implementation phases of the GIS. Every agency or department that has an interest in the final GIS product should be represented, and committee members should report regularly on progress to their own senior management.

✦ Begin with a thorough understanding of the organization. How big is it? What is its mission? Are all departments in the same location? Is there a central information technology (IT) department? Do individual departments currently maintain their own databases? In many organizations, maps are seen as presentation graphics and not as information, and it is not unusual—particularly in the private sector—to find that GIS capabilities exist only in a graphics department that serves organization-wide needs for marketing materials. Obviously, for GIS to be of full benefit to the entire organization, a graphics department that uses it only to make the occasional map is not the optimal organizational location.

✦ Once you have determined how information is collected, stored, and disseminated within the organization, make a preliminary judgment on whether the GIS database should be designed to fit into and be administered by the IT department, or whether GIS databases should be maintained by individual departments. As is often the case, this decision may be driven by management considerations, rather than by the most efficient organization of a GIS database.

✦ Contact upper management and IT staff, secure their involvement in the process, and discuss with them how this new technology will affect existing business processes.

✦ Identify and contact potential users and set up meetings to educate and listen, and to get an overall understanding of information needs, budgets, and staff capabilities. This is an important step. All levels of management and technical staff should have an overall understanding of the costs, benefits, and implications of introducing GIS into their current operations. It should be emphasized that GIS projects can be designed to be scalable—beginning with well-defined current applications and adding new functions and new users over time. Potential users should be encouraged to consider possible future GIS applications.

✦ Explain the development process to management and to potential users, and estimate the time required to produce a functioning GIS application.

+ Estimate costs and benefits. This issue will arise early on in the proposal process, and should be addressed at first in a general way, based on experience with other GIS projects, since actual cost data will not be available initially. It should be noted that not all benefits of a GIS can be quantified. For example, the ability to visualize information and to place objects and events in their geographic context results in better decisions. This is a major benefit of any GIS but one difficult to measure.

+ Initially, the most expensive part of a GIS is the purchase and development of data—an up-front cost. However, management should be aware that as time goes on, staff costs associated with using and maintaining the system begin to outweigh the original development cost. Initial, one-time expenditures can be amortized over a period of years. Staff time and maintenance costs cannot. Cost-benefit analysis, like project design, is an iterative process that will continue throughout the development cycle.

+ Once the coordinating committee is established, it should be empowered to make decisions or recommendations to senior management, depending on the organizational culture, and a budget should be established for a needs assessment. Initially, the committee should decide on the use of consultants to assist with the needs assessment, design, and development phases of the project. In addition to technical expertise, a consultant with experience in implementing GIS projects can play a useful role as facilitator as the GIS coordinating committee weighs the costs and benefits of various options. The consultant's knowledge of data sources and the relative costs of hardware, software, and data development can help the committee to evaluate these options, both in the initial stages and throughout the development of the project.

Project development

The development phase of a GIS application is best thought of as a cycle, because it is iterative in character, with continual back-and-forth between designers, users, and management. The development cycle begins with a needs assessment—organization-wide, preferably—and then proceeds, based on the findings of the needs assessment and on preliminary management decisions, to a preliminary design of the application and its spatial and attribute databases, including assignment of access rights and maintenance responsibilities to anticipated users. This process is then repeated and refined, and the completed application is periodically re-evaluated and adjusted. The costs and benefits of various options presented during development are assessed and re-assessed as development proceeds and as costs become more concrete. Issues that will need to be addressed, both at the beginning and throughout the development cycle, include:

+ Sharing of GIS development tasks
+ The type, number, and information needs of users and clients
+ Scalability of the application and flexibility of the design

✦ Preferred media for GIS output: paper maps, printed reports, Internet, or intranet

✦ Organizational issues: the application's cost center, its accessibility, and the cost and responsibility for data procurement and maintenance

Development cycle

The development of a GIS includes the following tasks. This list is courtesy of the Erie County (New York) Water Authority.

✦ Needs assessment

✦ Conceptual design of the GIS

✦ Survey of available data

✦ Survey of GIS hardware (H/W) and software (S/W)

✦ Detailed database planning and design

✦ Database construction

✦ Pilot study and benchmarks

✦ Acquisition of GIS hardware and software

✦ Systems integration

✦ GIS application development

✦ GIS use and maintenance

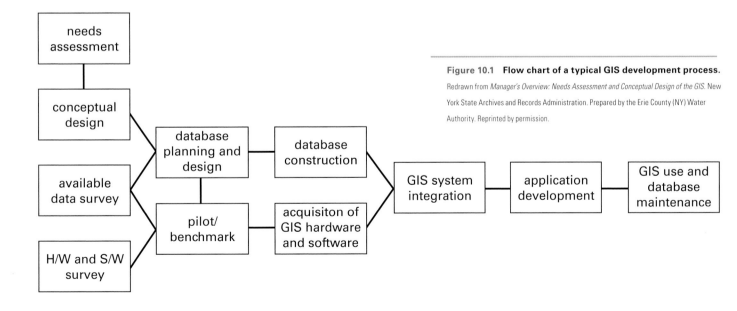

Figure 10.1   Flow chart of a typical GIS development process.

Redrawn from *Manager's Overview: Needs Assessment and Conceptual Design of the GIS.* New York State Archives and Records Administration. Prepared by the Erie County (NY) Water Authority. Reprinted by permission.

Needs assessment

The first task of the coordinating committee is to conduct a needs assessment, in effect, a survey that includes interviews and questionnaires. The needs assessment report identifies the users and uses of the GIS, the data requirements, the quantity and frequency of GIS output required by clients, and the GIS functions that will be used to produce this output. In addition to current user needs, the survey may identify future uses for the GIS.

The needs assessment report should serve as a performance-based outline specification for the GIS as the project proceeds to the design phase—a blueprint for what needs to be done, rather than how it gets done. The report should be circulated to upper management as well as to potential users, for a consensus on the scope of the project and a favorable decision to proceed with project development.

Conducting the needs assessment survey

The survey provides detailed information for the needs assessment, and the survey form should be carefully designed and explained at interviews with respondents. In addition to eliciting better responses from the survey, these interviews present a good opportunity to again explain the technology and its benefits—in effect, to repeat the process of educating and listening that was begun at the proposal phase. The completed survey forms should be discussed and refined at a follow-up interview with respondents. Design of the survey forms will vary in accordance with the type and complexity of the application and the technical sophistication of the respondents. When tabulated and combined with information from user interviews, they should provide answers to the following:

+ Who are the GIS users, and who are the potential clients for the GIS product? (management, the public, departments within the organization, other municipal agencies)
+ How will the information be shared?
+ What is the technical skill level of potential users?
+ What kinds of information will this application provide?
+ What specific spatial and attribute information will users need?
+ What are the data quality standards for this information?
+ What are the distribution requirements for the GIS output (for example, volume, frequency, and media)?

MASTER DATA LIST		
Entity	**Attributes**	**Spatial Object**
Street_segment	name, address_range	Line
Street_intersection	street_names	Line
Parcel	section_block_lot#, owner_name, owner_address, site_address, area, depth, front_footage, assessed_value, last_sale_date, last_sale_price, size (as of previous January 1st)	Polygon
Building	building_ID, date_built, building_material, building_assessed_value	Footprint
Occupancy	occupant_name, occupant_address, occupancy_type_code	None
Street_segment	name, type, width, length, pavement_type	Polygon
Street_intersection	length, width, traffic_flow_conditions, intersecting_streets	Polygon
Water_main	type, size, material, installation_date	Line
Valve	type, installation_date	Node
Hydrant	type, installation_date, pressure, last_pressure_test_date	Node
Service	name, address, type, invalid_indicator	None
Soil	soil_code, area	Polygon
Wetland	wetland_code, area	Polygon
Floodplain	flood_code, area	Polygon
Traffic_zone	zone_ID#, area	Polygon
Census_tract	tract#, population	Polygon
Water_district	name, ID_number	Polygon
Zoning	zoning_code, area	Polygon

The answers to these questions will provide a preliminary basis for project design decisions, cost estimates, choice of hardware and software, and choice of a distribution system for the GIS product (Internet, intranet, local or wide area network, or individual workstations).

The needs report should include:

✦ A master data list showing geographic features and their attributes, as required to meet user needs (figure 10.2).

✦ A list of applications, by application type and output requirements (see figure 10.3).

✦ Data flow diagrams describing data flow for each of the required GIS applications.

Figure 10.2 A sample of a master data list. From *Manager's Overview: Needs Assessment and Conceptual Design of the GIS*. New York State Archives and Records Administration. Prepared by the Erie County (NY) Water Authority. Reprinted by permission.

App#	Application Name	Type	Frequency
1	Zoning query	Query & display	85/day
2	Customer phone inquiry	Query & display	100/day
3	Fire dispatch map	Query & display	86/day
4	Fire redistricting map	Map analysis	1/year
5	Crime summary map	Query & display	12/month
6	Patrol dispatch summary	Query & display	133/day
7	Complaint summary map	Query & display	624/year
8	Subdivision development map	Query & display	No estimate
9	Counter query map	Query & display	85/day
10	Land use/land value	Map display	1/year
11	Assessed value map	Query & display	144/year
12	Grievance map	Query & display	2500/year
13	Comparable value map	Query & display	No estimate
14	Built/vacant map	Display	1/year
15	Water and sewer line map	Query & display	30/month
16	Hydrologic profile map	Spatial model	1440/year
17	Sewer system flow analysis	Spatial model	12/year
18	Emergency repair map	Query & display	110/year
19	Storm drainage map	Spatial model	700/year
20	Fire flow test map	Spatial model	260/year
21	Easement map	Query & display	520/year
22	Zoning map	Query & display	50/day
23	Floodplain map	Query & display	50/day
24	Youth league residency	Check query & display	3500/year
25	Mosquito control area map	Query & display	50/year
26	Site plan approval process	Query & display	200/year
27	Census data map	Display	48/year
28	Population density map	Map analysis	50/year
29	Land use inventory	Display	24/year
30	Retail space projection	Spatial model	24/year
31	Office space projection	Spatial model	12/year
32	Traffic volume map	Query & display	24/year

Figure 10.3 A sample of a list of GIS applications from a needs-assessment survey, from the town of Amherst, NY. From *Manager's Overview: Needs Assessment and Conceptual Design of the GIS.* New York State Archives and Records Administration. Prepared by the Erie County (NY) Water Authority. Reprinted by permission.

Design and application development

The designers of the GIS will translate the performance requirements identified by the needs assessment into detailed specifications for hardware, software, database design and content, system configuration, and GIS functionality.

Design and application development includes:

✦ Design of the GIS database, including the development of a data model that defines the relationships of spatial entities and attribute data.

✦ Planning for the life cycle of the GIS database, from the initial identification of required data to updating, maintenance, and archiving as the GIS is used and maintained.

✦ Hardware and software specifications.

✦ Data selection.

✦ Development of metadata for the selected spatial and attribute datasets.

✦ Look and feel of the screen interface and printed reports.

✦ Standard symbology for maps and graphics.

✦ Specifications for custom programming, if required. Custom program functions include designing user-friendly interfaces for operational personnel, automation of repetitive tasks, Internet map server functions, customized reports, dialog boxes for data input and database queries, and the programming of GIS functionality for specific user needs.

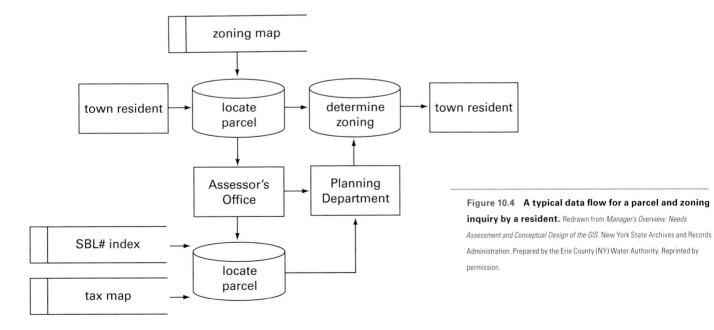

Figure 10.4 A typical data flow for a parcel and zoning inquiry by a resident. Redrawn from *Manager's Overview: Needs Assessment and Conceptual Design of the GIS.* New York State Archives and Records Administration. Prepared by the Erie County (NY) Water Authority. Reprinted by permission.

Survey of available data

The survey identifies sources of spatial and attribute data, evaluates data quality, and documents costs. Where vector spatial data is not available, the survey will estimate the cost of converting paper maps and raster data to vector format.

The data survey begins with an inventory of data that either exists in-house or is known to be available from other agencies. If a data clearinghouse exists for the area, this will be the next stop in the data search. Other relevant data may be found on the Web sites of government agencies and commercial data providers. The Internet has simplified the search process in recent years, and should make GIS data still easier to find in the future.

Availability and cost of GIS data

As a general rule, data acquisition is the most expensive part of a GIS, and this is particularly true of urban applications. Regional GIS maps (e.g., state and county boundary maps) that do not need to meet stringent requirements for accuracy and currency are easily obtained from government and commercial sources at little or no cost, and these maps are often linked to useful demographic and other socioeconomic attribute data. By contrast, detailed GIS mapping of the built environment is difficult to obtain. Unfortunately, urban applications typically require this kind of data, so the development budget for an urban GIS should allow ample time to look for detailed digital mapping and attribute data that may reside in city agencies and elsewhere.

Where vector mapping is not available, conversion to vector format from paper maps, photogrammetry, or remote sensing images will be required, as described in chapter 7, and this can be a costly process.

The data survey report will include cost estimates for data acquisition and conversion, and these estimates will go a long way toward determining the final cost of the project. If they exceed previous budget estimates, the coordinating committee will need management approval to increase the budget, to reduce the scope of the project, or to relax data quality standards—another example of the iterative character of the GIS development process.

Data collection for private sector GIS projects may also be complicated by a larger institutional issue: the public's right to access data that has been generated by federal, state, and local government. At the federal level, the Freedom of Information Act (FOIA) can be used to request geographic data and there are some similar statutes at the state level. Public agencies have time limits for response to such requests. It should be noted that FOIA requests for GIS data are not always successful and are usually time-consuming because public agencies can and do restrict access to public data. The public's right to know and the public agencies' duty to protect privacy and security are issues that remain unresolved. In general, local GIS data is harder to obtain for private sector uses than regional, state, and federal data. The data survey for a private sector application will have to assess the chances of obtaining the public datasets that were identified by the needs assessment.

Metadata and data quality

Relevant information from the metadata files that accompany the source data should be included in the data survey report.

In theory, data sources provide metadata files that allow you to judge whether the data meets the standards called for in the needs assessment report. In practice, source metadata is often incomplete, and it may be necessary to contact data providers directly to fill in missing information.

Choice of GIS hardware and software

Performance standards for GIS output having been defined by the needs assessment, the next step is a survey of commercially available hardware and software that will meet these standards. This survey will enable both GIS designers and management to evaluate the trade-offs among cost, quality, and scalability, and to make acquisition decisions.

In addition to cost, the selection criteria for hardware and software should include:

✦ Scalability of software: does the GIS software provider offer extensions to the basic software package that will accommodate future GIS functions such as network analysis or three-dimensional GIS?

✦ Scalability of database management software: is the RDBMS able to handle larger table sizes without reducing retrieval speeds?

✦ Scalability of hardware: can the hardware capacity for expansion meet increased user demands without reducing data retrieval times?

✦ Compatibility with existing hardware and database management software: this is a consideration when the project involves the expansion of an existing GIS, or when GIS functionality is combined with an existing RDBMS.

Database planning, design, and construction

Database design is concerned with how tables are organized, what attributes each table will contain, what spatial data will be linked to individual attribute tables to produce GIS map layers, and what key attribute columns will be used so that tables can be related to each other. Database planning is concerned with the life cycle of the database: how the database will be administered and updated, how it will be backed up and archived, who will update and maintain metadata, and how the data model will change as uses and users are added. Before addressing these questions, database designers must understand the mission, operations, and management style of the organization. The overall design and planning of the GIS database will require answers to institutional questions such as:

+ Will the GIS be used by only one department, or will it be shared by several departments or agencies?
+ What are the privacy and security restrictions on the spatial and attribute data? Who should have access to what?
+ Who will have editing privileges and be responsible for updating maps and attribute data?
+ Is there an existing IT department willing to administer the database, or is one of the proposed user groups willing to take this responsibility?

The planners and designers of the GIS database will need a clear understanding of these organizational issues.

Technical qualifications

Designing and administering all but the simplest databases requires technical expertise, and detailed database design should be carried out by consultants or by staff with database management training. Potential users should also be involved in database design discussions, since user access needs must be understood by the database designer, and users may be assigned responsibility for editing and updating individual tables and map layers.

Figure 10.5 (facing page)　Sample of a data dictionary from LotInfo, a commercially available GIS product for urban planning and real estate in New York City. Parcel level digital mapping is combined with property, land-use, and zoning information from the NYC departments of Finance and City Planning, and with transit data from the Metropolitan Transit Authority. LotInfo.

Data Sources

1) DOF-RPAD—Department of Finance New RPAD file, June 2004

2) DCP-PLUTO—Department of City Planning, Pluto disk, December 2003

Field Names Source Color Code

    Red   DOF-RPAD

    Blue DCP-PLUTO

Field Name	Type	Description	Source	Source Field Name
BBL	Char (10)	Borough, Block and Lot—Concatenated (unique lot ID) Concatenated by Space Track Inc. from borocode, block, lot.	*DOF-RPAD*	
BOROCODE	Decimal (1,0)	**1**=Manhattan; **2**=Bronx; **3**=Brooklyn; **4**=Queens; **5**=Staten Island	*DOF-RPAD*	**BORO**
BOROUGH	Char (2)	**MN**=Manhattan; **BX**=Bronx; **BK**=Brooklyn; **QN**=Queens; **SI**=Staten Island	*DOF-RPAD*	**BOROUGH**
BLOCK	Decimal (5, 0)	Tax Block number	*DOF-RPAD*	**BLOCK**
LOT	Decimal (4, 0)	Unique Tax Lot Number within Borough/Block	*DOF-RPAD*	**LOT**
ADDRESS	Char (28)	Lot Address	*DCP-PLUTO*	**ADDRESS**
ZIPCODE	Char (5)	5-digit zip code	*DCP-PLUTO*	**ZIP**
COMMDIST	Decimal (3,0)	Community district code	*DCP-PLUTO*	**CD**
CENSUSTR2K	Char (7)	US Census 2000 Tract in which lot is located	*DCP-PLUTO*	**CT2000**
CENSUSBL2K	Char (4)	US Census 2000 Block	*DCP-PLUTO*	**CB2000**
SCHOOLDIST	Char (2)	School District	*DCP-PLUTO*	**SCHOOLDIST**
COUNCIL	Char (2)	City Council District	*DCP-PLUTO*	**COUNCIL**
FIRECOMP	Char (4)	Fire Company	*DCP-PLUTO*	**FIRECOMP**
HEALTHAREA	Char (4)	Health Area	*DCP-PLUTO*	**HEALTHAREA**
HEALTHCTR	Decimal (2,0)	Health Center District	*DCP-PLUTO*	**HEALTHCTR**
POLICEPRCT	Char (3)	Police Precinct	*DCP-PLUTO*	**POLICEPRCT**
ZONEDIST1	Char (9)	Zoning District	*DCP-PLUTO*	**ZONEDIST1**
ZONEDIST2	Char (9)	Second zoning district—for lot with split zoning	*DCP-PLUTO*	**ZONEDIST2**
OVERLAY1	Char (4)	Commercial overlay	*DCP-PLUTO*	**OVERLAY1**
OVERLAY2	Char (4)	Second commercial overlay—for lot with split zoning	*DCP-PLUTO*	**OVERLAY2**
SPDIST1	Char (5)	Special purpose or limited height district	*DCP-PLUTO*	**SPDIST1**
SPDIST2	Char (5)	Second special purpose or limited height district—for lot with split zoning	*DCP-PLUTO*	**SPDIST2**
ALLZONING1	Char (21)	Zoning, commercial overlay, special purpose/limited height district	*DCP-PLUTO*	**ALLZONING1**

Constructing the database

The process of acquiring the spatial and attribute data that is needed for a GIS application, integrating this data into a single database format, and converting spatial data into vector format where necessary, are the most expensive parts of GIS development for most applications. The conversion of map data from raster format or hard copy to vector format can be both expensive and time-consuming, and this task is often outsourced to companies around the world who specialize in this work. It should be noted, however, that outsourcing requires a considerable amount of coordination and supervision. If data conversion is outsourced, in-house staff members or consultants should be designated for coordination, quality control, and contract administration. Other options include in-house data conversion using on-screen digitizing and rubber sheeting techniques, as described in chapter 7, and the traditional digitizing table. Recently available software for on-screen digitizing has made this an attractive option for data conversion.

Integrating and processing attribute data

Another time-consuming but important task is the integration and processing of attribute databases that have been collected from various sources. Upon completion of this task a data dictionary with verbal descriptions of each field in the various database tables should be compiled and made available to users (figure 10.5).

Pilot study

The first GIS product that will be seen by users and management will be the pilot study, and a realistic time frame should be established for producing pilot study output, in order to retain momentum for the project. The pilot study should also serve to demonstrate the capabilities of the GIS software for future applications, and it should be widely disseminated throughout the organization. Sample data can be used to demonstrate future applications that are called for in the needs assessment, but for which data is not yet available. The pilot study output may also elicit comments and suggestions that will affect project design.

Systems integration

A simple stand-alone GIS will require very little system integration. The standard PC set-up with Internet access and a printer or plotter will suffice and will most likely be installed by in-house staff. More complicated configurations should be installed by a consultant or by trained staff. Complex GIS configurations include server/client installations, local area networks (LANs), wide area networks (WANs) for remote users, and the applications service provider (ASP) model for the Internet. Finally, existing hardware and software may need to be integrated with the new system, and this may require outside help.

GIS use and maintenance

The coordinating committee should continue to meet regularly as the GIS is used and maintained. Once the application development phase is completed, the database maintenance and updating responsibilities of each department should have been well defined. However, adjustments will need to be made as time goes on to keep pace with staff and organizational changes. Feedback from clients may result in adjustments to the design and content of the GIS product, and new uses for the GIS may be proposed. In effect, the development cycle continues throughout the life of the application (figure 10.6).

Maintenance tasks include:

✦ Backup: The database should be backed up daily, preferably at an off-site location.

✦ Metadata: The updating of metadata files as the database is updated is an important, and often neglected, maintenance function. Ongoing staff responsibility for maintaining the metadata files should be clearly defined.

✦ Documentation: A common problem in municipal GIS departments is staff turnover. Trained GIS specialists are in demand. It is important, therefore, to budget time for full documentation of the GIS as it applies to each department, so that new staff can continue to operate and maintain the system.

✦ Response to feedback from users and clients: The coordinating committee should schedule periodic interviews with users and clients to get feedback, evaluate the system, and incorporate changes.

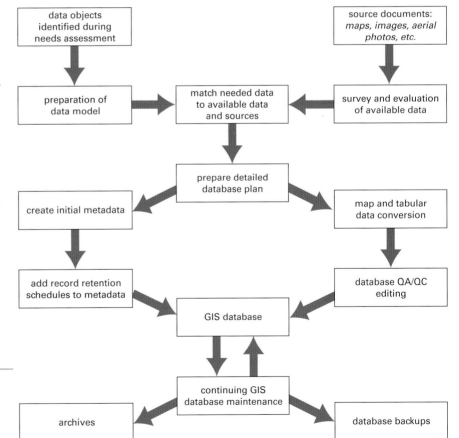

Figure 10.6 Redrawn from *Manager's Overview: Needs Assessment and Conceptual Design of the GIS.* New York State Archives and Records Administration. Prepared by the Erie County (NY) Water Authority. Reprinted by permission.

Management perspective

Management level questions that need to be addressed, especially when resources and benefits are shared among departments or agencies, include the following:

+ Who stands to benefit from the GIS, and to what degree?
+ Which of these groups
    ◇ are prepared to contribute to the initial project budget?
    ◇ are prepared to budget for on-going maintenance costs?
    ◇ have trained GIS staff?
    ◇ have and maintain spatial and attribute data that is relevant to the project?
    ◇ have an IT department that currently administers a GIS database?
+ How will costs, resources, and data be shared among departments or agencies?
+ Which groups will have access to which data?
+ How will maintenance and updating responsibilities, including editing of the spatial database, be shared among groups?
+ Is an enterprise database feasible?
+ If so, which department or agency will administer the database?
+ Will there be a lead agency or department with authority to choose data sources, hardware, and software?
+ Which group of participants in the GIS project will own the data?
+ Will agencies that own the data license it and charge for it? (Some city agencies charge other agencies for access to GIS data as a way of reducing the strain on internal budgets.)
+ Who will be responsible for data archiving and retention?

We have discussed some of these development issues in the context of project design and development, and stressed the importance of communication at the technical level. Management will address these issues from a different perspective: in the light of operational requirements, projected savings, available staff time, and budget constraints. It is important, therefore, to maintain communication between the GIS coordinating committee and management at each stage of the development process. Once the project is implemented, continuing communication among technical staff, the GIS coordinating committee, and management will be important for its continued success.

Written Exercise 10

Develop a project flow chart

Develop a flow chart for the proposal project you selected in the written assignment of chapter 9.

References and further reading

Aberley, Doug. 1993. *Boundaries of Home: Mapping for Local Empowerment.* Gabriola Island, B.C.: New Society Publishers.

Aronoff, Stan. 1993. *Geographic Information Systems: A Management Perspective.* Ottawa: WDL Publications.

Chrisman, N. R. 2001. *Exploring Geographic Information Systems.* New York: John Wiley and Sons.

Erie County Water Authority, National Center for Geographic Information and Analysis, GIS Resource Group. 1996. *Manager's Overview: Needs Assessment and Conceptual Design of the GIS.* Albany, NY: Local Government GIS Development Guides, New York State Archives and Records Administration.

Grimshaw, David. 1999. *Bringing Geographical Information Systems into Business.* New York: John Wiley and Sons.

Harder, Christian. 1998. *Serving Maps on the Internet: Geographic Information on the World Wide Web.* Redlands, CA: ESRI Press.

Huxhold, William E., and Allan G. Levinsohn. 1995. *Managing Geographic Information System Projects.* New York: Oxford University Press.

Jankowski, Piotr, and Timothy Nyerges. 2001. *Geographic Information Systems for Group Decision Making: towards a participatory, geographic information science.* London: Taylor and Francis.

Martin, D. J. 1996. *Geographic Information Systems: Socioeconomic Applications.* London: Routledge.

Masser, Ian. 1998. *Governments and Geographic Information.* London: Taylor and Francis.

Obermeyer, Nancy, and Jeffrey Pinto. 1994. *Managing Geographic Information Systems.* New York: The Guilford Press.

O'Looney, John. 2000. *Beyond Maps: GIS and Decision Making in Local Governments.* Redlands, CA: ESRI Press.

Reeve, D. E., and Petch, J. R. 1999. *GIS, organisations and people: a socio-technical approach.* London: Taylor and Francis.

Severance, Dennis G., and Jacque Passino. 2002. *Making IT Work: An Executive's Guide to Implementing Information Technology.* San Francisco: Jossey–Bass.

Tomlinson, Roger. 2003. *Thinking about GIS: Geographic Information System Planning for Managers.* Redlands, CA: ESRI Press.

Ethical Issues *in* GIS *and* Urban Planning

Many of the challenges and problems that we face in GIS stem from ethical issues, but they are not so different from those that have always affected mapping in general: The potential for maps to be used for propaganda or to purposefully mislead the public; the inherent subjectivity of mapped data and the implications of that subjectivity on data reliability and completeness; the responsibility of the mapmaker to be as honest and unbiased as possible in the presentation of data; and the ramifications of the role maps play in the worlds of planning, policy, and decision making.

A multitude of issues

Being a relatively new technology, GIS brings new issues to the fore, both general and specific ethical issues. These include the new ease of mapmaking with GIS and what that implies for the democratization and vernacularization of maps; the use of counter-mapping for advocacy and empowerment purposes by nongovernmental groups, usually in opposition to official maps; the magnitude of investment, both in capital, time, and human expertise to build a GIS, and the resulting differentials in GIS access between more affluent groups and governments, and those less affluent; the individual's loss of confidentiality and privacy in light of the detailed geographic, health, demographic, socioeconomic, and consumer data now available and mappable; the potential for "Big Brother" surveillance made possible with georeferenced data and GIS; the ownership of, and rights to, digital geographic and attribute data, especially data assembled with taxpayer money; and liabilities associated with data dissemination.

These issues must be considered in virtually every GIS project that we encounter in urban planning, and sometimes the project design itself is shaped by ethical and legal constraints or objectives.

The subjectivity of maps

Maps are powerful objects and can shape the way nations define themselves and the way individuals experience their environments. Maps are also very subjective documents, despite the common perception of maps as objective, even scientific, descriptions of reality. It is largely due to this widespread perception of maps as inherently objective and truthful that allows their subjectivity to be masked and concealed much of the time.

When information is portrayed graphically in the form of a map, it can take on an aura of truth that often goes unquestioned. Further, maps have the power to naturalize artificial, human-made constructions—to lead us to believe that existing conditions as portrayed on the map are inevitable and immutable. Yet we know that all mapmaking, by definition, is a process of generalization—a human being deciding what to put in and what to leave out—and therefore maps are, at best, only partial truths. The reality that is expressed on a map is very much biased by the mapmaker's view

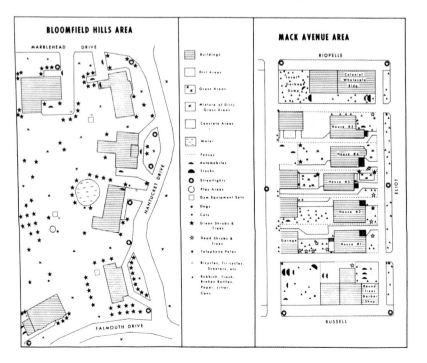

Figure 11.1 Maps of sections of Detroit, showing things the USGS normally overlooks: dead shrubs and trees, broken bottles, paper, litter, bicycles, scooters. From *The Power of Maps,* by Denis Wood. (From the Detroit Geographical Expedition and Institute, Field Notes: Discussion Paper No. 3, The Geography of the Children of Detroit, Detroit, 1971.) Reprinted by permission of Guilford Press, © 1992.

of reality, whether this bias is conscious or not. In analyzing or presenting data, urban planners must always be aware of any potential biases they or their agencies hold, and strive to portray a balanced picture. This is one reason why multiple views of the same data are desirable, as map readers can see the data presented in more than one way and have more resources with which to draw their own conclusions, as discussed in chapters 4 and 5.

Planners also must guard against omitting features or other information on maps that may carry special meaning to certain communities. Even the decision of which features are considered permanent enough to include on a map can be fraught with political overtones (figure 11.1). Urban areas are famous for their mix of disparate cultures, but there can be grave misunderstandings between planners and constituent (stakeholder) groups due to varying cultural perceptions. Since most often "official" mapping is handled by governmental agencies, features important to some groups may not be taken into account and may be excluded from the map. Planners are usually most interested in permanent features of the built environment, but this focus may not adequately represent the reality of the community's view of their environment. It is crucial to incorporate the community's point of view if planning is to be democratic, meaningful, and ultimately, successful.

Even the very idea that a map should record only "permanent" features of the environment can be called into question, as can the definition of a permanent feature (Wood 1992). A large-scale map of a community may include buildings and street trees, for instance, but not the location of a trash-filled lot, which may be equally permanent in terms of time scale, if not intention (figure 11.2). One of the main ethical issues in GIS is how to incorporate the community's viewpoint into maps that will ultimately be used to decide the future or fate of that community's environment.

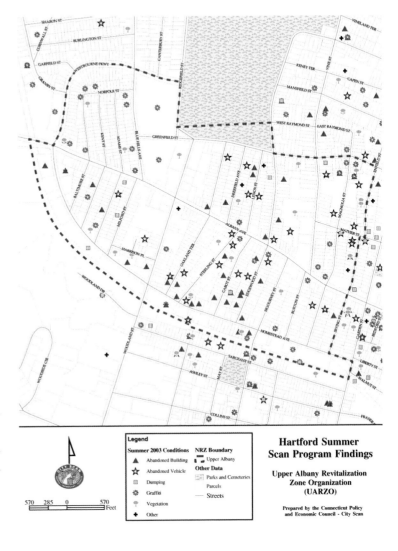

Figure 11.2 **Mapping quality of life issues—street level conditions. In the CitiScan Project in Hartford, Connecticut, neighborhood volunteers and high school students with GPS units, handheld computers, and digital cameras recorded data about physical conditions in their neighborhoods. The data was then entered into a GIS and mapped for use in resource allocation and prioritization.** (CPEC) Connecticut Policy and Economic Council, *www.city-scan.com.* © Portions of this document include property of the CPEC and are used with permission. CityScan™ is a trademark of the Connecticut Policy and Economic Council.

Maps as propaganda

Maps are sometimes used as political propaganda, an extreme example of the subjectivity of maps. Propaganda maps are constructed by using deliberately skewed or misrepresented data (either spatial or attribute) to elicit a particular desired response in the map reader. This response can have far-reaching effects on public opinion and resultant policy making, including justification for genocide and war, as demonstrated by Germany's propaganda maps in the 1930s (figure 11.3). We would like to be able to say that this type of mapmaking is confined to the past, but unfortunately it still appears with alarming regularity. Many maps today are produced with the intentions of inflaming emotions and inspiring action (figure 11.4). And, of course, the line can be fuzzy between propaganda maps and advocacy maps that espouse a particular point of view and are meant to increase public awareness (or indignation and outrage) about topical issues, such as in figure 11.5.

Many governmental agencies as well as community groups produce propaganda maps, although defining them as such often depends on who is looking at them. Sometimes, propaganda maps do not contain erroneous information *per se* but are viewed as propaganda maps due to the manner in which the data is portrayed. For example, a map showing a nuclear energy facility with concentric circles radiating outward in waves throughout the surrounding area, delineating the potential impact areas of various types of nuclear accidents and malfunctions could be considered, depending on labels, color, and graphic choices, as a map designed to alarm rather than inform. Likewise, a map depicting Superfund sites indicated by a skull and crossbones symbol is probably not attempting to show a balanced or objective view of the data. Planners, whether those working for government or for community or nonprofit advocacy groups, must be particularly vigilant that they do not produce maps designed to misrepresent or exaggerate information in order to present a certain bias, thereby misleading or prejudicing the public.

Figure 11.3   Typical propaganda map symbols: (a) arrows represent pressure on Germany from all sides; (b) circle signifies the encirclement of Germany before and after WWI; (c) pincers personify the pressure against Germany from France and Poland from the west and east.

From *Map Use and Analysis,* by J. Cambell. Reproduced with permission from the McGraw-Hill Companies, © 2001.

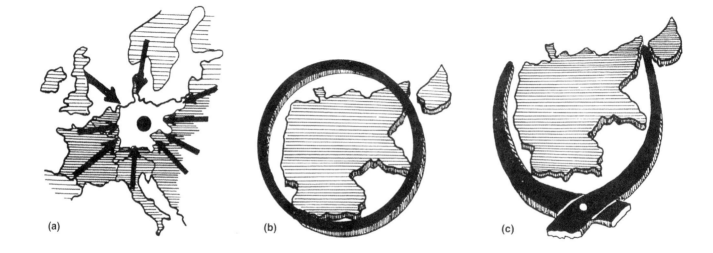

(a)   (b)   (c)

Figure 11.4   A poster displayed by the Army of the Pure, a militant organization based in Pakistan, shows missiles juxtaposed against a map of India and Pakistan, and suggests that the map of the subcontinent be changed to include "more Pakistans." Copyright © 2000 The New York Times Co. Reprinted with permission.

Figure 11.5   A map produced in the United States in the immediate wake of 9/11. Maps can express a viewpoint or attitude, such as patriotism, honor, resilience, or anger. Courtesy of Mike Smith published in the *Las Vegas Sun*, 9-16-01 © United Features Syndicate. Reprinted with permission.

Participatory GIS and the democratization of GIS

Many planning agencies and nonprofit planning organizations have employed or are beginning to consider Participatory GIS in their work with local community groups. Participatory GIS (PGIS) or Public Participation GIS (PPGIS) is the process of using GIS technology as "a way of enhancing local peoples' abilities to share and analyze their knowledge of lifestyles and conditions, thereby better enabling them to plan" (Chambers, quoted in Cinderby, 1998:3) and has proven to be an effective way to inject a degree of community input and control into the planning process.

GIS (and mapping in general) can be a powerful tool in community-based planning and PGIS. It helps give the community access to information and technology; enables the layperson to speak the same language as decision makers; enhances problem-solving and critical-thinking skills in community members; produces creative solutions to community problems; improves community leadership skills, and fosters a fuller understanding of urban issues, computer technology, the sciences, public policy, and community advocacy.

Practical applications of GIS for community-based organizations include: inventory of existing community conditions; community planning and economic development efforts; review of projects and developments proposed for their community; monitoring and verification of developer—and government—prepared environmental assessment of proposed projects and zoning and land-use amendments; evaluation of service delivery by governmental agencies; citizen complaint and resolution tracking; determination of community needs; preparation of testimony for public hearings;

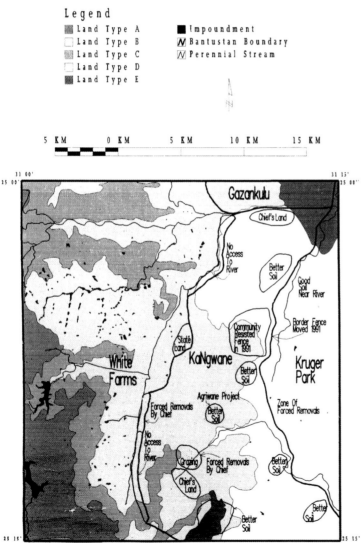

Legend

Land Type A
Land Type B
Land Type C
Land Type D
Land Type E

Impoundment
Bantustan Boundary
Perennial Stream

5 KM   0 KM   5 KM   10 KM   15 KM

Figure 11.6 Map of land types and composite mental map of Kiepersol, Eastern Transvaal, South Africa. Maps were developed based on participatory GIS projects for equitable post-apartheid land reform and land redistribution, and show areas of good soil for agriculture. From *Pursuing Social Goals Through Participatory GIS: Redressing South Africa's Historical Political Ecology*, by Harris, et al. in *Ground Truth: The Social Implications of GIS*, John Pickles, ed. Reprinted by permission. The Guilford Press, © 1995.

proactive efforts for pollution prevention strategies; coordination efforts with fire departments and police for public safety; vacant lot clean-up, drug dealer tracking and drug eradication; hazardous chemical storage conditions and locations; and resident emergency evacuation plans. NIMBY (Not In My Back Yard) and LULU (Locally Unwanted Land Use) conflicts are also well-suited to mediation with GIS and participatory planning, as are the controversies that pit environmental protection against economic development.

Although some of the potential uses of a community-based GIS are reactive (verification of government—or industry—prepared environmental assessments) many more uses are proactive. While this kind of use requires that community members reach some type of consensus about their future direction, it also permits them to realize success in shaping their community.

PGIS often uses perceptual mapping exercises by the community participants to denote features of interest. "Perceptual maps can help describe communities' knowledge of their local environment in a form intelligible both to members of the group and also outsiders.... Local environmental knowledge is of high quality when compared to data compiled by outside experts. It also holds numerous advantages when compared to traditional spatial datasets. Perceptual maps contain information unobtainable from other environmental data on the social settings for resource use" (Cinderby 1998). Perceptual maps are then integrated within the GIS framework, allowing for the accommodation of multiple viewpoints (figure 11.6).

There are a number of criticisms leveled against using GIS for participatory planning. Some question whether using the high-tech approach of GIS eliminates the true participatory nature of community-led planning, given the usual lack of GIS expertise within community groups. However, PGIS can still be effective if done in collaboration with GIS experts from outside the community, provided that the experts act as consultants to the community and not as the managers of the project. Another problem is the fact that GIS can be seen as imposing a quantitative rather than a qualitative view of space on the planning process, which runs contrary to many community-led planning endeavors and the associated conceptualization of community space. This is where the use of perceptual mapping and efforts to map the elusive experience of place can come into play. Counter-mapping can be used as a tool to facilitate communications within the group, among groups, and in presenting a cogent picture of their community to the outside world.

Using participatory GIS can raise a number of ethical issues, however, and these must be addressed when it is used in a new project. Perhaps the first important ethical issue pertaining to participatory GIS is the need to reconcile differing and conflicting definitions of the community. A prime benefit of PGIS is the inclusiveness it can afford and the legitimacy it bestows on the resultant planning efforts. However, a fundamental flaw in the process is that there is no predetermined set of individuals that comprise the community. What is this entity called *the community,* and who decides who is included in it? Who determines who speaks for the community, and how does one ensure adequate representation of all the voices of the community? How can we even identify the geographical extent of the community in terms of the boundaries of an affected area?

All too often, planners and other officials refer to the community as though it were a homogeneous, monolithic entity, when in reality communities are generally assemblages of many-faceted stakeholder groups with potentially conflicting interests. Within any community there are more powerful and more marginalized individuals or groups, as well as divisions based on gender, age, educational attainment, economic class, race and ethnicity, and personality traits. People who are self-confident in nature will probably make their opinions known more often (and more loudly) than shy people, which may put an undue emphasis on their views. A resident in a community may have different priorities than a business owner, and a homeowner may have interests that differ from those of a tenant. We can never be sure that there is any single community voice or community opinion. PGIS loses all claims to legitimacy if it is not as inclusive as possible and if it does not find ways to incorporate a wide range of all possible viewpoints.

A common trap for planners is to believe that, by definition, the people who take part in a PGIS are actually representative of the community, without putting some effort into making sure that this is so. The other side of the community representation coin is the equally serious charge of tokenism, or including one of each possible type of people, ostensibly to represent that group's position. These are problems common to all the public participation aspects of planning, but they are particularly evident in PGIS, because PGIS attempts to generate a graphic representation of the community's shared vision, resulting in a physical record of the community's ostensible consensus on an issue. Therefore, it is even more visible than usual when the viewpoint presented as the community's is one-sided or does not reflect the multiple perspectives of the whole community.

GIS for advocacy purposes: Counter-mapping

An emerging and popular use of GIS for planning is in the production of maps and spatial analyses for advocacy purposes. These so-called counter-mapping projects often challenge the prevailing official views of communities enshrined in urban planning maps and enterprise GIS, and are usually intended to offer some measure of empowerment to local areas trying to plan their futures (figure 11.7).

Counter-mapping can be defined as the efforts of a local community, group of indigenous people, or ethnic minority to assemble data, generate maps and other graphic representations, and disseminate these materials for the purpose of better understanding their environment, taking control of their space, directing how their space is interpreted by the official powers, and confronting the official view of their community in order to promote constructive change and equity (Maantay 1996, 33).

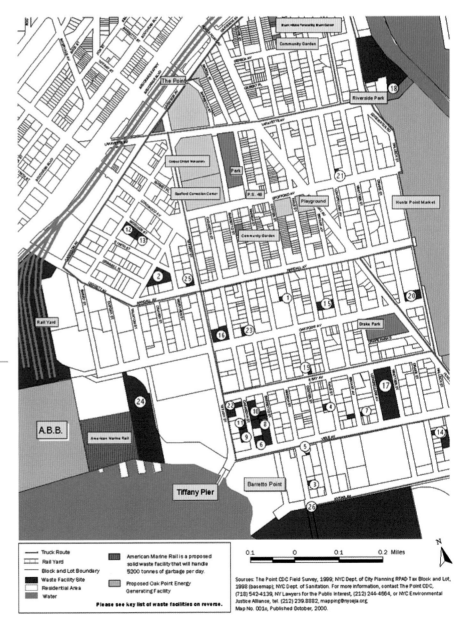

Figure 11.7 A map prepared by the Environmental Justice Alliance with information gathered in a field survey by the Point Community Development Corporation. The map was intended to demonstrate the over-saturation of waste-related facilities in the Hunts Point section of the Bronx, in response to the threat of additional waste-related facilities, such as the proposed marine/rail solid waste transfer station and the proposed waste-to-energy plant. © The Point CDC.

Often, counter-mapping activities involve indigenous groups' assertions of land rights, most frequently in tribal, rural agricultural, or hunter–gatherer societies. But counter-mapping has proven very effective as well in urban settings in developed countries, especially among environmental justice advocacy groups. Counter-mapping may be considered a subset of PGIS. A number of organizations promote the use of community mapping and the creation of local maps by the people. One such group, Common Ground, in Britain, states that "by making parish maps [community maps] and putting on them the places and features that you love and displaying them in a prominent position in the neighborhood, there is a better chance that these things will not only be recognized and enjoyed by others, but respected and protected as well" (Harmon) (figure 11.8).

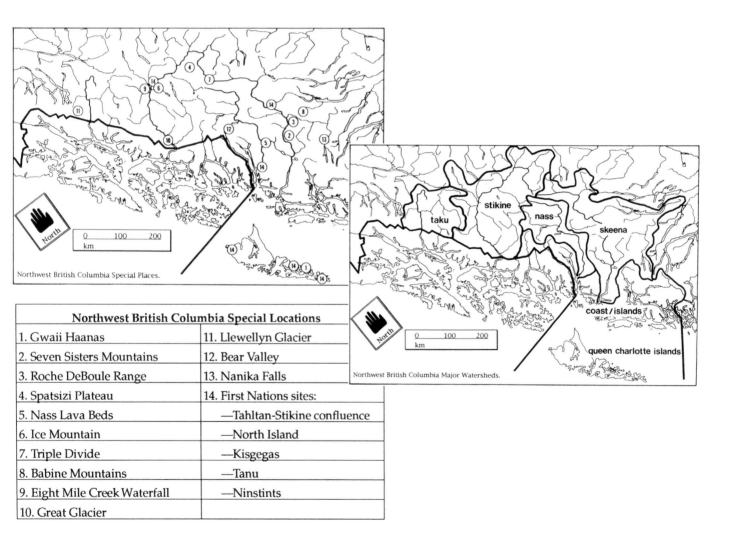

Northwest British Columbia Special Places.

Northwest British Columbia Major Watersheds.

Northwest British Columbia Special Locations	
1. Gwaii Haanas	11. Llewellyn Glacier
2. Seven Sisters Mountains	12. Bear Valley
3. Roche DeBoule Range	13. Nanika Falls
4. Spatsizi Plateau	14. First Nations sites:
5. Nass Lava Beds	—Tahltan-Stikine confluence
6. Ice Mountain	—North Island
7. Triple Divide	—Kisgegas
8. Babine Mountains	—Tanu
9. Eight Mile Creek Waterfall	—Ninstints
10. Great Glacier	

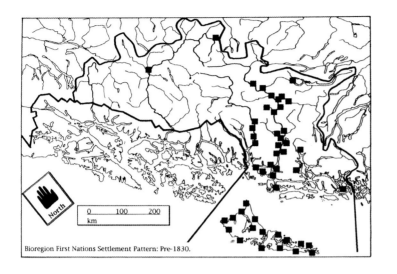

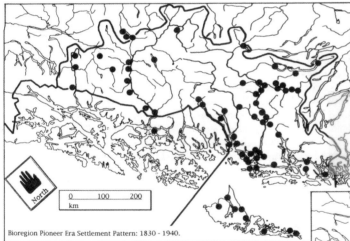

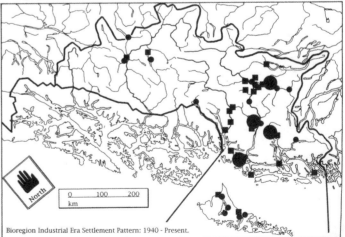

Figure 11.8   Bioregion Mapping Project for the First Nations of British Columbia. These demonstrate how mapping "Experience of Place" can be a way to understand your territory and aid in conflict resolution and community empowerment. This selection shows watershed boundaries and special places (on facing page), and historic settlement patterns. Other maps contained current and historic political and administrative divisions, physiographic regions, climate provinces, ecoregions, tribal territories, community use patterns, mining areas, and other economic activities.

Reproduced from: *Boundaries of Home: Mapping for Local Empowerment*, Doug Aberley (1993) New Society Publishers, *www.newsociety.com*.

Technology and the use of local knowledge bases

One of the critical ingredients of a community-based GIS is the incorporation of local knowledge into the database. Local knowledge typically differs from information gathered in official databases in that it is more particular, more detailed, more specific to the locale, and of importance to the community for educational or archival purposes. A local knowledge base may consist of something as simple as a community-led land-use inventory, at a level of accuracy and detail missing from official land-use databases, and including unique categories of land use or previously unrecorded variables of interest to the community. A local knowledge base might also consist of a rating of quality-of-life or safety issues by street, as informed by the life experiences of the residents, or it might include the spatial delineation of spheres of influence of various street gangs. Information like this would be virtually impossible for people outside the community to obtain. The addition of local knowledge can help identify historic, cultural, or traditional places of importance to the community, either to the lives of individuals or to the broader life of the community, which may have otherwise escaped official attention. A GIS enriched by local knowledge bases is an invaluable tool in community-based planning. Some would even argue that without local knowledge bases the GIS will not be used to its optimal planning capabilities.

However, with GIS and other geographic information technologies, there are certain difficulties in using data based on local knowledge, since it often does not readily fit the GIS format. Some spatial concepts cannot be adequately represented in a GIS, and they do not lend themselves to Cartesian representation. Local knowledge is fluid, and when translated into a two-dimensional geography, consists mainly of zones of transition. Boundaries between activities or functional areas are not usually sharply demarcated, but tend to be overlapping or gradual.

Local knowledge that needs to be part of a GIS may include extreme temporal variability. Spatial locations often are attached to attribute information that varies by different units of time. For instance, a property lot may be used by a certain group in one part of the day and a different group at night; a neighborhood has one description or a certain demographic characteristic during the weekdays and another during the weekends; an area has seasonal variations with one type of use in the summer and another in the winter. Some types of uses may have considerable temporal overlap. Temporal variability and fluidity are common in indigenous rural cultures, but are also prevalent in urban societies.

Local knowledge bases often reflect spatial information that is not geographically fixed in a way that can easily be portrayed in a static GIS. There are inherent problems representing the complex social/political/gendered construction of space in a strictly Cartesian geometry. It has been suggested that there is an element of dishonesty in trying to force local knowledge into the so-called rational–scientific mode of representation within a GIS. (Harris et al. 1995).

Also, the very term *local knowledge* assumes that there exists a homogeneous knowledge base within a given community. Clearly, as shown above, this is an oversimplification. The use of GIS has been based on its ability to facilitate and support decision making, which in turn is based on the underlying assumption of a single objective knowledge, a single set of facts. "With the inclusion of more than 'one knowledge,' it is likely that a GIS database will contain conflicting information and substantive fuzziness. . . . With the inclusion of locationally fuzzy local knowledge, many issues begin to arise as to how multiobjective goals, based on multiple criteria, and using spatially imprecise and possible conflicting data, might actually achieve what is assumed to be consensus decision making. Indeed, greater quantities of information may promote social conflict" (Harris et al. 1995, 196).

Some of these problems can be overcome by developing innovative applications. "What is challenging is learning how to put these techniques together in a process that fits each unique landscape, cultural group, and land-use system, and to use these western-derived techniques to create a map that is both scientifically accurate and a correct reflection of local perception" (Flavelle 1995, 72). As planners, it is our moral responsibility to try to acquire and use spatial data that reflects local knowledge, while maintaining scientific accuracy. Nevertheless, GIS remains a technology rooted in the rational–scientific approach, rather than being well-suited to the shifting dynamics of most local knowledge frameworks. Given the financial and temporal constraints of most governmental or nonprofit planning functions, incorporating local knowledge bases in the planning process remains an ideal that is not often achieved. However, it is important enough that we should aspire to include it.

Access to hardware, software, and skills

Another major issue regarding community-based GIS and PGIS projects is the ability to sustain high technology tools within communities that historically are not technically oriented. This has changed somewhat with more user-friendly computer systems and less expensive Windows-based, menu-driven GIS software, but a high level of computer expertise and sophistication about data use and analysis is required and not always available in many communities.

Typically, obtaining the hardware and software is less difficult for many community groups than getting the technical support and analytical expertise. Some of these problems can be solved by technical support staff developing standards for database updates and management, and creating macros or mini-programs for community users to access commonly required analyses. This reduces the need for community users to have advanced computer or analytical skills, and sets up a format for community members to use in maintaining their GIS.

Perhaps the best solution to the technology problem is to devote a sizable amount of the initial project time and funds to GIS training programs and community GIS education. It is unconscionable to start a community-based GIS project without having planned for its long-term sustainability. Lack of planning for sustainability will likely condemn the project to being useless. It will have wasted money and time, and have raised people's hopes under false pretenses. Broad-based GIS training will ensure not only community access to the GIS, but provide enough people within the community who are well-versed in the GIS to maintain it, perform basic trouble-shooting, update databases, and continue to educate others in its use. It will also prevent one group within the community from becoming a technological elite, with greater access to information and therefore to power. Well-designed training programs will also lead to a greater likelihood of the community GIS experts developing the innovative approaches necessary to effectively portray the complex reality of their own world.

The benefits of this approach are many and help give the community a sense of control over its own destiny and a sense of being on a level playing field with the official powers. At the beginning of any community-based GIS endeavor, it should be stressed to the community that the project will require a huge, ongoing commitment on the part of the community to create and maintain the GIS. The point should also be emphasized that GIS is not a panacea that will solve all of the community's problems. Often community-based planners and community members alike have unrealistic expectations of what GIS can do and the ease with which it can be done, and this can lead to disillusionment with and abandonment of the GIS. It is the responsibility of the experienced GIS user to present GIS to the public in a positive yet realistic light. It is also important for planners and GIS users to themselves understand the very real limitations of GIS, and to realize that GIS can only be a part of the solution to planning questions.

Vernacularization of GIS

Historically, mapping and cartography, being expensive and highly skilled technical endeavors, have remained the province of the elite and powerful. This has been true until recently for GIS and other geomatic technologies. With the advent of the personal computer, relatively inexpensive desktop GIS software, and readily available data served over the Internet, mapping with GIS is available to virtually anyone. Will this new understanding of mapping by the ordinary citizen lead to a new form of empowerment? Is the vernacularization and increased accessibility of maps potentially as subversive and far-reaching as the similar increase in accessibility of books five hundred years ago with the invention of the printing press and the subsequent dissemination and use of printed materials and news information?

It is a widely held belief that increased accessibility of printed information in the fifteenth century had profound impacts on the ascendancy of democratic ideals, due to the increased availability of knowledge to a larger audience. If knowledge is power, does it automatically follow that more open access to GIS will produce more democratic conditions? Or will the still-skewed access to information and technology mean there will still be haves and have-nots in the GIS realm, similar to the so-called digital divide?

To be a truly participatory vehicle for change, GIS must be approached as a bottom-up rather than a top-down technology. This is difficult to accomplish, since the government and other powerful entities still control most of the information utilized in the GIS databases and still retain most of the technical expertise in using these systems. "In order to incorporate a GIS successfully, participation will have to be broad-based, inclusive, gender-sensitive, and biased towards the interests of marginalized people . . . If participatory GIS is genuinely to empower the most oppressed social segments of a community, then it must become part of a political program explicitly aimed at restructuring social relations, and hence contesting local and state political structures" (Harris et al. 1995, 218).

Some disadvantages accompany the many advantages of the democratization of GIS. Now that making a map is quite easy and does not require a very high level of technical expertise, does that diminish the power of maps? Will accessibility of data and mapping software result in mapping overload? It is easier to produce a "bad" map than a "good" one, but because of desktop mapping software, they will both possibly look equally professional, and therefore appear to be equally correct or true. Will the average person be able to discern the good from the bad, and will there be a flood of badly constructed maps that, purposefully or not, are misleading in content? Will greater access to mapped data and the ability of user interface with the data lead to greater understanding and critical thinking about the information? Quite possibly these types of cautionary questions were asked five hundred years ago at the dawn of the printed book era: would increased access to knowledge lead to its corruption and misinterpretation, or would it lead to enlightenment? Time will tell if the average person will become more literate in map reading and interpretation because of the proliferation of maps in her daily life. A more educated public should be considered a desirable goal of planning, and we believe it is a responsibility of planners to promote map literacy and disseminate the fruits of their GIS expertise and knowledge. They can accomplish this through involvement in public education and training, participatory GIS, and pro bono work as GIS consultants to nonprofit community and environmental organizations.

Data access

Because it is usually government agencies that collect and maintain large databases, there are ethical and legal issues concerning public access to this data. Who does this data belong to? Who should have rights to use it? One of the implicit premises of democracy is the right, or perhaps even the requirement, for citizens to have full access to information necessary to evaluate policies that affect their lives. The federal Freedom of Information Act (FOIA), and its counterparts at the state level, mandate that public agencies make records available to the public, except as dictated by national security or privacy concerns. FOIA was promulgated in 1972 to reduce the possibility of governmental abuse of power inherent in secret record-keeping. FOIA and its local versions are an important avenue of data acquisition for the urban planner—even a planner working within a governmental agency—as agencies do not automatically share data, especially when going from one level of government to another, such as from federal or state to county or city.

Our governmental agencies are responsible for gathering and maintaining data, and these tasks are paid for by taxpayers' dollars. Yet very often this data is off limits to the public, or, in fact, even to other agencies. Indeed, in many cases, the public does not even know that a particular kind of data exists. How open should access to this data be? And in the many cases where governmental agencies charge individuals for data, there is a question of fairness to taxpayers, who have paid for the data through their taxes and are then required to pay an additional user fee to access that data.

From the point of view of the governmental agencies charged with collecting, maintaining, and distributing this data, however, these tasks can quickly become burdensome and even overwhelming in some short-staffed agencies struggling to keep up with requests from the public. In addition to data preparation, there are costs involved with disseminating data— equipment and material costs to find, select, and copy data, and then transfer files to disks or other media. Many agencies have decided to pass these charges along to the end user.

With such a policy, agencies can recoup some of the costs of these tasks, but at the risk of limiting data access to those who can afford to pay for it. In the case of GIS data, the problem of recouping costs is even more extreme because, as we have noted, GIS databases are expensive to construct and prepare, with raw data requiring considerable manipulation to make it useful to an end user within or outside the agency. Passing along to the public all these value-added costs incurred in developing a GIS database would likely make data less inaccessible to many less-affluent people.

Data ownership

Issues of data ownership arise most often when a commercial firm takes governmental data—available to them for free or at a relatively minimal cost as it is to any other member of the public—and performs its own value-added processes to the data. It then repackages the data as a new product, which it then sells to the public and makes a profit. Who owns a dataset that started as governmental (public) data, but has been transformed by private enterprise into a different, more useful, product? How much work must be done on a dataset before a company or individual can claim ownership of the material? GIS data is information, and information, by its very nature, is intangible; it is inherently difficult to establish its ownership. The end result of such questions is a growing number of legal conflicts over intellectual property.

Data liability

There are also liability issues involved in disseminating data. When data is distributed to the public, either directly from governmental agencies or via private companies, who verifies that the information is correct? Who is liable for incorrect information and for the legal and financial ramifications of using incorrect information contained in public (or commercial) datasets? Suppose a property owner or her contractor obtains a city spatial database of water and sewer mains, and bases a plan for new construction on their stated location. In the process of digging a foundation, a main is hit where the map shows there is none. Can the homeowner or the contractor hold responsible the agency that gave them the data? Similarly, we might have a scenario where a wetlands delineation map prepared by the state department of environmental protection shows the boundaries of a wetlands within a certain property, rendering most of the property off limits to development due to strict environmental regulations regarding wetlands. However, the map is out of date and does not reflect current wetland conditions. Nevertheless, the property owner will not be able to sell the property for its full worth as a residential parcel with the restrictions on development that are in place, nor can he obtain a loan. This type of error is likely to go uncorrected for years, with the burden on the property owner to prove the inaccuracy of the mapped data.

Other liability issues are surfacing in the area of 911 emergency infrastructure mapping, automobile navigation systems, airport and flight path data, and locational data obtained through global positioning systems. This is especially pertinent because much data within a government agency is shared with other agencies or departments. This increases the chances that data could be put to use for purposes other than the original ones. This could lead to fatal consequences such as when an ambulance or fire truck fails to find the source of an emergency call because emergency personnel are using an outdated or erroneous address database—one that may have been obtained from another agency and was not subjected to systematic updating and maintenance.

A different kind of liability concern arises when incorrect personal information is used in assessing an individual's eligibility for a loan, a home mortgage, or insurance coverage. How much does information compiled about us affect decisions made on our behalf, without our being able to refute or correct erroneous information, or often not even being aware that this erroneous information exists? There are a number of databases today compiled by commercial companies from census data and other publicly available information. These databases are developed primarily for marketing purposes, using geodemographic data to classify communities by socioeconomic indicators. Geographic data is used to delineate marketing zones, a practice that is of immense help to advertisers, real estate salespeople, insurance brokers, and members of other sales-related industries who are interested in targeting potential customers. With the availability of easy mapping, spatial data examination, and display with GIS, demographic data can be located with pinpoint accuracy, in a way never before possible. As has been amply demonstrated elsewhere, seeing information delineated on a map is much more powerful than looking at a table or list of addresses.

A larger issue relates to what has been called the ecological fallacy, which poses the question: how much does the average person represent any one individual in a given area? The ecological fallacy "is committed when the analyst or reader assumes that the average individual within an area represents any given individual therein. Because the ecological fallacy is a way of thinking with which anyone who publishes statistical research must contend, some might argue that there is no reason to single out the visual representations produced within geodemographics. But there is in fact a difference between the commission of the ecological fallacy in statistical work where results are displaying in, for example, tabular form, and its commission in the case of mapped data. This is because people tend to see maps as direct representations of reality in ways that tables and charts are not. There is ample evidence of this…[and] concerns arise just to the extent that it is possible to produce visual representations that any reasonable reader will directly read as associating characteristics of behavior or belief with individuals or members of households. In an important sense, such maps paint their subjects, and in colors that may be inaccurate or damaging, that may impute to those subjects behavior or beliefs which they take to be matters that should remain out of the public eye" (Curry 1998, 124).

Many of these questions of liability and ownership have not been thoroughly addressed by regulators or the courts yet; the jury is still out on deciding how to handle these new forms of intellectual property. As planners, we are likely to be on one side (the public) or the other (the government) in the debate, and often in any one planning career, we will find ourselves on each side at one time or another. Although there are few definite regulations to guide our actions, an awareness of some of the issues will at least prepare us to face the hard decisions as they emerge.

Privacy, confidentiality, and surveillance issues

The converse of the issue of full public access to data is the potential for invasions of individual privacy created by the wide-spread availability of such data. Geodemographic data now can pinpoint the geographic location of individuals and with it, previously intimate and private information attached to those locations and individuals. Much information about an individual based solely on an address can be gathered from public or private sources. Privately held data includes buying habits, income, vacation destinations, leisure activities, telephone and computer use, religious affiliation, banking and investment accounts, educational record, and medical information. Public data includes assessed value of a home, motor vehicle information, political contributions, and criminal record. Many people clamor for full access to data until they realize what that might entail for the dissemination of private data about themselves. Where do we draw the line? What constitutes legitimate use of private data?

GIS has often been promoted in trade magazines, software company material, and in planning courses for its various applications "that permit us to gain greater levels of clarity and control over the social and economic domain" (Pickles 1995, 16). Much of this clarity and control comes at the expense of the individual's right to privacy, and in many cases, the individual is not even aware that her privacy is being invaded. There is an enormous potential for abuse in the use of data about individuals and the concomitant surveillance possibilities.

"The prospect of socioeconomic application of GIS permitting efficiently functioning organization such as insurance companies to develop 'geodemographical' insurance rate schedules based on the identification of zones and localities of high risk, the targeting of civil rights groups (the 'politically militant') for particular police or vigilante attention, or the extension of direct-mail solicitation to exact-market targeting based on recorded purchasing and general expenditure records (already a reality, of course) . . ." is generally seen as objectionable when stated in terms of rights to privacy. All too often, however, these uses of GIS are seen as normal and neutral, as scientific uses of socioeconomic data (Pickles 1995, 16).

"The ease of computer mapping, combined with the increasing availability of datasets, has made maps more readily available and made the possibility of privacy infringement much more likely" (Curry 1998, 125). The construction of data profiles is particularly troubling in terms of privacy, since an individual can be profiled by someone utilizing individual-level data, such as that available from credit card companies, governmental agencies, and so forth, and combining these with publicly-available aggregate data, such as sociodemographic data from the census or other data providers at the tract or ZIP Code level. This type of data profile may be an eerily accurate portrayal of an individual, and the potential for abuse of this profile is enormous.

As planners and users of GIS data, we must be conscientious about ensuring that we apply our efforts to constructive and positive transformative uses of GIS, to improve people's sense of community, and to strengthen the community itself, while not contributing to the loss of independence and loss of control over one's life that computer technology and GIS in particular has been accused of promoting.

This chapter has done no more than briefly introduce some of the contentious ethical dilemmas facing GIS practitioners and planners today, has not done justice to the complexities of the issues, and has only touched on other important ethical concerns. Therefore, we encourage the reader to familiarize herself with the materials listed in the additional reading section of this chapter for further insights into this fascinating sphere of knowledge. Some of the case studies, such as Trevor Harris' participatory GIS in Kiepersol, South Africa, are particularly instructive.

In concluding the discussion on ethics and GIS, we repeat the caution given by Denis Wood in *The Power of Maps,* in which one of the chapter subheadings is "Maps are Heavy Responsibilities" (Wood 1992, 188). Maps and GIS are important inputs for decision making, and have the power to effect far-reaching changes for good or ill. We, as the mapmakers, must shoulder that responsibility and act with integrity and awareness of the consequences, even though the ethical choices are not always as clear cut as we might like. Planners should be cognizant of the eventual use of their GIS products and analyses and, before they begin work on a GIS project, make a commitment that they will, like physicians, first do no harm.

Written Exercise 11

Participatory community-based planning with GIS

In this exercise, you will be developing the framework for a participatory GIS project. You may use the problem scenario outlined below, or use a current community-based problem in your area. Then, answer the eight questions.

Problem scenario

The city owns many vacant parcels of land in a large part of the city, due to landlord abandonment and foreclosure on properties over a period of time. Many of these vacant lots have been turned into community gardens by nonprofit neighborhood gardening groups, and some have been active gardens for decades. The gardens serve as cultural, educational, and environmental focal points of many neighborhoods, which are typically poorer communities. The city would now like to sell some of these properties, since land value has gotten much higher in these neighborhoods, and the city could use the money that would accrue from the sale as well as the subsequent property tax revenue from new development. The properties may be sold to developers at auction and are generally used for new housing construction. This new housing is not necessarily affordable to the existing residents of the community and may serve to instigate the gentrification process and the displacement of the population.

A community group has contacted you about conducting a participatory GIS project to try to save the community gardens. To do that, the group is willing to prioritize the gardens that are to be saved, based on criteria to be determined. Your first task will be to help them prioritize the gardens with GIS. In addition to prioritizing the gardens, you are to figure out what other, different vacant properties are available near the community gardens, which can then be considered for development in their place. Additionally, you are to look for parcels with abandoned or underutilized housing, suitable for rehabilitation into affordable housing units. The goal is to have a viable alternative plan to present to the city council before they vote on the fate of the gardens.

You must gather all the data required for the project and then prepare for an intensive two-day planning charrette with twenty-five members of the community gardens advocacy group. During this charrette, you will be one of the facilitators for technical planning support and GIS, and the community members' viewpoints and perspectives are to be taken into account. Local knowledge bases are to be incorporated wherever feasible, and a vision plan for preserving the community gardens, as well as practical solutions, are expected.

Questions

1 What steps would you take to implement this event?
2 How would the participants be selected?
3 What data would be necessary?
4 How would you structure the participatory GIS project?
5 How would you incorporate the community group's participation into the final product?
6 What types of local knowledge bases would likely be used in the project?
7 What do you envision the final product would be?
8 How would this plan be implemented effectively?

References and further reading

Abbot, Jo, Robert Chambers, Christine Dunn, Trevor Harris, Emmanuel de Merode, Gina Porter, Janet Townsend, and Daniel Weiner. "Participatory GIS: Opportunity or Oxymoron?" *PLA Notes 33* (October 1998).

Barndt, Michael. "A Model for Evaluating Public Participation GIS Programs." Paper presented at the NCGIA Specialist Meeting, Santa Barbara, CA., October 1998.

Chambers, Robert. "The Origins and Practice of Participatory Rural Appraisal," *World Development 22,* no. 7 (1994):953–969.

Cinderby, Steve. "Participatory Geographic Information Systems (GIS): The Future of Environmental GIS?" Paper presented at the NCGIA Specialist Meeting, Santa Barbara, CA., October 1998.

Cassettari, S. 1998. *Introduction to Integrated Geo-Information Management.* London: Chapman and Hall.

Cho, G. 1998. *Geographic Information Systems and the Law.* New York: John Wiley and Sons.

Christopher, A. J. 1994. *The Atlas of Apartheid.* London: Routledge.

Convis, Charles, Jr., ed. 2001. *Conservation Geography: Case Studies in GIS, Computer Mapping, and Activism.* Redlands, CA: ESRI Press.

Curry, Michael. 1998. *Digital Places: Living with Geographic Information Technologies.* London: Routledge.

Earnshaw, Rae A., and John Vince. 1997. *The Internet in 3-D: Information, Images, and Interaction.* San Diego, CA: Academic Press.

Epstein, E. F. 1991. "Legal Aspects of GIS" in *Geographic Information Systems: Principles and Applications.* Harlow, UK: Longman.

Flavelle, Alix. 1995. "Community-Based Mapping in Southeast Asia" in *Cultural Survival, 18:4. Geomatics: Who Needs It?*

Greene, R. W. 2000. *GIS in Public Policy.* Redlands, CA: ESRI Press.

Greeves, T. 1987. *Parish Maps: Celebrating and Looking After Your Place.* London: Common Ground.

Harmon, Katherine. 2004. *You Are Here: Personal Geographies and Other Maps of the Imagination.* Princeton, NJ: Princeton Architectural Press.

Harris, Trevor, Daniel Weiner, Timothy Warner, and Richard Levin. 1995. "Pursuing Social Goals Through Participatory GIS: Redressing South Africa's Historical Political Ecology" in *Ground Truth: The Social Implications of GIS.* New York: The Guilford Press.

Harris, Trevor, and Daniel Weiner. "Community-Integrated GIS for Land Reform in Mpumalanga Province, South Africa." Paper presented at the NCGIA Specialist Meeting, Santa Barbara, CA., October 1998.

Howard, Daniel. "Geographic Information Technologies and Community Planning: Spatial Empowerment and Public Participation." Paper presented at the NCGIA Specialist Meeting, Santa Barbara, CA., October 1998.

Leitner, Helga, Robert McMaster, Sarah Elwood, Susanna McMaster, and Eric Sheppard. "Models for Making GIS Available to Community Organizations: Dimensions of Difference and Appropriateness." Paper presented at the NCGIA Specialist Meeting, Santa Barbara, CA, October 1998.

Maantay, Juliana. 1996. "Bringing Mapping (and Power!) to the People: Using Counter-Mapping to Change Conditions in the South Bronx." Discussion paper, The Center for a Sustainable Urban Environment, Bronx, New York.

McHaffie, Patrick H. 1995. "Manufacturing Metaphors: Public Cartography, the Market, and Democracy," in *Ground Truth: The Social Implications of GIS*. New York: The Guilford Press.

Monmonier, Mark. 1991. *How to Lie With Maps*. Chicago: University of Chicago Press.

Nyerges, Tim. "Linked Visualizations in Sustainability Modeling: An Approach Using Participatory GIS for Decision Support." Paper presented at the annual meeting of the Association of American Geographers, Los Angeles, March 2002.

Peuquet, Donna J. 2002. *Representations of Space and Time*. New York: The Guilford Press.

Pickles, John. 1995. *Ground Truth: The Social Implications of Geographic Information Systems*. New York: The Guilford Press.

Plewe, B. 1997. *GIS On-Line: Information Retrieval, Mapping, and the Internet*. Santa Fe, NM: OnWord Press.

Poole, Peter. 1995. *Indigenous Peoples, Mapping, and Biodiversity Conservation: An Analysis of Current Activities and Opportunities for Applying Geomatics Technologies*. The Biodiversity Support Program.

Steiner, Frederick. 1991. *The Living Landscape: An Ecological Approach to Landscape Planning*. New York: McGraw–Hill.

Szego, J. 1994. *Mapping Hidden Dimensions of the Urban Scene*. Stockholm: Swedish Council for Building Research.

Weiner, Daniel, T. S. Warner, T. M. Harris, and M. R. Levin, "Apartheid Representation in a Digital Landscape: GIS, Remote Sensing, and Local Knowledge in Kiepersol. South Africa." *Cartography and Geographic Information Systems* 22 no. 1 (1995):30–44.

Wood, Denis. 1992. *The Power of Maps*. New York: The Guilford Press.

Other Geotechnologies *and* Recent Developments *in* GIS

This chapter discusses advanced GIS applications and developing technologies related to GIS that are expected to have a major impact on the future quality and affordability of both GIS data and GIS analysis.

Three-dimensional GIS

The list of urban planning functions that require visual information in three dimensions is a long one, but the most common are topographic mapping, modeling of surface terrain, and modeling of the built environment. Planners can also benefit greatly from using 3D GIS in considering aesthetics, viewsheds, shadow studies for proposed structures, lines of sight for telecommunications towers, general urban design, infrastructure planning, zoning proposals, and emergency management.

Topographic mapping

Terrain elevations are represented in vector maps as contour lines, drawn at specified intervals. A contour line connects points on the ground of equal elevation, and the contour interval is the vertical distance between contour lines. As is the case with two-dimensional vector GIS data, points are the basic building blocks of a 3D GIS model. Three-dimensional points have x-, y-, and z-coordinates, with the z coordinate referenced to a base elevation such as mean sea level in a terrain map, or 0 in an architectural CAD drawing. Contours can be automatically generated from gridded arrays of 3D points, using interpolation methods that were discussed in chapter 9, or from irregular points on the map using a Triangulated Irregular Network model (TIN). GIS software with 3D capability automates the process of generating contours from grids and TINs.

Gridded arrays of 3D points on a terrain surface are known as digital terrain models (DTM), or digital elevation models (DEM), although DEMs commonly refer to specific mapping products of the same name, published by the U.S. Geological Survey. USGS DEMs are available nationwide in five grid sizes, or sampling intervals, and are described officially as "a digital file consisting of terrain elevations for ground positions at regularly spaced horizontal intervals . . . Although all are identical in the manner the data are structured, each (product) varies in sampling interval, geographic reference system, areas of coverage, and accuracy, with the primary characteristic being the spacing, or sampling interval, of the data." USGS DEMs can be downloaded from the USGS Web site.

The most refined of the USGS DEM products is the 30×30 meter DEM that is designed to generate contour mapping for 7.5-minute (1:24,000 scale) USGS quadrangle maps (see chapter 6). Remember that the accuracy of a contour map generated from a DEM depends on the density of the point grid that was used to generate it, as well as on the methodology used to interpolate points and generate the contours. The interpolation methods cited in chapter 9 allow you to generate a contour map using a grid that contains just a few points, but the resulting map will be useless—or worse, misleading.

USGS digital topographic maps for most areas in the United States use a contour interval of 10 feet or more, meaning that the contour lines represent elevations that are 10 feet apart from each other vertically. If a topographic map has a 10-foot contour interval, contour lines will represent elevations of 10 feet, 20 feet, 30 feet, and so on, with respect to a base elevation of zero, or sea level. Although a 10-foot interval is appropriate for general planning purposes, building and infrastructure design in an urban setting calls for terrain mapping that is reliable at intervals of 2 feet or less. Nonetheless, the USGS DEMs are useful for the preliminary planning of large-scale projects in three dimensions, and for an overview of the configuration of an urban landscape.

There are three ways to generate topographic maps at the level of detail needed for urban planning purposes: traditional on-the-ground survey techniques, aerial orthophotography, and a relatively new technology called lidar. The first two methods are expensive and are best suited to small geographic areas. Lidar promises to make detailed topographic mapping affordable in the future, even for large geographic areas.

Surface modeling with GIS

GIS software with 3D capability can generate surface terrain models that can be viewed from any point in space on the computer monitor, as if they were a physical model. Surface models, like contour maps, are generated from points—either irregularly spaced or arrayed in grid form as previously described. Three-dimensional points arrayed in grid form can be used to produce **wire frame** surface models. Surfaces defined by irregularly spaced points are modeled using a triangulated irregular network (TIN). The irregularly spaced points that are used by the TIN model (which also have x-, y-, and z-coordinates) become the vertices of three-dimensional triangles (figure 12.1).

The choice of wire frame model versus TIN model generally depends on whether the 3D points are available in grid form or are irregularly spaced. Assuming an equivalent density of points in the area to be modeled, a TIN model based on irregularly spaced points that respond to terrain characteristics will produce a more accurate representation of the surface.

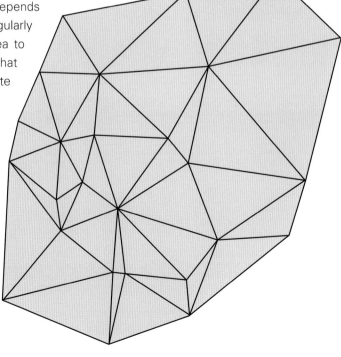

Figure 12.1 A simple Triangulated Irregular Network (TIN). Vertices of the triangles have x-, y-, and z-coordinates.

There are several other advantages to the TIN model. Colored and shaded raster images can be draped over the TIN model, providing realistic effects. GIS software with 3D capability lets the user view both TIN models and wire frame models in perspective—from any viewpoint chosen by the user—and manipulate light sources and, therefore, shading. For spatial analysis in three dimensions, surfaces that were generated using the TIN model allow the user to find the height of any point on the surface, determine lines of sight between points, calculate slope and aspect (the direction the triangular surface faces), find the steepest path along a sloped surface, generate profiles, and make engineering calculations for operations such as cut and fill. The accuracy of these calculations, of course, will depend on the accuracy and density of the 3D points that are used to create the model. TIN models can also produce realistic renderings of terrain for presentation purposes (figure 12.2).

Figure 12.2   A shaded terrain surface using the TIN model.

Environmental simulation: Modeling the built environment in three dimensions

Simulation models of the built environment—that is, 3D computer models of existing and proposed buildings and infrastructure—can be combined with GIS to create a powerful tool for both designers and planners. Simulation models are normally created in CAD because CAD editing tools are optimized for detailed definitions of form in three dimensions—although some GIS software vendors are developing editing tools to rival the capabilities of CAD. CAD drawings can be imported into most GIS software packages.

Nearly all CAD software and a number of GIS packages enable 3D spatial models of buildings and infrastructure to be superimposed on terrain models to simulate viewsheds, lines-of-sight for telecommunications equipment, and the aesthetics of new development in an existing urban context. These models can be viewed in perspective from any chosen point in space, and may be used with solar software—which generates shadows cast by buildings and other objects at a particular date and time of day—for studies of new development. In fact, shadow studies are often required as a part of the approval process for proposed high-rise structures. When detailed simulation models of the built environment are integrated into a GIS, nonspatial attributes such as land use and property values can be displayed thematically, as shown in figure 12.3a.

Digital environmental models can also be used with animation software to produce virtual fly-bys of proposed developments or any other model of the built environment. These can be an effective design tool and a persuasive presentation technique because of their cinematic attributes.

Environmental simulation such as this promises to bring the work of architects, urban designers, and urban planners closer, by involving all in the aesthetic, socioeconomic, and environmental decision process that is essential to the planning and design of new developments at any scale. The difficulty—and high cost—of environmental simulation stems from the time required to produce detailed 3D digital drawings of buildings and infrastructure. To date, the primary player in this arena has been the telecommunications industry, which uses it for line-of-sight studies necessary for the placement of cellular telephone towers. It should be noted that lidar technology, described in more detail below, significantly reduces the cost of simulating the existing built environment, leaving only new development to be modeled in three dimensions and combined with the lidar model. This may result in more extensive use of environmental simulation studies in the future.

In the absence of the detailed and costly data that goes into a simulation of the built environment which often requires detailed digital drawings of buildings, a simple but often effective 3D image can be produced by simply extruding two-dimensional features such as urban lots. GIS software with 3D capability allows the user to automatically assign heights (*z* values) to two-dimensional geographic features, using a value in the attribute database. Attribute databases describing urban lots often include a *number of stories* field for structures built on the lot. This field is typically included in property tax databases. By multiplying the *number of stories* value by an estimated story height (for example, 12 feet), you can arrive at an estimated building height for each developed lot, and produce a rough image that illustrates density, building heights, and the overall character of an urban area, as shown in figure 12.3b. If building footprint mapping is available, the two-dimensional outlines of the footprints can be extruded, instead of the boundaries of the lots, resulting in a more refined image.

Figure 12.3a **A three-dimensional GIS model of buildings in downtown New York, with attribute data displayed thematically by color. Colors show floors above the fourteenth floor, with a square footage of 8,000 or less. Red indicates floors in buildings that are 8,000 square feet or less; cyan indicates floors in buildings completed prior to 1945.**
Reprinted by permission of the Environmental Simulation Center, New York, NY.

Figure 12.3b **A lot-level 3–D image of midtown Manhattan, using building heights (from an attribute database) as the third dimension.**

Using the third dimension to express an attribute value

Finally, the z-axis can be used effectively to express nonspatial attributes of geographic entities, such as median income by census tract, showing how these conditions relate to each other geographically. The z-axis can be exaggerated as needed to make the point, since it bears no relationship to the x and y spatial dimensions of objects on the ground (figure 12.4).

Figure 12.4 Using the third dimension to represent the non-spatial attributes of geographic entities: assessed total value of lots in downtown Brooklyn, color-coded by land use.

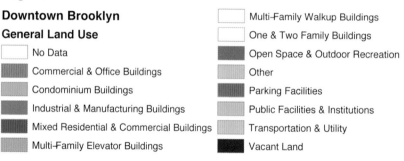

Extrusion

Tax lots are extruded by a z-value that represents total market value per square foot of lot area.

Legend

Downtown Brooklyn

General Land Use

- No Data
- Commercial & Office Buildings
- Condominium Buildings
- Industrial & Manufacturing Buildings
- Mixed Residential & Commercial Buildings
- Multi-Family Elevator Buildings

- Multi-Family Walkup Buildings
- One & Two Family Buildings
- Open Space & Outdoor Recreation
- Other
- Parking Facilities
- Public Facilities & Institutions
- Transportation & Utility
- Vacant Land

Advanced GIS analysis

Adding the time dimension

The position of objects can be tracked over time as well as in space by GIS tracking software, using time and location data provided by traffic sensors, GPS receivers, satellite remote sensors, or field crews. Adding the time dimension to GIS creates a powerful tool for managing municipal operations such as infrastructure maintenance and repair, delivery operations, traffic monitoring, and emergency response. For example, tracking the leading edges of a toxic plume over time can show its direction, growth, and rate of speed, enabling emergency responders to predict which additional areas will be in danger, and when. Other geoevents that must be tracked over time include flooding, locations of snow-removal trucks, wildfires, and the spread of diseases such as the West Nile virus.

Network analysis

GIS software is available for analyzing networks such as roads, rail lines, pipelines, and rivers. In addition to spatial information, the network analysis database stores attribute data such as speed limitations and one-way street designations. When network analysis is applied to a street or roadway network, users can find the shortest distance and the best route between points on the map, to optimize delivery routes, find the facility that is closest to a point on the map, calculate drive times, and create written driving directions. Popular commercial Internet sites such as MapQuest use GIS network analysis software.

Geostatistical analysis

Common uses of geostatistical GIS software include developing statistically optimal surface models, validating the statistical accuracy of surfaces, analyzing very large quantities of data such as are found in raster GIS models, and producing statistical graphics. Geostatistical analysis can also be used to make predictions about the behavior of spatial entities over time, and to examine the effects of different data values on the shape of surfaces—the spatial counterpart of the *what if?* queries that are performed on attribute data. Geostatistical analysis is most commonly used to analyze complex natural systems and, of course, requires the expertise of staff with both statistical and GIS training. In an urban context, geostatistical analysis is primarily used by environmental planners to predict future conditions and to estimate environmental impacts of natural and man-made systems. It can also be used to examine the spatial patterns of disease and other epidemiological and health care issues.

Business analysis

GIS software tailored for the kinds of geographic analysis relevant to commercial enterprises is available from a number of vendors. Spatially referenced business data includes information about some 15 million businesses in the United States, and demographic analyses of more than 100 million U.S. households. Retail chains and others use this software and data to optimize site selection, to predict the most profitable land uses for individual sites and define sales territories, and to make other decisions based on the geographic distribution of consumer demand. Some business software programs include applications of the gravity model—a spatial model that estimates the attraction of commercial centers based on their size and the distance that the consumer must travel to get there; equal competition area models based on Theissen polygons; and spatial analyses of concentric circles around a chosen site. GIS business software is also used by economic development agencies in large urban areas.

Remote sensing

Remotely sensed data is any data gathered at some distance from the earth, usually by airplane or satellite.

Satellite imagery

Satellite images are now available worldwide in digital raster format from several commercial and government sources. These images can be combined with vector mapping and used as basemap layers to add visual information and orientation to a GIS map. Satellite images also provide a check on human error in vector map layers, and on changes that have taken place on the ground since the vector maps were produced. In addition to current satellite imagery, older archived imagery is available at a much lower cost.

The satellites of the Landsat project have been in orbit since 1978, and archived imagery from Landsat and other satellites is a valuable data source for GIS applications that track development and environmental trends. Satellite imagery can be purchased at various processing levels, ranging from raw data to georegistered, graphically enhanced imagery. Of course, higher levels of processing cost more.

Spatial vs. spectral satellite imagery

In addition to visual information, the satellite's digital sensors can produce images that reveal the physical make-up of objects on the ground. Satellite images can be divided into two categories: **Panchromatic**—providing images of features on the ground, and **multispectral**—indicating selected physical characteristics of these objects such as temperature or chemical composition by color and brightness. These two categories are commonly referred to as **spatial images** and **spectral images.** Spectral images that indicate the temperature of objects on the ground are generally called **thermal images.**

The difference between spatial and spectral satellite data has to do more with content than technique. The satellite's digital sensors measure electromagnetic reflectance from the earth's surface in bands, or portions, of the electromagnetic spectrum. The digital sensor that collects the data for panchromatic (spatial) imagery measures reflectance across a wide band that includes the visible portion of the spectrum, to produce black and white images that resemble aerial photographs.

The digital sensors that collect data for spectral imagery, by contrast, measure reflectance in various bands of the spectrum (or in different parts of the same wavelength), and produce data for colored images that combine red, green, and blue at brightness levels corresponding to the reflectance values of objects on the ground for each individual band. Computers on the ground, using data values transmitted by the satellite, produce the actual images.

Since the reflectance values of materials in the various bands of the electromagnetic spectrum are known, the spectral image allows the user to identify materials, plant types and other characteristics of objects, either by eye or by using image processing software. Wetlands are commonly delineated by spectral imagery that identifies wetland plant life. The World Trade Center emergency response case study in part 2 provides further examples of these techniques.

Sensors that can measure reflectance in a large number of individual bands, known as *hyperspectral* sensors, are under development. National Oceanic and Atmospheric Administration (NOAA) satellites now carry Advanced Very High Resolution Radiometer (AVHRR) sensors that can operate in five bands. Spectral technology is developing and promises to permit in the future more subtle distinctions between objects on the ground.

Spatial resolution of satellite imagery

Satellite sensors divide the earth's surface into grid cells and record a single set of values for the electromagnetic reflectance of each cell. In other words, the grid cell is displayed on the screen as a square of uniform color or value, and no detail is visible for objects within it. Smaller grid cells produce more detailed images of features on the ground. The level of detail that can be displayed as a function of grid cell size is called **spatial resolution**. The smaller the grid cell, the higher the resolution. **One-meter resolution** means that the grid cell size in the image corresponds to one square meter on the ground. Figure 12.5 shows a 2.5-meter resolution image, so its grid cell size corresponds to 2.5 square meters.

Figure 12.5   A 2.5-meter resolution satellite image of an urban area: Tel Aviv, Israel. Courtesy SPOT Image Corporation. © CNES 2003.

The lateral dimension of a grid cell is referred to as the ground sample distance (GSD) of the image. (A 1-meter GSD corresponds to 1 square meter on the ground.) As spatial resolution increases, the image becomes both more detailed and more accurate. If you have an image with 1-meter resolution, your screen will be made up of squares, or pixels, each with a single color and brightness, and each corresponding to 1 square meter on the ground. 10-meter resolution means these squares will correspond to 10 meters on the ground, and you will not be able to distinguish features that are smaller than 10 square meters in size.

When choosing the spatial resolution of an image, it is important to know that file sizes and data costs increase exponentially as spatial resolution is increased. Your choice of spatial resolution will depend on the size of the smallest object that you will need to be able to distinguish in the image. This choice, of course, is also affected by the size of the geographic area to be covered, known as the coverage. For example, a spatial resolution of one kilometer might be appropriate for a global climate study. But in a small portion of a city, you would want to be able to distinguish buildings and streets, and you would probably require 1-meter resolution.

Temporal resolution of satellite imagery

In addition to the selection of the spatial resolution of a satellite image, decisions must also be made regarding which time scale, or temporal resolution, is appropriate to use. Again, this will depend on what purpose the image is intended to serve. Temporal resolution, by measuring how often the information on the map or image is updated to reflect current conditions, is also a measure of accuracy. For instance, the information on USGS quadrangles is updated infrequently on an irregular schedule, and so the temporal resolution could be considered to be very coarse, measured perhaps in years or even decades. By contrast, most satellite images have a much finer temporal resolution than this, measured in days or even hours. This makes satellite images much more useful for certain types of analyses, such as change detection studies and emergency response needs. It is obviously critical to have the most up-to-date information when trying to deal with natural or man-made disasters such as hurricanes, oil spills, floods, wildfires, volcanoes, earthquakes, toxic releases, or acts of war and terrorism. Fine temporal resolution is important in making comparisons over short periods of time, to understand how a natural or man-made process is progressing in order to provide appropriate response or mitigation in situations where events change rapidly. Good temporal resolution may also be necessary in order to evaluate change over time on a consistent time scale, for instance, forest clear-cutting or urban sprawl, by viewing a yearly or monthly snapshot of the area, with a consistent geographic extent, spatial scale, and time interval for valid comparison.

However, there is a trade-off between spatial resolution and temporal resolution. Generally, sensors that have a coarser spatial resolution have an orbiting schedule that allows them to return to a given geographic area more frequently, thereby improving the temporal resolution. To monitor events that are taking place on an hourly or daily time scale, it would probably be necessary to sacrifice spatial resolution in favor of better temporal resolution. This limitation is also partially based on the geographic extent of the area to be monitored, but advances and innovations in satellite systems may eventually reduce or eliminate this trade-off.

Table 12.1 shows how image resolution typically corresponds to the scale of a paper map.

Image resolution	Typical map
1000 meters	1:1,500,000
30 meters	1:80,000
20 meters	1:50,000
10 meters	1:24,000
5 meters	1:12,000
1 meter	1:2,000

Table 12.1   Courtesy SPOT Image Corporation. © CNES 2003.

Aerial photographs

At present, satellite imagery is not publicly available at resolutions higher than 1 meter, and images that require this level of detail are produced by aerial photography. Aerial scanners with multispectral sensors similar to those on satellites can be used with airplanes to produce spatial and spectral imagery at spatial resolutions higher than 1 meter, but at a higher cost than satellite imagery.

Synthetic Aperture Radar (SAR)

The satellite-mounted digital sensors that we have discussed so far are known as electro-optical sensors. These are passive instruments that measure electromagnetic energy reflected from the sun, and thermal infrared radiation produced by heat sources on earth. Cloud cover and fog impede these measurements, and electro-optical sensors cannot operate at night.

An alternative to the passive electro-optical sensor is Synthetic Aperture Radar (SAR). SARs are known as active imaging systems because they transmit a radar signal and measure the return signal after it strikes the face of the earth. SAR can acquire the data for images at night and through fog and cloud cover, and may be the best option for coastal regions and in emergency situations that call for spatial and spectral imagery on a 24-hour basis.

Processing remote sensing data

Commercial image processing software can be used to produce custom images, and some pre-processed GIS images are available from government and the private sector. These include:

✦ **Classification maps:** Thematic maps of land areas that have been classed as groups such as land cover type or land use.

✦ **Change detection:** Images that show changes in the same area, created by identifying changed values in individual pixels, as recorded by successive overflights. Changed pixels are usually indicated by color. Change detection is useful in identifying environmental changes and new features in the built environment such as buildings and roads.

✦ **Edge matching:** When separate satellite images are combined into a single mosaic, image processing and edge-matching software is available to match edges, balance colors, and create a seamless database.

✦ **Merging images:** Image processing software that enables the user to merge images of the same area created at different resolutions. For example, a high-resolution panchromatic (spatial) image can be merged with a lower-resolution spectral image to combine spatial detail with spectral information about objects on the ground.

The Internet is the best place to look for these products.

Global Positioning System (GPS)

The Global Positioning System satellites were originally designed for navigational and military uses, but now are increasingly used to generate civilian GIS data in the field, to update GIS data, and to verify the location of points on a GIS map. For example, a field crew checking the condition of power line poles can use a low-cost GPS receiver to enter the position of a pole in need of repair, show it as a point or symbol on a basemap displayed on a personal digital assistant (PDA) using software such as ArcPad from ESRI, enter attribute data about the pole, and transmit this information to a central database. Maps can also be created or corrected from points entered in the field, using a GPS receiver to record their lat–long coordinates.

GPS has the following three parts:

+ **The space segment** consists of a constellation of satellites that orbit the earth every 12 hours; at any given time there are at least 24 in orbit.
+ **The control segment** is a network of earth-based tracking stations, monitored by a master satellite tracking station located in Colorado.
+ **The user segment** is the many thousands of receivers that pick up signals from the satellites, process the signal data, and compute the receiver's geographic location in three dimensions. Values for the horizontal (x and y) dimensions are converted into latitude and longitude. Vertical dimensions are measured meters, referenced to a chosen baseline, such as mean sea level. Signals from four satellites are required to compute position and time. GPS satellite coverage is global, and ground receivers can receive signals from five or more satellites at any point on the earth's surface.

The U.S. military operates the system and provides two levels of satellite signal data to users. The first, known as the Precise Positioning Service (PPS), delivers a high degree of positional accuracy but is available only to U.S. government agencies and to some universities. The second, known as the Standard Positioning Service (SPS), is available to users worldwide without charge. Prior to May 2000, the Department of Defense provided only *Selective Availability* of GPS signal data to SPS users. Selective availability degraded the GPS signal data, resulting in horizontal positional accuracies no better than 100 meters. But in May 2000, the department removed this restriction, improving the horizontal positional accuracy of SPS receivers to 15 meters—an acceptable margin of error for many GIS applications—and making it possible for the first time for anyone, anywhere, to generate GIS point data in the field using a GPS receiver that costs no more than a few hundred dollars.

Since then, the positional accuracy provided by low cost GPS receivers in North America has been increased dramatically by a Federal Aviation Administration (FAA) funded system known as the Wide Area Augmentation System (WAAS). WAAS uses a precisely surveyed network of remote receivers to correct errors in the signals received by the user's

GPS receiver. All this, of course, takes place behind the scenes, giving the user access to highly accurate positional data, recorded by a single WAAS-enabled GPS receiver. WAAS testing in September 2002 confirmed accuracy performance of 1 to 2 meters horizontally and 2 to 3 meters vertically throughout the majority of the continental United States and portions of Alaska.

WAAS was designed to support in-flight navigation and precision landing approaches for aircraft, but it is available to the public; WAAS-enabled GPS receivers now cost little more than standard GPS receivers. Japan's Multi-Functional Satellite Augmentation System (MSAS) and Europe's Euro Geostationary Navigation Overlay Service (EGNOS) are similar to WAAS. The United States is working on international agreements to share these technologies. Positional data at the WAAS level of precision should eventually become available to users worldwide.

Differential GPS (DGPS)

Another way to achieve a high degree of accuracy in GPS readings is to use two GPS receivers—one stationary and one roving. The two receivers in combination are able to identify and correct errors and return positional data that is more accurate than readings from the stationary receiver alone. DGPS can deliver positional accuracies of 1 meter or less, depending on the receiver system. DGPS receivers include software to make the necessary corrections using data from both receiver locations. DGPS equipment costs several thousand dollars, requires experienced personnel in the field, and requires additional set-up time. Before the selective availability restriction was removed, DGPS was the only way to use GPS to generate positionally accurate GPS data points, and it is still used extensively where better than 1-meter accuracy is required.

Carrier Aided Tracking and surveying

The surveying profession has been revolutionized by an application of DGPS known as Carrier Phase Tracking that returns positional accuracies down to as little as 10 centimeters. Carrier Aided Tracking uses special equipment that tracks signals from the Satellite Carrier signal (the L1 and L2 bands of radio frequencies) for error correction. Carrier-aided tracking can also be applied to field-generated GIS point data where high levels of accuracy are required.

All these developments in GPS technology should significantly increase the use of GIS for field operations in the future, reduce the cost of generating new GIS spatial data, and provide a cost-effective means of verifying and updating GIS mapping from the field.

Lidar technology

The recent development of lidar (Light Detection and Ranging) technology promises to make highly accurate three-dimensional mapping available to GIS users at affordable prices. With lidar, lasers strike objects on the earth's surface, including buildings and trees, and produce "point clouds"—thousands of individual points, each with an accurate x-, y-, and z-coordinate. Normally, lidar equipment is mounted on airplanes or helicopters, although lidar images can also be created from ground level. Helicopters are used when highly detailed information is needed for a narrow band of terrain (figure 12.6). The Long Island Rail Road (LIRR), for example, used helicopter-based lidar to produce a detailed three-dimensional model of its tracks and equipment. After processing, railroad station buildings, signal towers and other facilities showed up in fine detail, but the width of the image was limited to the LIRR right-of-way. Airplanes are used to cover wider areas, such as the 3D lidar mapping of lower Manhattan that was produced daily to manage the response to the World Trade Center attacks.

Lidar points can be generated with a positional accuracy that is within inches horizontally and a little more than two feet vertically. Lidar can also be programmed to produce topographic contour at intervals of just over two feet. Lidar is far more affordable than other 3D mapping techniques such as aerial surveying, and as a result it is likely to be increasingly used in the future, particularly in urban areas. The difficulty of lidar use lies in the enormous number of points generated by the system. Heavy processing is required to produce a readable image, and this may double the price of the raw data. Even the processed data, however, is inexpensive when compared to older methods. A 3D file of processed lidar data may cost in the range of $1,000 per square mile, depending on the level of detail required. This is cheaper by an order of magnitude than traditional techniques. The processing of lidar data—reducing the point cloud to a clear image of terrain, vegetation, and structures—requires special software, and is perhaps best left to lidar system providers and consultants.

Figure 12.6   A simplified illustration of one-way, three-dimensional lidar data that is acquired and processed. Redrawn from an image courtesy of TerraPoint USA Inc.

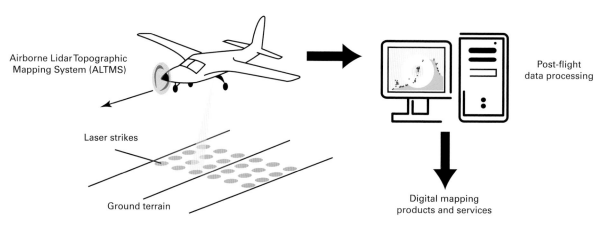

Airborne Lidar Topographic Mapping System (ALTMS)

Laser strikes

Ground terrain

Post-flight data processing

Digital mapping products and services

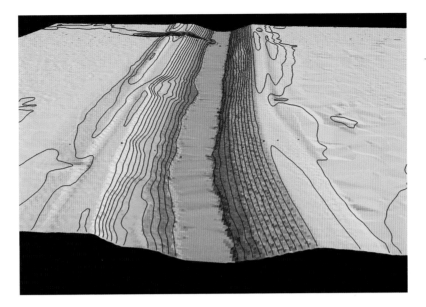

Figure 12.7 A three-dimensional image of a river basin with contour lines, processed from lidar data. Courtesy of TerraPoint LLC.

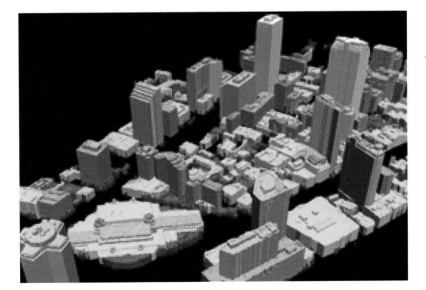

Figure 12.8 Buildings in downtown New Orleans, color coded by height. This image was extracted from lidar data. Courtesy of TerraPoint LLC.

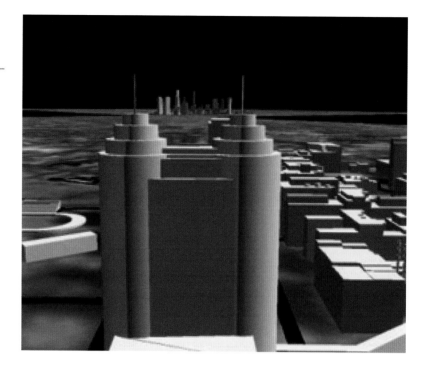

Figure 12.9 A presentation quality image of buildings in Houston processed from lidar data. Courtesy of TerraPoint LLC.

Mobile mapping

Mobile mapping brings GIS and GPS together with two new developments in computing and telecommunications: lightweight, hand-held computers and wireless communication. Mobile mapping enables field crews to both generate and receive GIS data, making possible interactive exchange and updating of GIS data between a centralized database and the field. Currently available GIS software for mobile mapping supports the display of both vector and raster images, and allows the user to edit and query the attribute data associated with geographic features and to download links to photographs, documents, and other images, with wireless access to the Internet. Mobile mapping is particularly well-suited to time-sensitive operations such as package delivery, maintenance and repair of equipment, and tracking geoevents in an emergency. For example, field crews can use mobile mapping to locate damage to buildings and infrastructure in an earthquake, and quickly relay this data to an emergency operations center. In addition, mobile mapping can significantly improve the quality of the GIS data that is stored in the central database by adding ground-truth corrections made on the spot and in real time by field crews, then uploaded.

Internet mapping software

GIS data can also be distributed over the Internet, or, for the internal use of a large organization, over an intranet. GIS data that resides on a server can be directly downloaded as a layer into the user's GIS display. Internet GIS applications can function interactively, using Internet map server software such as ArcIMS from ESRI. Map functions such as panning and zooming are programmed individually using the IMS software. Nearly all of the functions to be found in a desktop GIS software package can be found on a GIS client, but it is an expensive process, and speed (the time required to retrieve data on the client's machine) can become an issue, particularly where graphic images are concerned. The client request to the remote server must be returned and displayed on the client's screen. As a result, very large map files may make interactive use of the Internet impractical.

A second limitation for small organizations with limited budgets is the cost of programming GIS functions using IMS software. As IMS software evolves and high-speed access becomes more affordable, these limitations are likely to become less of a problem. Internet GIS use—both by GIS professionals and the public—is already extensive and will undoubtedly increase in the future.

Wide area networks

Like any application, GIS software and data can either reside on individual machines or be networked in any of the standard network configurations. In addition to the standard Local Area Network (LAN), GIS data and functionality can be distributed over a Wide Area Network (WAN), for use in remote locations. WANs have the advantage of making the full range of GIS functionality available to the remote user without the programming of two-way requests (client to Internet server and back) that characterizes interactive Internet GIS. WANs can be a good solution for joint use of a GIS by several agencies or offices spread around a municipality. However, WANs depend for speed on the telecommunications network, and in the past the speed of some WAN applications has been found to be unsatisfactory.

There are two basic WAN configurations. In the first, the software resides on the client machine and data is distributed over the WAN. In the second, both software and data reside on the central server, enabling system administrators to change and update software only once, rather than on each of the client machines. WAN software is currently being developed to optimize speed. As high-speed telecommunication systems become available, the WAN may become a viable alternative to the Internet for some interactive GIS applications.

Conclusion

In this chapter we have discussed advanced GIS techniques, together with recent developments in related disciplines that are beginning to make it possible for GIS users on a limited budget to produce sophisticated GIS analyses—taking both the third dimension and time into account, and using spatial data that approaches the level of positional accuracy normally associated with engineering drawings and CAD.

Computer hardware has come down in price, while processing speed and storage capacity have increased exponentially—a trend that should continue well into the future. These developments, together with recent advances in data compression software, allow the GIS user to incorporate remote sensing and other data associated with large file sizes into a PC-based GIS product. At the same time, accurate remotely sensed spatial data, in both two and three dimensions, is becoming more available, easier to find on the Internet, and less expensive.

These developments are especially germane to urban GIS use because, as we have discussed in preceding chapters, the urban built environment must be planned in three dimensions and the density of cities calls for large scale mapping with a high degree of spatial accuracy.

Finally, as the map editing capabilities of GIS software continue to develop, and as accurate spatial data becomes increasingly available to GIS users, we should see a wider use of GIS by CAD users in the fields of architecture and engineering. Planning and designing the built environment is an iterative process that ideally involves continued interaction among architects, engineers, and planners. We have emphasized GIS as an urban planning tool, but as the technology and its data sources evolve, GIS may serve in the future as a common language, used by all three professions to visualize data, evaluate alternatives, and present solutions to each other and to decision makers.

Written Exercise 12

Project documentation and policy recommendation report

Based on your findings from projects that you completed in lab assignments 3–12, prepare a report containing complete project documentation and policy recommendations. This report should include project background, the need for the project, data documentation, analysis methodology and project design, discussion of the findings, conclusions, and recommendations. Maps, graphs, and other graphic materials are an important part of the report. Advanced students can create a project flow chart for the GIS part of this project.

References and further reading

Birkin, Mark, Graham Clarke, and Martin P. Clarke. 2002. *Retail Geography and Intelligent Network Planning.* New York: John Wiley and Sons.

Brown, Judith R., Rae Earnshaw, Mikael Jern, and John Vince. 1995. *Visualization: Using Computer Graphics to Explore Data and Present Information.* New York: John Wiley and Sons.

DeMers, Michael N. 2001. *GIS Modeling in Raster.* New York: John Wiley and Sons.

Engenhofer, Max J., and Reginald Golledge, eds. 1998. *Spatial and Temporal Reasoning in Geographic Information Systems.* London: Oxford University Press.

Foody, Giles M., and Peter Atkinson. *Uncertainty in Remote Sensing and GIS.* New York: John Wiley and Sons.

Goodchild, Michael F., ed. 1996. *GIS and Environmental Modeling: Progress and Research Issues.* Fort Collins, CO: GIS World Books.

Hjelm, Johan. 2003. *Creating Location Services for the Wireless Web: Professional Developer's Guide.* New York: John Wiley and Sons.

Kelly, Richard E., J. Nicholas, A. Drake, and Stuart L. Barr, eds. 2004. *Spatial Modeling of the Terrestrial Environment.* New York: John Wiley and Sons.

Leick, Alfred. 1995. *GPS Satellite Surveying.* New York: John Wiley and Sons.

Lilles, Thomas M., Ralph W. Kiefer, and Jonathan W. Chipman. 2003. *Remote Sensing and Image Interpretation.* New York: John Wiley and Sons.

Longley, Paul A., and Michael Batty. 2003. *Advanced Spatial Analysis: The CASA Book of GIS.* Redlands, CA: ESRI Press.

Paine, David P., and James D. Kiser. 2003. *Aerial Photography and Image Interpretation.* New York: John Wiley and Sons.

Steede-Terry, Karen. 2000. *Integrating GIS and the Global Positioning System.* Redlands, CA: ESRI Press

Wilson, John P., and John C. Gallant, eds. 2000. *Terrain Analysis: Principles and Applications.* New York: John Wiley and Sons.

Part 2

Case Studies

We believe case studies should be an integral part of any introductory GIS book, for several reasons. Students and professionals embarking on the mission of learning GIS will benefit from studying how others have used GIS, and the myriad issues that faced others and will likely face them—how GIS projects are developed, the datasets required, the sources of that data, the methodology and analytical processes employed, the institutional framework of a GIS project, problems encountered, limitations of the data, and the results. As mentioned in the preface, we have found that students who pay close attention to how the case studies are constructed usually begin to understand GIS more quickly and easily. Understanding the structure and methods employed in the case study projects will help make the theoretical and technical concepts discussed in the first part of the book less abstract.

Since urban planning is an applied discipline, a strictly academic approach to GIS does not address many of the real-world concerns that urban planning students and professionals typically face when attempting to design and execute a GIS project. A cross-section of actual urban GIS projects can be very instructive in this regard by demonstrating in a concrete way the principles and techniques actually employed by GIS users. Additionally, GIS teachers may want to include a GIS project as part of the course requirements, and the case study examples should stimulate the imagination of students so they can better formulate their own GIS projects.

Because urban planning and operations almost always involve geographic location, there are literally thousands of potential applications of GIS for the urban environment. The examples we have chosen illustrate some of the major categories of GIS planning activities, although by no means all. However, these are all real-world projects that have been successfully implemented and they suggest the broad range of uses for GIS in an urban planning context.

These projects have been developed by many types of enterprises, including governmental agencies, nonprofit organizations, private-sector consulting firms, academic institutions, community-based groups, and collaborative partnerships and associations of all the above. Just as indicative of the growing influence and widespread use of GIS in the workplace is the diverse backgrounds of the project directors of these GIS efforts. The variety of the case studies, their sponsoring organizations, and their project directors again points up the breadth of scope and the far reach of the effects of GIS applications now in use for addressing contemporary urban issues. The case studies are summarized and listed below:

1   **Nonprofit Organization GIS for Strategic Planning and Public Outreach**   *323*
     Public outreach: Imagine New York
     The Lower Manhattan Preservation Fund
     The campaign for community-based planning
     Olympics 2012
     Micaéla Birmingham, Director, Planning Center, Municipal Art Society, New York City

2   **GIS Use by Urban Nonprofit Organizations for Housing Initiatives and Urban Services**   *335*
     Fordham Bedford Housing Corporation, University Neighborhood Housing Program, and the Enterprise Foundation
     Gregory Jost, Deputy Director, University Neighborhood Housing Program
     Patrick Logan, Director, Office of Policy and Planning, Fordham Bedford Housing Corporation
     Michelle Whetten, Deputy Director, The Enterprise Foundation

3   **Delivering Health Care Services to an Urban Population**   *341*
     Access to health care in Greenpoint–Williamsburg, Brooklyn, New York
     Zvia Segal Naphtali, President, Resource Mobilization, Inc. and Adjunct Clinical Professor, the Wagner School

4   Natural Habitat and Open Space Assessment   *349*
Long Island Sound Stewardship System
Jennifer R. Cox, Regional Plan Association

5   Urban Environmental Planning   *357*
Land-use classification for extensive vegetative roof acreage potential in the Bronx
Karen S. Kaplan, City University of New York (CUNY), Lehman College, Department of Environmental, Geographic, and Geologic Sciences, Urban GISc Lab

6   Emergency Management and Disaster Response   *367*
Using GIS at Ground Zero
John Ziegler, Space Track Inc.

7   Infrastructure Mapping for Planning and Maintenance   *377*
GIS mapping of the New York City sewer and storm drain system
Michael Crino, Baker Engineering
Karen Rutberg, AICP, Hunter College
Magdi Farag, PE, FASCE, New York City Department of Environmental Protection

8   Archaeology and Historic Preservation with GIS   *383*
Archaeology GIS project for the New York City Landmark Preservation Commission
Kenneth Mack, Technical Director, The Kenerson Group

9   Health and Environmental Justice   *393*
Mapping environmental hazards and asthma in the Bronx
Juliana Maantay, City University of New York (CUNY), Lehman College, Department of Environmental, Geographic, and Geologic Sciences, Urban GISc Lab

10   Crime Pattern Analysis   *409*
Exploring Bronx auto thefts using GIS
Christopher Hermann, New York City Police Department
Andrew Maroko, Lehman College, Urban GISc Lab

11   Community-based Planning   *415*
Graffiti in Community District Two, Queens, New York: Private expression in the public realm
Thomas Paino, Hunters Point Community Coalition, Queens, New York

12   Advocacy Planning and Public Information   *425*
Community Mapping Assistance Project for the Metropolitan Waterfront Alliance
Open Accessible Space Information Systems Internet map site
Steven Romalewski, Director, CMAP, New York Public Interest Research Group (NYPIRG)
Christy Knight Spielman, Senior GIS Analyst and Web Designer, CMAP

About the author

Micaéla Birmingham is the director of the Planning Center at The Municipal Art Society, New York, and the director of GIS projects there. She was previously employed in the GIS unit of the New York City Parks Department. She has an undergraduate degree in geography from McGill University in Montreal, and a Master of Science in City and Regional Planning degree from Pratt Institute in Brooklyn. Micaéla has been using GIS for planning applications for about six years and has taught basic mapping science courses in the GISc program at Lehman College, City University of New York.

Nonprofit Organization GIS *for* Strategic Planning *and* Public Outreach

Public outreach: Imagine New York
The Lower Manhattan Preservation Fund
The campaign for community-based planning
Olympics 2012

One

Micaéla Birmingham

The Municipal Art Society (MAS) is a private, nonprofit membership organization whose mission is to promote a more livable city. Since 1893, MAS has worked to enrich the culture, neighborhoods and physical design of New York City. MAS advocates excellence in urban design and planning, contemporary architecture, historic preservation, and public art. Among the organization's successes are the preservation of Grand Central Terminal, Times Square's bright lights and signage, and the temporary memorial called Tribute in Light.

MAS uses GIS to promote strategies to enhance quality of life in New York's many diverse neighborhoods. The Planning Center of MAS provides technical support to community-based organizations to create their own GIS projects and to develop unique neighborhood-level datasets for planning and analysis.

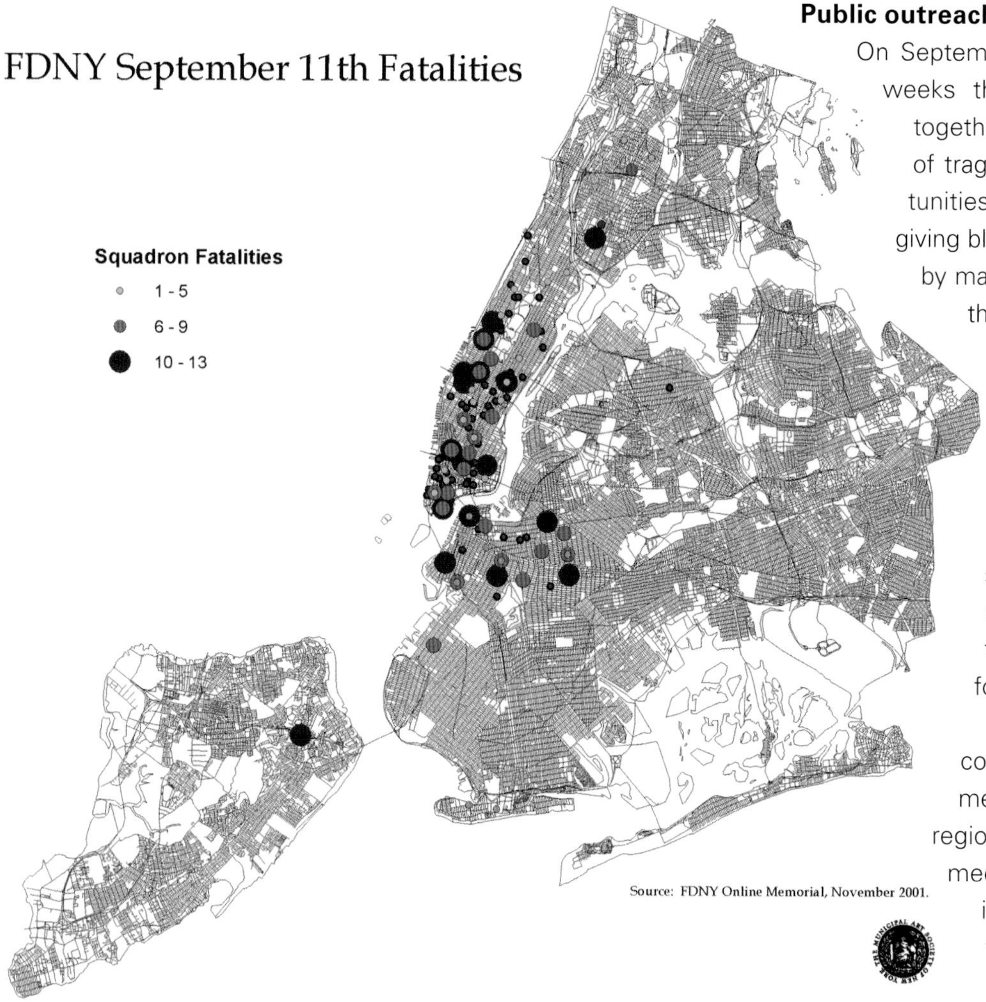

FDNY September 11th Fatalities

Squadron Fatalities

- 1 - 5
- 6 - 9
- 10 - 13

Source: FDNY Online Memorial, November 2001.

Figure 1

Public outreach: Imagine New York

On September 11, 2001, and in the days and weeks that followed, New Yorkers came together as no one imagined. In the face of tragedy, New Yorkers embraced opportunities to make a difference, whether by giving blood, by cheering on rescue workers, by making donations, or simply by making themselves available to help a neighbor in need.

In the continued spirit of this new civic engagement, MAS and a large network of partners sponsored *Imagine New York,* a project to encourage the public to share its ideas and vision for memorializing both the World Trade Center tragedy and the city's response, and for rebuilding downtown.

MAS approached cultural, civic, and community organizations, and government entities throughout the tri-state region, and asked them to host public meetings where people would discuss ideas for the future of the site and the region.

This initial call to action was met with a dramatic response, and the project grew rapidly from a 45-member steering committee at the beginning of February 2002, to nearly 200 project partners in a month. In all, more than 3,000 people participated in 230 public workshops and design charrettes, contributing to the collection of more than 19,000 ideas.

Because the goal was to include the ideas of as many people as possible from all walks of life in the metropolitan region, outreach strategy was critical. GIS allowed for the analysis of both the geographic extent of contact efforts, and the demographic characteristics of participants' communities. Using data that detailed the locations of fire and police stations which had lost significant numbers of their uniformed officers, impacted neighborhoods could be mapped, and outreach efforts focused on them (figure 1).

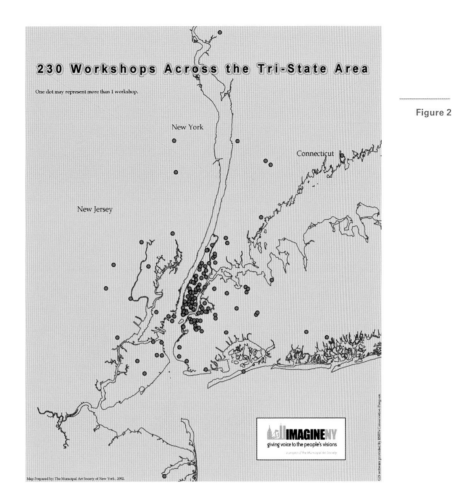

Figure 2

GIS also allowed for an analysis of the neighborhood demographics of *Imagine New York* participants. By overlaying the locations of local *Imagine New York* workshops with Census data, MAS was able to promote a diversity of public representation by ensuring that workshops were located in low-, moderate-, and high-income communities from a variety of ethnic backgrounds. The overall demographic distribution of *Imagine New York* participants closely compares to the demographic breakdown of the New York metropolitan region (figure 2).

GIS was also critical to the actual work of participants during workshops, when they shared in small groups their ideas about rebuilding New York City. To illustrate these ideas, each group was given basemaps of the World Trade Center site and of Lower Manhattan in which they could draw, sketch, and describe their ideas in real geographic terms, with the help of a trained facilitator. These basemaps were key to the envisioning process that led to the creation of thousands of brilliant site plans, transportation enhancements, tri-state link-ages, and pedestrian amenities.

The Lower Manhattan Preservation Fund

The destruction of the World Trade Center complex caused severe damage to buildings in the immediate vicinity and even to many historic buildings several miles away. In an effort to help repair and clean these affected historic structures, MAS created the Lower Manhattan Preservation Fund, which gave financial support for repairing affected buildings within the boundaries specified by the Federal Emergency Management Agency (FEMA). Using a database of building locations and survey information from preservation inspectors, and using GIS, MAS mapped all the historic structures in lower Manhattan. Having all of these

Figure 3

BUILDINGS THAT MATTER

structures identified on a map, with notations about their status and location, made fund allocation and field inspections efficient and effective.

A database of damaged, destroyed, and threatened structures was obtained from Preservation Fund administrators and geocoded by address to the building lot centroids. The geocoded theme was then overlaid with a polygon theme of building footprints. By associating the attribute information of the two themes a new shapefile of the affected buildings could be created. To date the Preservation Fund has supported twenty-three buildings (figures 3 and 4).

Figure 4

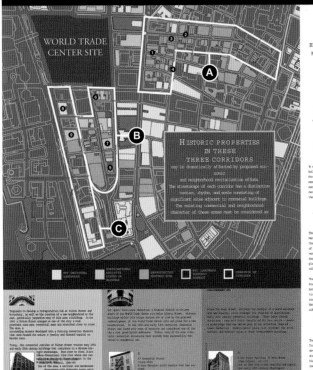

The campaign for community-based planning

Community-based planning describes the participation of local residents, neighborhood organizations, and advocacy groups to create a locally initiated urban plan for their community. The goals of the MAS campaign were to develop a platform to support the development and implementation of community-based planning activities, and to promote structural and legal mechanisms to institutionalize community plans within New York's planning framework—historically a top-down planning structure.

GIS was used to describe where community-based planning is taking place in the city as a whole, and also to show the intricate details of smaller site plans (figure 5). Borough-wide maps describe where plans fall along political boundaries. For communities that did not have the ability to create their own maps of planning regions, MAS created site plans using the ArcMap application, which community groups continue to use in their planning processes (figure 6).

Figure 5

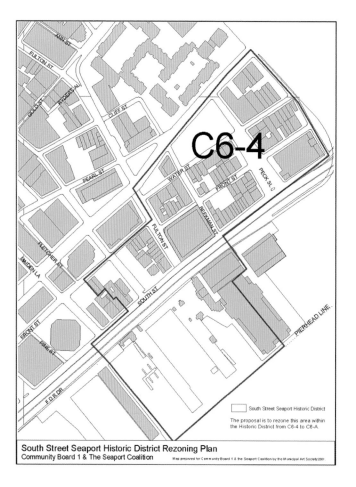

South Street Seaport Historic District Rezoning Plan
Community Board 1 & The Seaport Coalition   Map prepared for Community Board 1 & the Seaport Coalition by the Municipal Art Society 2001.

Map ID	Organization Name
Bronx	
BX1	The Bronx Center Steering Committee, The Urban Assembly
BX2	The Point CDC, Youth Ministries for Peace and Justice
BX3	Sustainable South Bronx/The Point CDC
BX4	Bronx CB 3
BX5	Nos Quedamos Committee, NYC HPD & DCP
BX6	The Residents of Claremont Village
BX7	West Farms Task Force, Phipps CDC
BX8	Jerome Park Conservancy
BX9	Bronx Borough President's Office
BX10	Cherry Tree Assoc.
BX11	Mid-Bronx Seniors Citizens Council
BX12	Mount Hope Housing Company
BX13	PROMESA
BX14	Banana Kelly Community Improvement Assoc.
BX15	Mid-Bronx Desperados Community Housing Corp. CCRP
BX16	Bronx CB 8
BX17	Design Trust for Public Space, Bronx CB 3, Trust for Public Land
BX18	Mothers on the Move
BX19	The Parks Council (now New Yorkers for Parks)
BX20	SOBRO & 138th Street Merchants and Professionals Assoc.
Brooklyn	
B1	Brooklyn CB 1
B2	Brooklyn CB 1
B3	Brooklyn CB 6
B4	Brooklyn CB 7
B5	Ridgewood Bushwick Senior Citizens Council
B6	The Gowanus Dredgers Canoe Club
B7	5th Avenue Committee, Brooklyn CB 6
B8	Technical Advisory Committee, NYC DCP & DOT
B9	Myrtle Avenue Revitalization Project LDC
B10	Brooklyn CB 2
B11	Brooklyn Bridge Park Development Corp.
B12	Brooklyn CB 3
B13	Astella Development Corp.
B14	Southwest Brooklyn Industrial Development Corp.

Map ID	Organization Name
Manhattan	
M1	Manhattan CB 1
M2	Manhattan CB 1
M3	Manhattan CB 2
M4	Hell's Kitchen Neighborhood Association
M5	Manhattan CB 4
M6	Friends of the High Line
M7	Manhattan CB 9
M8	Manhattan CB 11, CIVITAS
M9	CIVITAS
M10	CIVITAS
M11	Manhattan CB 11
M12	Manhattan CB 6, East Midtown Coalition for Sensible Development
M13	Friends of NOHO Architecture
M14	WEACT & Manhattan CB 9
M15	Manhattan CB 10
M16	E. 79th Street Neighborhood Assoc.
M17	Alliance for Downtown New York, Inc., Manhattan CB 1
M18	Manhattan CB 6
M19	Greenwich Village Society for Historic Preservation
M20	Manhattan CB 7
M21	Asian Americans for Equality
M22	Cooper Square CDC
M23	Hell's Kitchen Neighborhood Assoc.
M24	Institute for Rational Mobility
Queens	
Q1	Greater Jamaica Development Corp.
Q2	Queens CB 6
Q3	Queens CB 7
Q4	Queens CB 11
Q5	Queens CB 14
Q6	Office of the Queens Borough President
Q7	Hunters Point Community Coalition
Q8	Office of Tony Avella, 19th Council District
Staten Island	
SI1	Fox Hills Tenant Association
SI2	North Shore Waterfront Greenbelt
SI3	Staten Island CB 3
SI4	Staten Island Civic Associations (see plan)

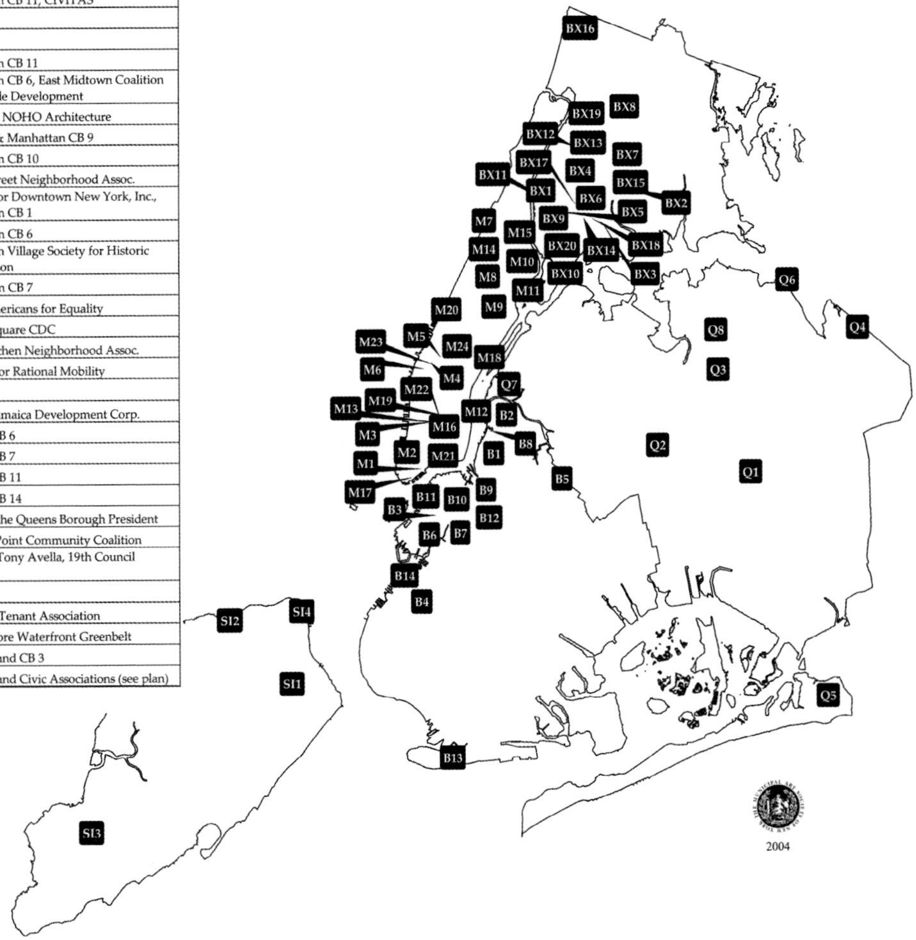

Figure 6 Community-based planning in New York City.

As part of this campaign, the Municipal Art Society published *Planning for All New Yorkers: A Briefing Book of Community-Based Plans*. It is the first-ever compilation of community-based plans from across the city, and includes plans from both discrete geographic areas and from citywide or boroughwide plans that focus on common issues. The *Briefing Book* is an ever-growing body of evidence of communities' successes in planning their own destinies by way of preserving, housing, revitalizing, and greening their neighborhoods.

GIS analysis of community-based planning recommendations

Not only does the *Briefing Book* itself add a critical planning issue to the political agenda and provide a wealth of information to community planners, GIS analysis conducted on the contents of the *Briefing Book* revealed trends in community-based planning across New York. The goal of this analysis was to examine common themes raised in the recommendations of plans contained within the book and to identify patterns in the geographic distribution of plans and their associated demographic character. This analysis provided the basis for spatial visualizations, which, through a series of maps, modeled ways that New York City would change in the event that these plans were implemented. All of the plans in the *Briefing Book* were entered into a database that recorded the plan location and community organization information, and identified the community issues raised by the plan, including

Figure 7

Community-Based Economic Development Recommendations & New York City Median Income

74%
of Economic Development Recommendations are from Low or Moderate Income Neighborhoods

Up to 26 of the 51 plans in the Briefing Book include economic development recommendations, ranging from policy recommendations such as industrial retention to increased job training activities and the strengthening of commercial corridors. The most common policy recommendation, found in 39% of the plans, is attraction and strengthening of compatible businesses in neighborhoods undergoing revitalization and / or transition from manufacturing to other uses. The promotion of high performance industry is a priority in some of these neighborhoods.

●   Reinforce Commercial Corridors

◉   Create an Economic Development/Revitalization Zone

●   Support Local Retail

○   Promote High Performance Industry

Median Income 1990 (in Dollars)

	0 - 13000
	13001 - 24000
	24001 - 35000
	35001 - 46000
	46001 - 70000
	70001 - 150001
	Council District Boundary

Sources:
US Census 1990. US Bureau of the Census, 1990.
Planning for New Yorkers: A Briefing Book of Community-Based Plans.
The Community-Based Planning Task Force, 2001.

For the purposes of this analysis, low- or moderate-income neighborhoods
refers to census tracts where the median household income is less than $24,000.

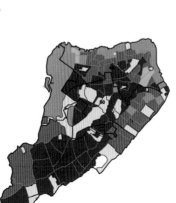
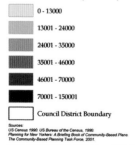

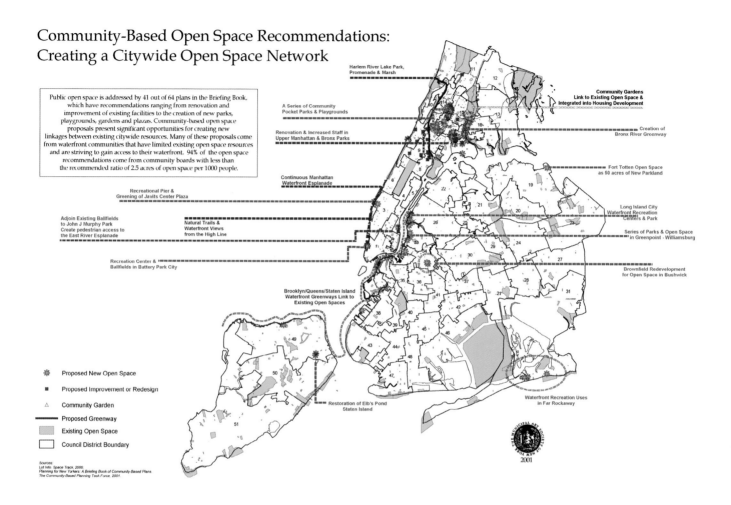

Community-Based Open Space Recommendations: Creating a Citywide Open Space Network

Public open space is addressed by 41 out of 64 plans in the Briefing Book, which have recommendations ranging from renovation and improvement of existing facilities to the creation of new parks, playgrounds, gardens and plazas. Community-based open space proposals present significant opportunities for creating new linkages between existing citywide resources. Many of these proposals come from waterfront communities that have limited existing open space resources and are striving to gain access to their waterfront. 94% of the open space recommendations come from community boards with less than the recommended ratio of 2.5 acres of open space per 1000 people.

Figure 8

environmental conservation, housing, transportation, and economic revitalization. The plan locations database was then geocoded to the New York City Linear Integrated Ordered Network file. With a shapefile of the plan locations, data queries were performed to select the geographic areas with common planning recommendations.

These queries allowed the depiction of neighborhoods with common local issues. These issue-based maps were then overlaid with income shapefiles from the Census Bureau to identify demographic trends in communities initiating their own plans (figure 7).

For issues such as open-space and waterfront access, parcels of land proposed for new park or waterfront development were digitized and geocoded (figure 8). These features were overlaid with a database of existing parks and open space provided by the NYC Department of Finance, in the Real Property Assessment Database (RPAD). This enabled the identification of relationships among existing and proposed waterfront and open-space access points.

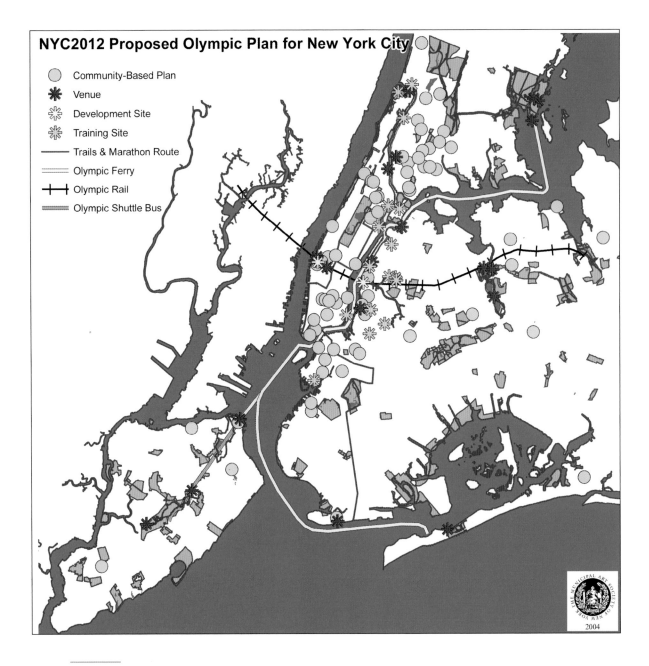

NYC2012 Proposed Olympic Plan for New York City

Community-Based Plan
Venue
Development Site
Training Site
Trails & Marathon Route
Olympic Ferry
Olympic Rail
Olympic Shuttle Bus

Figure 9

When viewed on a citywide basis, the community-based proposals created the framework for an overall comprehensive open-space plan.

Similarly, GIS enabled the visualization of the impacts of community-based planning on the green landscape of neighborhoods. By digitizing parcels of land that communities were proposing for new open space using ArcMap edit tools, estimates could be made of the total number of new acres of open space that would be created if planning recommendations were implemented.

By summarizing the total number of proposed park acres in each community board, and comparing this to the existing acres of open space, we were able to see the changes these recommendations would have on communities.

Interestingly, this GIS analysis concluded that implementing community-based plans would enable several low-income neighborhoods in the south Bronx to match or exceed the city-recommended standard for open space. Previously, they had had a substandard ranking.

Olympics 2012

In New York's bid for the 2012 Olympic Games, MAS used GIS to address the need for a link between large-scale Olympic planning and locally based initiatives. All of the proposed Olympic-related development sites, training sites, and athletic venues were digitized with ArcMap. Overlaid on this polygon layer was a point layer showing the locations of more than sixty-five community-based plans maintained in the MAS database. Potential Olympic development locations were linked to the locations where community organizations have been planning the future of their neighborhoods. Five clusters of overlap were identified and used as a basis for roundtable discussions between Olympic bid representatives and community planning organizations, with the goal of identifying commonalities in both the city's Olympic vision and the individual community's vision (figure 9).

These intimate workshops of ten to fifteen people allowed Olympic-bid organizers to get an insight into these neighborhoods in a way they previously had been unable to achieve from the traditional method of presentations at community board meetings. GIS maps helped community leaders from Jackson Heights and East Elmhurst in Queens to coordinate a plan with Olympic organizers to avoid closures of sections of Flushing Meadow Park that would have conflicted with the annual July 20 Colombian Independence Day festival, an important cultural event that brings in significant revenue for the community.

About the authors

Gregory Jost is the deputy director of University Neighborhood Housing Program (UNHP). He manages various projects and edits the newsletter and Web site (including the Community Resource Guide) for the organization. He also provides technical assistance to other nonprofits regarding housing development, research (statistical and property), and mapping. Before joining UNHP, he worked as the Housing and Relocation Specialist for Newark Emergency Services for Families, Inc. in New Jersey. He graduated from Fordham University with a Bachelor of Arts degree in Urban Studies and is working on a Master of Arts degree in Humanities and Sciences at Fordham.

Patrick Logan directs the Fordham Bedford Housing Corporation's Office of Policy and Planning. This office works to increase the Housing Corporation's impact on affordable housing, education, transportation and open space. He graduated from Fordham University with a Bachelor of Arts degree in Communications and a Master of Arts degree in International Political Economy and Development.

Michelle Whetten is the deputy director of The Enterprise Foundation and guides its neighborhood initiatives for the New York region. She manages child care, workforce development, community safety, organizational development, and neighborhood planning projects and oversees private and federal grants and contracts. Before joining The Enterprise Foundation, she was Program Coordinator at the Neighborhood Design Center, a Baltimore-based nonprofit organization. Michelle graduated from the University of California, Davis with a Bachelor of Science degree in Environmental Policy Analysis and Planning and from the University of Illinois with a Master of Science degree in Urban and Regional Planning.

GIS Use *by* Urban Nonprofit Organizations *for* Housing Initiatives *and* Urban Services

Fordham Bedford Housing Corporation, University Neighborhood Housing Program, and the Enterprise Foundation

Gregory Jost

Patrick Logan

Michelle Whetten

Two Bronx-based nonprofit organizations use GIS independently in a number of ways that previously would have required the services of expensive consultants. The Fordham Bedford Housing Corporation (FBHC) and the University Neighborhood Housing Program (UNHP), partners on a number of affordable housing initiatives over the years, have successfully deployed desktop GIS to further their goals and have done so on a limited budget.

FBHC, an owner and manager of seventy buildings with 1,800 apartments for low- and moderate-income families in the northwest Bronx, began mapping properties by street address in a GIS in 2001, with technical assistance from the Enterprise Foundation. These basic maps were clean and easy to update and to export as graphics. Subsequently, as technology improved and better data files became available through the City of New York's Department of City Planning, these maps were built on blocks and lots, rather than street addresses. This resulted in a better visual and spatial representation of the properties owned by FBHC and of other neighborhood attributes.

A primary function of these maps has been to help make local improvements to transportation infrastructure, green spaces, and affordable housing development. Maps that highlight long stretches of a busy eleven-lane roadway in need of crosswalks and traffic signals, that draw attention to pedestrian walkways in disrepair, and that plot known drug-dealing locations in a neighborhood.

UNHP began working with mapping technology in 2000 with the creation of the award-winning, interactive Community Resource Guide (CRG) on its Web site, *www.unhp.org.* The guide was created to assist community development staff, primarily in the Bronx but also in other areas of the city, to quickly locate useful community-specific data online. The mapping component of CRG provides users with maps that are otherwise not readily available online.

These early, rudimentary maps of ZIP Code areas and community districts were created with free ArcExplorer software from ESRI; because of their lack of detail, they were probably useful only to someone already familiar with the area. But in 2001, a technology grant from the Enterprise Foundation allowed UNHP to upgrade computer systems, mapping software, and geographic data files. The group purchased ArcGIS 8 in 2002 at a nonprofit discount from ESRI. This significant software upgrade, however, came with a considerable learning curve. Training by ESRI and months of experimentation and practice were necessary to gain a level of functionality with the software.

Despite the steeper learning curve, the benefits of the new mapping capabilities have been significant. The maps on CRG have been completely revamped, improved, and expanded. Currently, maps are available for each of the five boroughs as both image files and as PDFs, and they show each borough by ZIP Code, community district, census tract, school district, and police precinct, and by city council, state assembly, state senate, and congressional district. Users can identify their communities based on these various geographical boundaries and then locate the data online using CRG.

Another major benefit of the expanded mapping capabilities has been the impact on the joint effort between UNHP and FBHC in strategic community planning efforts, such as the Fordham Community Action Plan (FCAP). The FCAP combines the community-development and planning efforts of both organizations for a specific area, called the focus area, identified from mapping past, current, and future affordable housing and community development efforts. The current FCAP map (figure 1) based on a tax block and lot map file from the New York City Department of City Planning, linked to existing databases, includes buildings that FBHC owns and manages, affordable housing projects to which UNHP has lent money for acquisition or improvement, potential affordable housing projects that the groups have surveyed for feasibility, and current public infrastructure improvements that local organizations have initiated. Overlaid with these buildings are neighborhood attributes such as parks, schools, and other institutions.

In addition to in-house use, the maps are a critical element of presentations made to gain support for FCAP efforts. They complement before-and-after photographs of past development work, architectural drawings of current and future development efforts, and demographic and other statistical information about the focus area. The resulting presentation is a virtual tour through neighborhoods and through time, serving to preview an actual tour.

GIS maps are also helping make the case for more affordable housing. The two organizations are hoping to convince the Department of Housing and Urban Development (HUD) to allow nonprofits (with local government as a partner) to acquire foreclosed HUD homes at a discount. These homes can then be substantially renovated and sold to low- and moderate-income owner–occupants. As part of the process of analyzing the need and feasibility for such a program in New York City, the groups acquired a list of potential properties from HUD. Mapping this housing data showed a startling concentration of homes in the northeastern end of the focus area that were headed for foreclosure and HUD ownership. Other local groups have been using this data to demonstrate the need for the institution of a similar program.

GIS is also helping to create an early-warning system for housing. The proposed Multifamily Foreclosure Prevention Clearinghouse project incorporates a mappable database of Bronx multifamily properties with five or more units, as well as data downloaded from the departments of Finance, Buildings, and Housing Preservation and Development. The data contains common indicators of housing in danger of foreclosure, such as administrative violations and liens the city holds against the property. With this information, UNHP will be able to contact mortgage holders and attempt to resolve problems before foreclosure. This Building Indicator Project map also allows certain properties to be highlighted if they are in close proximity to other at-risk buildings.

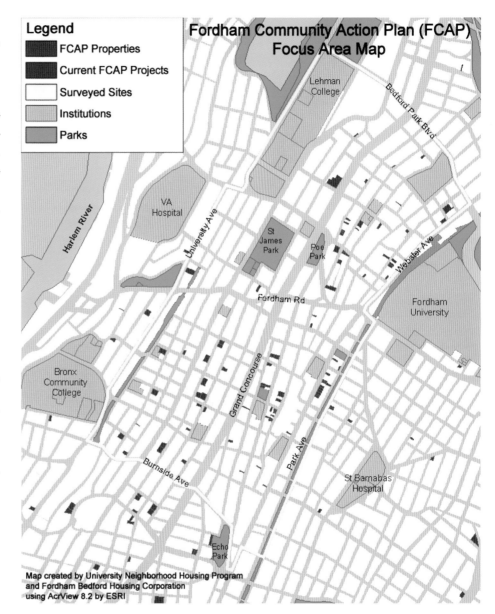

Figure 1

Being digital, the maps created today are a marked improvement from those sketched thirty years ago by community organizers who helped create the UNHP and FBHC. Those early maps, with their handwritten notations and idiosyncratic symbols contained valuable information about the local housing stock and were an essential tool in the community organizer's arsenal, employed for winning the millions of renovation dollars that helped save the community from the abandonment that had overcome south Bronx neighborhoods. This work continues with GIS mapping adding ever more dynamic data to make today's case for community development.

The Enterprise Foundation Programs

Since 1987, the Enterprise Foundation has developed more than 16,000 units of affordable rental and homeownership housing in New York City, with more than eighty-five community-based partners. With a map file of all of the housing and community facilities (some shown in figure 2) that we have helped finance or develop over the years—which incorporates LotInfo, a property-based GIS application for New York City; and Maptitude, a GIS software package—the foundation creates maps every week for a variety of purposes, including:

Figure 2

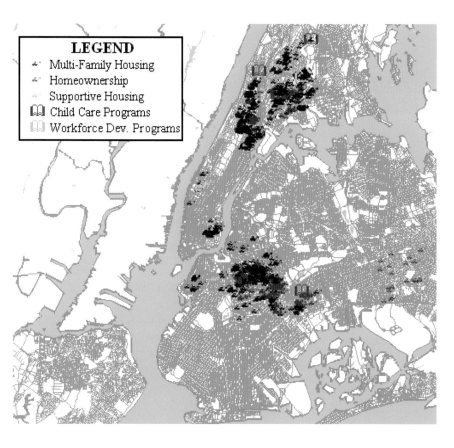

LEGEND
- Multi-Family Housing
- Homeownership
- Supportive Housing
- Child Care Programs
- Workforce Dev. Programs

Tour maps

The foundation frequently takes funders, board members, and staff members on tours of neighborhoods where it has made a significant investment. A map, such as the one in figure 3, is created for each tour identifying the bus route, location of Enterprise Foundation projects, and tour stop.

Public relations

Occasionally small maps in newsletters and brochures are inserted to show the location of a particular project or the location of all the foundation's buildings in the city.

Project investment committee packages

Foundation project managers prepare underwriting packages that are reviewed by an investment committee prior to a loan or equity investment being made in a housing project or community facility. For each project, a map is prepared showing the location of the proposed project in relation to other Enterprise Foundation buildings, and to other assets such as parks, hospitals, subways, and schools.

Board presentations

At an annual board meeting where accomplishments and programs are discussed, maps allow board members to see where work has been done the previous year.

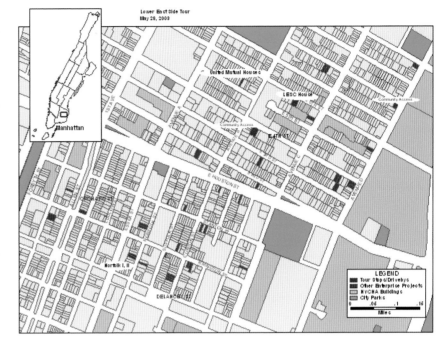

Figure 3

Neighborhood assessments

The Enterprise Foundation has developed neighborhood assessment software called Neighborhood Survey Pro. Using personal digital assistants (PDAs), residents and other volunteers use the software to inventory property and building conditions. The information is uploaded to a Microsoft Access database and then mapped in a variety of ways to show building vacancies, ownership, maintenance status, and other attributes.

Other projects

The Enterprise Foundation managed a welfare-to-work program that involved moving 750 individuals from substandard housing to new apartments. GIS software and geographic data showed where the participants originally lived and where they moved. By adding a foundation projects layer, maps could then identify community-based organizations in the vicinity that could be a resource to these new residents.

About the author

Zvia Segal Naphtali is an adjunct clinical professor of public administration at New York University's Wagner School and teaches courses on spatial analysis and statistical mapping, including "GIS and Health Care." She is the founder and president of Resource Mobilization Inc., a consulting firm established in 1982 that specializes in statistical and demographic data analysis and GIS mapping applications. She received her PhD in sociology from New York University and is the author of the ESRI Virtual Campus course, "Mapping for Health Care Professionals Using ArcView 3.x." She is a GIS presenter at conferences and universities in the United States and Israel.

Delivering Health-care Services *to an* Urban Population

Access to health care in Greenpoint–Williamsburg, Brooklyn, New York

Zvia Segal Naphtali

Background

In this case study, focusing on the Greenpoint–Williamsburg neighborhood of Brooklyn, New York, we will see that the location of health services, such as hospitals and clinics, is a key factor affecting accessibility to care. We will also see that GIS plays an important role in evaluating health care delivery. The *Dartmouth Atlas of Health Care* argues convincingly that in health care, geography is destiny. Variations in the supply of health-care resources, in the physician workforce, and in the quality of care have profound impacts on the health of the population. A study by the New York City Department of Health and Mental Hygiene called "Health Disparities in New York City," reported that substantial inequalities exist among New Yorkers of different racial/ethnic groups. Poor New Yorkers, as well as African-American and Hispanic residents, bear a disproportionate burden of illness and premature death (Karpati et al. 2004). The study noted that factors such as poor access to medical care, unhealthy behavior, and poor living conditions are common among certain economic and racial/ethnic groups. This case study is an example of the health-care research conducted by Resource Mobilization Inc.

As chronicled on the New York City Department of City Planning's Web site, Greenpoint–Williamsburg developed more than one hundred years ago during Brooklyn's great industrial age, when both sides of the East River were dominated by large factories, oil refineries, and shipyards. Along the waterfront, workers' homes and factories intermingled, setting a pattern of mixed use that still shapes these neighborhoods. Over the years, Greenpoint–Williamsburg grew and adapted to changing economic conditions. The refineries and shipbuilders disappeared, and new generations of businesses, entrepreneurs, artists, and residents emerged. Today, these neighborhoods are once again vibrant communities, from the bustling commerce of Manhattan and Bedford avenues to the many distinctive side streets. The waterfront, however, remains largely derelict, dominated by empty lots and crumbling structures, and almost entirely inaccessible to the public (New York City Department of City Planning). A comprehensive rezoning of Greenpoint–Williamsburg, approved in May 2005, set the stage for revitalizing a vacant and underused stretch of this Brooklyn waterfront. The plan created opportunities for new housing and further commercial development.

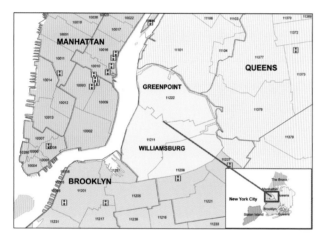

Figure 1 **Access to health care: Hospitals near Greenpoint–Williamsburg, Brooklyn.**

Hospital usage in Greenpoint–Williamsburg

In this case study, we sketched a health care profile of the Greenpoint–Williamsburg neighborhoods by focusing first on hospital usage. The Greenpoint–Williamsburg neighborhood is located within Community District 1 of Brooklyn. Greenpoint borders the Borough of Queens. Both Greenpoint and Williamsburg are located just across the East River from lower Manhattan. The analysis presented here is limited to two ZIP Codes (11211 and 11222) which include all of Greenpoint and most of Williamsburg. Williamsburg extends farther south and east and borders Bushwick and Bedford–Stuyvesant. Figure 1 shows there are no hospitals in the Greenpoint–Williamsburg area. There is a cluster of hospitals in Manhattan, across the East River. There are hospitals south of Greenpoint–Williamsburg in Brooklyn and farther away in Queens.

Although plans were undertaken to improve the quality of health services at one of the nearby hospitals—Woodhull Medical and Mental Health Center—redevelopment did not address the health-care needs of the people living in Greenpoint–Williamsburg. Furthermore, there were no plans to build new hospitals. Two satellite health-care clinics of the Woodhull Medical Center were established in the Greenpoint–Williamsburg neighborhood. They were the Manhattan Avenue Family Health Center in Greenpoint at 960 Manhattan Avenue and the Williamsburg Health Center/Health and Hospitals Corporation at 279 Graham Avenue in Williamsburg. Other health clinics in Williamsburg included Bedford Medical Family Health at 100 Ross Street, the Department of Health Child Health Station at 333 Roebling Street, and the Williamsburg Family Service Center.

Where do the residents of Greenpoint–Williamsburg go when they need a hospital? Working with GIS, we can investigate this question and others: Are people willing to cross the East River to Manhattan in search of a hospital? Or are the hospitals in Brooklyn the preferred choice? What happens in a disaster or in emergency conditions?

Transportation to hospitals

The next question to ask is "how accessible are the hospitals in other parts of Brooklyn, Manhattan, and Queens to residents of Greenpoint–Williamsburg?" Figure 2, a map of the transportation routes in the area, shows that Greenpoint–Williamsburg residents have easy access to other parts of Brooklyn, Queens, and Manhattan via subways and express highways and fairly good access to Manhattan via bridges and tunnels. Buses could be added to the map because there are many bus routes going through this area. A closer analysis of proximity-to-care issues would require a model of "travel to treatment" that goes beyond proximity as straight-line distance. This requires us to incorporate travel routes defined by public transportation networks and travel time, including transfers. However, accessibility measures based on travel time or distance offer only a partial view of access to services. People are willing to travel farther to go to a hospital that more closely meets their needs. Where their doctors practice and their health insurance coverage also affect hospital choice. This case study did not analyze travel time. However, it did analyze the health insurance coverage of residents in New York City and in ZIP Codes 11211 and 11222.

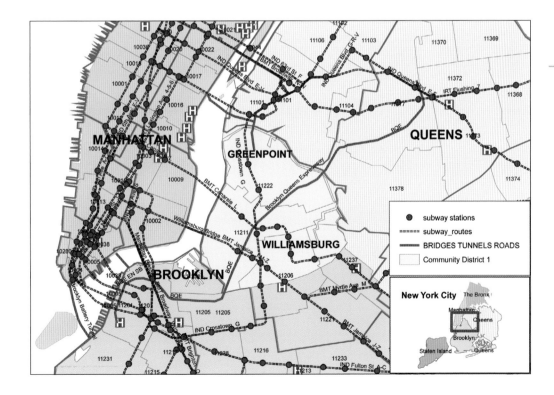

Figure 2   Transportation network Greenpoint–Williamsburg.

Hospital admission patterns

To determine which hospitals Greenpoint–Williamsburg residents go to for their health-care needs, we mapped 2004 hospital admission data for ZIP Codes 11211 and 11222. The data is available from the New York State Statewide Planning and Research Cooperative System (SPARCS). We decided to map each of the two ZIP Codes that make up the Greenpoint–Williamsburg neighborhood.

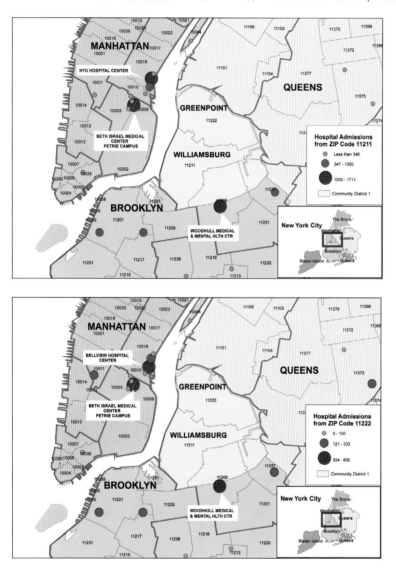

Figure 3   Hospital admissions in 2004 from Williamsburg ZIP Code 11211.

Figure 4   Hospital admissions in 2004 from Greenpoint ZIP Code 11222.

Figure 3 shows the top three hospitals in admissions from ZIP Code 11211 in Williamsburg. There were 1,711 admissions out of 11,447 to Woodhull Medical and Mental Health Center (14.9 percent), and 1,541 admissions to Beth Israel Medical Center's Petrie Campus (13.5 percent). The third hospital shown on the map, NYU Hospital Center, had 1,071 admissions from the 11211 ZIP Code, or 9.4 percent of the total admissions. Clearly residents of this part of Williamsburg preferred to go to Manhattan hospitals instead of those available nearby in Brooklyn, such as Maimonides Medical Center, Wyckoff Heights Medical Center, and Long Island College Hospital. We were surprised to see that a hospital on Staten Island, the Staten Island University Hospital North, also ranked among the top choice of Williamsburg residents with 582 admissions.

As shown in figure 4, the analysis of hospital admissions from ZIP Code 11222 in Greenpoint, which had much fewer hospital admissions than Williamsburg, came up with only slightly different results. Woodhull Medical and Mental Health Center again topped admissions. Total admissions to this hospital was 835 out of the 4,024 patients from ZIP Code 11222 (20.8 percent). The second highest ranking hospital was Bellevue Hospital Center in Manhattan with 370 out of the 4,024 admissions (9.2 percent), and the third was Beth Israel Medical Center's Petrie Campus, also in Manhattan, with 341 admissions (8.5 percent). One hospital in Queens—SVCMC (Saint Vincent Catholic Medical Centers) St. John's Queens Hospital—had 242 of the total admissions (6 percent).

Population distribution

An important geographic pattern to investigate in evaluating or planning health service delivery systems is the distribution of the population that will be receiving the care (Cromley and McLafferty 2002). In this analysis, ZIP Code 11211 had 2.8 times the admissions of 11222 (11,447 versus 4,024). Do the differences in population size or the age distribution between the two ZIP Codes explain the differences in hospital admissions? Answers come from mapping the population and the distribution of young and old residents.

Figure 5 shows that ZIP Code 11211 is more heavily populated and has a higher proportion of children under 5 than its neighboring ZIP Code 11222 and, therefore, it is not surprising that it has many more hospital admissions. (Note that the maps reflect population density.) Also notable are the unpopulated areas shaded with cross-hatch. These are the sites of the Brooklyn Union Gas Company and wastewater treatment plant. Along the western side is the East River waterfront and the piers.

Next, in figure 6, we compare maps of the population over age 65 and under age 5. The data on the population of Greenpoint–Williamsburg is from the 2000 Census and does not reflect the vast changes that have taken place since then. Nonetheless, the maps only partially explain the differences in hospital admissions between the two ZIP Codes. The over 65 population is more evenly distributed over the two ZIP Codes, but there is a marked concentration of children in the 11211 ZIP Code and, as shown in the previous map, large areas of 11222 are unpopulated.

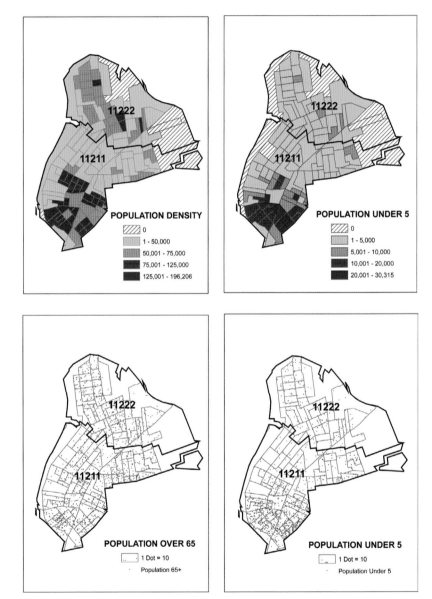

Figure 5 Total population by census tract and population under 5 (normalized by area).

Figure 6 Population by census tract and population over 65 and under 5.

Access to health care and health insurance

The last part of this case study examines health coverage for the population of Greenpoint–Williamsburg and also maps health coverage for New York City as a comparison. A 2003 report by the New York City Department of Health and Mental Hygiene, "The Health of Greenpoint: NYC Community Health Profiles," found Greenpoint residents had less access to health care than the citywide average (Karpati et al. 2003). The report said that 18,000 people (14.5 percent of the total population) had no health coverage, 11,000 did not get needed medical care in the past year, and 32,000 did not have a personal doctor.

Figure 7 examines the geography of health-care coverage in New York City, focusing on Greenpoint–Williamsburg. The map shows the distribution of different types of health-care coverage (private, Medicare, Medicaid, and no insurance coverage) in Greenpoint–Williamsburg and nearby parts of Brooklyn, Manhattan, and Queens. Pie charts show the relative distribution types of coverage.

Figure 8 examines different parts of New York City and contrasting patterns of health care coverage. The East Side of Manhattan (ZIP Codes 10021, 10028, and 10128) includes higher income neighborhoods. The pie charts show a very small segment designating Medicaid and no coverage whereas private coverage and Medicare dominate. Immediately to the north in Harlem, Medicaid and no coverage dominate. The same pattern is apparent in the South Bronx, areas with large Hispanic and African-American populations. Inspecting other parts of New York City would be interesting because neighborhoods with a high percentage of immigrants, such as the western parts of Queens, have a large number of people who lack health coverage.

Figure 7　Health care coverage in 2003 (Greenpoint–Williamsburg).

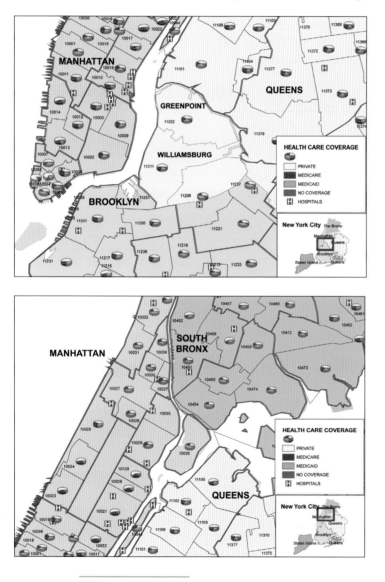

Figure 8　Health care coverage in 2003 (Manhattan, South Bronx, Queens).

Conclusion

This study could be followed up by spatial analysis of additional data. The geographic distribution of health-care access and utilization could be further explored. Hospital admission data covers many categories of illness and hospital choice often depends on the availability of medical and surgical services. Open-heart surgery, for example, is only available in some hospitals. Hospital capacity is another variable. Greenpoint–Williamsburg residents may visit neighborhood health clinics or local physicians' offices, which could be located and mapped. The spatial distribution of residents by demographic and socioeconomic characteristics could also be explored.

References and further reading

Cromley, E. K., and S.L. McLafferty. 2002. GIS and Public Health. New York: Guilford Press

Dartmouth Atlas of Health Care. 1996, 1998, 1999. The Center for the Evaluative Clinical Sciences at Dartmouth Medical School. Hanover, NH. *www.dartmouthatlas.org/atlases/atlas_series.shtm*

Karpati, A., et al. 2003. The health of Greenpoint. Community Health Profiles 21:1–12.

Karpati, A., et al. 2004. Health disparities in New York City. A report from the New York City Department of Health and Mental Hygiene.

New York City Department of City Planning. Projects and proposals, Greenpoint–Williamsburg. *www.nyc. gov/html/dcp/html/greenpointwill/greenplan1.shtml*.

New York City Department of City Planning. 2002. Community Board 1-Borough of Brooklyn, Williamsburg 197-A Plan.

New York City Department of City Planning. 2002. Community Board 1-Borough of Brooklyn, Greenpoint 197-A Plan.

New York City Department of City Planning. 2005. Community District Needs, Fiscal Year 2006.

About the author

Jennifer R. Cox is the Regional Plan Association's GIS manager. Jennifer joined RPA in 2000 to integrate innovative GIS and remote sensing technologies into regional research. She earned a Master of Arts degree in Geography from Hunter College, and a Bachelor of Arts degree in Geography from the State University of New York at New Paltz. Previously, she was an assistant GIS manager in the Division of Natural Resources, at the New York State Department of Environmental Conservation and a GIS specialist at the Town of East Fishkill, New York. Jennifer is a steering committee member for New York City's Open Available Space Information System (OASIS NYC) and is a coordinator of the NYC ARC Users Group. Currently, she is concluding a public access assessment for the Long Island Sound Stewardship System, starting the greenways/bikeways regional inventory for the Healthy Communities Initiative, and launching a regional build-out and alternative growth scenario for the NYC Metropolitan Region.

Natural Habitat *and* Open Space Assessment

Long Island Sound Stewardship System

Four

Jennifer R. Cox

Introduction

Long Island Sound is a place of recreation and relaxation for millions of people who live and work around it. People fish, swim, and boat in the Sound, and also use it as a destination to simply sit and relax. While about 10 percent of the nation's population—more than 20 million people—live within 50 miles of its shores, only 20 percent of that shoreline is accessible to them.

Regional Plan Association (RPA) is a partner with Audubon Society New York, Save the Sound, and the U.S. Fish and Wildlife Service to promote the development of the Long Island Sound Stewardship System. Our goal is to create a network of protected and reclaimed natural areas and parks stretching from New York City to Rhode Island, with the Sound at its heart—a system of protected sites that will preserve the Sound's upland and estuarine natural systems, while providing new recreation and public access opportunities for the people of New York and Connecticut. In order to identify the open-space and public-access priority areas around Long Island Sound, RPA—with the help of many stakeholders—compiled a public access inventory. The inventory identified areas where water-based activities such as fishing, swimming, boating, and hunting take place around the Sound, and secondarily the locations of open-space resources for hiking, biking, and environmental education within the Long Island Sound project area.

RPA also created an open-space and public-access assessment, designed to become a tool for the identification of priority areas and for the further inclusion and prioritization of other high-value areas in the system. The goal of the assessment was to provide a Sound-wide basis for evaluating public access and recreation sites proposed for inclusion in the stewardship system by local, state, and federal agencies, and by other interested organizations. The assessment process, which required changing vector datasets into raster datasets and then making composite maps, involves three steps: establishing the criteria to build the assessment, resolving the priority areas using the assessment as a tool, and accepting public comment.

Assessment methodology

The first step in the public access assessment was establishing the criteria to define high-value areas around the Sound. The criteria include a wide range of datasets, of which three were given high priority: existing parks and protected areas, unprotected open space, and existing private recreation facilities. These are the places that the public already uses to access the Sound.

Each criterion identified in the assessment becomes a map and a GIS layer. Through a series of regional workgroups, RPA and other stakeholders assigned specific values to each criteria map, which in turn dictated the data inventory, collection, and compilation of the geographic information.

The criteria maps are then grouped into four categories based on theme: *Public Access to the Water; Water Resources Protection; Open Space, Cultural, and Other Recreation Resources;* and *Recreational Need.*

Conducting a raster-based analysis, RPA added the weighted values of the criteria maps for each of the four categories, resulting in four summary maps, one for each category. The values for each summary map were then reclassified into an ordinal scale by RPA.

The *Public Access to the Water* category, as shown in figure 1, depicts a composite of high- to low-value areas, showing existing water-based recreational facilities, areas where potential recreational facilities might be suitable, and facilities located closer to the shoreline that had higher water quality and made best use of the water. The assessment summary map depicts a portion of the Connecticut River within the Long Island Sound study area. The criteria used for including areas on the *Public Access to the Water* map included:

✦ Within ¼ mile of the shoreline
✦ Within 100 feet of the shoreline
✦ Water quality based on state classifications for best use
✦ Degree of impairment toward meeting best use
✦ Presence of existing public recreational facilities (swimming, fishing, hunting, boating)
✦ Potential recreational facilities (swimming, sandy beaches, potential fishing access, hunting)
✦ Wetlands and mudflats, boating (bays and coves)

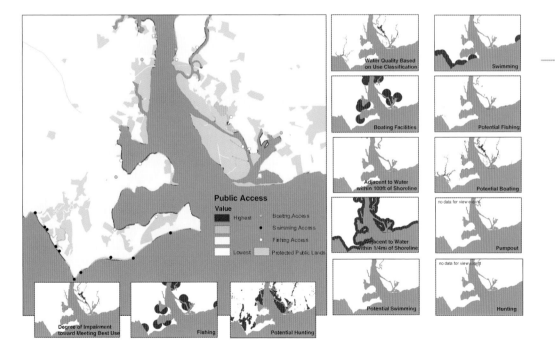

Figure 1

Figure 2

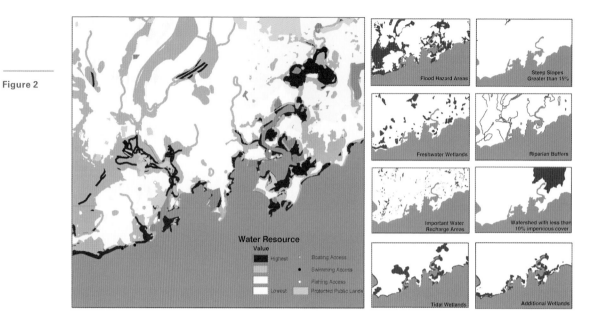

The *Water Resources Protection* category is also a composite of high- to low-value areas. The maps' highest valued areas contain locations of wetlands, riparian areas, and floodplains. Additional value was given to areas of steep slope, of watersheds of less than ten percent impervious land cover, and areas with soils important to groundwater recharge. The assessment summary map, as seen in figure 2, depicts a portion of the western Connecticut shoreline within the larger Long Island Sound project study area. The assessment summary map values are added, then reclassified, to depict the highest valued areas in the study area. The criteria used for the Water Resources Protection included:

+ Tidal wetlands (300-foot buffer) and freshwater wetlands (100-foot buffer)
+ Flood hazard areas
+ Riparian areas (streams and water bodies with 100-foot buffer)
+ Steep slopes (greater than 15 percent)
+ Watersheds with less than 10 percent impervious cover
+ Important groundwater recharge area (up to 3 percent slope and greater than 20 inches per year)

The *Open Space, Cultural, and Other Recreation Resources* category attempts to show additional open spaces in the Long Island Sound project study area. The summary map (figure 3) is a composite of high- to low-value areas made up of existing parklands, unprotected lands (such as agricultural areas, scenic views, and bluffs), and their adjacent areas.

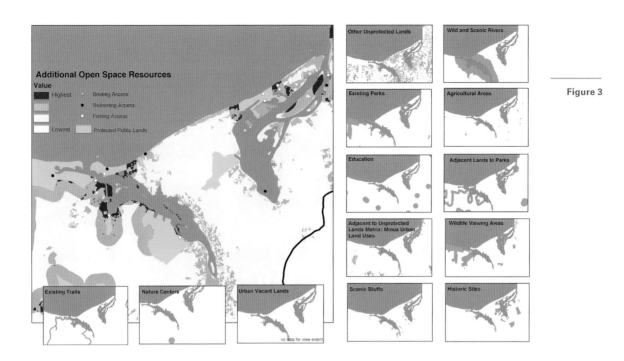

Figure 3

Additional values went to areas with greenways, trails, and bikeways that connect our open spaces, to places close to educational facilities, and to historical and cultural locations, including nature centers. The assessment summary map depicts a portion of the Nissequogue River shoreline along the New York side of the Long Island Sound project study area. The criteria used for the Open Space, Cultural Resources, and Other Recreation Resources included:

✦ Existing parks and public or private conservation areas
✦ Adjacent to existing parks and conservation areas
✦ Unprotected lands matrix
✦ Other unprotected lands
✦ Urban vacant lands
✦ Existing or proposed greenway or bikeway
✦ Designated wild and scenic river
✦ Important agricultural areas
✦ Proximate to nature centers and aquariums
✦ Proximate to schools
✦ Federal and state register historic sites
✦ Scenic bluffs and escarpments
✦ Recognized wildlife viewing areas

Figure 4

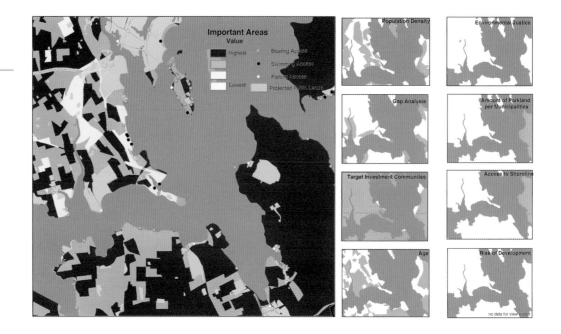

The *Recreational Need* category depicts a composite of high- to low-value areas. Locales with needs for additional public access based on population density and on open-space access per municipality appear as high-value areas. Additional value was placed on areas where a facilities gap was detected, or where rapid population growth and land development are occurring, as well as on special-needs communities, including places of environmental justice concern. Within the assessment, the summary map (figure 4) depicts a portion of the New York City shore, Queens, and the Bronx. The criteria used for the Recreational Need category included:

+ Population density by census block group
+ Current unrestricted access to Long Island Sound by municipality and county
+ Amount of parkland by municipality
+ Targeted investment communities (special funding)
+ Rapid pace of development in municipality (building permits, households, population change)
+ Gap facilities analysis
+ Special needs groups: age
+ Environmental justice community

Priority areas methodology

What are the high-value areas along Long Island Sound for recreational access? Using the public access assessment the stakeholders looked for approximately one hundred recreational hot spots along the Sound that could be included in a stewardship system. RPA set up four regional stakeholder workgroups to find the high-value areas along the Sound for each region. These same workgroups helped identify the assessment's criterion.

The workgroups were tasked with identifying of the high-value areas, comparing the high-value areas on each of the four summary maps with the existing open-space basemap. The basemap (figure 5) contained recommendations identified in other public programs or existing plans. The regional workshop group selected twenty to thirty priority areas, drawing ovals around them.

Once stakeholders identified the high-value areas of the public access assessment, the public was given the opportunity to voice its opinion of the two assessments—public access and ecological—at public hearings. Following the public hearings, the RPA, the U.S. Fish and Wildlife Service, other stakeholders, and the public would use the assessment as a tool to identify sites to be designated for the stewardship system.

Figure 5

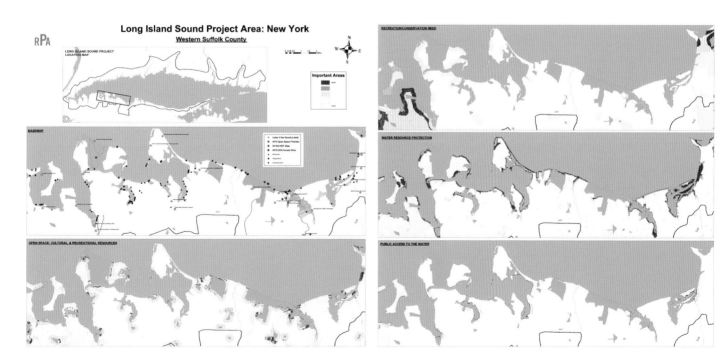

About the author

A native of Miami, Florida, **Karen Kaplan** earned a Bachelor of Arts degree from Smith College with a major in art and design, studied printing technology at Rochester Institute of Technology, and became a certified purchasing manager in 1988. After a career in manufacturing and four years with a global forest products consultancy based in Finland, Karen completed the Lehman College Graduate Certificate Program in Geographic Information Science in 2003, followed by a GIS internship in redistricting in the office of the New York state senate minority leader. She completed a Master of Arts degree in Social Studies education in 2004.

Urban Environmental Planning

Land-use classification for extensive vegetative roof acreage potential in the Bronx

Karen S. Kaplan

Background

The fifth façade—the roof—offers acreage for a range of urban uses: recreation, agriculture, rainwater harvesting, and micropower in the form of generators, windmills, photovoltaic, and fuel cells. Vegetative or green roofs, a long-established technology in Europe, can support many of these uses and also mitigate pressing environmental problems so effectively that, coupled with a paradigm shift in eco-effectiveness, establishing green roofs has become not a question of *if* or *how,* but *when* and *where,* a spatial element that can be answered with GIS. This project primarily used the historical method to accurately record land and roof areas, and classify them into categories meaningful to roof users, with no control of any variable or evaluation of environmental mitigation potential. The objective was to present EVRAP (Extensive Vegetative Roof Acreage Potential) research results in a manner useful to policy and planning decision makers for future roofing options by both the public and private sectors.

Lower levels of vegetation in urban areas have a notably detrimental effect on the environment. "On warm summer days, the air in a city can be 6–8 degrees hotter than surrounding areas. Scientists call these cities urban heat islands [creating in turn, the Urban Heat Island Effect, UHIE.] What causes this to happen? There are fewer trees, shrubs, and other plants to shade buildings, intercept solar radiation, and cool the air by evapotranspiration [the process by which] plants secrete or transpire moisture through pores in their leaves, cooling the [surrounding atmosphere.] A single mature, properly watered tree with a crown of 30 feet can evapotranspire up to 40 gallons of water in a day, which is like removing all the heat produced in four hours by a small space heater" (Pon and Akbari 2002).

There is now a substantial, accepted body of evidence that vegetative roofs contribute to significant mitigation of such pressing urban environmental problems as storm water runoff (SWR), combined sewage overflows (CSOs), the urban heat island effect (UHIE), and air pollution. Green roofs are "prescriptive opportunities for sustainable environmental solutions" (Schultink 1997). Green roofs also provide economic benefits by doubling roof life, and improving urban aesthetics for residents visually stressed by a roof-landscape of tarpaper, concrete, metal, and plastic.

A further impetus is provided by the concept of eco-effectiveness. "Eco-effectiveness is a broad strategy. It engages the idea of an effective economy producing profits for companies . . . while treating people fairly and well and respecting, even celebrating the natural world . . . Imagine a structure . . . that engages with the sun the way a tree does, with a photosynthetic connection, moisture transpiration, habitation by hundreds of species, transformation of microclimate, distillation of water, and production of complex sugars and carbohydrates; that sequesters carbon, fixes nitrogen, and changes with the seasons. Imagine a building like a tree, a city like a forest" (McDonough 2000). The New York City Parks Department supports street trees, but the high value of land precludes reforestation in the traditional sense. New York can join other cities in Europe and the United States, including Portland, Seattle, and Chicago, which are looking to the fifth façade of every building for the surface area needed to bring nature into the built environment. Urban rooftops are the new frontier.

Extensive vegetative roofs employ commercially viable, proven technologies, mostly from Germany, for both new construction and retrofitting, where structural strength and minimal maintenance are constraints. According to the product literature for one company, Hydrotech, "the soil mixture, composed primarily of mineral materials mixed with organic medium, can be very shallow (as little as three inches). The entire system is very light, weighing little more than a traditional ballast roof, allowing for safe installation on almost any existing roof."

While the exact composition of layers may vary by manufacturer, extensive roof gardens work best with low maintenance, drought-resistant succulents such as sedums and other low-growing plants, especially Mediterranean herbs and those adapted to alpine conditions of little soil, no water, high winds, and high sun exposure. Mosses are not recommended because they can create a fire hazard when very dry. Ornamental grasses and bamboos require much greater soil depths. Load restrictions are the main limitation, as is roof slope. "A flat roof requires an additional layer to drain excess water away from the root zone. A roof slope between 5° and 20° works best . . . Roofs with up to a 40° slope can be greened, but slopes between 20° and 40° require a wooden lath grid . . . to hold soil substrate until plants form a thick vegetation mat" (Scholz–Barth 2001).

Virtually all published studies using spatial analysis for solutions to urban environmental issues have neglected New York; they have focused on natural landscape analysis and on the UHIE and potential energy savings from alternate colors of roofs or increased ground level vegetation. Various nonspatial studies credit vegetative roofs with reducing as much as 75 percent of rainwater, mostly by greatly slowing the rate of runoff by several hours as rain percolates through the vegetative system (HUD/FHA 2002). Combined Sewage Overflow (CSO) mitigation and extended roof life are now considered the key economic drivers for vegetative roofs. In the case of the Ford Motor Company, these two rationales provided sufficient cost justification for the installation of a 450,000-square-foot extensive vegetative roof on an assembly plant in River Rouge, Michigan—one built without any government incentives or subsidies.

Green roofs can have a significant impact on the amount of Combined Sewer Overflow (CSO) and Storm Water Runoff (SWR), which become a serious nonpoint source of water pollution when rainfall exceeds one inch per twenty-four-hour period in New York City. The city Department of Environmental Protection is assessing, along with other technologies, the future potential of green roofs to mitigate CSOs. One DEP question is how much EVRAP is needed to avoid CSOs during a twenty-four-hour rain event in each of the two drainage basins that serve the Bronx.

This project could not assess potential for the entire range of buildings because each existing building would need to be defined in the database by criteria such as load-bearing capacity, roof slope, and access. Cumulative calculation of square footage by lot parcel and by land-use class provided the main output, creating a hierarchy of landscapes for further GIS analysis and providing supporting baseline data for current initiatives at the neighborhood and county level for the Bronx.

Figure 1

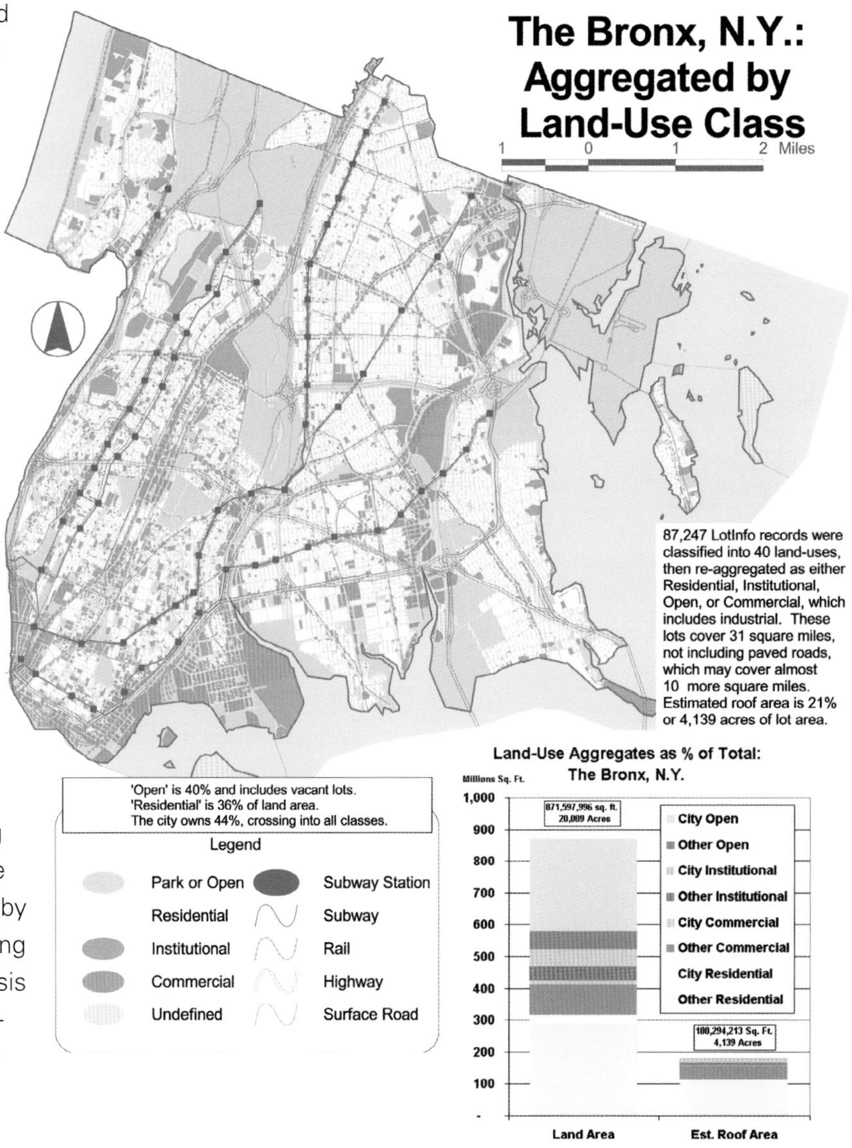

The Bronx, N.Y.: Aggregated by Land-Use Class

1 0 1 2 Miles

87,247 LotInfo records were classified into 40 land-uses, then re-aggregated as either Residential, Institutional, Open, or Commercial, which includes industrial. These lots cover 31 square miles, not including paved roads, which may cover almost 10 more square miles. Estimated roof area is 21% or 4,139 acres of lot area.

'Open' is 40% and includes vacant lots.
'Residential' is 36% of land area.
The city owns 44%, crossing into all classes.

Legend

	Park or Open	●	Subway Station
	Residential	∿	Subway
	Institutional	∿	Rail
	Commercial	∿	Highway
	Undefined	∿	Surface Road

Land-Use Aggregates as % of Total: The Bronx, N.Y.

Millions Sq. Ft.

871,597,996 sq. ft.
20,009 Acres

- City Open
- Other Open
- City Institutional
- Other Institutional
- City Commercial
- Other Commercial
- City Residential
- Other Residential

180,294,213 Sq. Ft.
4,139 Acres

Land Area   Est. Roof Area

Project data sources, and methodology

1   Basemap of the Bronx with water bodies, major surface streets, subway lines and stations, parkways, highway segments, ZIP Codes, and population density (Source: Census 2000 TIGER/Line files)

   a   The Shoreline TIGER/Line file was used to create a Bronx polygon.

   b   LotInfo subway and subway station files are for all five boroughs and were clipped to the Bronx polygon.

   c   There was no area data for highways or paved roads, which may add as much as 10 square miles to the 31 square miles represented by the LotInfo records. The TIGER/Line Street file had to be queried to create individual shapefiles which were then unioned to create interstate highway, parkway, and major surface road themes.

2   Water Pollution Control Plant (WPCP) Drainage Basins (Source: NYC DEP)

3   Land-use data for the Bronx (Source: LotInfo 2002)

   a   There are 87,251 records in the *Bx_lots.dbf* file, which was edited to facilitate queries through deletion of many unnecessary fields and of four other records due to lack of data.

   b   The 87,247 resulting LotInfo records were classified into forty land-use classes and then re-aggregated into four broad classes called *Open, Institutional, Residential,* and *Commercial.* Of these records, 72,488 were classed as residential. Additional queries were done to create files for subclass analysis, including city ownership. The research project generated eleven final layouts. Figure 1 shows the Bronx aggregated by these four broad classes.

   c   Each DBF file was opened in an Excel workbook for spreadsheet analysis and the creation of all charts and tables. The four broad classes were clipped for study areas: the Wards Island WPCP Basin and ZIP Code 10474 for the Hunts Point neighborhood. Figure 2 shows the Wards Island WPCP Basin.

   d   The LotInfo *grosssquare foot* field could not be used because it multiplied the building depth and width by the number of stories. A new field was added to calculate the square footage of the building's footprint.

   e   The actual classifications were based on the LotInfo *LNDUSE_G* and *LNDUSE_D* fields in order to facilitate queries and symbology.

   f   Sanborn maps are now linked by volume and sheet number to LotInfo 2002 and these fields were left in the DBF files. Actual data is separate and was not necessary although Sanborn's digital elevations would be highly desirable for future projects, for three-dimensional modeling of selected buildings to assess roof slope. It was outside the scope of this project to identify load-bearing capacity or replacement roof schedule by building, even if that data were available.

4 Aerial orthophotography to optimize analysis of surfaces and roofs, used in eight of the eleven final layouts (Source: NYCMap)

5 Locator map for Bronx County (Source: ESRI)

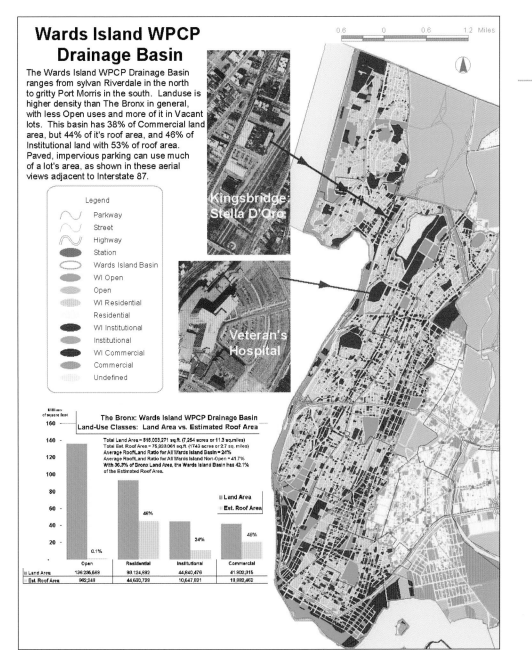

Figure 2

Results

Results show that GIS was an effective tool for the development of a land-use classification database to determine how much roof acreage is available for extensive vegetative roofs in public and privately owned buildings in the selected areas of the Bronx. All calculations of roof acreage are directionally valid estimates.

Spatial analysis provided reliable roof area data and maps that could be meaningful to private real estate development, as well as to public policy and planning initiatives.

Spatial analysis confirmed the intuitive conclusion that warehouses, industrial buildings, and big box retail would have the best EVRAP. The roof area distribution analysis was critical in determining that roof areas of 10,000-square-feet and greater are most meaningful, with a roof area of 1,712 acres, the equivalent of two Central Parks shown in detail in table 1 and in figure 3. Total EVRAP was 4,139 acres, detailed in table 2. Significant EVRAP building classes revealed through the analysis were private elevator buildings, walk-ups, mixed residential, schools, and vacant lots.

Estimated Roof Area 10,000 square feet or more								
	Land area	Land acres	Roof area	Roof acres	Ratio	Avg. stories	# Buildings	# Records
Total Bronx	201,539,643	4,627	74,559,061	1,712	37%	4.4	4,227	3,484
City owned	81,222,479	1,865	14,215,473	326	18%	5.1	612	357
Wards Island	87,265,894	2,003	41,172,478	945	47%	4.9	2,363	2,026

Table 1

Summary of Aggregate Data							
	Land area	Est. roof area	Land acres	Roof acres	Land SQMI	Roof SQMI	Buildings	Records
Open	349,063,707	1,740,253	8,013.4	40.0	12.5	0.1	572	7,870
Residential	314,712,024	113,331,136	7,224.8	2,601.7	11.3	4.1	72,488	69,244
Commercial	110,693,962	45,222,284	2,541.2	1,038.2	4.0	1.6	7,759	8,315
Institutional	97,127,303	20,000,540	2,229.7	459.1	3.5	0.7	2,207	1,818
Total	**871,597,996**	**180,294,213**	**20,009**	**4,139**	**31.3**	**6.5**	**83,026**	**87,247**
	Land area	Est. roof area	Land acres	Roof acres	Land SQMI	Roof SQMI	Buildings	Records
City open	291,637,026	973,609	6,695.1	22.4	10.5	0.0	339	1,866
City residential	24,496,831	2,857,704	562.4	65.6	0.9	0.1	849	462
City Commercial	14,030,780	1,922,921	322.1	44.1	0.5	0.1	155	266
City Institutional	52,120,740	10,823,508	1,196.5	248.5	1.9	0.4	494	389
Total	**382,285,377**	**16,577,742**	**8,776.1**	**380.6**	**13.7**	**0.6**	**1,837**	**2,983**

Table 2

EVRAP: Estimated Roof Area Analysis

Equal area classification analysis revealed that 3,484 lots theoretically have buildings, averaging 4.4 stories, with footprints (or roofs) of 10,000 square feet or more. Estimated roof area averages 37% of the land area of these lots. 55% of this estimated 1,712 acres is in the Wards Island Basin, and 19% is owned by New York City. One possibility would be to green these roofs, without distinction to land-use class.

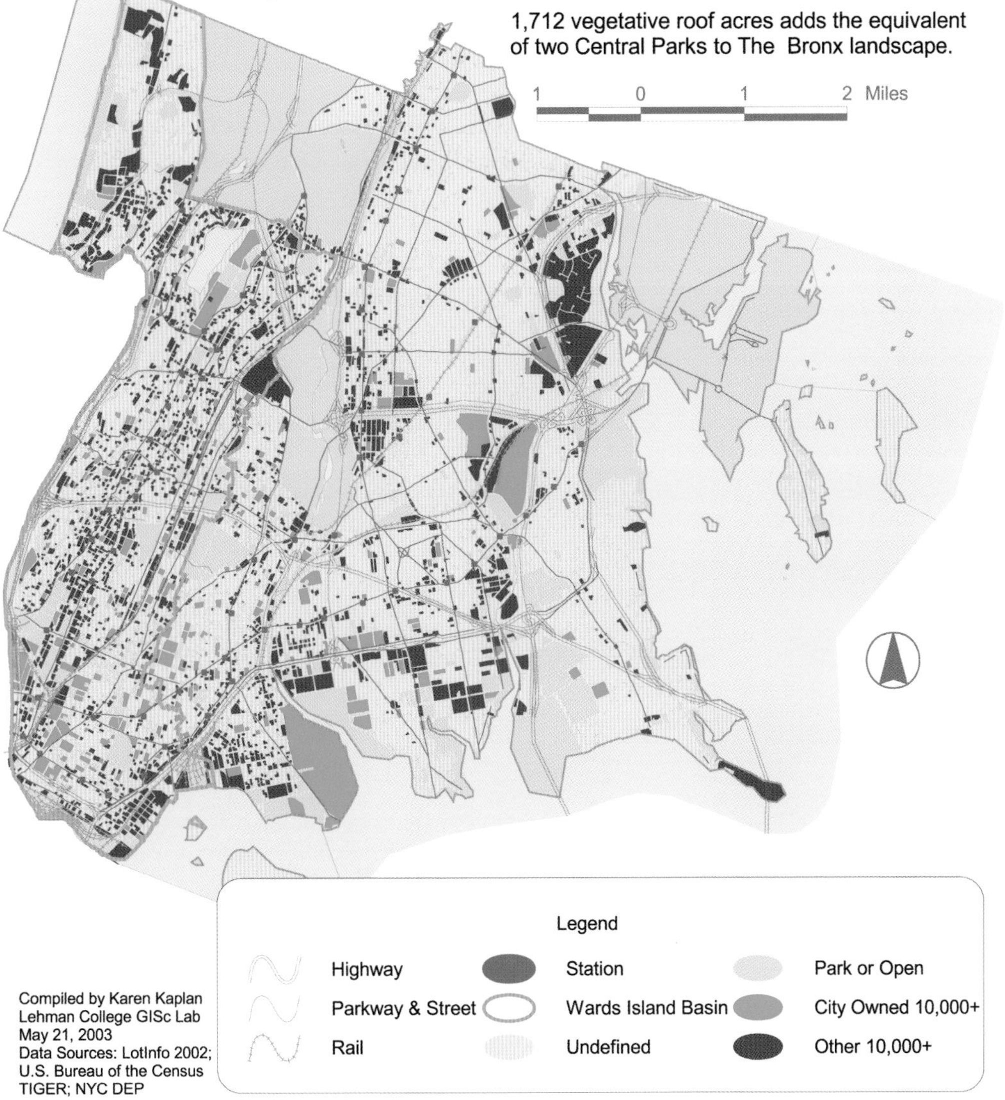

1,712 vegetative roof acres adds the equivalent of two Central Parks to The Bronx landscape.

Figure 3

Compiled by Karen Kaplan
Lehman College GISc Lab
May 21, 2003
Data Sources: LotInfo 2002;
U.S. Bureau of the Census
TIGER; NYC DEP

Legend

～	Highway	⬭	Station	░	Park or Open
～	Parkway & Street	◯	Wards Island Basin	▨	City Owned 10,000+
～	Rail	░	Undefined	⬤	Other 10,000+

Conclusion

There is good EVRAP in the Bronx. The fact that estimated roof area is less than 21 percent of land area means that environmental mitigation strategies need to be specific to areas such as Hunts Point. The best EVRAP classes identified by this project are six-story elevator apartment buildings, schools, factories, warehouses, and any building with a roof area of greater than 10,000 square feet.

The next step would be to use the same process on the other boroughs of NYC, and to refine this data further by using Sanborn roof slope data as well as thermal satellite imagery to identify the worst UHIE.

This project demonstrates the power of GIS spatial analysis to support an important eco-effective technology by providing the analysis to generate visual displays of the "minimum information needed to make informed [public policy] decisions" as Schultinck defined them in 1997.

References and further reading

McDonough, William D., and Michael Braungart. "Eco-Effectiveness: A New Design Strategy" in *Sustainable Architecture White Papers,* New York: Earthpledge Foundation, 2000.

Pon, Brian, and Hashem Akbari et al. 2002. "Existing Climate Data Sources and Their Use in Heat Island Research." Lawrence Berkeley National Laboratory Report 41973. 12 October 2002.

Scholz–Barth, Katrin. 2001. Green Roofs: Stormwater Management from the Top Down, *Environmental Design and Construction.* January/February.

Schultink, Gerhardus. 1999. Environmental Indices and Public Policy: A Systems Perspective on Impact Assessment and Development Planning, *International Journal of Environmental Studies, 56*: 237–258.

Low Impact Development (LID) Practices for Stormwater Management, PATH Technology Inventory (Washington, D.C.: United States Department of Housing and Urban Development, Office of Policy Development and Research, 27 November 2002).

About the author

John C. Ziegler is the founder and president of Space Track Inc., a New York City GIS consulting company founded in 1988. Space Track was one of many New York consulting firms that volunteered to provide GIS support to the disaster response team in the weeks following the 9/11 attack on the World Trade Center. Prior to founding Space Track, John was the New York regional director of the Federal Insurance Administration (FIA) and deputy regional director of the Federal Emergency Management Agency (FEMA). As the FIA's New York Regional Director, he implemented the riverine and coastal floodplain mapping of New York, New Jersey, Puerto Rico, and the Virgin Islands. John holds a Master of Urban Planning degree from New York University, a Master of Architecture degree from Columbia University, and a Bachelor of Arts degree from Harvard College. He is a charter member of the American Institute of Certified Planners (AICP), and a Registered Architect in New York State. From 1989 to 2001, he taught urban planning and GIS as an adjunct professor at New York University's Wagner Graduate School of Public Service.

Emergency Management *and* Disaster Response

John Ziegler

GIS played a vital role in the response to the attacks on the World Trade Center (WTC) in New York in 2001. Unusual demands are placed both on GIS technology and on GIS users when an urban center faces a disaster of the suddenness and magnitude of the WTC attacks. For example, major GIS applications are normally developed over a period of months or years. In the case of this and other disasters, there was an urgent need for GIS product within hours of the attack, and up-to-the-minute updates of GIS mapping were required by responders on a twenty-four-hour basis for many weeks thereafter. In addition, the information and map data needed by responders were extremely diverse and included building plans and satellite imagery, as well as attribute data collected from a wide variety of sources. These conditions posed both technical and institutional challenges that were overcome with remarkable success by the mostly volunteer GIS staff brought together immediately after the attacks. Without GIS, it would not have been possible to meet the ongoing demand for mapped information needed by recovery teams and command centers. Both technical and institutional lessons can be learned from this successful GIS response to an unforeseen disaster.

Background

In September 2001, the City of New York was nearing completion of a citywide GIS basemap project, begun in 1996, that contained more than 100 layers of mapped information, known as NYCMap. Since it was to be shared by all city agencies, the city had established a GIS coordinating committee composed of representatives of city agencies that required GIS products, chaired by the city's Department of Information Technology and Telecommunications (DOITT), which was also the lead agency and central repository for the NYCMap. By 2001, most city agencies were using GIS applications on a day-to-day basis, and had GIS data that had not yet been incorporated into the NYCMap.

Outside city government, an organization of some 300 local GIS users from government and the private sector had been formed in 1992. This organization, known as GISMO (GIS and Mapping Operations), had been meeting regularly to discuss GIS issues, and had maintained a contact list of members and their organizations. GISMO had proved to be an effective means of bypassing official channels to enable its members to share GIS solutions and to keep abreast of developments in the field.

By September 2001, therefore, the groundwork had been laid for a coordinated response to the emergency by the region's GIS professionals, technicians, and aficionados, including students from local universities.

The GIS response

The Mayor's Office of Emergency Management was located in the World Trade Center complex, and was itself destroyed in the attack—as was essential GIS hardware, software, and data. But within hours, DOITT had established a temporary mapping center, known as the Emergency Mapping Center (EMC), and it began to reassemble hardware, software, and data from city, state, federal, and private sector sources. Within a few days, the EMC was relocated to a Hudson River pier facility in Midtown, staffed by city agency personnel and volunteers. The EMC's people and information resources were augmented by personnel from the Federal Emergency Management Agency (FEMA) and other federal agencies, and from New York's State Emergency Management Office (SEMO).

At the new EMC, computers and plotters were connected into a local area network, and ESRI's ArcGIS desktop mapping software and its extensions were chosen as the core technology. The city had previously contracted for emergency response software from E Team, a California company whose products focus on disaster response support and recovery, and which uses Internet mapping service technologies such as ArcIMS software from ESRI.

E Team software had been scheduled for installation at the destroyed EOC on September 14; it was instead installed at the EMC. Verizon Communications provided emergency wireless access to the twelve command posts that were set up in the area, enabling two-way transfer of mapping and attribute data between the command posts and the new EMC. Wireless coverage of the area also enabled field crews to transfer data on field conditions to the command posts using handheld equipment.

Satellite and aerial photography

The next step was to collect current remote sensing images of the disaster area from a variety of local, federal, and private sector sources.

Two satellite systems—SPOT IMAGE and the Space Imaging IKONOS system—were used to show changing conditions at Ground Zero and the surrounding area. IKONOS provided continually updated photographs of the site at one-meter resolution, its images showing the extent of debris and dust and the condition of buildings and providing up-to-date basemapping for other GIS layers. These images, taken at 423 miles above the earth's surface, were supplemented by aerial photography that provided better detail.

Aerial photography became available only after September 13, when flight restrictions were removed for approved aircraft by the FAA; the area was then flown periodically by private firms and by the National Oceanic and Atmospheric Administration (NOAA). Most of these images were already georeferenced and could be overlaid on GIS mapping without further manipulation, although the level of spatial accuracy varied.

Lidar

Lidar (Light Detection and Ranging), an airborne mapping technique that is described in chapter 12, produces a three-dimensional computer model of objects on the earth's surface. Unlike a satellite photograph, a lidar model can be viewed from any angle. The vertical elevations of points on a lidar model are accurate to within 2½ feet, and horizontal accuracy is within inches. Lidar produced accurate three-dimensional mapping of the area around Ground Zero, and these images were updated by repeated flyovers that showed the changing condition of damaged buildings and debris.

Figure 1 Thermal Imagery by EarthData Solutions, LLC.

Table 1   Multidisciplinary Center
for Earthquake Emergency Research
(MCEER) © Huyck and Adams 2002.

Date	Source	Type of data	Comments
9/11/2001	SPOT	Multispectral and panchromatic imagery	Data available on Internet
9/12/2001	IKONOS	Multispectral and panchromatic imagery	Data available on Internet
9/13/2001	Fire Department Keystone	Digital photographs (oblique) Digital photographs (vertical)	
9/14/2001			
9/15/2001	EarthData IKONOS	Digital aerial photographs (vertical) Lidar imagery Multispectral and panchromatic imagery	
9/16/2001	AVIRIS EarthData	Hyperspectral imagery Thermal imagery	
9/17/2001	EarthData	Digital aerial photographs (vertical) Lidar imagery Thermal imagery	EarthData releases orthophotos and thermal imagery of Ground Zero
9/18/2001	AVIRIS EarthData	Hyperspectral imagery Thermal imagery	EarthData releases Lidar imagery; AVIRIS thermal data released
9/19/2001	EarthData	Digital aerial photographs (vertical Lidar imagery) Thermal imagery	
9/20/2001	EarthData	Lidar imagery Thermal imagery	EarthData releases orthophotos and Lidar imagery of Lower Manhattan
9/21/2001	EarthData	Digital aerial photographs (vertical) Lidar imagery Thermal imagery	Poor photography and thermal data due to cloud cover; EarthData releases orthophotos and Lidar imagery of Staten Island
9/22/2001	AVIRIS EarthData	Hyperspectral imagery Digital aerial photographs (vertical) Lidar imagery Thermal imagery	
9/23/2001	EarthData AVIRIS NOAA	Lidar imagery Thermal imagery Hyperspectral imagery Lidar imagery Aerial photography (vertical)	
9/24/2001			Poor weather conditions
9/25/2001	EarthData	Lidar imagery Thermal imagery	

Date	Source	Type of data	Comments
9/26/2001	EarthData	Digital aerial photographs (vertical) Lidar imagery Thermal imagery	
	NOAA	Lidar imagery Aerial photography (vertical)	
9/27/2001	EarthData	Lidar imagery Thermal imagery	
	Pictometry	Digital aerial photographs (oblique)	
9/28/2001	EarthData	Lidar imagery Thermal imagery	
9/29/2001	EarthData	Digital aerial photographs (vertical) Lidar imagery Thermal imagery	Poor data due to turbulence
9/30/2001	EarthData	Digital aerial photographs (vertical) Lidar imagery Thermal imagery	
October 2001	EarthData	Digital aerial photographs (vertical) Lidar imagery FLIR imagery	EarthData continues acquiring data; FLIR system is used from 10/17/2002 to 10/22/2002

Thermal and multispectral imagery

Temperatures at Ground Zero were a major concern in the first weeks after the attack. Underground fuel tanks in the World Trade Center complex posed a risk of explosion, and knowing the location of these tanks and their proximity to high-temperature areas of the debris pile was of critical importance. Thermal imagery was collected from both aerial and satellite sources. Imagery was continually updated as temperatures subsided during the first weeks. On September 16, aircraft from NASA's Jet Propulsion Laboratory began flying the area using AVIRIS hyperspectral equipment, which produces actual (as opposed to relative) temperature readings. The extent of dust and debris from building materials was estimated by the U.S. Geological Survey using satellite multispectral imagery. Table 1 shows the various kinds of remote sensing data that was assembled in the days after the attack, while figure 1 shows an infrared thermal image of hot spots in the giant debris pile, superimposed on orthophotography, taken on September 16.

Figure 2 3D Manhattan database by EarthData Solutions, LLC.

Computer-aided design (CAD) images

First responders needed detailed architectural plans of the WTC complex and of the surrounding buildings and infrastructure because of the need to locate underground fuel and chemical tanks and other hazardous objects. Like lidar images, CAD drawings can display the location of elements in three-dimensional space. Where available, two-dimensional and three-dimensional CAD drawings were imported into the GIS and superimposed on satellite images of the surrounding area. The need for detailed information about the built environment when dealing with emergencies of this kind suggests that future GIS applications for emergency planning should include links to available CAD drawings of buildings and infrastructure. Figure 2 shows in vivid detail the extent of the damaged buildings from the attack and density of the urban infrastructure.

GIS mapping and attribute data

In addition to the map layers available from the NYCMap, the Department of City Planning had an address-matched, block-level digital map of the city, known as LION, a refined version of the Census TIGER/Line file, as well as CAD-based lot-level digital mapping. These files, although useful, did not meet high standards for geographic accuracy. Also available was a file of alternate addresses for properties. Other city agencies had attribute data that could be matched to map features by block and lot number or by address. Street-level mapping at the regional level was available from the Internet.

The integration of needed mapping and attribute data was one of the first tasks of the mapping center team.

Paper maps and map distribution

Paper mapping was used when digital mapping was not available, or when it would take too long to obtain it. Paper maps were scanned into the computer and rubber-sheeted to conform to existing vector mapping. The Fire Department, for example, had hand drawn a 75-foot grid on a paper street map of the Ground Zero area for operational purposes. The map and grid were scanned and rubber-sheeted to overlay a vector block-and-lot map with additional information about Ground Zero conditions. The composite map was then plotted and distributed. Because of time constraints, available paper maps were often combined in this way with digital mapping.

In the first five weeks after the attack, some 2,600 hard-copy maps were produced by the GIS team at the Emergency Mapping Center and provided to responders, government agencies, and the public. These included:

✦ Restricted access areas
✦ Damaged and destroyed buildings
✦ Thermal mapping of Ground Zero
✦ WTC operations command posts and facilities
✦ Zones requiring safety gear such as hard hats, gloves, goggles, and respirators
✦ Building conditions in the vicinity of Ground Zero
✦ Facilities such as food stations, truck washes, toilets, and medical aid stations
✦ Red Cross shelters
✦ Fire districts and firehouses
✦ River crossing status
✦ Utilities status
✦ Telephone outages
✦ Water outages
✦ Pedestrian and vehicular access for residents, responders, and the public
✦ Multispectral images of downtown New York
✦ CAD mapping of buildings and infrastructure
✦ Transportation centers for responders
✦ Regional mapping of logistics staging areas, donor sites, military facilities, and hospitals

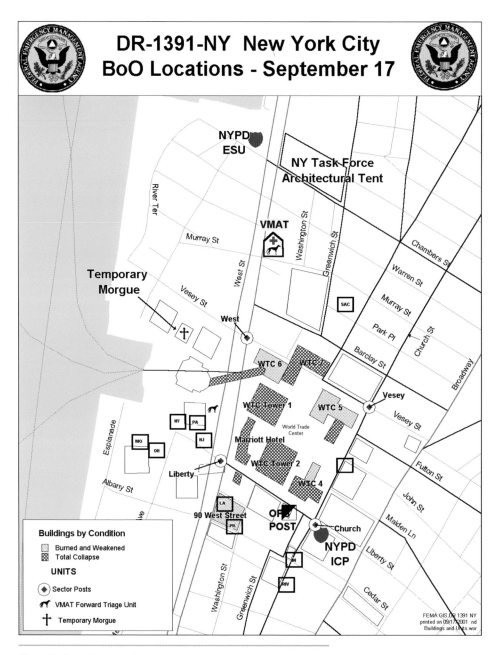

Figure 3   Courtesy Department of Homeland Security, Federal Emergency Management Agency.

As the underlying data changed, these maps were continually redrawn. The FEMA map in figure 3 is a typical example of a map produced for operational purposes.

The demand by responders was primarily for hard-copy maps, from basic letter-size handheld maps to 30- by 40-inch wall maps. The greater demand was for the larger maps, and large-scale plotters were in use around the clock at the Emergency Mapping Center. The public was kept informed of local and regional conditions by the Office of Emergency Management (OEM) Web site, which used Internet Map Server (IMS) technology from ESRI to display information. GIS data can also be transmitted by e-mail. With many telephone lines down, the Internet became the primary means of communication in the first days after the attack.

People

Immediately following the attack there was an urgent demand for maps showing current conditions—a demand that continued twenty-four hours a day for the next several weeks. A steady stream of responders came to the mapping center needing maps based on new information gathered from satellites, aerial photographs, lidar images—and field data reported by the responders themselves. Many of these maps called for knowledge of advanced mapping and imaging techniques such as CAD systems, 3D GIS, and superimposing (or draping) of imagery. In addition, many of the early maps needed by responders required georeferencing of raster data so that vector data could be overlaid on satellite and orthophoto images of building damage, areas of debris, and thermal and air pollution conditions.

Conclusion

People and interagency cooperation were keys to the success of the GIS response to the WTC attack. The continuing demand for maps required people outside government with GIS expertise as well as city, state, and federal GIS users who had access to needed geographic data. Another key was the pre-disaster network of local GIS professionals, GISMO, which spanned both government and the private sector and thus provided an unofficial but invaluable communication channel. When technical people from all agencies can be brought together in one place on short notice, institutional barriers fall and extraordinary things can be accomplished.

About the authors

Michael Crino is a GIS specialist and planner for Baker Engineering NY, Inc., and the principal data investigator for the New York City sewer mapping project. He holds a Master of Arts degree in Geography and Environmental Planning from Towson University and is an officer in the United States Naval Reserve. He is an active participant in a number of professional organizations and most recently helped start the New York State GIS/LIS Association.

Karen Rutberg, AICP, works at the Center for Advanced Research of Spatial Information (CARSI) at Hunter College, where she manages the quality assurance and program management contract for the NYC Department of Environmental Protection (DEP) Sewer Mapping Project. She holds a Master of City Planning degree from the University of Pennsylvania's School of Design and a Master of Arts degree in Geography from Hunter College. For the last fifteen years, she has worked as a city planner, information technology consultant, and geographic information systems specialist.

Magdi Farag, P.E., FASCE, is the director of engineering for the NYC Department of Environmental Protection (DEP) Bureau of Water and Sewer Operations. He also serves as the DEP Sewer Mapping Project director. He holds a Bachelor of Science in Structural Engineering and a Master of Science in Soil Mechanics and Foundation Engineering from Cairo University in Egypt. He also holds a Master of Science in Civil Engineering, Soil Mechanics and Foundation Engineering from the Polytechnic Institute of Brooklyn. He is recognized as a professional engineer in the states of New York and California and a fellow in the American Society of Civil Engineers.

Infrastructure Mapping *for* Planning *and* Maintenance

Case Study Seven

GIS mapping of the New York City sewer and storm drain system

Michael Crino

Karen Rutberg

Magdi Farag

Background

The New York City sewer system is a complex network of more than 6,200 miles of sanitary, storm, and combined sewers; 300,000 manholes; 14 wastewater treatment plants; and thousands of additional facilities. The system developed in a piecemeal fashion primarily in the late nineteenth century and early twentieth century, with some sections predating the Civil War (figure 1). The system has been administered by various jurisdictions and agencies. Changes in managing responsibility and improvements in engineering design techniques have produced a system infrastructure that varies considerably in age, materials, and structure. Records of the sewer system were prepared with different standards, and with different datum for each borough. Today, responsibility for planning, construction, and maintenance of this diverse system lies with the New York City Department of Environmental Protection (DEP).

The size of the system and the borough's different design and maintenance criteria have made it difficult to create and maintain a single map of New York City's entire sewer network. Each of the city's five boroughs has a unique method for storing and maintaining its own sewer maps and records. To date, no boroughwide or citywide sewer system map exists in a digital format.

The primary sewer records are the original as-built and contract drawings, on a variety of media: vellum, paper, Mylar, sepia, and even linen. The quality of these documents has deteriorated over the years and some original drawings were lost. The individual boroughs have made various efforts to transfer some of this data to more comprehensive system maps, but these boroughwide maps show only a small portion of the sewer system's vital attributes. Their spatial accuracy is inconsistent among the boroughs. Because of these limitations, these maps can only be used for graphic representation purposes or as an index to more efficiently locate the paper as-built and contract plans. In addition, because all of the source information is stored on hard-copy media, the maps are difficult to update, and in some cases not maintained at all.

GIS implementation

With limited ability to perform digital planning and analysis for the city sewer system, and with the prospect of losing additional vital information over time, the DEP investigated GIS as a long-term solution. The advantages of a GIS approach became clear. The DEP would benefit from reduced dependence on the fragile original-source documents, planning and engineering analysis would be enhanced, and data maintenance could be more centralized. In addition, a topological sewer network would allow the DEP to take advantage of advanced spatial modeling capabilities such as network traces and flow analysis.

Although the benefits of a GIS are abundant, several other considerations became part of DEP's evaluation process before a commitment to GIS was made. Foremost was the daunting task of the conversion itself. With the sewer system pieced together from more than 190,000 drawings of differing quality, it was difficult to estimate how much of the system could be properly converted into a GIS. There was also the commitment required of DEP after the conversion to maintain the newly created GIS; the ability to update the sewer network and to operate the custom GIS applications requires a continuous investment in staffing, training, software, and hardware long after the initial conversion is completed.

Recognizing that the existing mapping and data retrieval system was inadequate, the DEP initiated the sewer mapping project in 2002. The project was the realization of the first phase of DEP's vision to integrate, leverage, and extend existing resources into an agency-wide GIS. This system is envisioned to eventually contain the water system and document images database. Also part of the design is future interoperability between the sewer GIS and the Hansen database, which the DEP uses for work order tracking. DEP's vision for the GIS includes developing more than 100 different applications for 11 different bureaus within the DEP to help improve business processes. These applications include three-dimensional mapping, modeling and analysis, air quality modeling, floatables analysis, hydrant flushing support, permits tracking, siting, water quality pattern analysis, and many others. ArcGIS from ESRI is the software platform and Oracle is the RDBMS.

The benefit of the GIS extends beyond users at the DEP. Because the DEP GIS will be housed at New York City's central data repository, the sewer GIS will be available for integration with other GIS layers such as the land base, tax block and lot, orthophoto, and others as they come online.

Project partners

The project requires partnership and coordination among three main groups. The DEP is responsible for the city's sewer system and serves as the project's administrator. The DEP approves the final database design and methodology for data conversion, and performs an intensive review of all delivered data to ensure the quality meets DEP standards. The DEP is assisted by a quality assurance (QA) team represented by the Center for the Advanced Research of Spatial Information (CARSI) lab at Hunter College, and private-sector consultants PlanGraphics, Inc. and Camp, Dresser & McKee.

Baker Engineering NY, Inc. is the lead consultant performing the data conversion and is supported by a number of team members specializing in surveying, document scanning, and quality control. Baker is responsible for developing the final database design, procedures, and methodology for conversion, and for all data delivered to the DEP.

Pilot implementation

The sewer mapping project is structured into two phases, pilot and production. The pilot provides the opportunity to verify the data dictionary, data accuracy, data completeness, and appropriateness of processes to support the full-scale production conversion. The QA team also has the opportunity to test and refine the QA procedures and software, and to validate the sampling approach. Because significant differences exist among the source information for each borough, the pilot includes a full conversion of a small area of the Bronx, Brooklyn, Manhattan, Queens, and Staten Island. In addition, adjacent areas connecting Brooklyn and Queens are part of the pilot project in order to test edge-matching between two different boroughs.

Data collection and conversion

Data for the project is primarily collected through scanning and then interpreting the original as-built and contract source documents. More than 190,000 drawings from all five boroughs and numerous government agencies were scanned early in the project's process. Scanning the documents creates a permanent record, and because they are linked to the geodatabase, end users have access to the original information.

Interpreting all of these documents requires capturing updated data from various drawings and achieving outdated attributes (figure 2). Indexing, a procedure that associates a source document with its respective street centerline, is the first step in filtering the documents. In cases where multiple-source documents are indexed to a single street segment, the as-built document with the most recent date defaults as the source of information in the data conversion. In cases where missing dates and illegible text limit data utility, more extensive analysis of the source drawings is performed.

After completing the indexing process, the conversion team carefully studies each document to identify its content. In a process called scrubbing, each drawing is reviewed and the pertinent attributes are highlighted for inclusion in the database. The challenge of scrubbing stems from the inconsistencies within the source documents. Although drawings are primarily classified as as-built or contract drawings, the actual sources vary and include private sewer plans, Works Progress Administration (WPA) drawings, inspector tracings, and a host of bridge, tunnel, and highway projects all with their own datums, symbols, and annotation styles. Domain values updated throughout the pilot project help standardize the attributes, but the skills of those interpreting the drawings are required to recognize the many spatial and attribute anomalies across the source documents.

For sections of the city that have no original source document, or where the drawings cannot be interpreted because of their poor condition or missing information, the spatial location and associated attributes are collected as part of a field survey. GPS is the primary field collection method though total station is also used. The sewer features are digitized relative to the New York City land base, also known as NYCMap. Although

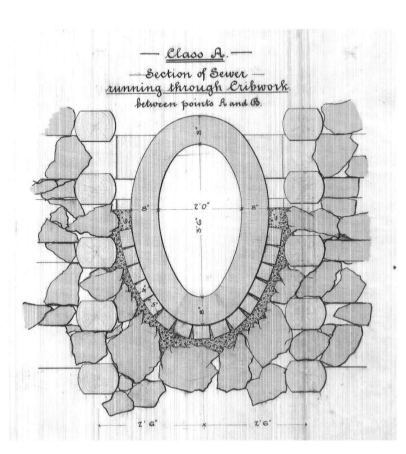

Figure 1   Cross section of sewer in Manhattan built in 1888. New York City Department of Environmental Protection.

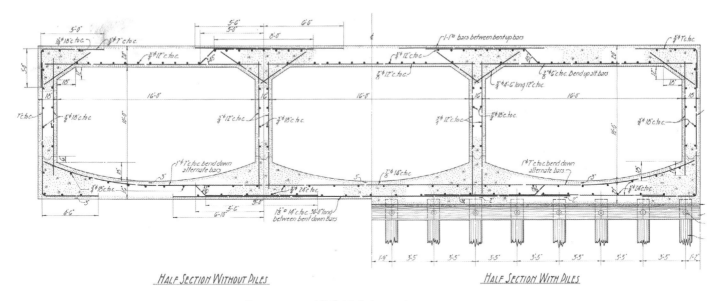

HALF SECTION WITHOUT PILES

HALF SECTION WITH PILES

TRIPLE BARREL 16'-0" x 10'-0" SEWER SECTION

Figure 2 Cross section of triple barrel sewer in Queens built in 1945. New York City Department of Environmental Protection.

most drawings overlay well with the NYCMap, a host of other steps are taken to ensure the converted data is within the project's accuracy specifications. Some drawings show offset distances to property lines and distances between manholes, helping to correctly position sewer system features. Field-collected data provides an excellent distribution of precise points from which other features can be geopositioned. A combination of automated mapping routines and manual digitizing connect the sewer system's features, and rules defined within the geodatabase help ensure correct topology within the newly converted data.

Project summary

The DEP faces numerous challenges implementing the GIS, including managing the logistics of a complex conversion; capturing appropriate information from various, sometimes conflicting documents that are in different formats, sizes, and styles. In all, there are about 444,000 features and approximately 10.3 million attribute values in the NYC GIS sewer map. Although the investment in GIS is considerable, its development was deemed essential to maintain and improve the integrity of the aging infrastructure, provide services and issue permits, and develop, repair, and extend the city sewer system. It also provides a reliable and permanent record of the sewer system that will not be subject to unforseen disasters.

About the author

Kenneth Mack is the technical director at The Kenerson Group, a Massachusetts-based consulting firm, where he oversees all GIS operations and is responsible for applications and data development, client training and services, Web site maintenance, and product documentation. Kenneth came to The Kenerson Group from Space Track Inc., a New York City GIS company, where he held similar responsibilities. Before that, he served as geographic specialist at the U.S. Census Bureau. His education includes a Bachelor of Arts degree from the Eugene Lang College at the New School, and three years of graduate studies in Archaeology and Geography at Hunter College.

Archaeology *and* Historic Preservation *with* GIS

Archaeology GIS project for the New York City Landmark Preservation Commission

Eight

Kenneth Mack

Project background

The New York City Landmarks Preservation Commission (LPC) is an agency of the New York City government charged with designating and regulating landmark-designated properties in the five boroughs. An additional mission is to track sites of archaeological sensitivity and significance for regulatory review and for potential landmark designation. Consequently, the Archaeology Department at LPC maintains a database of such sites and their associated descriptive and administrative information. This database had, prior to the completion of the project described here, been stored in Microsoft Access format as a group of related tables and stored queries, with a user form for viewing, entering, and editing the site data.

LPC staff decided to implement GIS technology in an effort to enhance Archaeology Department data creation, management, and analysis capabilities. The impetus for a GIS-based solution arose for several reasons. First, the data itself is intrinsically spatial in nature; the key identifying attribute of an archaeological site is intuitively its location. The database lends itself to being built around places and spatial relationships. Furthermore, interacting with a map interface, rather than simply a form of text boxes and buttons, offers a much more natural and dynamic portal to the site data. Additionally, a departmental GIS would enable staff to perform spatial analyses and queries on the data—for example, to select all sites with Native American archaeological findings that are located within 500 feet of properties owned by the city—and to include maps in printed reports. Perhaps most importantly, the GIS would also provide the means to digitize and integrate information currently on paper historical maps in the department's archives.

The initiative also would integrate the department in the general citywide movement toward the generation, utilization, and standardization of spatial data. This effort, spearheaded by the Department of Information Technology (DOITT), encouraged municipal entities to take advantage of and contribute to the city's growing central repository of spatial data.

Project overview

The pilot phase of this project took the form of three main tasks. The first was originally envisioned simply as the geocoding, or mapping, of the approximately 27,000 records in the sites table; however, preliminary attempts to do this revealed a significant number of site records with missing, inadequate, or erroneous locational information. Therefore, the task became a database clean-up and geocoding task, which itself required some rethinking of the database structure. The second task, application development, entailed adding custom functionality to the standard GIS desktop software, ArcGIS 8.3, so that staff members could easily view, add to, and edit the site data directly through the ArcMap interface. Additionally, querying and symbolizing the data, performing spatial analysis, and generating reports would all be facilitated, either through application development or GIS training. The third task, a historic map digitization pilot, would fulfill the Archaeology Department's need to study the feasibility and utility of digitizing and georeferencing available historic maps; several pilot maps would therefore be digitized and georeferenced, and the process evaluated. Space Track Inc., a GIS consulting firm specializing in municipal data integration and application development, was hired to develop the GIS.

Implementation

The Archaeology Department's Access database of projects and sites is known as the Universal Site File (USF). The main table is actually the projects table, in which each record is a project—the administrative unit of work and study which consists of one or more sites, each of which is associated with a tax lot and is defined by the tax lot number where it is located. (A very small number of sites represent other types of locations, such as street beds, and have no tax lot identifier.) The projects table contains a number of attribute fields with descriptive and administrative information about the projects, such as start and end dates, review status, findings, time period, and comments. The sites are kept in their own table; they cross-reference the project number and contain two types of locational information: borough, block, and lot numbers (which, when concatenated, become the unique identifier for tax lots in NYC), and an address. This cross-referencing is efficient because the more elaborate project attributes don't need to be repeated throughout the sites table. The sites table is much larger than the projects table, because it contains at least one, and often multiple sites for each record in the project table. The USF contains a number of other cross-referenced tables as well; time period categories, for example, have a many-to-many relationship with the sites, because there can be multiple time periods associated with one site.

The borough, block, and lot numbers found in the sites table represent the key locational data for the proposed GIS. Every tax lot in New York City is uniquely identified by its borough-block-lot (BBL) number, a ten-digit number formed by concatenating the borough code (one digit), the block code (five digits), and the lot code (four digits).

Working with New York City data sources

In a data layer created and maintained by the Department of City Planning, New York City's tax lots are mapped and carry the BBL number attribute. This layer, known as the Bytes of the Big Apple tax lot map, was therefore chosen as the key basemap layer in the LPC GIS, with the BBL providing the link between the map and the sites table. Additionally, using the tax lot map as the key layer would give the GIS the ability to integrate all kinds of property- and building-based data keyed to the BBL attribute, available throughout city agencies. However, two factors prevented the simple linking up of sites to the map via the BBL.

First, there were the data quality and currency issues with the USF mentioned earlier, issues common to legacy data migration projects. The BBL information in particular was problematic. Human error in data entry accounted for some of the problems, but most were due to the fact that many of the USF archaeology projects were years old, and because of lot subdivisions and aggregations, their lot numbers were no longer valid. In fact, archaeology projects often were initiated at locations precisely because rezoning, development, and construction was occurring there.

The second overall obstacle was more complex. The tax lot map, a massive and complex graphical representation of the city's nearly one million properties, is not in fact the official database of BBL and address data in NYC—it was never meant to be. It was originally created using CAD software as a schematic—rather than geographic—representation of the city's lots, and only later converted to native GIS formats. The official BBL and address database is on a mainframe system known as Geosupport, maintained by a different division of City Planning. City Planning is committed to keeping the tax lot map as current as possible, but there is still a discrepancy between the mapped lots and reality as officially represented by Geosupport—not surprising given the scale of the city's geography and the speed with which lots subdivide, aggregate, or renumber.

In any case, in the interests of agency and citywide data integrity and standardization, the sites' BBLs needed to be validated against the Geosupport database before being geocoded to the tax lot map. Furthermore, this requirement applied both to the initial geocoding of existing sites and to the future, ongoing entry of new sites. Ultimately, two separate solutions were devised.

For the initial effort, a flat ASCII file was exported out of the Geosupport mainframe system; this file contained all valid BBLs, along with all valid addresses associated with those BBLs, including alternate addresses. This Geosupport file was imported into ArcGIS and linked to the tax lot map via the BBLs, and thus converted into a mapped data layer. This layer of valid BBLs and associated addresses was then used, after some additional modifications to field structure, as the reference data against which the sites table was matched. Of course, some valid BBLs did not find their mapped counterparts and were consequently not represented in the Geosupport data layer; these valid but unmapped lots from the Geosupport file were set aside in a separate file as a potential quality-assurance resource; they represented, and in fact quantified, the discrepancy between the valid Geosupport data and the map.

Once the existing sites were geocoded and set up as a feature class within the GIS database, functionality had to be implemented to allow for the ongoing validation and entry of new sites by LPC staff. Fortunately, an additional resource became available for this application programming task during the planning of this project, when the City Planning department released its new Geosupport for Windows product, which provided a Windows API to Geosupport data and functionality.

Geocoding methodology and results

First, the BBL was calculated for each record in the original sites table based on the values in the individual borough, block, and lot fields. Then, a field called *matchcode* was added to the sites data, to hold values indicating to what degree of precision the record was geocoded.

For the first pass, the BBL value was used to geocode the records against the Geosupport data layer; for those that were successfully matched, the BBL remained as the validated BBL. For those that could not be matched, the BBL was recalculated as NULL. The unmatched records were then taken through a second and third pass against the Geosupport data, this time using the address; if a match was found for the address, the Geosupport BBL corresponding to that address was used to update the BBL field for that record. In a small number of cases the address was adjusted if the error was apparent (for example, adding "St." to "130 Suffolk" or correcting "720 Greewich St" to "720 Greenwich St"). All records up to this point, having been successfully matched against Geosupport data and assigned a valid BBL, were given a matchcode value of 1; these level 1 records ended up accounting for 66.1 percent of the total database.

The remaining unmatched records therefore possessed invalid (or incomplete) BBL values; the strategy at this stage was to get as many of these sites as possible on to the map, unvalidated but at an approximate (say, block-level) location where they could be researched, validated, and moved into their correct location during later quality control by LPC staff. To accomplish this, they were geocoded against the Department of City Planning's LION street centerline data layer, using the address; or against the block layer, using the block number on its own. Records that were matched in these fourth and fifth passes were updated with a NULL BBL to preserve the validity of the sites' BBL data, and assigned a matchcode value of 2 or 3. These ended up accounting for 29.1 percent of the database. In most cases, these records represent the ones mentioned earlier that had obsolete lot numbers—lots having been renumbered due to rezoning, subdivision, or aggregation. When they are viewed on the map, their proper location is often apparent and easily corrected during quality control. Figure 1 shows two examples of this kind of discrepancy and how the surrounding data helps.

Figure 1

The remaining 4.8 percent of the site records (1,396 total records) could not be matched on any of the geocoding target layers, due to missing BBL and address data, or to the fact that they were in multiple boroughs. These records were detached from the database, to be reviewed by LPC staff and re-entered individually using the tool provided by the application. Once a site is entered and assigned its project ID number, it will automatically link up to its project information. Some of these records actually do have some useful locational information that could not be processed by the geocoding engine or the target data, including odd-appearing but valid addresses such as "Battery Park City North" or "Sheridan Square," neither with a block or lot number. Other records have information in the associated projects table that can be used to help locate the site, for example, a site that has no block or lot number, the address is *Battery to 59th St.,* and the associated project name is *Hudson River Park.* Some of these records represent street beds or other kinds of sites which simply can't be described using lot numbers or addresses; these need to be reviewed and entered manually. A matchcode of M was reserved for later use of the application itself, to be assigned to site records that are manually placed in point locations that are not within lots—such as street beds—and which therefore do not possess BBL numbers. Figure 2 provides an overview of all the geocoded sites, symbolized by their matchcode.

Figure 2

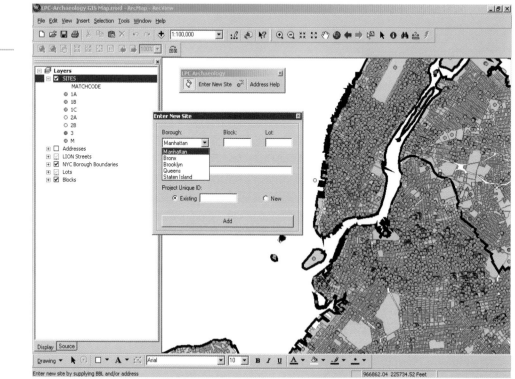

The modified USF: Toward a departmental GIS

The USF remains an Access database, but also includes the additional table components that allow it to function as an ArcGIS geodatabase. The USF contains the new sites data layer, or feature class—it is no longer just a standalone table—as well as the projects table and all the other tables, forms, and queries that existed in the prior Access version. Some necessary modifications have been made to certain tables, such as renaming fields or tables, or adding an autonumber field. Also, the tables containing the remaining unmatched sites (one for each borough and one for multiborough records) have been added to the USF. The USF is stored in a shared directory on the network.

The most interesting aspect of this process is the shift from a project-based database—in which descriptive archaeological data is attributed to archaeological projects—to a site-based geodatabase, in which it is the individual sites that are assigned archaeological attribute data. This shift will remove a number of limitations, such as the inability to record which specific locations contained archaeological findings within a large, multilot project.

Customizing the application

A set of custom tools was added to the LPC ArcMap application, using Visual Basic 6 and the ArcObjects component library, which allowed the user to view, edit, and add new USF data directly from the ArcMap interface. For example, clicking an archaeological site on the map with the Identify Site tool selects it and opens the Identify Site dialog box, and displays the site's associated data pulled from multiple tables in the USF. The Identify Site dialog, designed to provide access to all the relevant data fields in the project and sites tables, floats above the map; clicking a new site repopulates the dialog with the new site data.

Figure 3

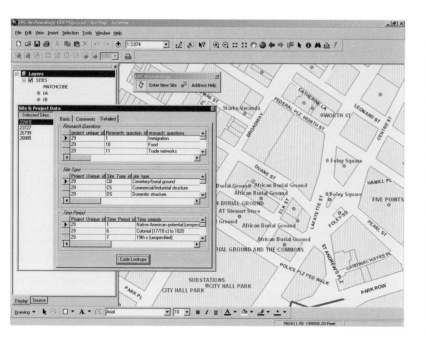

Figure 4

Some properties have multiple sites associated with them, which appear to be stacked at the property centerpoint. In such cases, a mouse click will select all, and the dialog will list all the selected sites on the side pane and display the first in the list. Clicking the different site ID numbers in the list will repopulate the dialog with the corresponding data for each site, as is being done in figures 3 and 4.

Adding new sites to the USF is accomplished by either supplying a BBL number and letting the GIS geocode it, or by clicking the location on the map. In the latter case, clicking the map will pick up the BBL number of the lot containing that point, if it falls within a lot. Both methods include BBL validation procedures, by implementing the Geosupport for Windows data and API provided by City Planning. With either tool, if the BBL fails the validation check, the user is notified and the site record is not created. If the BBL is returned as valid, the new site record is added to the database and initialized with locational data and default values in other fields.

The ArcMap query capabilities will allow LPC staff members to examine workflow issues, to review and research potential new sites, to produce other analyses, and to answer questions, historical and otherwise: What projects were reviewed in a particular neighborhood, and did those projects require subsurface work? How many eighteenth-century colonial sites are located in a particular geographic area? Additionally, all kinds of spatial analyses can be performed, allowing the user to examine patterns of certain types of findings, or to do overlay analysis with other types of data, or even to do tracking over time.

Georegistering the historic maps

To georegister historic maps, several sheets from paper-based historical maps were scanned, and then registered using the ArcGIS Georeferencing tools, which allow on-screen, point-and-click matching of images to vector data, with excellent results, as seen in figures 5 and 6. The map images overlaid with the vector basemap and archaeological site data provide great opportunities for site research and analysis. Beyond this, LPC staff envisions the ability to initiate a project to create vector data layers from the raster map images—for example, historical properties or streets—either by direct on-screen digitization, or with the use of the ArcScan extension, a newly developed technology that automatically detects and builds vector data from raster data.

Figure 5

Figure 6

About the author

Juliana Maantay has been active in environmental justice advocacy and research for more than ten years, and much of this work has involved GIS. Some aspects of the asthma and air pollution project described in this case study were started while she was Associate Director of the Center for a Sustainable Urban Environment at Hostos Community College in the South Bronx in the mid-1990s. Since then, she has continued her Bronx-based environmental and community focus from the Lehman College Urban GISc lab, examining spatial issues such as the proliferation of waste-related facilities, changes to industrial zoning, the urban community garden movement, the use of city-owned vacant land, and sustainable construction practices. She is also particularly interested in developing effective models for community-based participatory GIS.

Currently, she teaches GISc and Environmental Geography at Lehman College in the Bronx, where she also directs the GISc program. Her previous professional experience includes five years as Vice President of Architecture, Planning, and Real Estate at Paine Webber; senior environmental analyst at the New York City Department of Environmental Protection, responsible for environmental impact analysis; and senior environmental planner at the New York City Department of City Planning, where she was involved in urban planning and policy assessment. She holds a PhD and Master of Philosophy degree in Geography from Rutgers University, a Master of Urban Planning degree from New York University, a Master of Arts degree in GIS from Hunter College, and a Bachelor of Science degree in Environmental Analysis from Cornell University. She has been an avid GIS user since 1989 and a lover of maps since childhood.

Health *and* Environmental Justice

Mapping environmental hazards and asthma in the Bronx

Juliana Maantay

This project investigates possible explanations for the serious problem of high hospitalization rates for asthma in the Bronx, especially among poor and minority populations. It was undertaken by a consortium of researchers from Montefiore Medical Center, Albert Einstein College of Medicine, the South Bronx Clean Air Coalition, and Lehman College GISc Laboratory, under the aegis of the South Bronx Environmental Justice Partnership. The National Oceanic and Atmospheric Administration (NOAA) also provided significant support for the project.

The project investigated the possible spatial correspondence between the incidence of asthma hospitalization and the locations of environmentally burdensome land uses and activities. To test this hypothesis, we used GIS to map and model the major mobile and stationary sources of air pollutants in the Bronx. We concluded that there is such a correspondence between areas having high rates of asthma hospitalization and areas with high concentrations of major air emission sources, and therefore high levels of air pollution.

Background

Recent studies have linked high concentrations of known air pollutants to respiratory disease, demonstrating that increased air pollution is a serious public health and environmental concern. Reduced air quality and respiratory health problems, particularly asthma, have been associated with a number of sources, including emissions from industrial processes, particulate matter (PM) and increased levels of NO_x and SO_2 from truck traffic, and increased pollution from other land uses, such as waste-related facilities, medical institutions, and power plants.

These conclusions are of particular interest to the Bronx community, because residents of the Bronx, specifically children under the age of 14, suffer from one of the highest rates of asthma hospitalization in the country, according to the New York City Department of Health's 1999 publication *Asthma Facts*. The New York City Department of Health (DOH) reported a rate in the Bronx of 16.6 asthma hospitalizations per 1,000 people under the age of fourteen in 1997, a rate nearly 50 percent higher than the rate in the city as a whole. The 1997 rate for the Hunts Point–Mott Haven areas of the South Bronx is even higher: 23.2 per 1,000, nearly 250 percent above the citywide average (DOH 1999). DOH data also shows a marked increase in asthma hospitalization since 1988.

The Bronx also has a disproportionate number of solid waste transfer stations, Toxic Release Inventory (TRI) facilities, and other major stationary point pollution sources (Maantay 2002), and these are mainly concentrated in the South Bronx, where the asthma rate is relatively high. Figure 1 shows some of these pollution sources. The South Bronx also has one of the highest volumes of vehicular traffic in the nation, another factor in substandard air quality (Jackson 1995). There have been no major geographic studies of New York City that have addressed the potential connection between noxious land uses and asthma. Therefore, we are investigating the possible correlations that may exist between high rates of asthma and multiple sources of air pollution in the Bronx. We are also exploring the connection between the location of environmental hazards and the concentration of low-income populations and communities of color.

Figure 1

TRI Facility

Sludge Pelletization Plant (SPS)

Emissions from Housing Complex

Highway/Truck Route

In Phase One of this three-phase project, we asked the following questions:

1 What are the major sources of air pollution in the Bronx? What land uses and activities are typically associated with these sources?
2 For each source, what are the quantities and characteristics of the pollutants emitted?
3 What is the geographic extent of pollutant dispersion—the impact zone?
4 What are the characteristics (socioeconomic, racial, and ethnic) of populations potentially most affected by these pollution sources?
5 Is there a spatial correspondence between areas having high asthma hospitalization rates and areas in close proximity to major sources of air pollution?
6 Is there a connection between age, income, race and ethnicity, high rates of asthma hospitalization, and environmental burdens?

Phase One methodology

The Lehman College GISc team:

✦ Conducted a literature review of asthma research, specifically studies that examine the relationship between traffic, air pollution, and asthma hospitalization incidence. In addition, we collected information on basic epidemiological concepts, air quality issues, methods of assessing environmental health, and current air quality models.
✦ Compiled all federal- or state-monitored facilities that are regulated because of their potential to adversely affect health and the environment. These include TRI facilities, solid waste transfer stations, medical waste treatment sites, sludge processing plants, and wastewater treatment plants.
✦ Created an attribute database containing detailed information for each facility.
✦ Located and mapped each facility in the Bronx.
✦ Created a half-mile buffer around TRI facilities, and a quarter-mile buffer around major stationary point sources (SPS) facilities and compared population characteristics (and asthma rates) of areas within the buffers to the Bronx as a whole.
✦ Compiled a database of all major vehicular traffic routes, both limited access highways (LAH) and major truck routes (MTR).
✦ Mapped the locations of the limited-access highways and truck traffic routes.
✦ Created a 150-meter buffer around the limited access highways and truck traffic routes, and compared the population characteristics and asthma rates of areas within the buffers to the entire Bronx area. Buffer distances were based on standards used in environmental assessment, reflecting the distances that air pollutants typical for each source may travel.
✦ Compiled a database of rates of asthma hospitalization by census tract.
✦ Mapped the asthma hospitalization incidence data, both as rates and as cases.

+ Compiled a database of all major areas zoned for industrial land uses in the Bronx. These manufacturing (M) zones are legally permitted to contain polluting facilities, including those whose emissions are too small to require listing with the federal or state governments.

+ Using Census Bureau information, we characterized the potentially impacted population within census tracts by socioeconomic and demographic indicators such as race, ethnicity, and poverty status, then plotted these as choropleth maps.

+ Interpolated asthma rates, poverty levels, and percent minority population from data linked to census tract centroids, then plotted these as grid layers and classified by standard deviation.

+ Created dot density maps of asthma hospitalization cases.

Datasets used in this study.

+ Toxic Release Inventory Facilities (TRI) Source: U.S. Environmental Protection Agency (2000). (TRI locations geocoded and plotted by Lehman GISc team)

+ Local and through-truck route network Source: New York City Department of Transportation/Traffic Rules and Regulations (2002) (Truck routes digitized and plotted by Lehman GISc Team)

+ National Emission Trends (NET) Source: U.S. Environmental Protection Agency. (Stationary Pollution Source (SPS) facility locations geocoded and plotted by Lehman GISc Team)

+ Major manufacturing zones Source: New York City Department of City Planning, Citywide Industry Study: Geographical Atlas of Industrial Areas (1993) (Major M zones digitized and plotted by Lehman GISc Team)

+ Demographic and socio-economic data by census tract Source: U.S. Department of Commerce, Bureau of the Census; Census of Population and Housing, Summary Tape File 3a (2000)

+ Street segments Source: U.S. Department of Commerce, Bureau of the Census (2000) TIGER Files (Limited access highways selected and processed by Lehman GISc Team)

+ County shorelines, water bodies, parks, census tract boundaries, census tract centroids Source: U.S. Department of Commerce, Bureau of the Census (2000)

+ Permitted waste-related facilities Sources: New York City Department of Sanitation (2000); New York State Department of Conservation (2000) (Facility locations geocoded and plotted by Lehman GISc Team)

+ Zoning and land use, by tax lot Sources: LotInfo by Space Track (2002); New York City Department of Finance, RPAD (Real Property Attribute Data) (2002)

+ Digital orthophotos of New York City Source: New York City Department of Environmental Protection, NYCMap (2000)

+ Asthma hospitalization data per census tract Sources: InfoShare (2000); New York City Department of Health, Asthma Facts, brochure (1999).

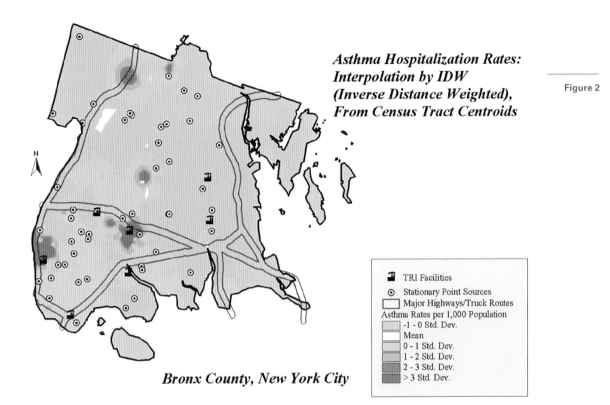

Asthma Hospitalization Rates:
Interpolation by IDW
(Inverse Distance Weighted),
From Census Tract Centroids

Figure 2

Bronx County, New York City

TRI Facilities
Stationary Point Sources
Major Highways/Truck Routes
Asthma Rates per 1,000 Population
-1 - 0 Std. Dev.
Mean
0 - 1 Std. Dev.
1 - 2 Std. Dev.
2 - 3 Std. Dev.
> 3 Std. Dev.

Phase One preliminary findings

Phase One preliminary analyses suggest that there is a spatial correspondence between the rates of asthma hospitalization by census tract and the locations of environmentally hazardous land uses and activities in the Bronx. Figure 2 shows asthma hospitalizations rates in the borough interpolated from census tract centroids.

Our analyses suggest that proximity to hazardous land uses is also associated with areas of high concentrations of minority and low-income residents, as the maps in figures 3 and 4 (next page) demonstrate.

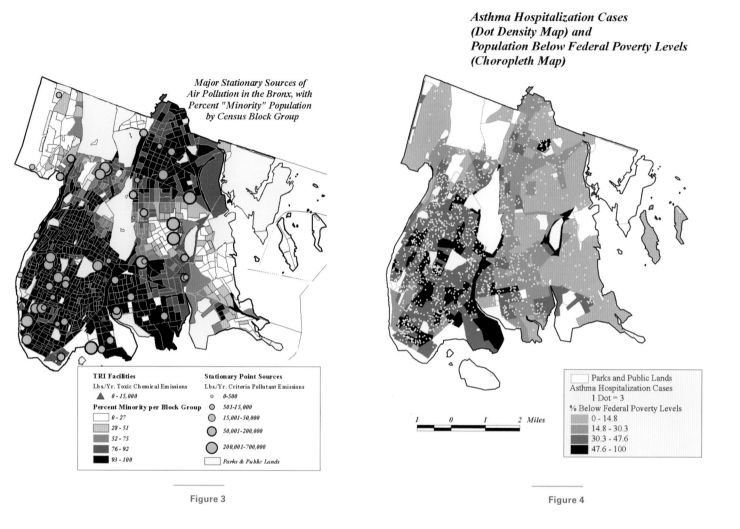

Major Stationary Sources of
Air Pollution in the Bronx, with
Percent "Minority" Population
by Census Block Group

Asthma Hospitalization Cases
(Dot Density Map) and
Population Below Federal Poverty Levels
(Choropleth Map)

TRI Facilities
Lbs./Yr. Toxic Chemical Emissions
▲ 0 - 15,000
Percent Minority per Block Group
☐ 0 - 27
▨ 28 - 51
▨ 52 - 75
▨ 76 - 92
■ 93 - 100

Stationary Point Sources
Lbs./Yr. Criteria Pollutant Emissions
○ 0-500
○ 501-15,000
○ 15,001-50,000
○ 50,001-200,000
○ 200,001-700,000
☐ Parks & Public Lands

1   0   1   2   Miles

☐ Parks and Public Lands
Asthma Hospitalization Cases
1 Dot = 3
% Below Federal Poverty Levels
▨ 0 - 14.8
▨ 14.8 - 30.3
▨ 30.3 - 47.6
■ 47.6 - 100

Figure 3

Figure 4

Phase Two methodology and data

The preliminary findings helped to give us direction for the more detailed analysis in Phase Two. This was made possible by obtaining a record-level database of asthma hospitalizations in Bronx hospitals from the New York State Department of Health SPARCS (Statewide Planning and Research Cooperative System) database. This database was customized for our purposes by including only those fields relevant to our research project. The database contained a separate, unique record of every hospital admission for the medical diagnostic codes associated with asthma during 1995–1999.

The records we received did not include patient names and addresses in order to protect patient confidentiality, but each record retained its unique patient ID number. Instead of patient address, the latitude and longitude coordinates of each address had been substituted, so that we could geocode and accurately map patients' residential locations without knowing the street address. Researchers signed affidavits agreeing to restrictions on their use of the data: that they would not manipulate data to reveal actual patient addresses, that it would not be shared with other researchers, that it would not be used beyond a three-year research period, and other limitations. Because of its highly confidential nature, it took more than six months to receive this data. However, the data was worth the wait. It allowed us to analyze the cases at a much higher resolution than previously could be accomplished using the cases and rates aggregated at the census tract level.

Phase Two of the project consisted primarily of using the SPARCS data to map the actual locations of patients who had been hospitalized for asthma. These locations were mapped in a variety of ways: separately by year, by month, by season, for all patients admitted to the hospital for asthma, for children 15 years and younger who had been hospitalized for asthma, and for all adults 16 years and older. We mapped cases (counts) as well as rates (counts divided by susceptible population), as shown in figures 5, 6, and 7. We then recreated the four types of buffers around the various pollution sources, similar to those used in Phase One (TRI, SPS, MTR, and LAH). Because the asthma hospitalization data was now mappable as points rather than as rates or as absolute numbers aggregated by census tract polygons, our maps could more accurately show exactly which cases were within the buffers and which were outside. From this we hoped to compare rates within and outside the buffers, and to ascertain whether people living in close proximity to a major pollution source are more likely to be hospitalized for asthma.

Asthma Hospitalizations
- ○ Outside Combined Buffers
- ● Inside Combined Buffers
- ▨ Combined Buffers
- ☐ Bronx, NY

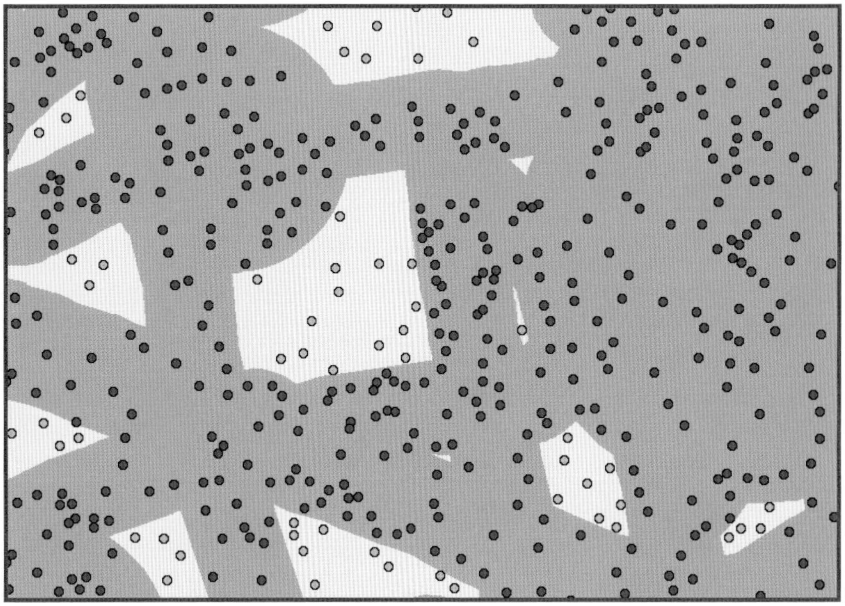

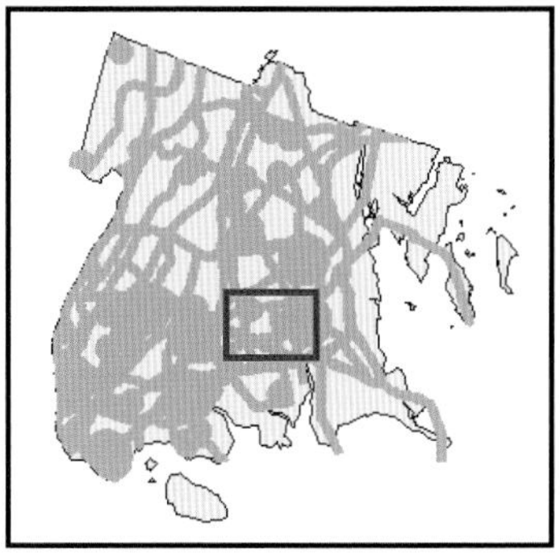

Each dot represents the residence of one Bronx person admitted to the hospital for asthma in 1999. Some dots represent multiple admissions of the same person, or multiple people admitted from the same address. The multiple cases are not shown as individual dots on the map, but have been included in statistical calculations. There were 8,188 hospital admissions for asthma in 1999: 5,876 of them from within the areas of the combined buffers, and 2,312 of them from areas outside the buffers. Overall in 1999, a Bronx resident was 27% more likely to be admitted to the hospital for asthma if living within a buffer area than if living outside a buffer area.

IMPORTANT NOTE The patient address locations shown on this map are derived from hypothetical data and do not represent actual addresses. Due to patient confidentiality requirements, the actual address locations could not be shown in a document for public dissemination, and this map is intended only to be illustrative of the methods used in the analysis. The actual address locations were, however, used by the researchers in the spatial analysis to derive the in- and out-of-buffer rates, odds ratios, and other statistical tests. The researchers were only permitted to show aggregated data (as opposed to record level data) in any maps available to the public.

Figure 5

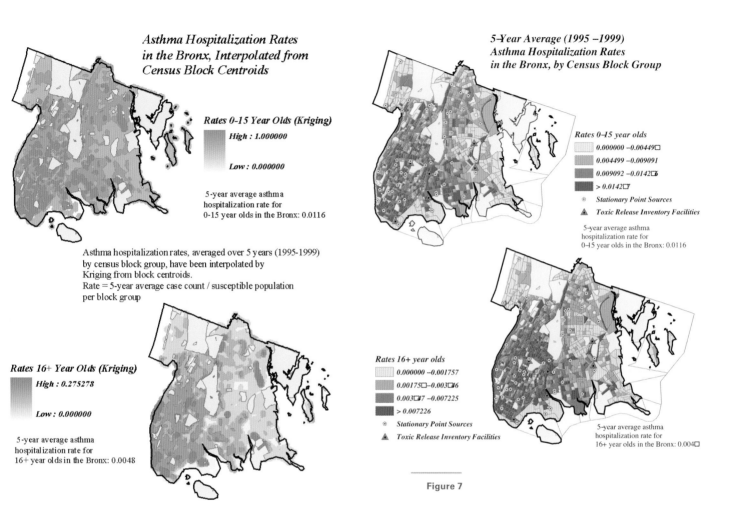

Asthma Hospitalization Rates in the Bronx, Interpolated from Census Block Centroids

Rates 0-15 Year Olds (Kriging)

High : 1.000000

Low : 0.000000

5-year average asthma
hospitalization rate for
0-15 year olds in the Bronx: 0.0116

Asthma hospitalization rates, averaged over 5 years (1995-1999)
by census block group, have been interpolated by
Kriging from block centroids.
Rate = 5-year average case count / susceptible population
per block group

Rates 16+ Year Olds (Kriging)

High : 0.275278

Low : 0.000000

5-year average asthma
hospitalization rate for
16+ year olds in the Bronx: 0.0048

5-Year Average (1995 –1999) Asthma Hospitalization Rates in the Bronx, by Census Block Group

Rates 0–15 year olds

0.000000 –0.00449
0.004499 –0.009091
0.009092 –0.0142
> 0.0142

⊛ *Stationary Point Sources*
▲ *Toxic Release Inventory Facilities*

5-year average asthma
hospitalization rate for
0–15 year olds in the Bronx: 0.0116

Rates 16+ year olds

0.000000 –0.001757
0.00175 –0.003 6
0.003 7 –0.007225
> 0.007226

⊛ *Stationary Point Sources*
▲ *Toxic Release Inventory Facilities*

5-year average asthma
hospitalization rate for
16+ year olds in the Bronx: 0.004

Figure 7

Figure 6

To accomplish this, we had to identify the ID number of the census block group that each hospitalization case was located within so that numbers of cases could be summarized by block groups, and so that rates based on population of the smallest reliable geographic unit of aggregation could be calculated. For this we performed a spatial join (point-in-polygon analysis) in order to append a block group ID number to each asthma hospitalization record. Then, we summarized the number of cases within each block group and calculated rates for each block group based on the number of cases divided by the population of the block group. We plotted the rates for each block group by year, then averaged the cases for each block group over the five years of the data, and calculated rates based on the average number of cases divided by population per block groups. We created an overall rate, a rate for children 15 years and younger, and a rate for adults 16 years and older, all based on the five-year averages.

We then clipped the block group files and the asthma hospitalization case files for each of the five years by each of the four buffers. This allowed us to see only the cases that fell within each type of buffer.

The next major step involved calculating the asthma hospitalization counts and rates of the portions of the block groups that were within each of the four types of buffers. We calculated and summarized the number of cases within each portion of a block group within a buffer. We then performed areal weighting to determine how much of a block group's area was within a specific buffer. The areal weighting script returned a new area (in square meters) for each polygon, representing the part of the block group within the buffer, and from this we calculated a ratio based on the block group's original area in square meters versus the new area—the part of the block group within the buffer. This ratio was then applied to the total population within each tract to calculate the approximate number of people residing within the portion of the block groups within the buffer. We used this estimate and the actual number of cases to develop a more precise rate for asthma hospitalization cases within the buffers. The rates within the buffers were then compared to the rates outside the buffers. In addition, we also looked at poverty rates and percent minority rates inside and outside of the buffer areas, as shown in figure 8.

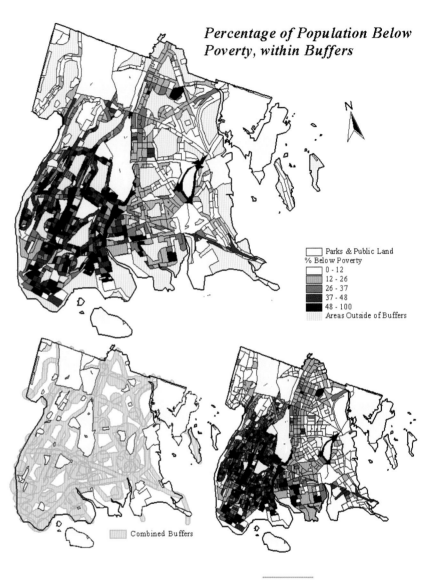

Percentage of Population Below Poverty, within Buffers

Parks & Public Land
% Below Poverty
0 - 12
12 - 26
26 - 37
37 - 48
48 - 100
Areas Outside of Buffers

Combined Buffers

Figure 8

Buffers Used in Estimating Impact Extent of Air Pollution Sources

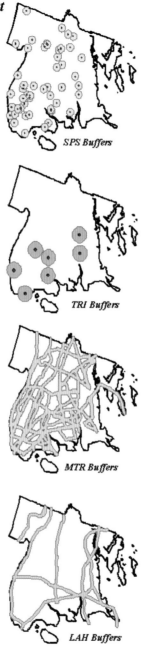

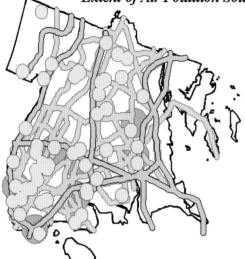

☐ *Stationary Point Source (SPS) Buffers*
■ *Toxic Release Inventory (TRI) Facility Buffers*
☐ *Major Truck Route (MTR) Buffers*
☐ *Limited Access Highway (LAH) Buffers*

SPS Buffers

TRI Buffers

MTR Buffers

LAH Buffers

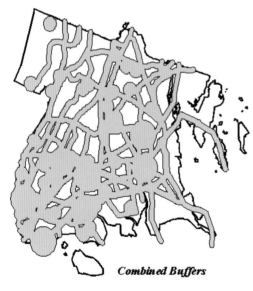

Combined Buffers

Figure 9

Using the point data of asthma hospitalization cases was much more informative than using the rates calculated from data aggregated by census tract. For one thing, we were able to pinpoint where the patients lived in relation to the buffers, rather than assuming that the rate was the same throughout the geographic extent of each census tract, as had been done in Phase One. For another, total population data was aggregated at the block group level, a finer resolution than tract level. Both these factors ensured that the rates we developed in Phase Two were more reliable, and so more accurately reflected reality.

These steps were repeated for each of the four types of buffers, and in each of the five years of SPARCS data, first for the total asthma hospitalizations, and again for children 15 years and younger, and then again for adults of 16 years or older. We also combined all four types of buffers into one spatial file, and calculated the numbers and rates of cases within the cumulative buffers as compared to outside the buffers. After combining the buffers, it was clear that there is actually very little area in the Bronx that falls outside the buffers, as seen in the figure 9 maps—an indication of how environmentally burdened the Bronx is.

Phase Two preliminary findings

Bronx-wide asthma hospitalization rates for children aged 0–15 years old were generally about 2.5 times higher than for the population 16 years and older based on individual year data and the five-year average by block group. These results are shown in table 1 and in figure 10.

In statistically comparing the rates of asthma hospitalization inside and outside the buffered areas, we discovered that people living within the buffers were approximately 29 percent more likely to be hospitalized for asthma than people living outside the buffers. Children under the age of 16 living within the buffers were approximately 17 percent more likely to be hospitalized for asthma than children living outside the buffers. The TRI buffers and the SPS buffers are particularly high-risk areas.

The odds ratios for children under 16 and adults 16 and older for asthma hospitalizations for each year from 1995–1999 were calculated. These compared the risk of asthma hospitalization for those who live in buffer areas surrounding TRI and SPS facilities and along MTR and LAH areas, with the risk of asthma hospitalization faced by Bronx residents living outside these buffer zones. Patterns of increased risk within the buffers were consistent from year to year.

Five-year average rates	
Age cohorts	**Rates**
0–15 years old	0.0116
16 years and older	0.00048

Table 1

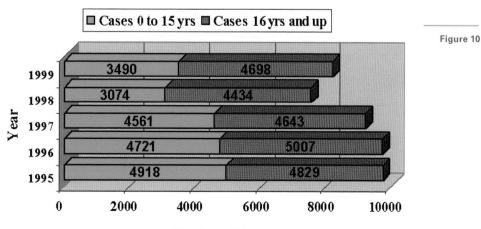

Asthma Hospitalization Cases in Bronx County

Figure 10

For adults living near TRI facilities, the increased risk for asthma hospitalization ranged from 29–60 percent, for SPS sites from 26–66 percent, along MTRs from 7–17 percent, and in all buffer zones combined from 28–30 percent. The increased risk for children living near SPS sites ranged from 14–30 percent, for TRI facilities 16–31 percent, and for combined buffer zones 11–17 percent. No increased risk was found along LAHs and none for children living along MTRs. The increased risk for the total population living in all buffer zones was 25–29 percent. Most of the odds ratios, as shown in table 2 and figure 11, were statistically significant at the $p < 0.01$ level.

Table 2

Buffer type	Adults	Children	Total population
Combined	*1.28–1.30	*1.11–1.17	*1.25–1.29
TRI	*1.29–1.60	*1.14–1.30	*1.33–1.49
SPS	*1.26–1.66	*1.16–1.31	*1.23–1.32
MTR	*1.07–1.17	1.00–1.09	*1.10–1.15
LAH	0.90–0.93	0.83–0.99	0.86–0.93
* indicated results are statistically significant at p<0.01			

Next steps

We intend to fine-tune the analysis with actual modeling of air pollution sources. Using mathematical and cartographic models, such as NOAA's Areal Locations of Hazardous Atmospheres (ALOHA) and EPA's Industrial Source Complex (ISC), will allow us to more accurately estimate the extent of the environmental impact, so that we can tailor each buffer distance and buffer shape to the actual pollutants and environmental conditions at each source. We can also model ambient air quality using air monitor data, and incorporate the use of remotely sensed data, meteorological data, and topographic data to refine the analysis, and to determine more definitively if asthma hospitalization rates are correlated with air quality.

Limitations of the data and analytical methods

There were some limitations of the data and methodology. In the SPARCS data, only 80 percent of cases had been successfully georeferenced; 20 percent of the addresses could not be assigned latitude and longitude coordinates for various reasons, such as incomplete or incorrect addresses or mismatched streets and ZIP Codes. Some of the lat–long coordinates were clearly incorrect—about 2 percent of the coordinates fell within census tracts where there was no population listed in the census. This could also be due to incomplete reporting of population by the Census Bureau or to the fact that, especially in urban areas, many groups such as undocumented aliens are substantially

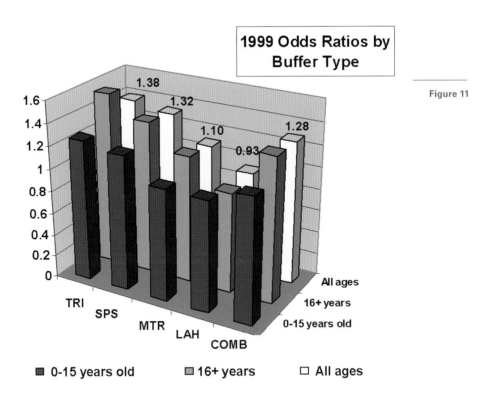

Figure 11

underreported in census counts. These people may be more likely to live in areas zoned for nonresidential uses.

Additionally, data on race and ethnicity and income was not reliable in the SPARCS database. Since the data was self-reported, many records had no entry for these variables. This information was essentially not useable, and we relied on census data to obtain race–ethnicity and income information by block group. This data could not be directly related to asthma hospitalization case records.

Because we had only the census data with which to derive hospitalization rates, based on population numbers at the block group level, we were restricted to the 1990 and 2000 population figures. The asthma hospitalization data covered the years 1995–1999. We decided to use only the 2000 census data, because block group boundaries changed from the 1990 census to the 2000 version, making averages of population figures difficult. However, rates based on 2000 population data may be inaccurate, especially for the years further away in time from the census year (the earlier years in our study time period). Rates for the years closer to the time the census data was collected tend to be more accurate, since case counts and census data reflected more similar populations.

The areal weighting method used to help estimate rates within buffers works well for calculating what proportion of a block group's area is within a buffer, but applying that ratio to population figures or other variables will result in an oversimplification. Population is not necessarily distributed evenly throughout a block group—just because 25 percent of the area of a block group is within a buffer does not mean that 25 percent of its population is within the buffer. This is a common fallacy in spatial analysis, but one needs to take into account that the derived rates may be incorrect due to possibly incorrect population base figures. This could particularly affect rates in industrial areas (where many of the polluting facilities are located) because population counts may actually be much lower than estimates based on areal proportions would indicate. Industrial areas tend to be relatively less densely populated than residential areas. Thus, inflated population counts based on areal weighting algorithms would result in lower rates than actually exist. Lower population counts based on reality would result in increased rates, because asthma hospitalization counts were based on actual georeferenced locations, not estimates.

Another area of uncertainty in the analysis is the establishment of the buffer distances. As mentioned earlier, these distances were based on standards commonly used in environmental assessment, reflecting the distances that air pollutants typical for each source are expected to travel. Altering the buffer distances in the analysis can potentially give quite different results, by increasing or decreasing the counts and/or rates within the buffers. We did, however, conduct a sensitivity analysis. When we altered the buffer distances for the MTR and LAH (using both 100 meters and 200 meters, rather than 150 meters) and the TRI (using ¼ mile, rather than ½ mile), we found no significant differences in rates within these various buffer distances for each pollution source. Establishing more precise impact and exposure extents by modeling is one of the tasks of the next phase.

Another significant shortcoming of the asthma data was that only hospital admissions for asthma are recorded in the database, not emergency room visits for acute asthma symptoms, or cases of known chronic asthma treated on an outpatient basis. Unfortunately, this is the most consistent data available. Doctors and clinics are not required to track or report asthma cases, so at this time, only hospital admissions are reported to the state department of health. In the future, emergency room visits will also be reported, and this will make the dataset much more reliable and comprehensive for analysis, showing a more realistic picture of the extent of the problem.

Acknowledgments

Funding for this project was provided in part by Montefiore Medical Center, Institute for Community and Collaborative Health; PSC–CUNY Faculty Research Awards; the George N. Shuster Fellowship; and NOAA–CREST (Cooperative Remote Sensing Science and Technology Center) Educational Partnership Program with Minority Serving Institutions (EPP/MSI).

This project benefited from the valuable work of the following: Dr. Hal Strelnick, director of the Institute for Community and Collaborative Health, Montefiore Medical Center; Holly Porter–Morgan, PhD candidate at the City University of New York Graduate Center;

Juan Carlos Saborio, Dellis Stanberry, and Susan R. Miller, Lehman College Geography students and NOAA–CREST research interns; Carlos Alicear, PhD candidate at Rutgers University and environmental analyst at the South Bronx Clean Air Coalition; and Jason Fletcher, statistical analyst at the Albert Einstein College of Medicine.

Dr. Maantay also gratefully acknowledges the guidance and support received from Dr. Arthur Getis, Dr. Michael Goodchild, and Dr. Don Janelle during the 2001 summer workshop on Spatial Pattern Analysis with GIS, held at the Center for Spatially Integrated Social Science (CSISS), University of California, Santa Barbara. Thanks are also due to Dr. Bill Bosworth, professor emeritus of Lehman College and director of the Bronx Data Center, for sharing his extensive knowledge of the census data for the demographic analysis.

References and further reading

Ciccone, G., et al. 1998. Road traffic and adverse respiratory effects in children. *Occupational and Environmental Medicine 55*:771–778.

Edwards, J., S. Walters, and R. C. Griffiths. 1994. Hospital admissions for asthma in pre-school children: relationship to major roads in Birmingham UK. *Archives of Environmental Health 49*:223–227.

English, P., R. Neutra, R. Scalf, M. Sullivan, L. Waller, and L. Zhu. 1997. Examining associations between childhood asthma and traffic flow using a geographic information system. *Environmental Health Perspectives 107*:761–767.

Guo Y., et al. 1999. Climate, traffic-related air pollutants, and asthma prevalence in middle-school children in Taiwan. *Environmental Health Perspectives 107* (12):1001–1006.

Jackson K., ed. 1995. *The Encyclopedia of New York City*. New Haven: Yale University Press.

Livingstone, A. E., G. Shaddick, C. Grundy, and P. Elliot. 1996. Do people living near inner city main roads have more asthma needing treatment? Case control study. *British Medical Journal 312*:676–677.

Maantay, J. A. 2002. Mapping Environmental Injustices: Pitfalls and Potential of Geographic Information Systems in Assessing Environmental Health and Equity. *Environmental Health Perspectives 110* (Supp. 2):161–171.

New York City Department of Health. 1999. *Asthma Facts*.

Oosterlee, A., M. Drijver, E. Lebret, and B. Bunekreff. 1996. Chronic respiratory symptoms in children and adults living along streets with high traffic density. Occupational and Environmental Medicine 53: 241–247.

Studnicka, M., et al. 1997. Traffic-related NO_2 and the prevalence of asthma and respiratory symptoms in seven year olds. *European Respiratory Journal 10*:2275–2278.

Van Vliet, P., M. Knape, J. de Hartog, N. Janssen, H. Harssema, and B. Brunekreef. 1997. Motor vehicle exhaust and chronic respiratory symptoms in children living near freeways. *Environmental Research 74*:122–132.

Wilkinson, P., et al. 1999. Case-control study of hospital admission with asthma in children aged 5–14 years: relation with road traffic in north west London. *Thorax 54* (12):1070–1074.

About the authors

Christopher Herrmann is a crime analyst with the New York City Police Department and an adjunct professor at John Jay College of Criminal Justice, City University of New York (CUNY). Chris holds a Bachelor of Science and a Master of Arts degree from John Jay College of Criminal Justice where he is a PhD candidate in crime mapping and spatial data analysis. He also earned an advanced certificate in GIS from Lehman College, CUNY. Chris's current academic interests include the development of spatial models within urban law enforcement environments, point data analysis, and the relationship between land use and criminal events.

Andrew Maroko has worked on funded research from the National Oceanic and Atmospheric Administration's Cooperative Remote Sensing Science and Technology Center (NOAA-CREST), and the National Institute of Environmental Health Sciences (NIEHS) in addition to acting as a consultant in both the private and public sectors. He focuses on environmental justice, biogeography, epidemiology, and public health. Andrew completed his undergraduate work in biology at Rutgers College and received an advanced certificate in GIS from Lehman College, CUNY. He is pursuing a PhD in earth and environmental science, specializing in geography and GIS, from the Graduate Center of CUNY.

Crime Pattern Analysis

Case Study

Exploring Bronx auto thefts using GIS

Ten

Christopher Hermann

Andrew Maroko

The New York City Police Department (NYPD) revolutionized the way GIS and crime statistics are used to pinpoint crime and deploy police officers. NYPD's Compstat process has received national and international recognition for its advanced methods of identifying areas of criminal activity. GIS continues to play a vital role in the crime-reduction strategies implemented throughout New York City. The Compstat process identifies four steps to crime reduction that have been critical to the department's crime-fighting success (NYPD Compstat 1999).

+ Accurate and timely intelligence
+ Rapid deployment
+ Effective tactics
+ Relentless follow-up and assessment

Spatial patterns and crime analysis

GIS helps police analyze spatial and statistical relationships within this "accurate and timely intelligence" framework. In order for police to construct successful crime-reduction strategies, departments need to identify:

+ what crime is happening (robbery, burglary, auto theft, murder).
+ where crime is happening (inside or outside, residential or commercial, specific address, street intersection, type of location).
+ when crime is happening (day or night, day of week, specific time of day).
+ why crime is happening (drug-related, gang-related, domestic).

Identifying spatial patterns and analyzing statistical trends is essential because it affords the GIS analyst an opportunity to develop spatial models and, to some degree, to forecast criminal events. By predicting crime, departments can allocate resources to address both sudden changes and future crime conditions.

Auto thefts in New York City

In 1990, 146,925 autos were reported stolen in New York City, and in 1995, that number was down to 71,543 (Giuliani 1996). In 2000, auto thefts were reduced even further to 37,231 (NYPD 2005). This case study will illustrate how GIS was used to analyze reported auto theft in the Bronx, New York City, in 2000. There were 6,977 auto thefts reported in the Bronx in 2000. Figures 1a and 1b illustrate the distribution of auto thefts by day of the week and by hour of day. While there appears to be little variation with respect to the day of the week, there is considerable variation regarding the time of day. This temporal information, when coupled with geographical analysis, would assist police in targeting "hot spot" locations during specific times of the day.

Figures 1a and 1b

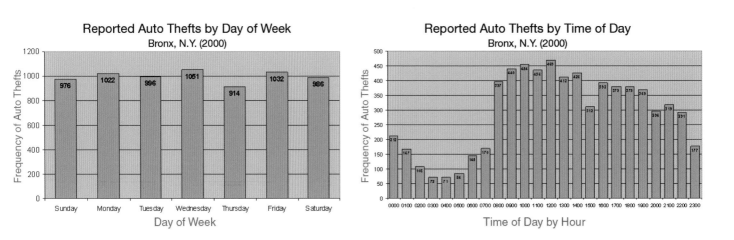

Figure 2 is a dot density map created with ArcGIS software to illustrate the spatial distribution of auto thefts throughout the Bronx in 2000. Spatial distributions and relationships are not always perceptible. It is very difficult to recognize distinct patterns within this map layout due to its large geographical area, the extensive spatial distribution of auto theft locations, and the sizeable quantity of points. When straightforward recognition of patterns is not feasible, GIS analysts can turn to the ArcGIS toolbox to better visualize and conceptualize spatial patterns and processes.

Methodology: Clustering and density analysis

The analysis in this case study combined clustering and density methodologies to describe the spatial distribution of Bronx auto thefts. A valuable feature of ArcGIS is its Spatial Statistics Toolbox which allows GIS users to analyze patterns, map clusters, and measure geographic distributions. There are numerous ways to analyze spatial and statistical data, and they vary according to the geographical extent of the study data. Combining statistical and spatial methodologies allows for more comprehensive explanations of the phenomenon being studied and for more robust models to be developed.

The clustering and density combination has become a popular toolset in the spatial analysis of crime. Clustering or "hot spot" maps illustrate those areas that contain concentrations of crime within a particular geographic region, over a specific period of time (Sherman and Weisburd 1995). In this case study, the clustering process identified 23 locations in the Bronx that contained the highest frequencies of auto thefts for the year 2000. For the clustering and density routines, police analysts used the CrimeStat III software package developed by Ned Levine and Associates. CrimeStat is a free, stand-alone spatial statistics program that analyzes crime incident locations using centrographic statistics, distance analysis, hot spot analysis, and spatial modeling.

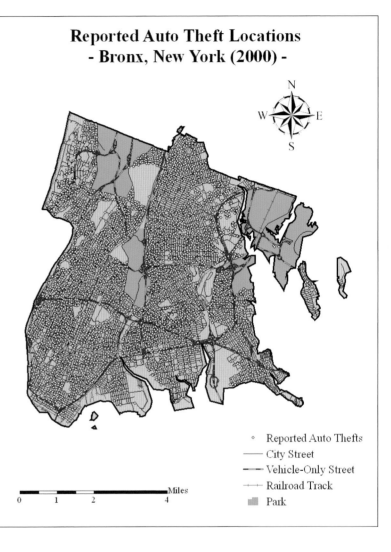

Figure 2

A nearest neighbor hierarchical (Nnh) spatial clustering routine was performed within CrimeStat. There are several advantages to using the hierarchical clustering routine. Nnh models have the ability to target concentrations of criminal events within a small or refined geographical area. The Nnh technique identifies groups of spatially close incidents and clusters points based on the nearest neighbor (next closest point). The technique generates ellipses or convex hulls, or both, around clusters of incidents. The parameters used for this routine included a randomized threshold distance (p<.05), minimum number of auto thefts per cluster (n=10), and export of ESRI shapefile data into both standard deviational ellipse format and convex hull format. Since ellipses are standardized, they will show more regular patterns if the analysis is repeated, and are best visualized in small-scale maps. Convex hulls create a polygon containing each point within the cluster and corresponds directly with the shape of the cluster, and they are best visualized using large-scale maps.

The Nnh spatial clustering routine returned twenty-three first-order clusters and two second-order clusters. The auto theft points that are closer to n or more other auto thefts (in this case n=10) within the specified threshold distance, were grouped into first-order clusters. In addition, several first-order clusters can sometimes be "linked" to form second-order clusters. These different levels of clustering provide the police with information that can be utilized by the police officer (first-order clusters), the precinct supervisor (second-order clusters), and perhaps a borough commander (third-order clusters).

Kernel smoothing

Kernel smoothing, also known as kernel density estimation, is typically considered a more refined statistical hot spot identifier when compared to cluster analysis. "Kernel smoothing involves placing a symmetrical surface over each individual point, evaluating the distance from that point to a referenced location based on a mathematical function, and summing the value of all the surfaces for that referenced location" (Levine 2004). Kernel density estimation continues to be one of the prominent forms of spatial data analysis in large police departments. Hot spot and clustering techniques provide statistics for the incidents themselves, whereas the interpolation techniques utilized by CrimeStat generalize those events based upon the geographical study area. When completed, a density map provides the GIS analyst with an estimation (z-score) for every location throughout the study area.

The Nnh spatial clustering routine differentiates between those auto thefts that belong to a cluster and those that do not. Since the Nnh routine employed a randomized distance parameter and a standardized minimum of ten auto thefts per cluster, some thefts were not allocated to a cluster. Conversely, the kernel density interpolation process estimates the density of all auto thefts within the region over a continuous grid surface. This method allows for visual inspection of both high- and low-density areas, while taking into consideration contextual knowledge of the built environment (street networks, land use, census data) and the natural environment (parks, lakes, rivers).

CrimeStat contains several kernel functions. The quartic function was selected because it weights incidents that occur closer to the center of the bandwidth more heavily. Bandwidth is the radius of the search area to be interpolated around each auto theft point, which gradually decreases with distance until the radius (0.1 mile) is reached. Relative density estimation simply divides the absolute densities by the area of the grid cell. This estimation process allows interpretation of the kernel density interpolation based upon the unit of measurement within the map, in this case, auto thefts per square mile.

Figure 3 illustrates the kernel density estimation with both the first-order and second-order clusters overlaid. Each of the first-order clusters corresponds well with the "hottest" parts of the density map. The second-order cluster of auto thefts in the northeastern section of the Bronx identifies a group of clusters, one of which has the highest frequency (n=48). Further examination of this second-order cluster hot spot reveals that this neighborhood ("Co-Op City") is one of the most densely populated areas within the Bronx. In addition, land-use data (LotInfo 2001) demonstrates that the highest frequency clusters were commercial parking garages located within close proximity to a busy shopping plaza ("Bay Plaza").

This case study demonstrates the utility of GIS for analyzing crime in the dynamic urban environment of New York City. The maps produced in this study begin to answer many of the questions necessary to develop a successful crime-control strategy.

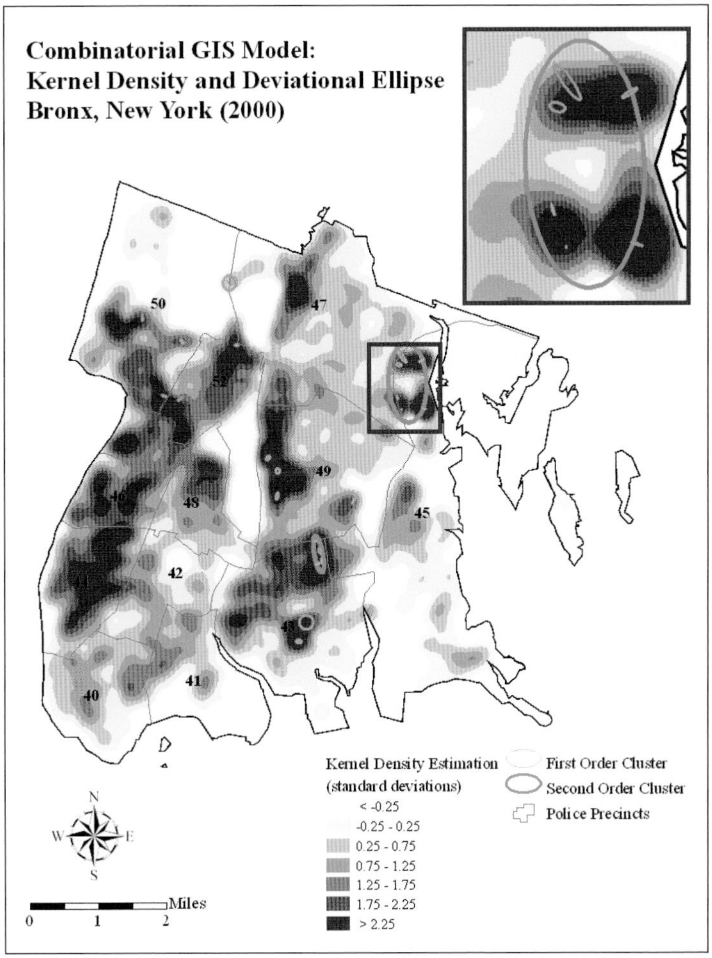

Combinatorial GIS Model: Kernel Density and Deviational Ellipse Bronx, New York (2000)

Figure 3

References and further reading

Giuliani, R. 1996. Archives of Rudolph W. Giuliani. Hearing on Insurance Commission. Senate Hearing Room, New York. *www.nyc.gov/html/rwg/html/96/insure.html*.

Levine, N. 2004. CrimeStat III: A Spatial Statistics Program for the Analysis of Crime Incident Locations (version 3.0). Ned Levine & Associates: Houston, Texas/ National Institute of Justice: Washington, D.C.

LotInfo. 2001. LotInfo, LLC, New York. *www.lotinfo.com*

New York City Police Department. 1999. CompStat: Leadership in Action.

Sherman, L. and Weisburd, D. 1995. General deterrent effects of police patrol in crime hot spots: A randomized controlled trial. Justice Quarterly 12: 625–648.

About the author

Thomas Paino, an architect for 25 years, earned an advanced certificate in Geographic Information Science from Lehman College. He currently uses GIS as an organizing tool in managing facilities for the New York City Department of Environmental Protection. It is his role as chair of the Hunters Point Community Coalition, a not-for-profit neighborhood preservation group, that led him to the graffiti survey described in this case study.

The Coalition had benefited from early use of GIS software by employing the services of Space Track Inc. to create a visual simulation of waterfront proposals using verifiable mapping criteria as a basis; this led to greater community involvement in the planning process.

He is a graduate of City College, CUNY, and a registered architect in New York.

Community-based Planning

Case Study

Eleven

Graffiti in Community District Two, Queens, New York: Private expression in the public realm

Thomas Paino

Background

It has been estimated that $750 million per year is spent in New York City to rid its buildings of graffiti (Austin 2001). According to *Newsday,* the Metropolitan Transportation Authority spends $3 million a year to control *scratchiti*—a graffiti hybrid produced with carbon steel implements against glass resulting in permanent markings on train and bus windows. These figures do not include the cost of enforcement and other police activity associated with graffiti control. Despite these substantial expenditures, one would be hard-pressed to find any two continuous blocks in New York City without some form of graffiti. This is particularly true in the boroughs, outside Manhattan tourist districts, where the phenomenon is ubiquitous.

Besides the fact that New York has one group of people creating graffiti (risking incarceration to do most of it) and another devoting enormous resources to getting rid of it, there exists a huge range of opinion about this cultural phenomenon that affects the daily visual experience of millions of New Yorkers. National graffiti magazines offer $1,500 fees to anyone submitting a photo of a recent piece the magazine deems laudable, while police detectives lament the fact that the Constitution prevents them from pursuing such publishers. Some property owners invite well-known local graffiti artists and artists from abroad to express themselves on the walls of their buildings, while vigilante groups form in other residential neighborhoods for the purpose of intimidating taggers, as graffiti artists are called. Clearly, the freedom to express oneself in this manner is a matter of very heated debate. At one extreme of opinion, such expression is considered a cultural art form, while at the other, graffiti is considered a sign of the breakdown of society, whose appearance causes entire neighborhoods to spiral down into crime-ridden wastelands.

Taking into account these extremes and contradictory behavior, it was decided to use GIS to describe, record, locate, and analyze the phenomenon and research the topic within an urban context that already has a host of descriptive data available. By placing a graffiti database layer over existing environmental descriptors and converting the mixed data to a graphic map image that could be manipulated, a much broader, yet also more precise view of this cultural and visual curiosity could be examined. No moral judgment of graffiti was made during this research.

Figure 1

Geographic study area

The graffiti phenomenon is international and this research process could probably be applied to any geographic area. In New York it is a citywide phenomenon. The district of Community Board 2 in Queens was chosen for several reasons. The district, shown in figure 1, covers approximately 500 blocks or about five square miles. It includes a wide range of land uses, from heavy industrial to high-rise commercial to low-density housing. Almost 20 percent of the land area consists of cemeteries and rail yards. Another 35 percent is industrial, including very large-scale, super-block entities and mid-block artisan studios. Census data indicates the rest is residential, accommodating approximately 100,000 residents, 25 percent of whom are described as youth. This variety allows a broad range of possibilities for linking land use to the presence of graffiti.

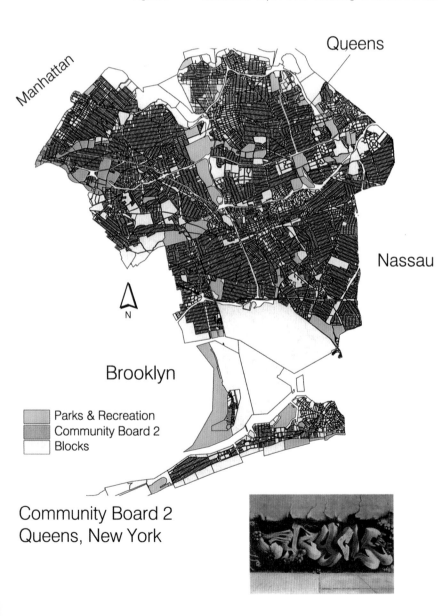

Manhattan

Queens

Nassau

N

Brooklyn

Parks & Recreation
Community Board 2
Blocks

Community Board 2
Queens, New York

The 108th Police Precinct covers the district and has the second highest arrest rate for graffiti infractions in Queens, according to police data. Police use a variety of techniques, including mounting specialized patrols, cooperating with community groups to eradicate graffiti from the area, and providing anti-graffiti technical assistance to property owners, resulting in additional anti-graffiti measures such as razor-ribbon fencing and paint applications which make graffiti removal easier. But the district also is home to at least four property owners who encourage graffiti; one of them invites artists from around the world to paint up the walls of his two-block industrial complex. The land-use mix, the substantial level of graffiti activity, and the extreme range of the phenomenon's acceptance made Community Board 2 a good place to conduct the study, as figure 2 shows.

Data and software

In order to analyze patterns of graffiti by location and by urban criteria such as land use, building type, or building age, data and software associating data to the smallest urban entity would be required. LotInfo software was chosen since its smallest denominator is the individual property lot keyed into the database by a unique identifier derived from the New York City tax lot and block number. The software, using the State Plane coordinate system, combines New York City Department of Finance data for each property lot with the Department of City Planning's Bytes of the Big Apple tax lot basemap. Each tax lot is its own polygon associated with a data table giving lot size, zoning, use, building size, building type, building age, property value, property owner, mortgage, and tax information. Each lot is geocoded by property address and street name. These descriptors provided a basis by which the graffiti information could be keyed to a specific address and lot number, and to all other data associated with that lot number. Photographs of specific graffiti could be linked to a specific property lot.

LotInfo has been designed to work in tandem with GIS software and mapping programs such as ArcView 3.x and ArcGIS.

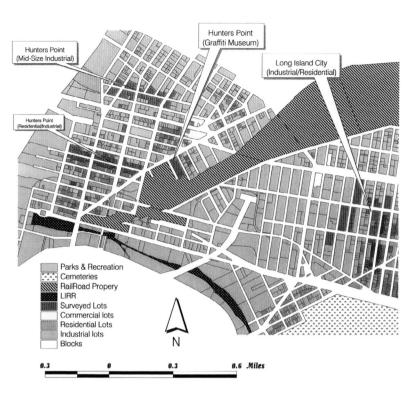

Legend:
- Parks & Recreation
- Cemeteries
- RailRoad Propery
- LIRR
- Surveyed Lots
- Commercial lots
- Residential Lots
- Industrial lots
- Blocks

Hunters Point (Mid-Size Industrial)
Hunters Point (Graffiti Museum)
Long Island City (Industrial/Residential)
Hunters Point (Residential/Industrial)

0.3   0   0.3   0.6 Miles

Pilot Study Area Context (Western Half)

Figure 2

Mural

Bubble letter

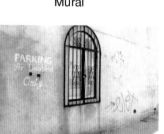

Message

Wildstyle

A/C fin compression

Tag

Figure 3   Graffiti types.

Determining graffiti type

Perusing some of the hundreds of Web sites featuring graffiti quickly provides the style characteristics of graffiti. When combined with a walk-though of Community Board 2, these styles predominated: tag, bubble letter, wildstyle, mural, and message. The glossary at the end of this case study and figure 3 explain and illustrate some of these styles. The mural is often an involved piece, using letter and symbol imagery mixed with stylized illustration. It is often colorful and meaningful to the immediate community and accomplished through cooperation with a building owner. The police oppose murals since they could contain imagery and messages offensive to a rival street group and so could possibly provoke additional crime. The message category found in this study, unlike the politically influenced version common outside the United States, is limited to profanity, and to actually useful information, such as "No Parking."

Pilot study area

As Community Board 2 is made up of five distinct neighborhoods and three basic land-use types, determining whether the degree of graffiti incidence is affected by neighborhood alone or by land use would require that both criteria be considered. The following combinations were established for potential survey areas: Residential/Commercial, Residential/Industrial, Large-Scale Industrial, Mid-Size Industrial. The Hunters Point neighborhood, a dedicated mixed-use area with an industrial zone underlayer, had its own unique category that required laborious, tedious table editing and a ground-truthing survey to designate individual properties accurately.

To establish the exact location that would serve as a model for the pilot area, a virtual walkthrough of the LotInfo map database was done, using the identify tool for potential lots to be surveyed; this revealed the exact use and condition of the building on that lot. This process was applied to each pilot model range and helped to reduce on-foot scouting in the field since much of the contextual information could be established beforehand. A scouting expedition of each area was also made to check for anomalies that might skew the total count. This was important if the results of the model areas were to be extrapolated to cover the entire study area.

Conducting the survey

Using the LotInfo software, grayscale maps of the model blocks for each study area were printed, with the lots to be surveyed consecutively numbered. A survey form was created that allowed the entry of information for each lot that quantified graffiti per type. Other information columns included building surface, a factor that could affect graffiti incidence; public exposure, to quantify whether a building faced the street directly or was obstructed by features such as fences, gardens, or other vegetation; and anti-graffiti to record if the building surface was treated with an anti-graffiti substance. One column was added for comments and another for recording the unique number for each lot, which is a combination of the borough code, block, and lot numbers. This last column was extremely important, because it was the link between surveyed information and the software-provided database for each lot. Photographs of representative graffiti were taken and keyed to the subject property.

The survey took several days over the course of three weeks. A total of 494 street-facing properties were surveyed on 65 blocks, which computes to about 7 percent of the total area of Community Board 2. A total of 123 photographs were taken on 5 rolls of film.

Data analysis

Completed survey forms were added to the existing LotInfo tables for each property. This enabled pertinent queries to be made and then a determination of which criteria should be mapped. Each property surveyed was keyed to its unique number through the use of the identify tool. This allowed the joining of the surveyed graffiti data to the standard data describing each property. However, since the standard tables have almost 65 fields, it was necessary to eliminate those fields not useful to this study, such as tax valuation and market value ratios. One large database table was created combining graffiti fields and property information for all property lots in Community Board 2.

A review of the tables allowed for an assessment of the criteria most likely to have a relationship with graffiti incidence. Zoning, land use, building type, and building owner all appeared to be related to the incidence of graffiti. Perhaps the most important finding in the study was the connection between graffiti incidence and the criteria of public exposure. Consistently, those buildings with an object such as a fence, plant, or ornaments placed between the building façade and the sidewalk were hit less frequently. Conceivably, evidence of a well-maintained building sends a message to the tagger to go elsewhere. Curiously, the application of anti-graffiti surface coating did not seem to have any effect whatsoever.

Data Source: LotInfo
Visual Survey

Graffiti Count
- 0
- 1 - 3
- 4 - 32
- 33 - 70
- 71 - 217
- Parks & Recreation
- Cemeteries
- RailRoad Propery
- LIRR
- Property Lots
- Blocks

0.3   0   0.3   0.6 Miles

N

Number of Graffiti Per Surveyed Lot (Western Half)

Figure 4

Data manipulation and map production

Once familiarity with the data was attained, its manipulation for the desired mapping results could be accomplished. To establish the Community Board 2 basemap, a query of all properties within the district was made and isolated from the master Queens County database, and a highlighted location map was created, as seen in figure 4. To isolate the 494 surveyed properties, a query of the anti-graffiti field was performed. Every surveyed property had either a "Y" or "N" in this field, guaranteeing its inclusion. These were then highlighted and shown within the zoning context. Neighborhood names were identified. Also, since from the survey, lots facing railroad properties seemed to be particularly vulnerable to graffiti, it seemed prudent to call out railroad lots with a unique symbol.

The next challenge was to look at the total numbers of graffiti incidents throughout Community Board 2. Each incident

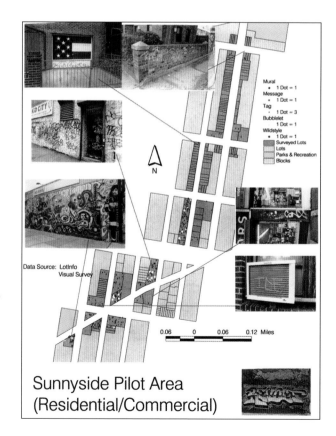

Mural
· 1 Dot = 1
Message
· 1 Dot = 1
Tag
· 1 Dot = 3
Bubblelet
· 1 Dot = 1
Wildstyle
· 1 Dot = 1

Surveyed Lots
Lots
Parks & Recreation
Blocks

N

Data Source: LotInfo
Visual Survey

0.06   0   0.06   0.12 Miles

Sunnyside Pilot Area (Residential/Commercial)

Figure 5

was quantified by type in the project table so that a choropleth map could be produced. A natural-breaks classification was edited so that properties with no graffiti would be placed in one category and those with extremely high quantity in another. By referencing both the choropleth map and the contextual map it became apparent that industrial properties are more likely to have graffiti, and to have more of it than other building types.

The defining blocks for each of the five pilot models were highlighted and promoted to form a new basemap. A dot matrix graph for each graffiti type per surveyed lot was created. In addition, photographs of representative graffiti were linked to the property where they were located. To guarantee accuracy in the placement of the source photographs, the unique lot and block number was keyed to each photo. In the table, this field actually houses the directory path where the corresponding photo can be found so that a hyperlink to the graphic output can be established. This is how the example of the latest graffiti form—air conditioner fin compression—was accurately identified to the correct building. These maps, three of which are shown in figures 5, 6, and 7, clearly show that tagging is the most prevalent graffiti type, but they also indicate some interesting anomalies such as relatively clean properties in the midst of buildings with numerous hits, and buildings with murals that have no tags.

Figures 6 and 7

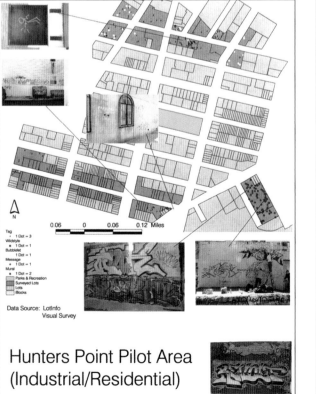

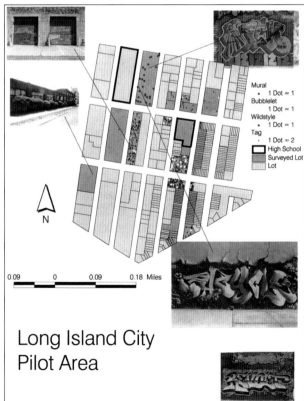

Conclusion

In some ways this research project has confirmed the obvious: graffiti is everywhere—not one block in the pilot models is completely clear of graffiti. Tagging is the most prevalent form of graffiti, and industrial buildings are the most likely to get hit. On the other hand, the project also raised new questions: Why are some buildings that are adjacent to heavily graffitied properties themselves clean? Why are residents the most outspoken about graffiti, yet residences are the least likely building type to get hit? Why are buildings covered with murals not buffed? And to return to an earlier point, why is it that so much money spent on getting rid of graffiti has had such little effect?

Differences in graffiti intensity do appear to be associated with areas identified with different ethnic groups, but such cultural connections are difficult to map. For example, in one personal interview for this study, a Chinese immigrant landowner in Queens commented on a rumored phenomenon that Chinese-owned buildings tend to have more graffiti. He agreed, saying that Chinese owners are more concerned with the income from a building than its appearance. "Only Italians borrow money to improve a building," he said.

Such a statement helps to support the observation that Chinese property owners tend to look solely at the income from property ownership whereas the Italians place a high value on the property's appearance as well. This finding would be significant to a municipality engaged in an anti-graffiti campaign, and might influence its decisions about translating its anti-graffiti informational literature.

The data for this research was collected in a manner that permitted many other analyses to be done. For example, the amount of graffiti by building type could be plotted, as could the amount of graffiti by owner type. This could answer questions about graffiti presence on subsidized housing as contrasted with private ownership. Cultural differences could also be explored further.

The study could also be applied to other forms of visual expression that confront the public. Most of the visual cacophony of the urban experience is caused by signage and advertising, much of it misplaced and in violation of zoning ordinances. A comprehensive GIS study of visual defacement and quality of life could reveal much.

Mapping will not resolve these issues. However, an overall picture of the phenomenon is the first step in deciding how much money, time, and effort one will devote to eradication of this uniquely urban form of self-expression.

Glossary

Bite	To copy another's style or tag.
Bubble Letter	A bold style of graffiti with letters inflated like balloons almost to the point of illegibility.
Buffing/Ragging	Cross out or go over an existing tag with your own.
Hit	The successful completion of making a graffito.
Racking	Stealing the items necessary to create graffiti.
Tag	Signature in the form of name, graphic, or mark, usually done hastily with spray paint.
Wildstyle	A powerful style of graffiti expressed through colorful angular and crisscrossed letters and arrows.

References and further reading

Austin, Joseph. 1998. "A Framework for Development." New York City Department of City Planning. 2001. New York: Columbia University Press.

Newsday, Man Charged in Subway Graffiti, April 5, 2002.

www.sixcentz.com

www.graffiti.org

About the authors

Steven Romalewski joined the New York Public Interest Research Group (NYPIRG) in 1984 as an environmental researcher and advocate, and in 1994 brought GIS to NYPIRG as a component of its environmental work. Steven was awarded a Revson Fellowship at Columbia University in 1995, and continued on at Columbia to receive a Master of Science degree in Urban Planning in 1998. He expanded his expertise in GIS and urban issues while at Columbia, and launched CMAP while completing his graduate studies.

Christy Knight Spielman joined the Community Mapping Assistance Project staff in November 1999. She holds a Master of Arts degree in Geography from the University of Kansas and a Bachelor of Science degree in Geography from Kansas State, and has extensive experience with computer mapping and graphic design. She is a Peace Corps veteran, having worked for two years in Burkina Faso in West Africa, where she taught English and cartography to university students.

Advocacy Planning *and* Public Information

Community Mapping Assistance Project for the Metropolitan Waterfront Alliance Open Accessible Space Information Systems Internet map site

Steven Romalewski

Christy Knight Spielman

Advocacy GIS mapping for the Metropolitan Waterfront Alliance

The Metropolitan Waterfront Alliance (MWA) is a network of organizations and concerned individuals dedicated to helping the New York City metropolitan region reclaim and reconnect to its greatest natural resource—the harbor, rivers, and estuaries of New York and New Jersey. The MWA project focus fills in a gap in GIS data and services created by city and government agencies that are more focused on collecting information on land resources. MWA shares its knowledge of waterfront resources with NYPIRG's Community Mapping Assistance Project (CMAP), which provides mapping services and geographically accurate data that nonprofit organizations can put to use in a variety of maps. For example, MWA has worked for several years to improve ferry service in and around New York Harbor through the New York–New Jersey Harbor Loop Ferry Plan, which connects mass transit and waterfront access points. The harbor, where the Hudson River meets the Atlantic Ocean, is home to international landmarks such as the Statue of Liberty and Ellis Island; it is also a hub of the region's shipping and transportation network. Dozens of development projects are underway, encompassing more than seven million square feet of office space, three thousand housing units, one thousand hotel rooms, and the expansion of thirteen different cultural attractions, all totaling nearly two billion dollars worth of new construction. At the same time, access to other notable harbor sites, including Governors Island, Homeport/Stapleton, and the Military Ocean Terminal at Bayonne is hampered by poor transportation access.

MWA wanted to create a plan to add new ferry stops. These would enhance regional mobility, improving the quality of life in communities underserved by transit and creating new work, living, and recreational opportunities for residents.

Figure 1

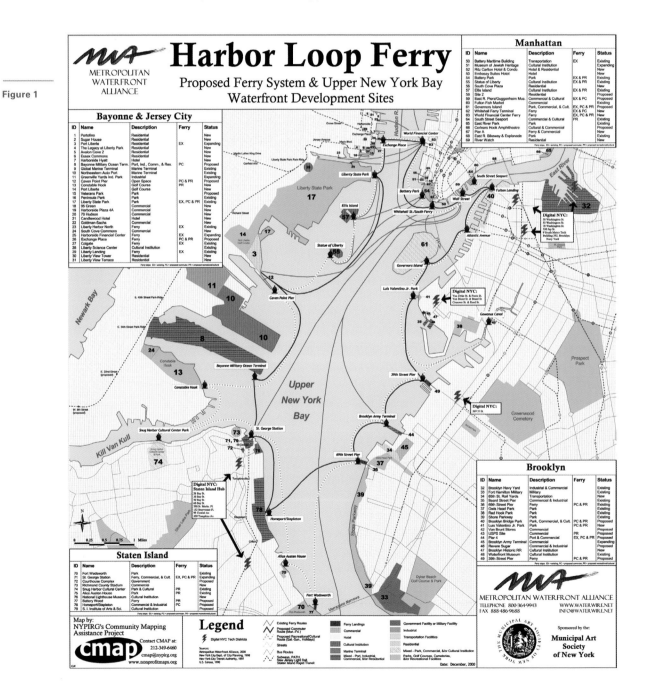

MWA came to CMAP to help visualize its Harbor Loop proposal. MWA provided a list of ferry stops and routes, and CMAP created digital versions to match existing basemap data. A series of maps was then created to show existing and proposed ferry routes, illustrating the potential for a revitalized waterborne transit system. These data and maps helped MWA secure a grant from New York's city council to implement the Harbor Loop system. As the Harbor Loop grows, MWA and CMAP continue to update the maps as needed (figure 1).

Several other examples show how GIS mapping and analysis help MWA fulfill its mission.

Maps for educating policy makers

By juxtaposing waterfront resources and New York City council districts, in 2001 MWA helped to educate political candidates about waterfront issues that were of concern to their constituents. CMAP and MWA converted these constituent concerns into powerful maps, such as the one shown in figure 2, that quickly conveyed a great deal of information to busy politicians.

Fliers for educating the public

After the tragedy on September 11, 2001, commuting conditions in New York changed drastically and frequently. MWA and CMAP quickly responded by providing maps with updated ferry route information, and distributed these maps by e-mail and by hand from the piers of lower Manhattan.

Figure 2

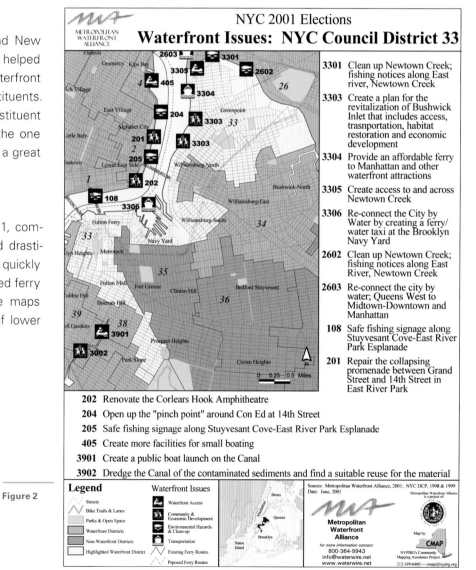

Data for interactive map applications

MWA shares the data it has collected with other interested organizations and institutions to help spread awareness of the importance of the waterfront to the region, and to help them with their own projects, such as interactive maps on the Internet to make waterfront information more accessible to the public. These organizations include the Open Accessible Space Information System (OASIS) and the Municipal Art Society's Community Information Technology Initiative (CITI).

The versatility of the data collected by MWA and digitized by CMAP is in large part due to careful matching of the data to the New York City basemap that is used by other institutions and organizations doing geographic data collection, digitizing, and analysis.

OASIS

The Open Accessible Space Information System (OASIS) (*www.oasisnyc.net*) is a partnership of more than thirty federal, state, and local agencies, private companies, academic institutions, grassroots groups, and nonprofit organizations focused on creating a one-stop, interactive mapping and data analysis application via the Internet, designed to enhance the stewardship of open space for the benefit of New York City residents. OASIS is a community-based undertaking, whereby local organizations can design and test a citywide, Web-based, open-space mapping resource for New York City. OASIS facilitates and focuses the delivery of GIS resources to provide timely and accurate information about the green infrastructure of NYC.

The vision for OASIS was to harness the growing power of online mapping tools so average citizens, neighborhood groups, and others could better appreciate and care for open spaces—whether those are community gardens, wetlands, or parks. The project was spearheaded by the U.S. Forest Service's Urban Resources Partnership, whose leaders realized that communities in urban centers were most in need of such tools. No place is more urban than New York, and so the city was chosen as the project launch site.

OASIS has two components. First, it is a collaborative partnership, bringing together disparate agencies, organizations, and individuals who share an interest in open-space issues. The OASIS partnership includes federal agencies, including the EPA and USGS, and local agencies including the New York State Department of Environmental Conservation and New York City Department of Parks and Recreation. Nongovernmental participants include firms such as ESRI, institutions such as Columbia University, and nonprofit agencies including New Yorkers for Parks, the Municipal Art Society, and the Environmental Justice Alliance.

A second component is collaboration—these groups have come together to pool their resources and create an online repository and delivery vehicle for open-space data. No other single source provides access to this much information in New York—almost four dozen layers of spatial data about the city's green infrastructure.

The information provided by OASIS is delivered to the public on the Web using ESRI's ArcIMS technology, allowing for interactive mapping by users, letting them query mapped data, and create custom maps of areas of interest. The maps that users create in response to their spatial queries can then be printed out or customized on-screen (figure 3).

Other OASIS Web site capabilities include:

✦ Interactive maps of open space by neighborhood.

✦ Identification of elected officials whose areas include parks and community gardens.

✦ High-resolution aerial imagery to let users locate trees and recreation areas.

✦ Detailed land-use data, including potential open space such as vacant lots.

✦ Wetlands, wildlife areas, and historic landmarks.

✦ Layering of census demographics with open space land-use patterns.

The OASIS Web site lets the public:

✦ Create maps of open space by ZIP Code, borough, tax block and lot, and/or neighborhood.

✦ Identify key open-space resources within or near a user-defined area.

✦ Locate these resources by name, type, and other attributes in addition to geographic-based searches.

✦ Identify other natural resources and landmarks near or adjacent to open spaces in the city.

✦ Calculate statistics based on open-space patterns by ZIP Code, borough, tax block and lot, and neighborhood.

✦ Undertake what-if scenarios, such as "What would my neighborhood look like if these vacant lots were turned into community gardens?" or "How would new bike lanes or bus routes improve my access to a park in the Bronx?"

✦ Use other mapping and data analysis tools.

Figure 3

The OASIS partnership is intended to address several problems. For community organizations and local residents, it makes green infrastructure data accessible to groups that are without GIS resources, strengthening these groups' abilities to participate in government decision making, since they are armed with accurate and detailed information and analysis about open-space issues in their neighborhoods. For example, community gardeners use the OASIS Web site to locate vacant lots that are in close proximity to their gardens which they then map and print out for meetings with local legislators.

For government, the Web site helps reduce the costs and redundancy of open space mapping efforts across agencies. It helps these agencies identify incomplete or inaccurate mapping data, especially at the neighborhood or community scale.

For all sides, the Web site helps reduce the need for independent mapping efforts that were duplicative and inadequate because of incomplete data. Further, it enables communities to nurture open space by creating maps of open space on demand, and by identifying key open-space resources and stewardship activity near a user-defined location.

New Yorkers for Parks and the New York City Audubon Society have compiled an inventory of natural areas in New York's parks, which has been incorporated into OASIS. Both groups participate in OASIS so that the general public can access information from their inventories using OASIS maps; in turn, each group benefits from sharing strategies and information resources with other OASIS partner organizations.

In addition, OASIS provides links to other Web sites, such as the Municipal Art Society's Community Information Technology Initiative (CITI), located at *www.MyCITI.org*. MAS worked with NYPIRG, CMAP, ESRI, and Space Track Inc., to integrate open-space data with CITI's maps, basing its own interactive mapping Web site for Community Boards on the OASIS system. This approach also helped MAS reduce expenses by using the online mapping platform already created for OASIS.

Other projects leveraging the success of OASIS include open-space stewardship organizations that are developing a system to display existing stewardship efforts and future stewardship needs throughout the five boroughs. There is also the Forest Service's Living Memorials Initiative that uses OASIS to display the locations of more than three dozen living memorials in the region created in response to the events of September 11, 2001.

The model of the OASIS partnership, the Web site itself, and these community applications are transferable resources that are being shared among partners in the planning community. Organizations in Washington, D.C. (the Casey Trees Foundation), Baltimore (the city and Maryland natural resources agencies), Boston (the Urban Ecology Collaborative), and Detroit (the city's planning agency) are all following closely the developments with the OASIS project to develop similar initiatives in their cities.

Part 3

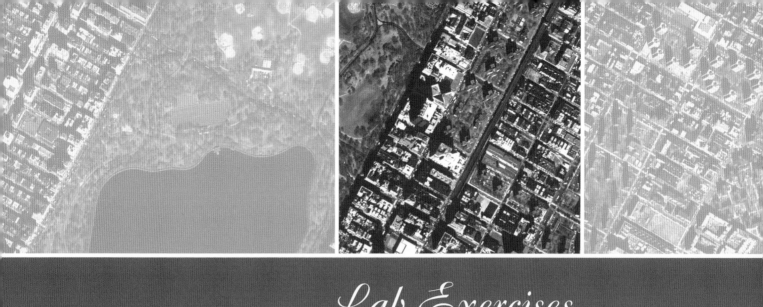

Lab Exercises

As noted in the preface, we feel these lab exercises will be of optimal benefit to the student or reader in a GIS lab setting where ArcGIS 9 software is installed and where an instructor experienced in ArcGIS is within easy reach to guide, answer questions, or respond to cries for help. While no previous experience with ArcGIS is required, some familiarity with this technology will likely moderate the steepness of the learning curve. It is expected, however, that students will know their way around the Windows 2000 or Windows XP environment, the operating system required to run ArcGIS. However, this book can also be of benefit as a self-study workbook for the astute reader who has ArcGIS installed on a home or work computer.

Note: These exercises were developed with ArcGIS 9.0. If you are using a newer version you may notice some slight differences between the graphics and what you see on your screen.

Each lab exercise is divided into a series of tasks and a task may contain several steps. Task direction and explanation are more detailed in the early exercises; the more advanced exercises later in the section assume that learning has taken place and the necessity for step-by-step explanation has diminished. Often, instructions for that operation in an exercise will be found in the assignment where they were originally given.

Extra Challenges are given in many of the exercises. These can be done, time permitting, by students who have completed the basic tasks of the lab exercise and who want to develop their GIS skills further.

Advanced Challenges are provided in many of the lab exercises for those students wanting an additional challenge. These tasks are generally given without detailed instructions and are intended for those students who easily and comfortably completed the basic tasks and want more in-depth experience with GIS.

Our thanks to Andrew Maroko of the Lehman College Urban GIS lab for concept and development work on lab exercise 8, and to Greg Studwell of the Municipal Art Society for concept and development work on lab exercise 12.

Bronx demographic data is courtesy of the Bronx Data Center at Lehman College. Our thanks to Professor William Bosworth, director, for his assistance.

A CD–ROM included with this book contains the exercise data. Installation instructions are on page 579.

1   **Making Maps of Urban Data** *439*

2   **Exploring Basic GIS Functionality** *443*

3   **Thematic Mapping: Dot Density Maps** *459*

4   **Thematic Mapping: Choropleth Maps** *471*

5   **Integrating Graphs and Maps, and Designing Map Layouts** *483*

6   **Developing an Attribute Database from an Internet Source** *499*

7   **Geocoding and Labeling** *505*

8   **Working with Relational Databases** *519*

9   **Generating Buffers and Using Selection for Proximity Analysis** *531*

10   **Geoprocessing Operations and Joining Tables** *541*

11   **Data Exploration and Geostatistical Analysis** *549*

12   **Advanced Layout Techniques** *555*

Making Maps *of* Urban Data

Many municipalities maintain interactive Web sites for planning and public information uses, as well as to support municipal decision-making and operations. These Web sites might contain, for instance, information about demographics, environmental conditions, topography and physiographic setting, land uses, zoning, crime incidents, properties and real estate, open space and parks, landmarks and visitor attractions, and municipal services, among many other sets of data.

In this exercise, you will access urban GIS Web sites that feature interactive digital maps showing various aspects of the physical and social characteristics of a city or cities. Some representative Web sites maintained by cities, towns, and metropolitan areas are listed below. (There are also many others that you will be able to find on the Internet.) Select one or more sites and explore the information contained therein. Each Web site has a section for interactive maps, which allow you to choose an area of interest and manipulate the map with simple tools. Using the navigation and display tools, create two basic maps that are illustrative of the data available on the Web site, then print them out.

The following Web addresses are subject to change.

San Diego, California

www.sangis.org

The Web site of SanGIS, a joint powers agency of the City and County of San Diego, responsible for maintenance of and access to regional geographic databases.

Baltimore County, Maryland

www.co.ba.md.us/Agencies/infotech/geographic_information_systems

The county's enterprise geographic information system offers online interactive maps. Internet map service (IMS) technology allows Web users to access the same map data used by county agencies and departments.

Miami–Dade County, Florida

gislab.fiu.edu/gisrsal/knight/intro.htm

The Miami–Dade County Interactive GIS Mapping is a product by Florida International University Geographic Information Systems/Remote Sensing Center sponsored by Knight Foundation. Data contributors include Miami–Dade County, FIU Metropolitan Center, and U.S. Census Bureau.

New York City, New York

www.oasisnyc.com

The New York City Open Accessible Space Information System (OASIS) cooperative is a partnership of more than thirty federal, state, and local agencies, private companies, academic institutions, and nonprofit organizations to create a one-stop, interactive mapping and data analysis application via the Internet to enhance the stewardship of open space for the benefit of New York City (NYC) residents.

Orlando, Florida

www.cityoforlando.net/gis

The city's Web site includes an interactive mapping function for Orange County and a gallery with dozens of standard maps of Orlando available to view and print.

Erie County, New York

www.erie.gov/environment/planning_ecdev/gis.asp

The county's Web site provides access to geographic data and technology, allowing users to visualize and query geographic data, including parcel boundaries, hydrology, NYS DEC and federal wetlands, floodplains, and demographics.

Ithaca, New York

www.ithacamaps.org

The city's Web site shows interactive views of properties, infrastructure, zoning, and other features and boundaries. This service is used daily by city staff, elected officials, consultants to the city, other agencies, citizens, and students.

Exploring Basic GIS Functionality

Two

In this assignment, you will be exploring basic GIS functions by working with a global dataset and focusing on the countries of West Africa. By using spatial and attribute data, you will change the geographic extent, zoom in and out, pan, make selections, sort the dataset, and perform simple queries. You will also label features, symbolize the data, and make a basic thematic map. In subsequent lessons you will build on the tasks you learn here, adding legends, scale bars, north arrows, and inset maps to your map layouts.

Unless otherwise noted, all mouse clicks are single and done with the left mouse button. However, double-clicks with the left mouse button and single-clicks with the right mouse button are also frequently used in ArcGIS.

TASK 1

Open ArcMap and add data

1   Start ArcMap by double-clicking the ArcMap icon on your computer desktop. (Alternately, click the Start button on the Windows taskbar, point to Programs, point to ArcGIS, and click ArcMap.)

2   In the ArcMap dialog, click the option to Start using ArcMap with a new empty map, then click OK.

3   On the Standard Toolbar click the Add Data button.

In the Add Data dialog, navigate to your **C:\UrbanGIS** folder—or the folder your instructor directs you to—and double-click the World.mdb geodatabase. Hold down the Shift key and click on all of the feature classes listed (cities, latlong, rivers, country, lakes, world30), then click Add.

Once a feature class is loaded into ArcMap it is referred to as a layer. The order of the layers in the table of contents (TOC) box on the left side of your screen will determine the order in which the data layers are drawn. You will want point data and line data placed above polygon data layers in the TOC, so that point and line data will also be viewed as above polygon layers in the map display. You can change the order of the data display by clicking the data layer to be moved and dragging it with the mouse to the top of the list, while holding down the mouse button.

TASK 2

Toggle layer visibility

When a layer is added to ArcMap, it is automatically turned on, which means that it is visible in the map display. You can toggle a layer's visibility on and off by checking the box to the left of the layer name in the ArcMap TOC. Toggling a layer's visibility is usually referred to as turning a layer off and on.

4 In the ArcMap table of contents, click the checkmark in the box next to the latlong layer to turn the layer off. Notice the change in the map display. Check the box next to the latlong layer to make it visible again.

By turning layers on and off you can control what is visible in your map, which is useful for reducing clutter in the map display and keeping the map focused on a particular set of features or set of layers.

TASK 3

Review the components of ArcMap

In ArcMap your map can be viewed in Data View or Layout View. Data View is used to explore, edit, and analyze spatial data. Layout View is your cartographic environment, where you build your map by assembling map frames, tables, charts, and other elements such as north arrows and scale bars.

5 From the View menu, click Layout View.

The tools and menus available in Data View are also available in Layout View, but the advantage of Layout View is the ability to see the map on a virtual page, or layout page, that reflects the size and orientation of the page chosen in the Page and Print Setup dialog box.

6 From the View menu, click Data View.

You can also toggle between Data View and Layout View by clicking the two small buttons (one has a globe icon, the other a page icon) in the lower left corner of the map display.

Other than Data View and Layout View, the default ArcMap interface is made up of the following components:

Menu bar ——————————→

Standard toolbar ——————————→

Table of contents (TOC) ——————————→

Map display ——————————→

Drawing toolbar ——————————→

Status bar ——————————→

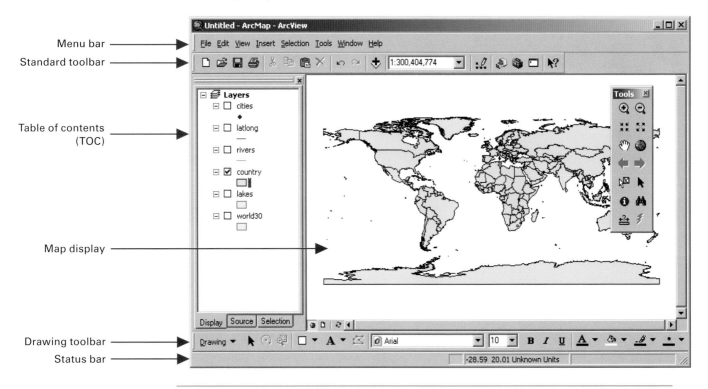

Menu bar: Contains options typical of other Window-based programs such as Microsoft Word or Excel, along with several others specific to ArcMap.

Standard toolbar: Contains a set of buttons used for common tasks such as opening a map, saving a map, printing, copying, pasting, and other more specialized tasks. If you are unsure of any button's function, hold the mouse pointer over the button and the button's name will appear in a ToolTip box, along with a short description in the Status bar of the button's function. For further information about a button's function, use ArcGIS Desktop Help, accessible from the Help menu.

Tools toolbar: Contains a set of buttons and tools used interactively with the map display for tasks such as zooming, panning, and selecting features in the map display. Can be moved or docked to any location you like.

Table of contents (TOC): Acts as a container for all of the data frames and layers added to ArcMap. Data frames are containers that hold a set of layers.

Map display: Displays a layer's features.

Drawing toolbar: Contains a set of buttons and tools used for adding text, graphics, and labels to the map display.

Status bar: Reports messages related to specific tasks and functions. The status bar also reports the coordinate location of the mouse cursor as it moves over the map display.

Because the ArcMap toolbars and TOC are dockable (i.e., you can click and drag them to anywhere on your desktop), your ArcMap interface may look slightly different from the one in the graphic.

TASK 4

Examine the attribute table

7   In the TOC, right-click the cities layer and click Open Attribute Table.

Look over the contents of the attribute table. Notice how it's organized into rows and columns (records and fields).

8   Close the Attributes of cities table by clicking the close button ☒ in the upper right corner of the table window.

9   Using the same method you used to open the Attributes of cities table, take a few minutes to open and explore the attribute tables for the other layers in your map. Close the attribute tables when you are finished looking through their contents.

TASK 5

Zoom and pan

10   On the Tools toolbar click the Fixed Zoom In button ⛶ three times.

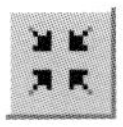

   After each click, wait for the map to redraw before clicking again.

11   On the Tools toolbar, click the Pan tool. 🖑

   Place your mouse pointer in the map display over the center of Africa. Click and hold down the mouse button and drag Africa to the center of the map display, then release the mouse button.

If you accidentally change the map extent, and want to return to the previous extent, or you just want to zoom quickly between map extents, you can use the Go Back to Previous Extent and Go to Next Extent buttons.

12 On the Tools toolbar, click the Go Back to Previous Extent button.

After the map redraws, click the Go to Next Extent button
to return to the extent in which Africa is centered in the
map display.

ArcMap also has two zoom tools that allow you to define the extent to which you zoom.

13 To reduce the clutter in your map, turn off (uncheck in the TOC) the latlong
layer. On the Tools toolbar click the Zoom In tool.

In the map, place your mouse cursor off the northwestern coast of Africa,
then click and hold down the mouse button while you drag a box around the
countries of West Africa—from Mauritania to Cameroon—as shown in the
graphic below (the area doesn't have to be exact). Once you've defined the
map extent, release the mouse button.

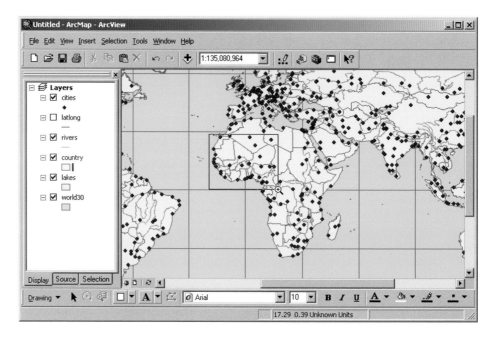

TASK 6

Identify features

14   On the Tools toolbar, click the Identify tool ,

then click any country in West Africa. Review the attributes that appear in the window.

At the top of the Identify Results window is the Layers drop-down list, which is used to choose the layer in which you want to identify features. Below the Layers drop-down list, the Identify Results window is split into two parts. The left portion shows which layer the feature identified belongs to, while the right portion contains a complete list of the identified feature's attributes. (Just below the layer name in the left side of the window, you will also see the primary display field value of the feature that was identified. The primary display field is defined by the user in the Layer Properties dialog box and usually contains values which are useful for labeling.)

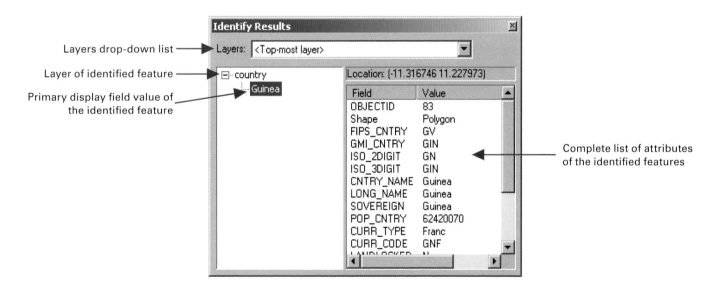

15   In the Identify Results window, choose cities from the Layers drop-down list. Try clicking on a county first, then click on any of the city points in the map display.

16   In the left-hand side of Identify Results window, click the primary display field value. While doing this keep an eye on the map display—you should see the associated feature flash in the map display when it is clicked in the Identify Results dialog.

17   Close the Identify Results window.

TASK 7

Select features

18 At the bottom of the TOC, click the Selection tab and uncheck all the layers in the TOC except country.

(If the Selection tab is not visible, from the Tools menu, click Options. In the Table Of Contents tab of the Options dialog box, check the box next to Selection in the Table of Contents Tabs list.)

The layers list in the Selection tab of the TOC allows you to control which features can and cannot be selected in the map.

19 On the Tools toolbar, click the Select Features tool.

Place your mouse pointer on the map in the ocean just off Mauritania and click and drag a box around all of West Africa to Cameroon, as shown in the following graphic. After you define the selection extent, release the mouse button.

Tip: If you aren't sure about country names, use the Identify tool. But again, the box you draw doesn't have to be exact.

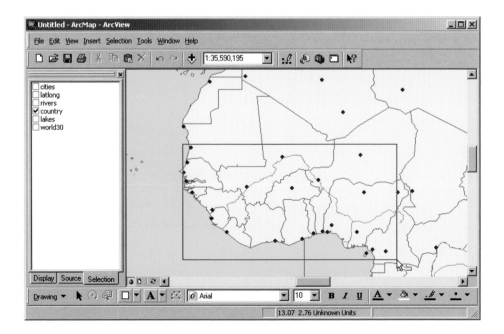

ArcMap lets you know which features have been selected by highlighting them in the map display.

You can also select features one by one by holding down the Shift key on your keyboard while you click on the features with the selection tool.

20   Using the Select Features tool, click Mauritania. (Remember to use the Identify tool if you need to.) Then hold down the Shift key and click on each country in West Africa until you have selected the sixteen countries. These are shown in the graphic. To make it easier to see where to click on the screen to select, try using the pan and zoom tools.

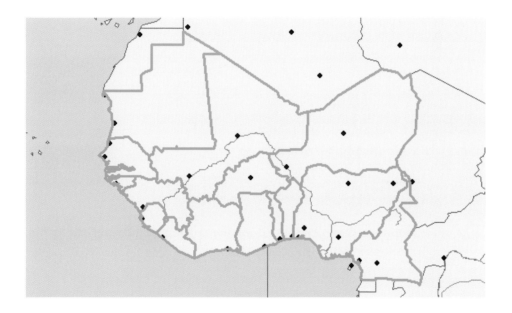

TASK 8

View and sort selected data in a table

21   At the bottom of the TOC, click the Display tab.

22   View the attributes of the country layer. (Right-click the country layer in the TOC and choose Open Attribute Table.)

The records of all the countries you have selected should now be highlighted in aqua color in the table. To see them you will have to scroll through the table to find each selected record.

23   At the bottom of the attribute table, click the Selected button to show only the selected records.

Now the entire table is highlighted because it's showing only the selected country records. Notice also that along the bottom of the attribute table, the number of selected elements is reported.

24   Scroll across the table until you locate the CNTRY_NAME field, then right-click on this field name and choose Sort Ascending.

All the country names are now arranged alphabetically.

25   At the bottom of the attribute table click the All button to show all the features again.

26   Close the attribute table.

TASK 9

Bookmark your geographic extent

When you create a bookmark, ArcMap remembers the extent of the map display defined by the bookmark. This feature lets you return to it anytime.

27   From the View menu, point to Bookmarks, then click Create. Name the bookmark **West Africa** and click OK.

This will save your West Africa extent in the map display. You can now return to it anytime, without having to use the zoom and pan tools.

28   On the Tools toolbar, click the Full Extent button.

29   From the View menu, point to Bookmarks and click West Africa.

TASK 10

Perform a simple query

  30   From the Selection menu, choose Select By Attributes.

The Select By Attributes dialog box allows you to enter a logical expression to find countries meeting specific criteria. In this case, you want to find all the countries from the current selection in West Africa that have populations of 10 million or more.

  31   At the top of the Select By Attributes dialog box, click the Layer drop-down arrow and choose **country** from the list of layers. Click the Method drop-down arrow and choose **Select from current selection.** In the list of fields, double-click **POP_CNTRY,** then click the greater-than or equal-to operator (>=). In the expression box at the bottom of the dialog, type **10000000** after this operator.

  32   Make sure your dialog box matches this graphic; click Apply, then Close.

There should be six countries in West Africa that meet this criteria.

33   From the Selection menu, choose Clear Selected Features.

Selecting features with an attribute query is a matter of choosing which layer you want to select features from, then creating a logical expression which defines the criteria of your desired selection. You will learn more about attribute queries in lab exercise 8.

TASK 11

Label features and edit the layer's symbolization

Now you are ready to label the map and change the symbology of some of the layers.

34   In the TOC, right-click the country layer and choose Label Features.

Because the name field is the primary display field of the country layer, the countries are labeled with their name attributes.

You may want to change the colors used to symbolize the country features. You can quickly do this by selecting a new color for a layer directly from the TOC.

35   In the TOC, right-click the color chip directly below the country layer. After the color palette appears, choose a color you like.

Another way to symbolize data is to categorize the features in a layer based on a set of attribute values and to then choose a unique color for the features in each category.

36   Right-click the country layer in the TOC and choose Properties. In the Layer Properties dialog box, click the Symbology tab. In the Show box, click Categories, then click Unique values. Click the Value Field drop-down arrow and choose CNTRY_NAME. Click Add All Values. From the Color Scheme drop-down list, choose a color scheme that you like.

Layer Properties

General | Source | Selection | Display | Symbology | Fields | Definition Query | Labels | Joins & Relates

Show:
Features
Categories
 └─ Unique values
 ├─ Unique values, many
 └─ Match to symbols in a
Quantities
Charts
Multiple Attributes

Draw categories using unique values of one field.   Import...

Value Field
CNTRY_NAME

Color Scheme

Symbol	Value	Label	Count
✔	\<all other values\>	\<all other values\>	
	\<Heading\>	**CNTRY_NAME**	
	Afghanistan	Afghanistan	?
	Albania	Albania	?
	Algeria	Algeria	?
	American Samoa	American Samoa	?
	Andorra	Andorra	?
	Angola	Angola	?
	Anguilla	Anguilla	?
	Antarctica	Antarctica	?
	Antigua_Barbuda	Antigua_Barbuda	?

Add All Values | Add Values... | Remove | Remove All | Advanced ▾

OK | Cancel | Apply

37 Click OK.

EXTRA CHALLENGE

Create a proportional symbol map of city populations

This task shows how to adjust the size of the city points to match population size.

EC 1   At the bottom of the TOC, click the Selection tab. Check cities and uncheck the other layers. Return to the Display tab of the TOC.

Because the cities layer contains major cities throughout the world, the class breaks will reflect the minimum and maximum populations of all the world's major cities, not the minimum and maximum populations for the major cities of West Africa only. To correct this, you must select the cities of West Africa and make a separate layer from them. Then, when you symbolize them by graduated symbol, classes based on the data range of just the West African cities will appear in the TOC.

EC 2   Using the Select Features tool, select all the cities in the West Africa countries. Once the selection is made, right-click the cities layer in the TOC, point to Selection, and click Create Layer From Selected Features.

The new layer is added to the top of the TOC. This is a temporary layer that only exists within your current map; no new files were created on your computer.

EC 3   Open the properties of the cities selection. If necessary, click the Symbology tab. In the Show box, click Quantities, then click Proportional symbols. In the Fields box, click the Value drop-down arrow and choose POPULATION. Click OK.

EC 4   To make the West African cities stand out better, turn off the cities layer.

Now you should have a map showing the cities symbolized according to population—the larger the city's population, the larger the symbol representing that city. On the next page is an example of how your current map could look. Your colors, symbols, and display extent will vary.

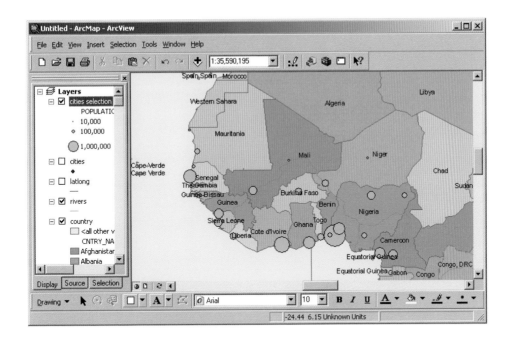

TASK 12

Save your work and exit ArcMap

Your work in this exercise is complete.

Saving your work in ArcMap also means saving your map. When you save a map, you save it to an ArcGIS map document file, which has the .mxd file extension. Map documents preserve the state of the map at the time that it was saved. For example, in this lab you added several layers to the map, zoomed to West Africa, created a spatial bookmark, made a selection layer from the cities, and defined the symbology for the cities selection and countries layers. When you save the map now, all of these settings will be preserved in the map document and regenerated in ArcMap the next time you open it.

42 Click the File menu in ArcMap, and click Save As. Navigate to **C:\UrbanGIS\ MyData** or to a directory your instructor has told you to use. Name the file **my_lab2.mxd** and click Save. Close ArcMap.

Thematic Mapping: Dot Density Maps

Three

In this assignment, you will create dot density maps showing aspects of

the population of the Bronx by census tract. As an option, you may show

in a second map the locations of various ethnic and racial populations as

a dot density map by census tract.

It is strongly recommended that you save your work every ten minutes or

so while working, in order to prevent computer tragedies.

TASK 1

Start ArcCatalog and create a folder connection

1   Start ArcCatalog by double-clicking the ArcCatalog icon on your computer desktop. Alternately, click the Start button on the Windows taskbar, point to Programs, point to ArcGIS, and click ArcCatalog.

ArcCatalog contains tools for managing your geographic data. It has a similar appearance to Windows Explorer and one of its primary uses is file management. From ArcCatalog you can connect to specific folders; preview and browse your geographic data and its associated attribute tables; manage data by creating, deleting, and renaming folders, files, and datasets; and create, view, and update metadata.

As a rule, use ArcCatalog whenever you copy, move, or delete any geographic dataset. ArcCatalog is specifically designed to handle the unique file architecture of geographic datasets, whereas standard operating system tools such as Windows Explorer are not. Your GIS data may be damaged if you use Windows tools to manage it.

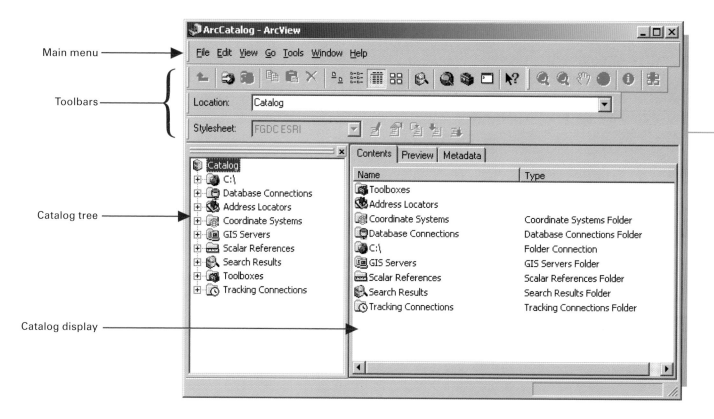

2   On the Standard toolbar, click the Connect to Folder button.

In the Connect to Folder dialog box navigate to and select the **C:\UrbanGIS** folder. (If your instructor is using a different location, follow the instructor's directions.) Click OK.

The direct connection that you made to the UrbanGIS folder provides quick access to your data from ArcCatalog or ArcMap.

From this point on, you will work with Bronx data in all of the lab exercises. Most datasets are located in the Bronx geodatabase in the UrbanGIS folder, while the others can be found in the UrbanGIS\SourceData folder. (All of the Bronx spatial datasets used in this book have been projected into the State Plane coordinate system, specifically NAD 83, State Plane New York, Long Island Zone, U.S. Feet.)

Main menu and toolbars: As in ArcMap, ArcCatalog contains a Main menu, a Standard toolbar, and several other toolbars containing buttons and tools grouped together based on functionality.

Catalog tree: Used to browse the content on your hard drive or network.

Catalog display: Used to view data, with three display options: Contents, Preview, or Metadata. These are activated by clicking the appropriate tab at the top of the display.

Content mode: (shown in the graphic) Displays the files and folders of the drive or folder that's selected in the catalog tree.

Preview mode: Displays the feature class that is selected in the catalog tree as it will appear when displayed in ArcMap. In this mode you can also preview a feature class attribute table or a stand-alone table.

Metadata mode: ArcCatalog has a set of tools for creating and maintaining metadata. In this mode you can view, edit, or create new metadata for your GIS data.

TASK 2

Open ArcMap, and add and symbolize layers

You will now switch back to ArcMap and load the data you will use throughout this project.

3   From the ArcCatalog Standard toolbar, click the Launch ArcMap button.

Click the option to Start using ArcMap with a new empty map, then click OK. After ArcMap opens, close ArcCatalog.

4   On the ArcMap Standard toolbar, click the Add Data button. In the Add Data dialog box, click the Look in drop-down arrow and choose Catalog from the top of the list. (This will set you to the highest level in your folder hierarchy.) Using the connection you made in the previous step, navigate to your **UrbanGIS** folder, then double-click the Bronx geodatabase, which is a file called **Bronx.mdb**. Add the following feature classes from the Bronx geodatabase to ArcMap: **SHR36005**, **WBD36005**, **LAN36005**, and **Bx_demog**. Use the Ctrl key on your keyboard to select multiple feature classes.

These layers in your map represent features in Bronx County, New York. The names of these layers are a bit cryptic. Here's what they contain:

◆   SHR36005—shorelines

◆   WBD36005—water bodies

◆   LAN36005—landmarks

◆   Bx_demog—census tracts and demographic data about the people who live in them

The numeric portion of these layers is based on a U.S. Census code scheme; 36 is the state code for New York, and 005 is the county code for the Bronx.

Naming conventions used for GIS data are often based on internal organizational needs for consistency and reusability, but they often make no sense to people outside the organization. For example, someone in the data division of the Census Bureau might easily recognize shr36005 as the shoreline data for the Bronx, but most other people will have no idea what this name represents. You may want something more helpful to the map reader.

5   Click **Bx_demog** to highlight it in the TOC. Click the layer name again to activate the text cursor and type **Census**, then click Enter. In the same manner, rename the other three layers to shorelines, water bodies, and landmarks. Make sure you're giving the descriptive name to the right layer.

Note: You just renamed the layers in the map document; you did not change the actual feature class names.

6   Look at each layer individually in the map display by turning off all the other layers. Make sure the layers are in the following order in the TOC: Shorelines, Water bodies, Landmarks, Census. Turn on all four layers.

In this exercise, you will be symbolizing the census tract data based on different demographic values. Right now the shoreline layer is obscuring the layers below it. Until you change the symbology of the shoreline layer to a hollow fill, you will not be able to see the features below it.

7   In the TOC, click the shoreline color chip. On the left-hand side of the Symbol Selector, click the Hollow fill symbol. On the right-hand side, increase the outline width value from the default .40 to 2. Then click OK.

In the map display you can now see the features in the other layers.

8   In the TOC, right-click the Census color chip and select a light brown color from the palette. Repeat this process to change the color of Landmarks to a shade of green and Water bodies to a shade of blue.

TASK 3

Examine the attribute table

9   Open the Attributes of Census table. (In the TOC, right-click the Census layer and choose Open Attribute Table.)

This table contains a record for each of the geographic entities in the spatial database, and holds values for several fields. The fields, which represent different attributes or variables present at each location, are shown along the top of the table. In the case of the *Census* table, there is a separate record for each census tract, and many fields, each pertaining to a different aspect of the demography of the tracts. For each census tract, the different fields give us different information about who lives in that tract. This information is aggregated at the tract level and typically reflects sums, means, medians, or percentages of the variable in that tract. Fields include such information as total population, non-Hispanic White population, percentage of population that is non-Hispanic White, number of households, median age of residents in the tract, and so forth.

10   Explore the fields contained in the attribute table. Note the number of records in the table (shown at the bottom of the table). This number represents the number of census tracts in the Bronx, and each record (a record is a single row in the table) corresponds to a census tract. Determine the field you want to use for your dot density map. Remember, a dot density map uses absolute numbers, not ratios or percentages, so make sure the field you choose contains absolute numbers. If you cannot tell what the values in a field represent based on the field name, refer to the **Metnames.txt** file that is within your **UrbanGIS\SourceData** folder.

11   Close the attribute table.

TASK 4

Prepare a data layer for symbolization and plot a dot density map

In this task you will create a dot density symbolization from your demographic data.

12 Open the properties of the Census layer. In the Layer Properties dialog box, click the Symbology tab. In the Show box, click Quantities, then click Dot density. In the Field Selection box, click one of the fields you chose to use for a dot density map, like TOTPOP, then click the arrow that points to the Symbol frame. Click Apply, then drag the Layer Properties dialog box to a location that does not block the map display.

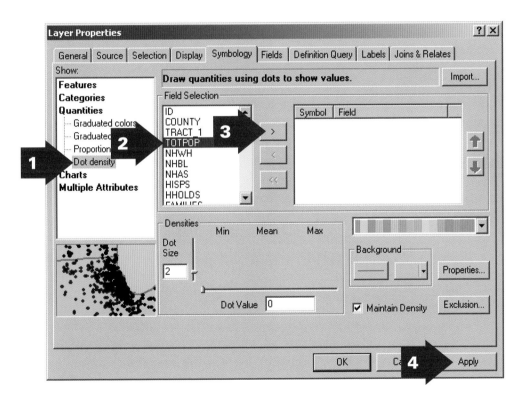

In the TOC, the legend for the Census layer reflects the dot symbology, reports the dot density value, such as 1 Dot = 2000, and shows which field(s) the symbology is based on.

ArcMap automatically chose the dot density value, and the size and color of the dots, but you can change these settings to match your needs.

13   In the Densities box of the Symbology tab, experiment with changing the Dot Value. (You can either type it in or use the horizontal slider bar.) Choose different values, click Apply, then look again at your map. You can keep doing this to examine the visual effect of the dot value. When finished, close the Layer Properties dialog box.

Tip: You may need to click the Field name again to activate the Apply button.

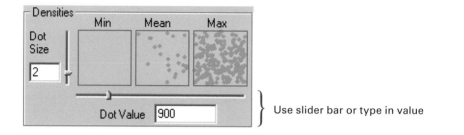

As you learned in chapter 3, there is a more objective approach for determining the dot value, which considers the distribution of the attribute values you are mapping. In the next step you will generate the information you need from the attribute table to determine an appropriate dot value.

14   Open the Attributes of Census table. Scroll through the table to find the field that you based your dot values on, then right-click that field name and choose Statistics.

Use the statistics that are generated to calculate the best dot value for your map, then close the statistics and attribute tables. (Refer to chapter 3 for a description of how to calculate a dot value.)

15   Open the properties of the Census layer, and input the Dot Value you calculated in the previous step. (If you want to change the color of the dot, right-click the dot in the symbol list and choose a new color from the palette.) Click OK to apply your changes and close the dialog box.

ADVANCED CHALLENGE

Mask a data layer

One of the problems with the algorithm used in determining dot placement on a dot density map is that it typically will not take into account conflicts with other datasets. For instance, dots, representing people, are often placed in the water or in parks because the census tracts might include those areas, and the algorithm randomly assigns dots within the entire area of the census tract. However, we know that is not logical to show population living beyond the land areas of the Bronx, or significant numbers of people living within parks or on lakes. The masking function will filter out areas that are inappropriate for dot placement. With masking you can select a data layer to use as a mask, so that ArcMap does not place dots in those areas.

AC 1   In the TOC, drag the Census layer to the top of the list. You may notice some dots in places you wouldn't expect, such as those designated as landmarks.

AC 2   Open the Census properties and, if necessary, click the Symbology tab. In the lower right-hand corner of the dialog box, click the Properties button.

AC 3   In the Dot Density Symbol Properties dialog box, check the Use Masking option, then experiment with the Masking options to see how it affects your map. When you are finished, close the properties.

TASK 5

Save the map and create layer files

16   From the File menu, choose Save As. Navigate to your **UrbanGIS\MyData** or to the folder your instructor has directed you to use. Name the file **DotDensity.mxd** and click Save.

As you learned in the previous exercise, saving a map document preserves all the properties of the layers in your map so that the next time you open the map document everything will appear as you had it when you last saved. Saving map documents is a great way to preserve the properties of your layers, but you could run into a situation where you want to save the symbology of a layer for use in other maps. This can be accomplished with layer files.

A layer file is used to preserve the properties that have been defined for a layer so that when it is loaded into ArcMap all of its properties are already defined. A common use of layer files is for preserving symbology.

You will now save some of your datasets as layer files so that you can use them again in other maps or data frames without the need to re-symbolize everything.

17   Right click the Census layer and choose Save As Layer File. Navigate to the **UrbanGIS\MyData** folder. Keep the default name (Census), and click Save. Layer files are saved to disk with the .lyr extension. Repeat this process for the remaining layers in your map.

Now that the layer files are created you can test them.

18   On the Standard toolbar, click the New Map File button.

Click the Add Data button. Navigate to **UrbanGIS\MyData** and add the layer files that you created in step 17.

Adding a layer file is similar to adding any other dataset to the map; the only difference is that when it is loaded into ArcMap its properties are already defined.

A layer file stores the layer properties assigned to a feature class and the path to its source data—it does not store the actual geometry used to draw the features. When a layer file is added to ArcMap it uses the path it stores to find its source data so that it can draw the features. If you move or delete a layer file's source data, it will not draw when added to ArcMap, and a red exclamation point will appear next to its name in the TOC.

If you encounter a layer file that's lost its connection to its source data, you can repair the connection by clicking the red exclamation point that appears next to its name in the ArcMap TOC and use the dialog box that opens to direct the layer file back to its source data.

EXTRA CHALLENGE

Create different dot density maps

After you complete one dot density map (of the Total Population field, for instance) you can create another one showing different data. You will need to add a new data frame to your current data view, which will let you create a new map with the same or different data, while staying within the same map document. The map you've already created will still be there and available for your use, but now you will have the option of adding another map to your layout. You will be able to display and work with only one map at a time in your data view, but you can toggle back and forth between any number of maps by highlighting the data frame name in the TOC, right-clicking, and choosing Activate. Thus, a new data frame can be created in order to show details of the original data frame, or an overview of the geography, or a different variable of the data with the same geographic extent. In this task, you will be using the same geographic extent (the Bronx) but showing a different variable (Latino population, for instance).

EC 1   From the Insert menu, choose Data Frame.

Your map display goes blank because the new data frame is automatically activated and it has nothing in it.

EC 2   In the TOC, click once on the name of the new data frame to select it, then click it a second time—but don't double-click it—to activate the text cursor. When the cursor appears, type in a new name for the data frame that reflects the type of demographics you are going to map with a dot density symbology.

EC 3   Click the Add Data button and add the four layer files that you created in Task 5.

EC 4   Open the Census properties and, if necessary, click the Symbology tab. Experiment with various dot density maps by selecting different attributes fields. For an added challenge, try adding multiple fields for different population counts and experimenting with the colors and symbology used for the dots. If you want, feel free to add more data frames to your map and to create different dot density maps within each that are based on a different set of demographics.

EC 5   When you are finished, save the map document and close ArcMap.

Thematic Mapping: Choropleth Maps

In this assignment, you generate a choropleth (graduated color) map showing by census tract the minority population of the Bronx as a percentage of the total population. Additionally, as an advanced optional portion of the assignment, you may also create a map of any other demographic variable that is suitable for choropleth mapping.

TASK 1

Open an existing project and add data

1. Open ArcMap. Choose to open an existing map, then navigate to and open the **DotDensity.mxd** map document that you created in the previous exercise. (If you did not complete the previous exercise, you will need to do so before beginning this one.)

You will be creating a new map using some of the same data as in lab exercise 3, but displaying different variables. In order to do this, you will need to add a new data frame to your data view. If you did the extra challenge task in lab exercise 3, you already have two data frames in your map document.

Data frames are containers for holding a set of layers. A new data frame can be created in order to show details of the original data frame, or an overview of the geography—an inset map—or to show a different variable of the data in the same geographic extent. You can create as many data frames as you want in a map, but you can only view one data frame at a time in Data View. In Layout View you can view all the data frames in a map at once. This capability is useful for making side-by-side comparisons among different maps in your data frames, for plotting maps with multiple views, or for making inset maps using your data frames. Note: The graphic below is an example; your map document will look different.

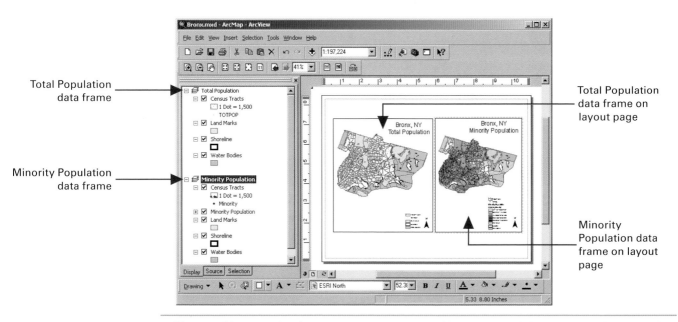

You can add as many data frames to a map as you want. Whether you are in Layout View or Data View all of your data frames and their contents will appear in the TOC, but you can display multiple data frames only when in Layout View. You will learn to do this in the next exercise.

To toggle between frames in a data view, right-click the frame's name in the TOC and select Activate. In this task, you will be using the same geographic extent (the Bronx) as before, but a different variable (Percent Minority) and a different thematic mapping technique (choropleth rather than dot density).

2   From the Insert menu, click Data Frame. Right-click the new data frame, and choose Properties. In the Data Frame Properties dialog box click the General tab, then click in the Name text box and replace the current name with one that is more relevant, such as Minority Population. Click OK.

Next you will save your changes to this map document.

3   From the File menu click Save As. Navigate to your **UrbanGIS\MyData** folder, name the map document **BronxPop.mxd**, and click Save.

You will now add data to the new Data Frame, and create a new map.

4   Click the Add Data button, navigate to your **UrbanGIS\MyData** folder, and add the four layer files that you created in the previous lab (Census, Landmarks, Shorelines, and Water bodies). After you load the layer files, click the Add Data button again and load **Bx_demog** from the Bronx geodatabase.

TASK 2

Create a new field and calculate values for it

In the **Bx_demog** dataset, you will have to add new fields to the theme table in order to calculate the percentage of minority population in the Bronx. The Bx_demog table itself does not have a field totaling the minority population in each census tract, so you must add such a field. Then you will need to create another new field to calculate the minority percentage from that number. The numbers given in the table are absolute numbers of people in each census tract, and you need the percentage of the total population that is considered to be a minority group.

5   Open **Attributes of Bx_demog**. Near the bottom of the table, on the lower right-hand side, click the Options button, then choose Add Field. (If you cannot see the Options button, you may need to increase the size of the attribute table window.)

A dialog box opens, prompting you to define a name and set the properties of the field. In this case you will give it a name of your choice, and define its type as long integer—a field type suitable for long whole numbers with no decimal points or fractions.

6   Give the new field a name (Minority, for instance). From the Type drop-down list, choose Long Integer. Make sure your dialog box matches the following graphic, then click OK.

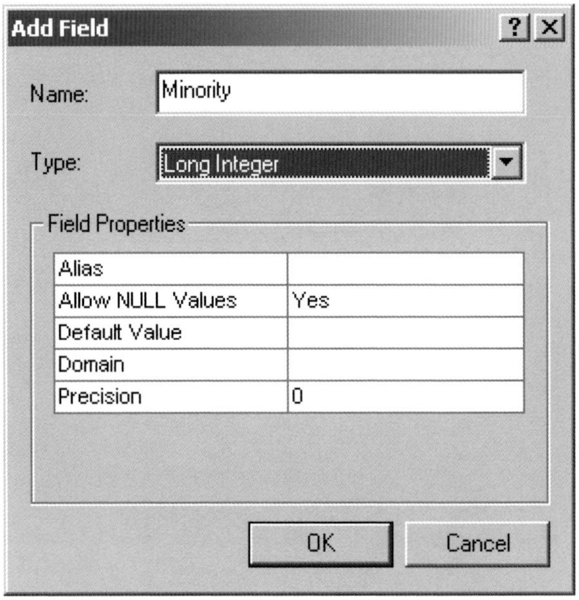

7   In the attribute table, scroll to the extreme right of the table. The new field exists in the table and contains all null values. A null value indicates the absence of a recorded value for a geographic feature; it is not the same as zero.

You will populate the new field by providing a mathematical formula for the values. For instance, Minority Population, for the purposes of this assignment, will be considered as the sum of the fields of Hispanics, non-Hispanic Asians, non-Hispanic Blacks, and others. In the attribute table, these field titles are HISPS, NHBL, NHAS, and OTHERS.

8   In the attribute table, right-click the name of the field you just created and choose Calculate Values; click Yes to the subsequent caution message. The Field Calculator appears.

With the Field Calculator, you calculate new values for the individual records in a field. It is like any other calculator but in this context, it lets you incorporate the fields from your attribute table into the equations you build.

When you open the Field Calculator, the name of the field from the attribute table where you right-clicked appears; it can be found above the expression box, to the left of an equals sign. Creating the expression is a matter of choosing fields from the Fields box and then applying the appropriate mathematical operators or functions. For example, in the upcoming steps you will add together the minority population values for each census tract and place the results in the new field you just created.

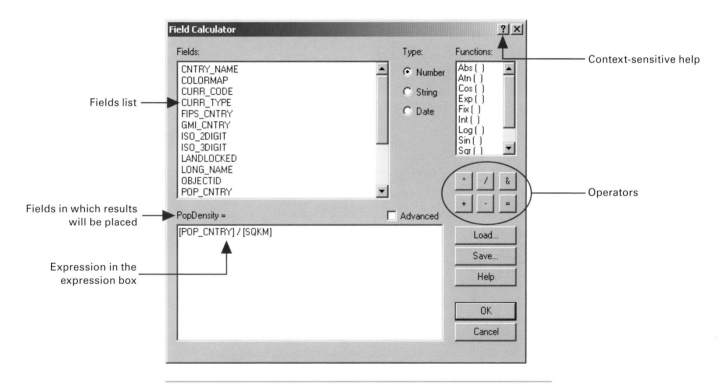

The Field Calculator is used to calculate values which are placed directly into a table field. In this case, population density (PopDensity=) is being calculated by dividing the population values in the POP_CNTRY field by the area values in the SQKM field. The results of this calculation are being placed in the PopDensity field. For additional context-sensitive help, click the question mark **?** in the upper right corner of the dialog, then click on the appropriate section of the dialog box. (Context-sensitive help is available for many of the ArcMap tools.)

9　Create an expression in the Field Calculator that will add together the values stored in the HISPS, NHAS, NHBL, and OTHERS fields. To do this, click one field needed in the expression, then click the addition button. Repeat this process until you have completed the expression. It should match the one shown below.

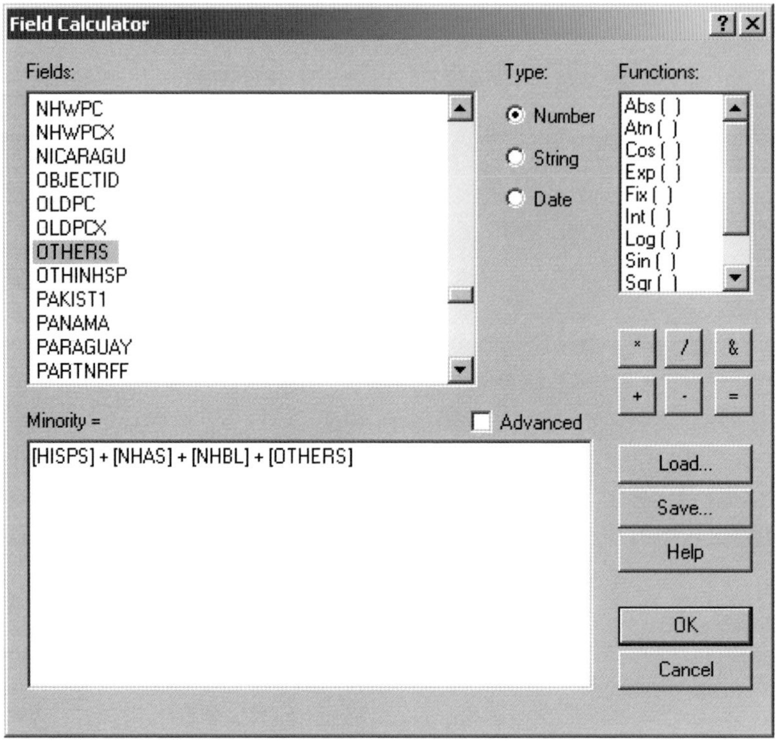

10　In the Field Calculator, click OK.

11　Check to make sure that you set up the equation correctly in the Field Calculator by manually performing the same calculation on a selected record. If your results don't match the results produced by the Field Calculator, you will need to recalculate the values.

Now that you have a field containing the total number of minorities, you can calculate the percentage of minorities in each census tract. First, you need to create a field that can store percentages. Unlike the integer field you created, the field that stores the percentages will need to store decimal values, not whole numbers. There are two field types you can choose from to store decimal values, Float and Double. In this exercise, use the Float field.

12   From the Attributes of Bx_Demog table, open the Add Field dialog box. (Refer to Step 6, above, to review how to do this.) Name the new field **Percent_Minority**, and set its Type as Float. Click OK.

To calculate these percentage values you will divide the values in the Minority field by the values in the TotPop field, and multiply the results by 100. There are a few census tracts that have no population because they are principally parks and water bodies. If you try to calculate the percentages of minority population on these tracts, the Field Calculator will generate an error because a number cannot be divided by zero. To prevent this, you can exclude the census tracts with zero population.

13   In the Attributes of Bx_Demog table, click the Options button, and choose Select By Attributes. Make sure the Method drop-down list is set to Create a new selection. From the Fields list double-click TOTPOP. Click the *greater than* operator   >   and type the number **0**. Make sure your expression matches the one shown in the graphic below. Click Apply and close the dialog.

All the selected tracts have a population greater than zero. When records are selected, the Field Calculator will apply the expression only to the selected set, so you can now calculate the percentage values without generating a logic error. (When no records are selected, the Field Calculator will by default apply the expression to all of them.)

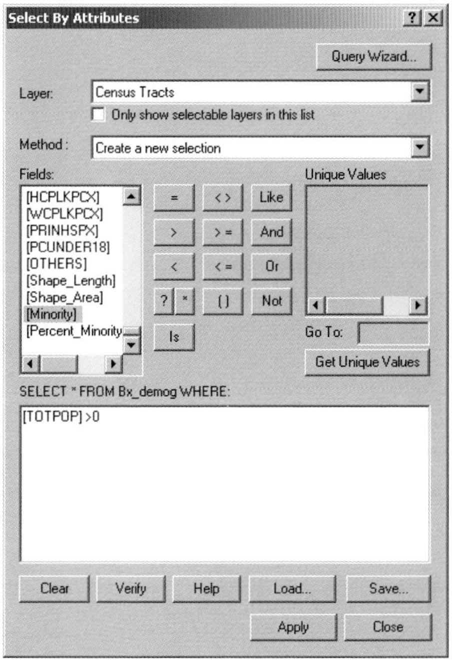

14   Open the Field Calculator for the Percent_Minority field (again, click Yes
     to the warning). Create an expression that will generate the percentage
     of minorities in each census tract. Assuming you named your minority
     population field Minority, your expression should match the one below. Click
     OK and examine the contents of the new field.

15   Close the Attributes of Bx_demog table, and clear the selected features.
     (To clear the selection, from the Selection menu choose Clear Selected
     Features.)

TASK 3

Classify the data and create a choropleth map

16 From your active data frame, open the Bx_demog properties layer and click the Symbology tab. In the Show box, choose Graduated colors from the Quantities category. In the Fields box, choose the Percent_Minority field from the Value drop-down list (it will be at the end of the list). Then, make any adjustments in the numbers of classes or the color ramp by changing the appropriate bars and pull-down menus. Experiment with different symbology settings. While experimenting with the symbology and classifications, click Apply each time you make a change so that you can see the effects without closing the Layer Properties. When you are finished defining the symbology, click OK.

You now have a choropleth map showing the percentage of minority population in the Bronx by census tracts. As you may have noticed, the tracts that have no data are not shown; this is because they contain null values and these fall outside the symbology classes.

TASK 4

Reclassify the data

When you choose to symbolize your data using the quantities classification, ArcMap automatically chooses Natural Breaks as the classification method, and breaks the data values out into five classes. You will change the classification method and the number of classes.

17 Open the properties of Bx_demog. On the Symbology tab, click the Classify button, and then select another method of classification from the Method drop-down list. (For information about each type of classification method, use the context-sensitive Help menu.) Use the Classes drop-down list to change the number of classes. When you're done, click OK in the Classification dialog box, then OK in the Layer Properties dialog box.

After you finalize your classification method and number of classes, you should change the class range labels to be more informative. Your class breaks should be labeled with a two-digit percentage (87%), not proportional decimals or percentages with decimal places (.870 or 87.32%). There are two ways to do this. The first is to manually type in the percentage values for each category. The other is to format all the labels at once using the Number Format dialog box.

18   Open the properties of Bx_demog, and if necessary, click the Symbology tab.
In the list of categories, click the Label column heading and choose Format
Labels (shown below). In the Number format dialog box, click Percentage,
then click Numeric Options. In the Rounding box, change the number of
decimal places to **0**. Click OK on all the open dialog box to apply the changes
and return to your map.

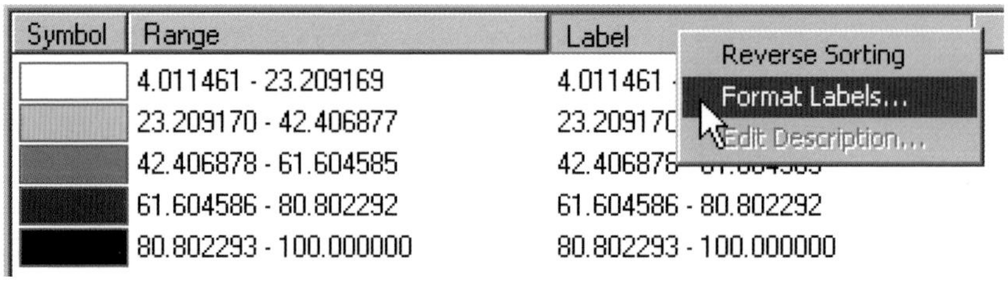

Symbol	Range	Label	
	4.011461 - 23.209169	4.011461 ·	Reverse Sorting
	23.209170 - 42.406877	23.20917C	Format Labels...
	42.406878 - 61.604585	42.406878	Edit Description...
	61.604586 - 80.802292	61.604586 - 80.802292	
	80.802293 - 100.000000	80.802293 - 100.000000	

19   In the table of contents, rename the Bx_demog layer. Finally, save the
Bx_demog layer (which now has a new name) as a layer file in your
UrbanGIS\MyData folder.

TASK 5

Save your project

20   Save your map document. (On the Standard toolbar, click the Save button.) If
you're not going to complete the challenge task, close ArcMap.

EXTRA CHALLENGE

Create a choropleth map with a dot density theme on top

An interesting technique that can reveal patterns in data that would otherwise be invisible is to overlay a choropleth map with a dot density map. For instance, you can use your choropleth map of percent minority, then place a dot density map of the absolute numbers over it, so both can be viewed on the same map.

EC 1   Add the Bx_demog feature class from the Bronx geodatabase to your map. If necessary, drag Bx_demog to the top of the data frame. Use the Layer Properties dialog box to symbolize Bx_Demog as a dot density map, then draw the dot values from the minority population field you created in Task 2. Be sure to use a color for the dots that makes them stand out against the background of the choropleth map.

EC 2   Save your map document. If you are continuing to the Advanced task, leave ArcMap open, otherwise close ArcMap.

ADVANCED CHALLENGE

Create new maps with the demographic data

AC 1   Create a choropleth map display showing the percentage of non-Hispanic White population. You will need to insert a new data frame, add the basemap data as well as the demographic dataset to the new data frame, then symbolize the demographic data with graduated colors based on the field that represents percentage non-Hispanic White. Give the data frame and the layers it contains meaningful names. If you have extra time, try creating a choropleth map showing the percentage of non-Hispanic Blacks in another new data frame.

AC 2   Save your map document and close ArcMap.

Integrating Graphs *and* Maps, *and* Designing Map Layouts

Five

In this assignment, you will create a bar graph (column graph) of population characteristics. To create graphs, you will need to make new tables and add them to your project. As an optional activity, you will create a pie graph. Then you will be using the maps and graphs that you have created so far to produce a map layout in the Layout View window of ArcMap. You will incorporate all the required map elements into your design, and then print out the final composition.

TASK 1

Open an existing project and choose appropriate information for the graph

    1   Open ArcMap. Choose to open an existing map, then navigate to your **UrbanGIS\MyData** folder and open the **BronxPop.mxd** map document that you created in the previous exercise.

    2   From the file menu click Save As. In the Save As dialog box, navigate to your **UrbanGIS\MyData** folder, name the map document **BronxPopLayout.mxd**, then click Save.

Saving to a new map document allows you to work with and change all the information contained in the original ArcMap document without altering the original project file. It is a good habit to do this at the beginning of a project.

    3   From the Minority Population data frame, open the Attributes of Percent Minority.

The selection of fields for your bar graph will depend on what information you are trying to communicate. Your graphs should show, at minimum, the minority population of the Bronx compared to the non-Hispanic White population, shown as totals aggregated from the 355 census tracts. Alternatively, you can also make a graph showing finer categories of the race and ethnicity breakdowns listed in the demographic table, such as non-Hispanic White, non-Hispanic Black, non-Hispanic Asian, and Hispanic. You will make the table for the bar graph first.

    4   Examine the fields in the table and decide which ones you want to include in your graph. Consult the **Metnames.txt** file if you are uncertain what specific field names in the Bx_demog table refer to. (If there are fewer than 50 people within a census tract category, the numbers are suppressed for confidentiality reasons.)

You will need to work with the absolute number fields, rather than the percentage fields because you must total the population values for all the census tracts (the records) for the Bronx, and you cannot add percentages for each census tract and obtain a meaningful number. You can, however, create Bronx-wide percentages for any of the ethnic/racial categories.

TASK 2

Aggregate the values for the fields to be graphed

Because there are 355 records in the demographic table (one record for each of the census tracts, far too many for a graph) you must aggregate the census tracts and get totals for all the fields. This will enable you to make a table showing total populations in the various racial and ethnic categories. Each field that you intend to include as a graph element will need to be totaled in this way.

5   In the attribute table, locate the first field you want to use in your graph, right-click its field name, and choose Statistics.

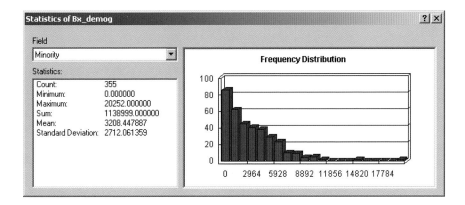

The Statistics dialog box includes summary statistics for the field you choose as well as a histogram that shows the distribution of the values.

6   Note and write down the Sum value for the field. Leave Statistics of Bx_
     demog open.

The value you just recorded will be used to populate a table, which will then be used to
generate the graph.

7   Click the Field drop-down list in the Statistics of Bx_demog dialog box and
     select the next field that you want to use in your graph and record the Sum
     statistic for that field. Repeat this process for the other fields you want to use
     in your graph. When you are finished, close Statistics of Bx_demog and the
     attribute table. At this point, you should have a written record of the sum
     statistics for all the fields you intend to include in your graph.

TASK 3

Create a table to store the values you will graph

8   Start ArcCatalog. In the Catalog tree, navigate to your Bronx geodatabase.
     Right-click the Bronx geodatabase, point to New and click Table. In the New
     Table dialog box, click in the Name box and type **Graph_Stats**. Click Next on
     the second and third panels, then click Finish.

9   Close ArcCatalog. Make ArcMap your active application. Click Add Data,
     navigate to your **UrbanGIS** folder and load the **Graph_Stats** table from your
     Bronx geodatabase.

Your newly created table will appear on the source tab of the TOC.

TASK 4

Add fields and records to your new tables

10   Open the Graph_Stats table (right-click and choose open). Click the Options
     button in the lower right corner of the table and choose Add Field. Click in
     the Name box of the Add Field dialog box and type a name that matches
     the type of statistics the field will hold. (For example, if you summarized the
     Minority field you might name the field MinorityPop.) Click the Type drop-
     down list and choose Long Integer.

(Alternatively you can create the new fields and define them in ArcCatalog, by opening the table's properties and clicking the Fields tab.)

11 Click OK on the Add Field dialog box. Repeat this process to create the remaining fields needed to hold the statistics for your graph. Be sure to choose Long Integer for type.

With the fields now in place you can populate them with values. You will need to start an edit session.

12 If the Editor toolbar is not already open on your ArcMap interface, on the Standard toolbar, click the Editor Toolbar button.

On the Editor toolbar, click the Editor drop-down arrow, then choose Start Editing.

13 In the Attributes of Graph_Stats click in the first cell associated with one of the fields you added and type in the corresponding value that you noted in task 2. Repeat this for the remaining fields in your table. (Do not enter a value for the OBJECTID field.)

14 When you are finished entering the values in the table click the Editor menu and choose Stop Editing. When prompted, click Yes to save your edits.

TASK 5

Create a bar graph

15   In the Attributes of Graph_Stats click Options, and choose Create Graph. (Alternately, you could also access the Graph Wizard from the ArcMap main menu by clicking Tools > Graphs > Create.) In step 1 of the Graph Wizard, choose the Column Graph type, then click Next. In step 2, uncheck Use selected set of features or records; check all the fields in the list of fields; and, for the Graph data series using option, choose Fields. Click Next. In step 3, give the graph a name of your choice and check the option to Show the Graph on Layout. After you complete the three steps in the Wizard, click Finish.

The new column or bar graph will appear on screen. Close the table. Drag the graph to a different location on your layout if you want.

EXTRA CHALLENGE 1

Use advanced options for design refinements

EC 1.1 Check out the design options available when you click Advanced Options in step 3 of the Graph Wizard dialog box. To access these after the graph is created, right-click the title bar of the graph window and choose Properties. Try changing the title of your graph, the font style and size, and other options.

ADVANCED CHALLENGE

Create a pie graph

Pie graphs conceptually are more complicated to construct than bar graphs because they are intended to represent the entire universe of categories that contain the variables being graphed. A pie graph compares parts to the whole. In other words, all the pie segments must add up to the whole, whereas a bar graph can contain disparate elements, such as hierarchically nested subsets of the whole. The fields that make up the pie graph must equal 100 percent of the population universe. For instance, you cannot have the field TotPop in a pie graph of population numbers, because all the constituent wedges of pie must total the whole, which is TotPop. Likewise, you could not include the field Minority in a pie graph showing finer categories of minorities, such as non-Hispanic Blacks, non-Hispanic Asians, and Hispanics. In a bar graph, you do not have this restriction.

To complicate matters further, the ArcMap graphing function will only allow pie graphs to be constructed from fields, not records. Only one field can be graphed at a time. All the values you want to graph, therefore, have to be located in the same field. If your table is not set up this way, you will not be able to create a meaningful pie graph. You should create a Pie_Demog table so that one field (called Population) contains all the necessary numbers you want to appear in the pie graph, and the records.

AC 1   Again open ArcCatalog and create a new table in your Bronx geodatabase. This table will need two attribute fields: a text field to store the name of the population type (such as non-Hispanic Black) and a long integer field to store the population statistic.

Hint: You won't be able to add fields in ArcMap until you close ArcCatalog.

AC 2   Once the schema of the table is created you will need to add a separate record (row) for each population statistic you want to appear in the graph. (Use the Percent Minority attribute table for values, and remember—you must be in an edit session to add rows.)

AC 3   After the table is constructed, open the Graph Wizard and run through the steps in the wizard to create a pie graph. Use the advanced options to access additional design selections and make sure that in the third step you check the option to show the graph on layout.

Suitable fields for a pie graph in this case might be one of the following sets of variables: (1) Non-Hispanic Whites, Minorities; OR (2) NHW, NHB, NHA, Hisp, Others. Both of these sets of variables represent a 100 percent universe of the population data, with no gaps or overlaps of categories. You could also create a pie graph of only minority categories—NHB, NHA, Hisp, Others—as long as you label the pie graph "Minority Population." This pie graph, then, would include all the minority population categories, with no gaps or overlaps, again displaying a 100 percent universe of the minority population.

TASK 6

Save your graphs and map

Graphs are saved inside a map document, which means you can close the graph window, then reopen it from the Graphs option in the Tools menu. You can also save the graph as a graph file (.grf), which enables you to use these graphs in other map documents.

16   From the Tools menu, point to Graphs and choose the graph you created in task 5. Right-click the title bar of the graph and choose Save. In the Save As dialog, navigate to your **UrbanGIS\MyData** folder, type in the name by which you want to save the graph, then click Save. If you created any other graphs, save them as well.

The graph you just saved as a .grf file can now be used in other map documents. Now that you are done creating the graphs, it's a good idea to clean up the map and save your work.

17   Close all the open tables and graphs in your map, then, if necessary, close the Editor toolbar by clicking the Editor toolbar button. From the File menu, click Save.

TASK 7

Layout preparation

Examine all the data frames, layers, and graphs you have created in the lab exercises so far. You should have at least two data frames and one graph if you have completed the basic tasks, and several more maps if you have done the extra credit or advanced tasks. Decide which of these would make an interesting and informative layout, understanding that the maps and graphs selected should relate to each other and support the data shown, not be disjointed and disconnected. For instance, if you are including the South Bronx demographic data in a detailed map, you would probably want to have a graph also showing South Bronx information, not a graph of the total Bronx population. Or, if your layout shows two maps, such as Percentage of Non-Hispanic Whites, and Percentage of Non-Hispanic Blacks, then your graph should feature those breakdowns, as well as the remainder of the categories needed to make up 100 percent of the population. The key is to think in advance and use common sense.

18   Determine which data frame you will use to create a map layout, then activate that data frame if necessary. Make sure the layers in the active data frame have descriptive names. If necessary, rename the data frame.

EXTRA CHALLENGE 2

Incorporate your extra challenge maps into your layout

EC 2.1 More than four maps on an 8½-by-11-inch layout is not recommended. However, depending on the selection of maps you have created in the previous extra challenge tasks, you may want to create a layout with maps other than those required in the previous task. You may also want to include more than two maps and one graph. You can be creative, but make sure the maps and graphs you select for your layout tell an informative and cohesive story, and are legible on an 8½-by-11-inch page.

TASK 8

Plan your layout format

Before you begin, make sure the layer names and legend text read the way you want them to appear in your layout.

19   Make sure you are in Layout View.

When you work in Layout View you work on a virtual layout page which shows the boundaries of the paper on which the map will be printed. You will notice that ArcMap has automatically placed all the data frames on the layout page, and that any of the graphs that you chose to show on the layout when you created them also appear on the layout page. At this point, all of the data frames in your layout might be stacked on top of each other on the page. You do not have to use all of them, and you will probably need to resize the elements and rearrange their location on the layout page. You will also need to add other map elements such as a north arrow, scale bar, title, and date.

20   Decide whether you want a portrait or landscape orientation for your layout, then from the File menu choose Page and Print Setup. In the Paper frame of the Page and Print Setup dialog box, click the appropriate orientation option for your map. (Note: You can also open the Page and Print Setup dialog by right-clicking the virtual page.)

21   Ask yourself the following questions to help you compose the page: Where should maps be placed, and how big should they be? How much of the page will be devoted to maps, how much to graphs, and how much for the other necessary information? How can all these elements fit on the page while still appearing large enough to be legible? It may help to sketch out the approximate layout of your elements. Do not forget to leave room for other elements such as legends, north arrows, scale bars, titles, and margins.

Your final layout must include at least two maps, one graph, and all necessary map elements. Large-scale inset maps, if used, must be referenced to a small-scale map with the inset area outlined or highlighted. The more you can visualize the final layout before you start inserting elements, the better.

TASK 9

Compose your layout

All your maps, legends, and graphs have been positioned automatically on the page. In this task, you will resize and reposition the data frames and add a legend. Before you begin, keep in mind:

- ✦ Everything in Layout View is treated as a map element, including text, graphics, data frames, legends, north arrows, and scale bars. All of the map elements have properties associated with them. To access these properties, right-click an element with the Select Elements tool and choose Properties.

- ✦   Maps and graphs that you do not want can be removed by using the Select Elements tool to click and drag an element off the layout page.

- ✦ When you select a data frame in Layout View with the Select Elements tool it becomes active. You must activate a data frame in order to add map elements to it. For example, if you wanted to add a scale bar to your layout showing the scale for a data frame named Minority, you would first need to activate the Minority data frame, then add the scale bar from the Insert menu.

- ✦ Use the Delete key with caution. Deleting will remove unwanted maps from the layout, but if you select a data frame with the Select Elements tool and press the Delete key, the entire data frame will be removed from your map document, not just from the layout page. If you want to keep a data frame in the map document, but don't want it to show on the layout page, just drag it off the virtual page, but don't delete it. This also holds true for graphs in your map—delete the graph from the layout and you actually delete the graph from the map document. (If you do accidentally delete a data frame, you can get it back by clicking the Undo button on the Standard toolbar.)

✦ While in Layout View, do not confuse the Layout toolbar with the Tools toolbar. Both toolbars contain zoom and pan tools, but they differ in function. The zoom and pan tools on the Layout toolbar allow you to zoom and pan on the virtual page without changing the map scale or spatial extent of the data in the data frame. The zoom and pan tools on the Tools toolbar will change the map scale and spatial extent of the data in the data frame, even if used in Layout View. For example, if you wanted to zoom in to a legend on your layout, you would use the zoom tool on the Layout toolbar, but if you wanted to change the scale of the data as it appears in the data frame on your layout, you would use the zoom tool on the Tools toolbar.

The Layout toolbar contains tools for zooming and panning on the virtual page without affecting map scale.

The Tools toolbar contains tools for zooming and panning within a data frame. Map scale is affected by these tools.

22 Click the Select Elements tool on the Tools toolbar, then click and drag each data frame off your layout page. Once they are all off your page, click and drag the data frame(s) that you want to show on your map back onto the virtual page. After you have positioned the data frames, resize them by clicking and dragging one of the square, cyan-colored handles of the frame with the Select Elements tool.

Once you get the data frame(s) positioned on the virtual page, you should add the legend(s).

23 From the Insert menu, click Legend. Go through each panel in the Legend Wizard, defining the settings as you go along. In each panel you can click the Preview button to see how the settings you made have affected the look of the legend. When you get to the final panel and are satisfied with the settings, click Finish. If you have other data frames in your layout, make them active and insert a legend for them as well. (Keep in mind that any settings you make can be changed after you close the wizard, by opening the legend's properties from the virtual page.)

The legend has a link to the data frame it was created from, so if you make any changes to the layer symbology or add layers to the map, these changes will appear in the legend.

TASK 10

Add other map elements to the layout.

Besides the data frames and their associated legends, there are a host of other map elements that you can add to your map, including north arrows, scale bars, neatlines, and textual information such as the map title. In the upcoming steps you will add a north arrow, a scale bar, a title, and a date to your map.

First you will add a scale bar. When adding a scale bar to a layout, you must choose the scale bar style you want to use, and define its properties. In this case, the only property you will change is the division units used for the scale bar.

The display units of the data frames in your map are currently decimal degrees (degrees of latitude and longitude). When initially added to a data frame, the units shown on the scale bar will reflect the display units of the data frame. Since decimal degrees are typically irrelevant units of measure for the casual map reader, you should switch the division units of the scale bar to something more meaningful, such as miles or kilometers.

24   From the Insert menu, click Scale Bar. On the left-hand side of the Scale Bar Selector dialog box, select one of the scale bar styles. On the right-hand side of the dialog box, click the properties button and, if necessary, click on the Scale and Units tab, then the Division Units drop-down list and choose the units that are appropriate for your layout. Click OK in the Scale Bar properties, then click OK in the Scale Bar Selector. After the Scale Bar is placed on the map, drag it to its intended location on the layout.

The next element to add is a north arrow.

25   From the Insert menu, click North Arrow. Choose a north arrow that you like from the left-hand side of the dialog, then click OK. After the north arrow is placed on the map, drag it to its intended location on the layout.

In some cases, you may find that a map element is obscuring another map element. You may want a legend, for instance, to appear in the foreground on top of other elements in the actual map space. You can control the drawing order of your map elements by right-clicking the map element with the Select Elements tool. Point to Order, then choose the appropriate drawing order option. You may have to play with this a bit in order for things to display properly if you are superimposing text or map elements onto your map.

Text elements, such as the map title, data source, and information about the map compiler (you) can also be added to the map from the Insert menu.

26   From the Insert menu, click Title. In the text box that appears, type a title for your map, then drag the title text to its intended location. From the Insert menu, click Text, enter today's date in the text box, then position the text on the layout.

As you probably noticed, the only difference between choosing to insert a title or text is that the Title option automatically places the text near the top of the layout and uses a large font size, whereas the Text option places the text at the middle of the layout and uses a relatively small text size. Once added to the map, however, you can change the size, location, or font style of either the text element or the title element.

27   Add the remaining text or map elements. At this point, what you place on the layout is up to you. You may want to add data source information for the map, which for this project would be the U.S. Bureau of the Census, Census 2000.

TASK 11

Finalize all layout elements and layout composition and preview at full scale

Your layout should be composed using techniques of good graphic design, as outlined in chapter 5. Consider the visual structure of the layout, including:

  ✦   proper balance of the map elements;

  ✦   alignment and centering of elements in relation to one another;

  ✦   placement, size, and style of text;

  ✦   legend placement and size;

  ✦   effectiveness of color selections, color contrast, and line weights;

  ✦   visual hierarchy and grouping style;

  ✦   legibility of elements;

  ✦   appropriate placement and relative sizing of elements to depict the desired emphasis.

Your finished layout should look orderly, neat, and clear, and not present the map reader with a confusing jumble of elements. Assume that the map reader has no previous knowledge of the subject matter being displayed in the layout. Your job is to present the information with clarity and in a way that piques the interest of the prospective map reader enough to look at the map and understand it. The best thematic maps are comprehensible at a glance.

28   Click the Zoom to 100% button on the Layout toolbar to display the layout at true (printed) 8½-by-11-inch page size. After clicking the Zoom to 100% button, use the Pan tool on the Layout toolbar to look over your layout page. This will enable you to determine if text and other elements will be actually legible on the printed page. Fix any problems with the legibility of elements.

TASK 12

Print the layout

29   From the File menu, click Print Preview. Make sure the page is set up the way you want it to print, then click the Print button in the upper left corner of the preview window. If you need to make any changes to the printing setup do so now from the Print dialog box; otherwise, click OK.

Be aware that printed colors often differ greatly from screen colors, and you may have to make significant alterations in your color choices after you see the first printout, in order to display the maps and graphs to optimal effect.

30   When you have printed successfully, save your map document and close ArcMap.

Developing *an* Attribute Database *from an* Internet Source

You have been asked by a community environmental organization to make a map showing the location of the Toxic Release Inventory (TRI) facilities in the Bronx. TRI facilities are those that manufacture, store, or use certain toxic chemicals in excess of 25,000 pounds per year of any of 650 listed chemicals. The Environmental Protection Agency (EPA) requires annual reports to be submitted from such facilities, and stores the data from these reports on its Web site, for public access in accordance with federal law. Using the EPA Internet site, you will access the list of current TRI facilities in the Bronx, and create a new table with actual street addresses for each TRI facility. In the next lab exercise, you will then use this list to create a spatial database of TRI locations by geocoding.

Note: This exercise requires Microsoft Excel.

TASK 1   .

Access the EPA's Web site and download information on TRI facilities

    1   Using your Internet browser, navigate to *www.epa.gov/enviro.*

The page you are taken to is the EPA gateway page from which you will navigate to and download the TRI facilities data.

    2   At the EPA EnviroFacts home page, click the Queries, Maps, and Reports link located on the left margin of the page. From this page, click the Toxic Release Inventory link, then TRI Customized Query Engine.

    3   Read through the information that describes the steps required to perform your query. Then, from the table below, click the Facility Information link. On the next page, click Step 2: Retrieve Tables for Selected Subjects.

After clicking the button the site produces a list containing the tables available to you based on your selections in Step 1. (So far, there should be only one table listed.)

    4   Check the box next to tri_facility, then click Step 3: Select Columns.

At this point, you must choose which fields (columns) you want to include in the table.

    5   In the list of columns, check TRI Facility Id, Facility Name, Street Address, County Name, and Zip Code. Scroll to the bottom of the page and click Step 4: Enter Search Criteria.

    6   Scroll down the page to the Geography Search section. In the County text box, enter **Bronx**, then, in the State text box, enter **NY**. At the bottom of the page, click Search Database.

In a few seconds you will have a table of all the TRI facilities in the Bronx. It should contain about 26 entries, one for each TRI facility in the Bronx. (The number of facilities listed will vary depending on when you access the EPA Web site, since periodically the site is updated.)

    7   Scroll to the bottom of the list and click Output to CSV File. Near the bottom of the following page click the name of the CSV file that was generated for you to download. (The name of the file will differ for each user.) Name the file **TRI.csv** and save it in your **C:UrbanGIS\MyData** folder.

8   Start Windows Explorer and navigate to the TRI.csv file that you downloaded in the previous step. Double-click the file to open it in Microsoft Excel. Within Excel, increase the widths of the columns so you can see the full names of all the records. From the File menu, click Save As. Near the bottom of the Save As Dialog click the Save As type drop-down list and choose DBF 4 (dbase IV) (*.dbf) as the file type, then click Save. When prompted, click Yes on the message box. Close Excel.

ADVANCED CHALLENGE

Add other fields to the Table from the EPA Web site

AC 1   Go back a couple of pages in your browser window, then click Form R Reports from the menu options on the left. Running a Form R query gives information about the chemicals emitted from each facility. This gives you a much longer list, since each facility is listed multiple times, one entry for each chemical emitted from the facility. By checking the appropriate columns, you can get information about whether the chemicals are known carcinogens and what specific toxic threat they pose to human life. Experiment with this option.

TASK 2

Open ArcMap, add data layers and the TRI table

9   Close your Internet browser, then start ArcMap with a new empty map.

10   Click the Add Data button, navigate to your **C:\UrbanGIS\MyData** folder and load the following layer files: Census, Landmarks, Shoreline, and Water Bodies. (Hint: Use the Ctrl key.)

Note: This assignment assumes successful completion of lab exercise 3. Since you symbolized these data layers before you saved them as layer files in lab exercise 3, they should be appropriately symbolized, but feel free to alter them.

11   From the File menu choose Save As. Name your map **BronxTRI.mxd** and save
      it in your **C:\UrbanGIS\MyData** folder.

You will now add the TRI table to your project.

12   Click the Add Data button. In the Add Data dialog box, navigate to the
      C:\UrbanGIS\MyData folder, click the TRI.dbf file, and click Add.

As soon as the table is loaded into ArcMap, the Source tab becomes active in the TOC.
Tables can only be seen in the TOC when the Source tab is chosen, so when a table is initially loaded into ArcMap the TOC automatically switches to the Source tab.

You will now open the TRI table and review its contents.

13   In the TOC, right-click the TRI table and choose Open.

Notice that the original field names became truncated when you converted the CSV table to
a DBF table. To make the field names more understandable, you will assign aliases.

14   Close the Attributes of TRI. In the TOC, right-click the TRI table and choose
      Properties. In the Table Properties dialog box, click the Fields tab.

All of the fields are listed in the Fields tab along with their properties. Notice the Alias field
is the same as each field's name. You will overwrite the current Alias values with more
descriptive names.

Alias values listed and entered here —————

Field names listed here —————

Table Properties   ? | ×

Fields | Definition Query | Joins & Relates |

Primary Display Field: [▼]

Choose which fields will be visible. Click in the alias column to edit the alias for any field.

Name	Alias	Type	Length	Precision	Scale	Number Format
☑ OID	OID	Object ID	4	0	0	
☑ V_TRI_FACI	V_TRI_FACI	String	39	0	0	
☑ V_TRI_FA_1	V_TRI_FA_1	String	44	0	0	
☑ V_TRI_FA_2	V_TRI_FA_2	String	36	0	0	
☑ V_TRI_FA_3	V_TRI_FA_3	String	33	0	0	
☑ V_TRI_FA_4	V_TRI_FA_4	Double	28	28	0	Numeric ...

[Select All]   [Clear All]

                                        [OK]   [Cancel]   [Apply]

The aliases you create are a property of the feature class or table, that is, they do not change the actual field name, just the name that appears when the table is viewed from within ArcMap.

15 In the Fields tab of the Table Properties, click the Alias name to the right of the V_TRI_FACI field and type **TRI ID.**

16 Create aliases for the remaining fields in the table. Use the table below to determine what alias to assign each field. When you are finished, click OK. (Do not create an alias for the OID field.)

Existing Field Name	Alias
V_TRI_FA_1	Name
V_TRI_FA_2	Address
V_TRI_FA_3	County
V_TRI_FA_4	Zip

17 Open the TRI table, verify that the aliases are working, then close the table.

The table is now ready to be linked to a spatial database through the geocoding function, which will be undertaken in the next lab assignment.

18 Rename the data frame **TRI Locations**.

19 Save your map. Close ArcMap.

Geocoding *and* Labeling

In this lab assignment, you will create a spatial database with the TRI information you downloaded from the EPA Web site in the previous exercise. In order to create a spatial database from a nonspatial attribute database, you will geocode—matching the street addresses given in the TRI table with a spatial database of street segments based on the TIGER/Line files. This spatial database lists streets by segments, typically streets between intersections, with ranges of left-side and right-side addresses. The address-matching process will locate an address generally within a half block of the actual location. Once you have successfully geocoded the TRI facility locations, you will be able to map their locations and perform spatial analysis on this data in later lab exercises.

TASK 1

Open existing project and add street data layer

1   Start ArcMap and open the **BronxTRI.mxd** map document you created in lab exercise 6.

2   Click the Add Data button, navigate to your **Bronx** geodatabase and load the **Bx_streets** feature class. (Bx_streets is made up of thousands of features which can take a long time to draw in the ArcMap display. You may want to uncheck it in the TOC to avoid slowing down your work.)

3   Open the Attributes of Bx_streets table. Scroll across the attribute table and review its fields. During your review, examine the FROMLEFT, TOLEFT, FROMRIGHT, TORIGHT fields. Together these fields store the address range for each street feature in the layer. Besides the range values, the table also contains fields for each street's name, type, and prefix direction (such as N. or S.); these are labeled, respectively, STREET_NAM, STREET_TYP, and PRE_DIR. When you are finished reviewing the table, close it.

Bx_streets is the reference data you will use to locate the TRI facilities in your map. Each address stored in the TRI table will need to be located on the street network based on the values in these fields.

Now you will examine the addresses stored in the TRI facilities table.

4   Click the Source tab at the bottom of the TOC, then open the TRI table.

Currently these addresses are stored in tabular form. In the geocoding process, the software will read each address in the TRI table and locate these addresses along the Bx_streets layer. The result of geocoding will be a new point feature class with one point placed at each TRI location.

TASK 2

Create an address locator

The geocoding process begins with the creation of an address locator, which in turn creates a service that ArcMap uses to match the records in a table to your reference data—in this case, Bx_streets. The address locator you create will store a set of properties that define how the address values are stored in the reference data.

5   Start ArcCatalog.

6   In the Catalog tree, locate and expand the Address Locators folder.

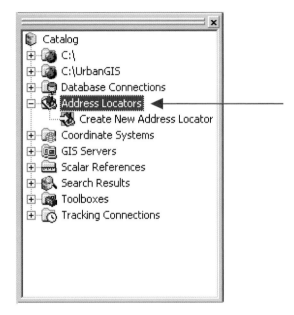

7   In the Address Locators folder, double-click Create New Address Locator. In the Create New Address Locator dialog box, scroll through the list of address styles, then locate and click the US Streets (GDB) style.

Styles with the (GDB) designation are intended for use with geodatabase files. Bx_streets is a geodatabase feature class, so you must select the geodatabase version of the US Streets style. Styles names followed by (FILE) are intended to be used with shapefiles.

8   Click OK.

The options presented to you in the Address Locator dialog box reflect the address style you just selected. You will use the left-hand side of this dialog box to give the address locator a name, to assign the reference data, and to define how the address data is stored within the reference data. The right-hand side of the dialog box lets you define additional properties for the address locator. You will leave all the options on the right side of the dialog box set to their defaults.

Name the address locator ——————

Define the reference data ——————

Define which fields in
the reference data store ——
the address attributes

9   In the New US Streets (GDB) Address Locator dialog box, click the Name text box and replace the default name with **TRI Facilities**. Click the Browse button next to the Reference data text box, navigate to your **C:\UrbanGIS** folder, double-click the Bronx geodatabase, then click the Bx_streets feature class, and click Add.

You may have noticed that the new US Streets (GDB) Address Locator dialog was intelligent enough to recognize which fields in the reference data were storing the address attributes. This was because the address fields in Bx_streets use standard naming conventions for address data. Without this capability, they would have to be defined manually.

10 Click OK.

The new address locator is now available for use. You can verify the existence of the new locator by looking for it in the Address Locator folders within the Catalog tree.

11 Close ArcCatalog.

TASK 3

Geocode TRI facility addresses

In this task you will geocode the TRI facilities. You must define which field in the TRI table stores their addresses. In the previous exercise, you created aliases for the fields in the table. These aliases will not be available to you when you work with the geocoding tools. Before you begin you should double-check which field stores the addresses.

12　In the ArcMap TOC, right-click the TRI table and choose Properties. If necessary, click the Fields tab. In the Alias column, find the Address alias, then look in the Name column to determine what the underlying field name is; this is the field you will choose when you set up the geocoding process. Close the properties.

13　In the ArcMap TOC, right-click the TRI table and click Geocode Addresses. In the Choose an address locator to use dialog box, click Add. In the Add Address Locator dialog box, click the Look in drop-down arrow and choose Address Locators, then click the TRI Facilities locator and click Add. Click OK to apply your selection of the Address Locator. (If you cannot find the TRI Facilities address locator, ask your instructor for help.)

After you define which address locator you will use, the Geocode Addresses dialog box opens. This dialog box is used to define the following three options:

+ which table stores the addresses;
+ which field in the address table contains the addresses;
+ where and how the results will be stored.

Since you activated this dialog box from the table instead of the Tools menu, the first option is already defined, but you must define the other two.

14　In the Address Input Fields frame, click the Street or Intersection drop-down list and choose the field that contains the addresses—the field you looked up in step 11. In the Output frame, click the Browse button. In the Saving Data dialog box, click the Save as type drop-down arrow and choose personal geodatabase feature classes. Name the new feature class **TRI_Locations** and save it within your **Bronx** geodatabase. Click Save, then OK.

When the geocoding process is completed, the Review/Rematch Addresses dialog box appears, containing statistics describing how successful the address locator was in matching addresses in the table to their locations within the reference data.

Keep in mind your results may differ from the graphic.

These statistics indicate how successful the address locator was in geocoding the addresses stored in the table. In this example, all but one of the addresses were successfully matched.

It's likely that at least one record in your TRI table was not matched. To resolve this problem you will use the Match Interactively function to locate the addresses that did not geocode with the initial batch of addresses.

TASK 4

Correct unmatched addresses and rematch

Before you try the Match Interactively function, reducing the spelling sensitivity of the Address Locator and running the process in batch mode may increase your success rate.

15 On the Review/Rematch Addresses dialog box, click the Geocoding Options button. Click in the Spelling Sensitivity text box and type a number lower than the current value, then click OK. Click Match Automatically to run the process again. If you still have unmatched addresses, go to the Rematch Criteria frame, click the Unmatched Addresses option, and click Match Interactively.

The Interactive Review dialog box is used to explore and fix the unmatched addresses. The upper portion lists the unmatched records; the lower portion contains the tools for fixing them.

Depending on when you downloaded the TRI file, and what type of geocoding options you set, your list of unmatched records may differ from the one used in the following steps. Regardless of any difference, the process to find and fix them is still the same.

16   In the upper portion of the Interactive Review dialog box, examine the highlighted record and write down the address that appears in the Arc_Street field.

You must now determine why this address did not geocode. To find the answer, you will return to the actual street data.

17   Click Close, then Done.

In the TOC you will see the Geocoding Result: TRI_Locations layer. This layer contains the results of your geocoding work so far, stored as points.

18   Open Attributes of Bx_streets and scroll across the table. Locate the STREET field. Right-click on the STREET field's name and choose Sort Ascending. Scroll through the list of street names and locate the streets with a name that matches or is a close match to those that were your non-matched addresses. Once these records are found, compare them to your unmatched addresses. Identify the problem.

19   From the Tools menu, point to Geocoding, then to Review/Rematch addresses, then click Geocoding Results: TRI_Locations. When prompted, click Yes to start an editing session, and again open the Interactive Review dialog box. Review the address of the record that's currently highlighted, then click the Modify button. In the Edit Standardization dialog box, correct the incorrect portions of the address, then press Enter, then close the dialog. (Examples of possible errors include misspelled street names: *Bulter St.* instead of *Butler St.*)

This process should produce several matching candidates in the candidate list. Review the address ranges in the candidate list, find and click the candidate that's a match for the current unmatched record, then click Match. Repeat this process for any other unmatched records until they are all matched. It's possible you will be unable to match all the records.

1. Select the record from the list of unmatched records. 2. Click the modify button. 3. Change the address as appropriate and press Enter. 4. Select the appropriate candidate, then click Match.

20   When you are finished matching the records, close the dialog box. The
      statistics in the Review/Rematch Addresses dialog box are updated to
      reflect the currently matched and unmatched addresses. Review these, then
      click Done. Close the Attributes of Bx_streets. If necessary, open the Editor
      toolbar, and from the Edit menu, choose Stop Editing and click Yes to save
      your edits.

The TRI_Locations layer now contains points for each TRI facility in the Bronx.

21   In the TOC, rename the Geocoding Results: TRI_Locations to **TRI Facilities**.
      Change the symbol and color used to represent the TRI Facilities to
      something of your choice, then save it as a layer file in your **MyData** folder.

TASK 5

Label the TRI facilities

In this task you will label each TRI facility with a unique ID number to replace facility names
that are long, difficult to read, and tend to clutter your map. You will create a new field in
which to store the ID values, then manually enter the values.

22   Open Attributes of TRI Facilities. Add a new field called **Facility_ID**. Define it
      as a Text field, with a length property of 2.

23   Start an Edit session. (From the Editor menu, choose Start Editing. When
      prompted, choose to edit the data in the personal geodatabase.) In Attributes
      of TRI Facilities, sort the table by the name field in ascending order. In
      Attributes of TRI Facilities, scroll across the table until you can see the
      Facility_ID field. Scroll to the top of the table, click in the first empty record in
      the Facility_ID field and type 1; in the next record, type 2; continue working
      your way down the table, adding the ID values in consecutive order until you
      reach the end of the table. Add ID values only for those facilities that have
      been matched in the geocoding process. You can tell this by looking for the
      letter *M* in the status field, designating a match; *U* designates facilities that
      have not been matched. When you are finished adding the values, stop your
      editing session and save your edits. Close the attribute table.

Now that the IDs are assigned, you can use them to label the TRI facilities.

24 First turn off Bx_streets, if necessary, so you'll be able to see your labels. Open the properties of the TRI Facilities layer. In the Layer Properties dialog box, click the Labels tab. Check the option to label features in this layer. In the text string box on the Labels tab, click the Label Field drop-down arrow and choose Facility_ID from the list.

You can use the options on the Labels tab to control how the labels will look on the map, how they are placed, and which features in the layer are labeled. For more information, click the context-sensitive Help button.

25 Experiment with the label properties. Each time you change a property (such as font, size, bold, and so forth), click Apply to preview the labels. When you are satisfied with the look of the labels in your map, click OK.

If you want individual control over the TRI Facilities labels you must convert them to Annotation. To do this, in the TOC right-click the TRI Facilities layer and choose Convert Labels to Annotation. Once this is done, you can reposition and change the style of individual labels. More information is available from ArcGIS Help.

TASK 6

Prepare the map for presentation

26 Now that you've geocoded as many of the facilities as possible and added labels, switch to Layout View and work on your map's appearance and information content—adding a title, legend, symbolization, and other elements you think would help the map reader.

TASK 7

Display TRI theme with the percent minority theme

You will now create a second map, using all the data layers from the first data frame, and adding the layer file showing the Percent Minority population by census tract that you previously created.

27   Press and hold down the Ctrl key on your keyboard while clicking on each layer name in the TOC. (Do not click the TRI table.) After selecting all the layers, release the Ctrl key, then right-click in the TOC and choose Copy. From the Insert menu click Data Frame. Right-click the title of the new data frame and choose Paste Layer(s).

28   Click the Add Data button. In the Add Data dialog box, navigate to your **C:\UrbanGIS\MyData** folder and add the Percent Minority layer file. Turn off the TRI Facilities labels. (Right-click TRI Facilities and click label features to unselect the option.) Name the new data frame **Percent Minority**.

ADVANCED CHALLENGE TASK

Make an inset map

AC 1   In Data view, zoom in on the area where most of the TRI facilities are located, the South Bronx. Switch to Layout View and prepare a layout showing both data frames in your current map. Make sure the Percent Minority data frame is active. Make the focus of the layout the Percent Minority data frame and use the TRI Locations map—which shows the full extent of the Bronx—as a smaller map that shows which part of the Bronx the Percent Minority map is focused on. Add a rectangle to the TRI Locations data frame that represents the extent of the Percent Minority data frame. Do this by opening the properties of the TRI Locations data frame, then clicking the Extent Rectangles tab.

TASK 8

Finalize and print a layout

29   Finalize your layout for presentation purposes.

Make sure you have all required map elements included in the layout. If necessary, review lab exercise 5 for more detailed instructions on creating a layout. If you want to enter text citing data sources, the data sources are the U.S. Environmental Protection Agency (EPA) for the TRI data, and Census Bureau, Census 2000 for the demographics and other layers in your map.

30   When you are satisfied with your layout, save your map, then print the layout.

31   Close ArcMap.

The Fire Department is conducting a study aimed at minimizing the response time for evacuating high-rise residential buildings in the Bronx, and it has asked you to produce a GIS map showing city blocks where these buildings are located. The Fire Department is especially interested in those high-rise buildings that are near TRI facilities, because these facilities often use or store hazardous materials. These can pose a hazard to nearby residents if there is an accidental release of toxic materials.

In this lab assignment, you will be creating a map that locates city blocks containing high-rise multi-family residential buildings, using two database tables, and examining the relationship between the locations of the high-rise buildings and TRI facilities. The Department of Finance (DOF) has supplied you with a parcel level ASCII (text) file of Bronx tax lots whose land-use category for these buildings is called multi-family

Working *with* Relational Databases

elevator buildings, and the Department of City Planning (DCP) has supplied you with a digital map of Bronx city blocks in ArcGIS format. The DCP file contains no attribute data fields except block number. (Do not confuse city blocks with census blocks; they are not the same.)

You will import the DOF text table into the attribute database of your GIS, then execute Structured Query Language (SQL) queries to select those fields and rows that are relevant to your task. You will use a summarize function to group the selected parcel data by city block, create a new table with the data grouped by block, then join this table to the mapped DCP table of Bronx city blocks, using the block number field as the key field. After you find the locations of the high-rise buildings in the Bronx, you will make a visual inspection, comparing them to the locations of the TRI facilities.

TASK 1

Open a new project and add layers

1　Open ArcMap with a new empty map. Click the Add Data button, navigate to the Bronx geodatabase in your **UrbanGIS** folder, and add the Bx_blocks feature class. Click the Add Data button again, this time adding the **bx-DOF.txt** from the **UrbanGIS\SourceData** folder.

You will now see the blocks layer and the bx-DOF table in your TOC.

2　Open the DOF table. Examine the contents, then close it. In an upcoming task, you will use the Join function to attach the bx-DOF table to the Bx_blocks layer. Once this is done, you will be able to map the attributes contained in the table. From the File menu choose Save As. Name the map **ResidentialHighRise** and save it to your **UrbanGIS\MyData** folder.

TASK 2

Create a definition query for the DOF table

3　Open the properties of the bx-DOF table. In the Table Properties dialog box, click the Definition Query tab, then click the Query Builder button.

A definition query defines which features (or records) will appear and be available for analysis when a feature class (or table) is used in ArcMap. Using a Query Builder dialog box, you will construct a definition query for the bx-DOF table to show only records that contain high-rise buildings, defined as those thirteen floors or higher.

4　In the fields list of the Query Builder, double-click 'NUMFLOORS'. Click the greater-than operator (>). Click in the expression box, and type **12** at the end of the expression. Make sure your expression matches the one in the graphic on the next page, then click OK in the two open dialog boxes.

TASK 3

Generalize the DOF table

5   Open the bx-DOF table.

Because you have refined your list with the definition query, the records now appearing in this table are considerably fewer than the last time you opened the table.

The bx-DOF table is aggregated by lots, whereas the Bx_blocks feature class is aggregated by blocks, a larger unit. To successfully join the two tables, you must first generalize, or group, the DOF table by blocks to ensure that you do not lose any data in the join process. To perform this generalization, you will use the Summarize function to group the records together based on common block numbers.

6   In Attributes of bx-DOF.txt, right-click the BLOCK field name and choose
    Summarize. Make sure that the Select a field to summarize option is set
    to BLOCK. For the second option, scroll to the bottom of the list of fields,
    then expand the UNITSRES field (number of residential units). Under the
    UNITSRES field, check the Sum option. For the third option, click the Browse
    button. In the Saving Data dialog box, choose Personal Geodatabase
    Tables from the Save as type drop-down list, then navigate to your Bronx
    geodatabase. Name the table DOF_sum and click Save. Make sure your
    settings in the Summarize dialog box match those shown in the graphic
    below, then click OK. Click Yes when prompted to add the table to the map.

Based on the settings you made in the Summarize dialog, ArcMap merged records with the
same BLOCK value together, while also summarizing the UNITSRES values for the merged
records. For example, if during the Summarize process, ten records were found with a
BLOCK value of 1, and each of these records had a UNITSRES value of 2 in the output table,
there will be one record with a BLOCK value of 1 and its UNITSRES value will be 20.

TASK 4

Join the DOF summary table to the blocks layer

7   Close the Attributes of bx-DOF table, and open the DOF_sum table.

Notice the data is now aggregated by BLOCK and there is a Sum_UNITSRES field that contains the number of residential units within each block. The Count_BLOCK field was automatically added, and it contains the number of the records found in the input table that have the same block ID as were merged into one record during the Summarize process.

8   Close the attribute table. In the TOC, right-click the Bx_blocks layer, point to Joins and Relates, then choose Join. Make sure the topmost drop-down list is set to Join attributes from a table. For the first option, choose TAXBLOCK from the drop-down list. For the second option, choose DOF-sum from the drop-down list. For the third option, choose BLOCK from the drop-down list. In the bottom right corner of the dialog box, click the Advanced button, choose the option to Keep only matching records, and click OK. Make sure your settings match the graphic below, then click OK. Click Yes to create an index.

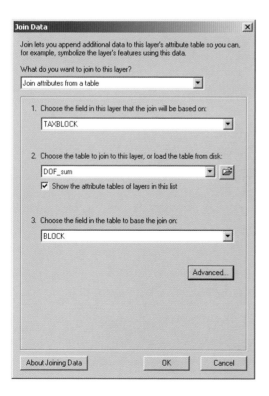

Using the settings you made in the Join Data dialog box, ArcMap joined the attributes in the DOF_sum table to the attributes in the Bx_blocks feature class. To create the join, ArcMap read through the values in the TAXBLOCK field of the Attributes of Bx_blocks and the values in the BLOCK field of the Attributes of DOF_sum, then joined records that matched in the respective fields. Records in the Bx_blocks layer, for which no matching record was found, were dropped from the results.

9   Open the Attributes of Bx_blocks table and explore its contents, then close it when you're done.

The Attributes of Bx_blocks also contains the attributes from the DOF_sum table. You may have noticed that there are fewer records (rows) in the table than before the join. This is due to the fact that not all blocks are represented in the DOF table, only the ones that have certain building types. This can also be seen in the map, because only the records preserved in the table are drawn as features in the map display.

TASK 5

Export the selected records to a new feature class

10 In the TOC, right-click Bx_blocks, point to Data, and choose Export Data. In the Export Data dialog box, choose to export all features using the same coordinate system as the layer's source data. Name the output **high_rise** and save it in your Bronx geodatabase. (Be sure to choose Personal Geodatabase feature classes from the Save as type drop-down list when browsing to the output location.) Check your Export Data dialog box against the graphic below, then click OK, and click Yes when prompted to send the exported data to the map.

This export process created a new feature class in the Bronx geodatabase named high_rise, with the joined attributes from the DOF table now a permanent part of the new feature class.

11 Save the high-rise layer as a layer file in your **UrbanGIS\MyData** folder.

Now that the layer has been created, the join can be removed from the Bx_blocks layer to restore it to its original state.

12 In the TOC, right-click Bx_blocks, point to Joins and Relates, then to Remove Join(s), and choose Remove All Joins.

This will complete the restoration of the layer; you may have to click the refresh button on the bottom left corner of the map screen to see the changes.

13 Click the Display tab at the bottom of the TOC, then drag the high_rise layer above the Bx_blocks layer. Change the symbol colors to make them distinct.

The high_rise layer in your map displays the blocks containing the buildings that meet the Fire Department's criteria.

TASK 6

Create a new data frame to display residential units as proportional symbols

14 Insert a new data frame in your map document. Name the new data frame **Residential Units**, then rename the original data frame **High-rise Buildings**.

15 In the High-rise Buildings data frame, right-click Bx_blocks and choose Copy; then, right-click the Residential Units data frame and choose Paste Layer(s). Using the same procedure, copy and paste the high_rise layer into the Residential Units data frame.

At this point, the new data frame should have the same layers as the old one.

In order to make the high_rise layer display the total number of residential units as a proportional symbol map, you must change its symbolization.

16 From the Residential Units data frame, open the properties of the high-rise layer, then click the Symbology tab. In the Show box, click Quantities, then click Proportional Symbols. In the Fields frame, click the Value drop-down list and choose sum_UNITSRES. Accept the defaults for the remaining settings, or, if you want, experiment with the other options. When finished making settings, click OK.

17 Save this symbolized high_rise layer as a Layer File named **Residential Units** in your **UrbanGIS\MyData** folder.

TASK 7

Conduct a visual inspection comparing the locations of high-rise buildings with TRI facilities

18 Add the TRI_Locations feature class that you created in lab exercise 7. Change the color of the TRI points if necessary, to make them stand out. Examine the TRI locations in relation to the largest concentrations of high-rise buildings. Using the Measure tool, compare the various distances between the TRI facilities and the blocks containing high-rise buildings.

TASK 8

Prepare map for presentation

19 Review your map's appearance and information content. Make sure that each layer is symbolized with different colors to differentiate the high-rise blocks from the other Bronx blocks. Removing the outlines from the high_rise layer may make the map clearer.

ADVANCED CHALLENGE

Create two additional maps showing relationships among high-rise buildings and other selected data

AC 1 Insert two new data frames and add the High-rise and Residential Units Layer files to all of them. In the first data frame, add the Percent Minority Layer File that you created in lab exercise 4; in the second data frame, create the dot density layer of total population as you did in lab exercise 3.

AC 2 After symbolizing the layers appropriately, create a layout containing maps from both data frames. Comparing the maps in the layout view, make a visual evaluation of the spatial correspondence between high-rise residential buildings, and areas with a high percentage of minority population, as well as areas with a high density of total population. What spatial generalizations can be formed based on the visualization of these maps?

TASK 9

Create and print a layout

20   Create a layout with two maps of the Bronx, the first displaying blocks containing high-rise residential buildings, and the second showing the number of residential units in the blocks containing high-rise buildings, along with the locations of the TRI facilities. If you completed the Advanced Challenge Task, then you will also have the option to include up to two additional maps showing the relationships among high-rise buildings and minority population or total population.

Make sure you have all required map elements in the layout. Review lab exercise 5 for detailed instructions on creating a layout. The data sources are the New York City Department of Finance, the New York City Department of City Planning, the EPA, and Census Bureau.

21   When you are satisfied with your layout, save your map.

22   Print your layout, then exit ArcMap.

Generating Buffers *and* Using Selection *for* Proximity Analysis

In this assignment, you will be creating a map showing the locations of the TRI facilities in the Bronx, as well as determining the demographic characteristics of the populations potentially affected by the TRI facilities. You will be examining the populations within a half-mile radius of the TRI facilities. This half-mile buffer has been determined to be the likely area of potential impact from the facilities. You will compare the characteristics of the populations within the half-mile buffers of the TRI facilities to the general population of the Bronx (the reference population). As an advanced task, you will create a graph comparing the demographic information.

TASK 1

Open your BronxTRI.mxd map document and save it as a new map

1 Start ArcMap and open the **BronxTRI.mxd** map document that you saved at the end of lab exercise 7. Switch to Data View if necessary. From the File menu, click Save As. Name the new ArcMap document **BronxBuffers.mxd** and save it in your **C:\UrbanGIS\MyData** folder.

TASK 2

Generate buffers

2 If necessary, activate the Percent Minority data frame, and zoom to the full extent.

3 Open ArcToolbox. (On the Standard toolbar, click the Show/Hide ArcToolbox Window button.)

Inside ArcToolbox, expand the Analysis Tools toolbox, then expand the Proximity toolset. In the Proximity toolset, double-click the Buffer tool.

Creating buffers is a matter of choosing the layer that contains the features you want to buffer, assigning the name and location of the output, and defining the buffer distance around the input features. In this case you are going to buffer the TRI facilities at a distance of a half-mile.

4 In the Buffer tool dialog box, click the Input Features drop-down arrow and select the TRI Facilities layer. Click the Browse button next to Output Feature Class. Save the output in your Bronx geodatabase and name it **TRI_Buffers**. Make sure that Linear unit is selected for the Distance option. Click the Distance units drop-down arrow and select Miles, then click in the Linear Unit text box and type **.5**. Leave the remaining options set to the defaults. Check your settings against those shown in the results graphic, then click OK. When the process is completed, click Close in the progress window.

When the buffer process is finished, the output feature class is automatically added to your map. You can now visualize the half-mile buffer zones around each TRI facility. At this point, you may begin to notice a clustering effect in the buffer zones.

5   Close ArcToolbox. (Click the Show/Hide ArcToolbox Window button.)

By default, ArcMap assigns a solid fill to the buffer features along with a randomly selected color. This works well for the initial visualization, but you also need to see which features the buffers overlay.

6   In the TOC, click the TRI_Buffer symbol that appears just below the layer name. In the Symbol selector, choose the Hollow fill symbol, and change the Outline Color to something that stands out on your map, such as Mars Red or Solar Yellow. You may also want to change the outline width to 2. (As with your work in the other lab exercises, how you symbolize layers in your map is up to you.)

TASK 3

Select by location

In this task, you will select the census tracts within the buffers. This selection process will require some subjective judgment, since the buffers are circular and census tracts are generally irregularly shaped rectangles that may be partially in and partially out of the buffers—that is, the buffers and the tracts are not spatially coincident. You will try a few different methods to determine which tracts should be considered within the buffers.

In a later task, you will use the selected tracts to determine the percentage of minorities living within the TRI buffer zones, so the selections you make in the current task will affect your demographic analysis.

7   From the Selection menu, click Select By Location. Read through the options in the Select By Location dialog box, then construct a selection statement to select features from the Percent Minority layer that intersect the features in the TRI_Buffers layer. Make sure your settings match those in the graphic below, then click Apply, and Close.

The outlines of the selected census tracts are now highlighted. This selection set consists of all the census tracts that intersect, even slightly, the half-mile buffers around the TRI facilities.

8   Open the Attributes of Percent Minority and determine how many census tracts are intersecting the TRI buffers.

Based on the number of selected census tracts, you must now decide if you think too many tracts are selected. You can decide based on visual or numeric evidence. You probably will want to narrow the selection so that it more accurately portrays census tracts in close proximity to a TRI facility. Through visual inspection, you will notice that many tracts are quite large, and even though they do intersect a buffer, only a small part of the tract (if any) is actually near the TRI facility.

9   Open the Select By Location dialog box. As you did in step 7, create a query statement to select the census tracts based on a spatial relationship to the TRI buffers, but this time use a selection method other than Intersect. Selection methods valid for the datasets include: Are contained by, Have their center in, Are completely within, and Are crossed by the outline of. Think of the reasons why the other options would not work for the datasets. After completing each selection, open the Attributes of Bx_demog, and determine how many census tracts were selected. Then, look at the selected tracts on the map, and determine if the selected tracts adequately represent the census tracts within the buffers. When finished, close Attributes of Bx_demog.

By now, you should have an idea as to which selection method best isolates the census tracts within the TRI buffers. Your next step is to create a new layer from the selected features.

10   If necessary, run your chosen selection method again, then close the Select By Location dialog box. Once the selection is made, right-click the Percent Minority layer, point to Selection, and click Create Layer From Selected Features.

The layer is created for you, and is automatically added to the top of the TOC. If you want to make it a permanent layer that you can use in other maps, save it as a layer file named **Buffer Tracts** in your **C:\UrbanGIS\MyData** folder.

If none of the selection sets produced by the methods you experimented with seem appropriate, try the Extra Challenge Task.

EXTRA CHALLENGE

Select features interactively

After you run a Select By Location query, remember that you can manually add or subtract individual census tracts from your selection. This means you can run the selection method that worked best, then add tracts or remove tracts to your selections.

EC 1   Turn off the selection layer that you created in step 10, then run the Select By Location method that you chose at the end of step 9. From the Selection menu, choose Set Selectable Layers. In the Set Selectable Layers dialog box, uncheck all the layers except the Bx_demog layer, then close the dialog box. Click the Selection menu again, point to Interactive Selection Method, and choose Add to Current Selection. Using the Select Features tool, click the census tracts you wish to add to your selection set. (You may need to zoom in to the areas from which you're choosing features.) To select the census tracts you want to remove from the selection set, change the Interactive Selection Method to Remove from current selection, then click the census tracts you wish to remove. When you are finished with the manual selection, convert your selection set to a selection layer, then save it as a layer file in your **C:\UrbanGIS\MyData** folder. Clear the selected features in your map (Selection menu, Clear Selected Features), and reset the options so that all the layers in your map are selectable.

TASK 4

Demographic analysis

To determine if there are differences in the demographic characteristics between the general population of the Bronx and the people living within the half-mile buffers of the TRI facilities, you must conduct a statistical analysis of the overall population of the Bronx and of the population within the buffers.

11   In the active data frame, open Attributes of Percent Minority. In the table, generate statistics for the Minority field. (Right-click the field name and choose Statistics.) Write down the Sum value for the Minority field. Now generate and record the Sum statistic for the TOTPOP field using the same method you used for the Minority field. Calculate the overall percentage of minority population for the Bronx. Record the results (Minority/TOTPOP). When finished, close the Statistics dialog box, then close the attribute table.

12 Open the attributes of your selection layer. Using the same process as in the previous step, generate the sum statistics for the Minority and TOTPOP fields, then use the Sum values to calculate the overall percentage of minority population within the buffer areas.

You now know what percentage of the total Bronx population is minority, and what percentage of the population living near the TRI facilities in the Bronx is minority. Compare the two percentage values.

You can repeat the steps in this task for any other attributes in your census data to investigate other ways the demographics differ between the Bronx as a whole and the areas near TRI facilities.

ADVANCED CHALLENGE

Create a graph of demographic comparison data

AC 1 Create a bar graph of the information you derived in the previous task. This will require you to create a new table containing these figures. Refer to exercise 5 if you need help.

TASK 5

Create a layout

Now that your analysis is complete, prepare your data for presentation and layout.

Your layout should contain the following: a map showing the TRI facilities, the half-mile buffers, and the census tracts symbolized by the Percent Minority field. If you made a chart in the advanced challenge, add that to the layout.

13 In your active data frame, symbolize the layers appropriately and, if necessary, give the layers meaningful names.

14 Switch to Layout View and add all the required map elements. Add the information you derived from task 4, your demographic analysis, as text to your Layout. (Click the Insert menu and choose Text.) An example of how the text can be worded is given below. Fill in the blanks with your own statistics, because everyone's numbers will differ slightly, depending on how you selected the census tracts included in the buffer zones.

Sample Text

Of the 355 census tracts in the Bronx, (number) tracts are within a half-mile of a TRI facility, and of the (number) people in the Bronx, an estimated (number) live within a half-mile of a TRI facility. Within the half-mile buffer zones, approximately (percentage) of the population is considered minority, while (percentage) of the total population of the Bronx is minority, according to the 2000 Census.

The data sources are the EPA, from 2002, and Census 2000.

15   Save your map document and close ArcMap.

Geoprocessing Operations *and* Joining Tables

Ten

Rarely is data presented to you in a neat package as it is in this book; much of the time you must track it down and enter it into your project yourself. In this assignment, you will create a table of children's asthma hospitalizations in the Bronx from data in a New York City Department of Health publication, then map this information. This health data is aggregated by ZIP Code to protect patient confidentiality. You will join the asthma table to a ZIP Codes feature class. The ZIP Code spatial data will need some geoprocessing before it is ready for mapping asthma rates. After mapping asthma hospitalizations, you will create a layout showing asthma hospitalization rates with an overlay of TRI facilities. As an advanced step, you will create a dot density overlay of asthma hospitalization cases.

TASK 1

Create a new table

In this task you will use ArcCatalog to create a new table inside the Bronx geodatabase to store the asthma hospitalization data. After creating the table you will use ArcMap to add fields and data.

  1   Start ArcCatalog.

  2   In the Catalog tree, navigate to your **UrbanGIS** folder, right-click the Bronx geodatabase, point to New, then click Table. In the New Table dialog box, name the table **Asthma,** click Next, then click Next again to bypass the configuration panel. In the third panel click Finish. Close ArcCatalog.

An empty table now exists within the Bronx geodatabase.

TASK 2

Add fields to the Asthma table

  3   Start ArcMap. Open the **BronxBuffers** map document that you created in lab exercise 9. In the map document, create a new data frame named **Asthma Rates**. Switch to Data View if necessary and collapse the other data frames in the map document. From the File menu, click Save As, name the new map document **BronxAsthma.mxd** and save it in your **UrbanGIS\MyData** folder.

The Asthma table must contain three fields, one to hold the ZIP Codes and two to store the asthma statistics.

  4   Add the Asthma table that you just created to the Asthma Rates data frame, then open it. From the Options menu in the Attributes of Asthma dialog box, click Add Field. Create a text field named **ZIP** with a Length property of 5, then click OK. Add a second field named **RatePer1000** and set its type as Float. Create a third field named **Hospitalizations** and set its type as Short Integer.

TASK 3

Populate the Asthma table

With the Asthma table now built, you can populate it with the available asthma statistics. You will add one record for each Bronx ZIP Code.

5   If necessary open the Editor toolbar, then start an edit session. (From the Editor menu, choose Start Editing.)

6   Populate the Attributes of Asthma with the Bronx asthma statistics contained in the AsthmaFacts.pdf report accessible from *www.nyc.gov/html/doh/ downloads/pdf/asthma/facts.pdf* in your **UrbanGIS\SourceData** folder. The AsthmaFacts.pdf contains tables and statistics for several areas in New York City. Go to table 6 (page 43 of 50), identify the relevant values, and enter only the values that apply to the Bronx, where ZIP Codes begin with 104. (The Number field in the document corresponds to the Hospitalizations field you just created.) When you are finished, choose Stop Editing from the Editor menu, close the Attributes of Asthma, then close the Editor toolbar. There should now be twenty-five ZIP Code records in your Asthma table with number of hospitalizations and rate per 1000 values for each ZIP Code.

TASK 4

Clip the ZIP Code features to the Bronx

In your SourceData folder there is a shapefile named nyczip that contains ZIP Code areas for New York City. In this task you will use a geoprocessing operation known as Clip to convert the nyczip shapefile to a feature class in your Bronx geodatabase that contains only the Bronx ZIP Code districts.

7   Click the Add data button and load **nyczip.shp** from your **UrbanGIS\ SourceData** folder. Click the Add data button again; this time add the **Shorelines.lyr** from your **UrbanGIS\MyData** folder.

The Clip tool uses the boundary of one layer to clip out a portion of another layer, like a cookie cutter. You will use the Clip tool to cut out the ZIP Code polygons from the nyczip shapefile that are within the boundaries of the Shorelines layer. (To get more information about the Clip function, consult Help.)

8   Open ArcToolbox and expand the Analysis Tools toolbox and then the Extract toolset, and double-click the Clip tool. Using the drop-down lists in the Clip dialog box, set the Input Features to nyczip, and the Clip Features to Shorelines. Name the output feature class **Bx_zip** and save it in your **Bronx** geodatabase. Make sure your settings match the graphic below, then click OK. Click Close when the process completes.

When the process finishes, Bx_zip is automatically added to your map. The extent of Bx_zip falls within the borders of the Shoreline layer.

9   Close ArcToolbox. Remove the nyczip layer from the Asthma Rates data frame. (In the TOC, right-click nyczip and click Remove.) Zoom to the full extent.

You now have a feature class containing the Bronx ZIP Code districts, and a table containing the Asthma hospitalization rates within the Bronx. You will join the Asthma table to the ZIP Code districts in a moment.

TASK 5

Format fields to prepare for a table join

When joining tables, the fields you join must be the same type—for example, an integer field with an integer field or a text field with a text field. Before you join the Asthma table to the attributes of Bx_zip table, determine the type of fields the ZIP Codes values occupy.

10   If necessary, click the Source tab at the bottom of the TOC, then open the Properties of the Asthma table. Click the Fields tab, look at the Type property, then close the dialog box. Do the same thing to determine the Type property of the ZIP Code field in the Bx_zip layer. In the Asthma table, the ZIP Codes are stored as text strings, while in Bx_zip they are stored as long integers. You will have to convert the format of the ZIP Code values in one of the tables to match the format used by the other. In this case, you will reformat the values in Bx_zip.

11   Open the Attributes of Bx_zip table. Add a new field to this table named **ZIP_Codes**, set its Type to Text and its Length property to 5.

12   In the Attributes of Bx_zip table, right-click on the name of the ZIP_Codes field and choose Calculate Values. (Click Yes to the warning message about calculating values outside of an edit session.) In the Field Calculator, click the ZIP field in the Fields list to add it to the expression. Make sure your expression matches the graphic below, then click OK.

The ZIP_Code field will be populated with the same values stored in the ZIP field, except now they are stored as text values. You can tell that they are text values because of their default alignment to the left; numeric values align to the right.

When you are done, you should remove the original ZIP field from the Attributes of Bx_zip table, because it's considered bad practice to have duplicate values within a table.

13 In the Attributes of Bx_zip, right-click the ZIP field and choose Delete Field, then click Yes on the warning message. After the field is deleted, close the table.

TASK 6

Join tables

You are now ready to join the Asthma table to Attributes of Bx_zip, allowing you to then create maps based on the asthma statistics.

14 In the ArcMap TOC, right-click Bx_zip, point to Joins and Relates, then click Join. In the first drop-down list at the top of the dialog box, make sure that Join attributes from a table is selected. Using the three numbered options, define a table join that will join the Attributes of Asthma to the Attributes of Bx_zip. (Remember, you must base the join on the fields storing the ZIP Code values.) When the drop-downs have been filled in correctly, click OK, and click Yes to create an index.

15 Open the Attributes of Bx_zip. If the join was successful the asthma statistics will now appear as additional fields. Close the table.

TASK 7

Map asthma hospitalization rates in the Bronx

You will now make a choropleth map of asthma hospitalization rates of children in the Bronx.

16 Open the Properties of the Bx_zip layer and click the Symbology tab. Using the options on the Symbology tab, create a graduated color (choropleth) map based on the RatePer1000 field. Choose an appropriate color ramp, then click OK to close the Layer Properties dialog box. In the TOC, change the name of the Bx_zip layer to **Asthma Hospitalization Rates of Children**.

17 Save the Asthma Hospitalization Rates of Children as a layer file in your **UrbanGIS\MyData** folder.

18 Copy the TRI Facilities layer from the Percent Minority data frame and paste it into your Bronx Asthma data frame. Add any other layers to the Bronx Asthma data frame necessary to convey the information to best effect. Create as many data frames as you think are necessary—for example, you may have one data frame showing the TRI facilities on top of the hospitalization rates layer, and another data frame showing hospitalization rates below the minority population layer, symbolized with a dot density scheme. You could also have a percent minority choropleth map shown beneath a dot density map of asthma hospitalization cases.

TASK 9

Create a presentation layout

19 Using the appropriate data frames, create a presentation layout of childhood asthma hospitalization rates in the Bronx. Include any charts, text, or other maps that might help explain the aspects of the data you want to present.

The Data Source for the asthma data is the New York City Department of Health (DOH), 2003. The source for the NYC ZIP Code data is Census 2000. The other sources to be listed in your layout will depend on what information your maps use. All other data sources have already been given in the earlier lab exercises.

20 When you are satisfied with your layout, save your map, then print the layout. Close ArcMap.

Data Exploration *and* Geostatistical Analysis

Eleven

In this advanced assignment, you will continue working with

the asthma hospitalization data to conduct data exploration and

geostatistical analysis. You will create a graduated symbol map

of rates based on areal aggregation at centroids, an interpolated

surface of rates derived from centroids, and add histograms,

scatterplots, Thiessen polygons, and other data exploration graphs

to your layout. It is strongly recommended that you have read and

understood chapter 9 before trying this lab exercise.

Note: This exercise requires ArcGIS Geostatistical Analyst.

TASK 1

Create a graduated symbol map

In addition to plotting hospitalization rates as a choropleth map based on ZIP Code polygon boundaries, you can also join the asthma statistics to the ZIP Code centroids (the geometric center of each ZIP Code district's polygon). Then they can be used to generate a graduated symbol map, and the values can be interpolated to create a continuous rate surface.

1 Start ArcMap and open the **BronxAsthma.mxd** map document that you saved in the previous lab. After the map opens, run the Save As command, name the map **Interpolation.mxd**, and save it in your **UrbanGIS\MyData** folder.

In this exercise, you will use the Geostatistical Analyst extension for ArcGIS. Before you can use the extension, you must turn it on in your current map.

2 From the Tools menu, click Extensions. In the Extensions dialog box, check the box next to Geostatistical Analyst, then click Close. Turn on the Geostatistical Analyst toolbar. (From the View menu, click Toolbars and check Geostatistical Analyst.)

3 Add a new data frame to your map. Name the new data frame **ZIP Centroids**, and, if necessary, make it active. Click the Add Data button and add the Bx_zipcentroid feature class and the Asthma table from your Bronx geodatabase.

4 Join the Asthma table to Bx_zipcentroid based on the ZIP field in the respective tables.

If you have not already, you should develop the habit of reviewing statistics for the datasets that you are working with, especially in the data exploration phase of mapping.

5 Open the Attributes of Bx_zipcentroid. Right-click the name of one of the fields containing the asthma statistics, and choose Statistics. After reviewing the summary statistics, choose the other Asthma field from the Field drop-down list and review its statistics. Close the statistics dialog box, then close Attributes of Bx_zipcentroid.

As in lab exercise 8, you can now symbolize your feature data based on the asthma statistics that you joined to it. In this case, your features are points rather than polygons, so instead of creating a graduated color map, you will create a graduated symbol map.

6   Open properties of Bx_zipcentroid and click the Symbology tab. In the Show box, click Quantities, then click Graduated symbols. In the Fields box, click the Value drop-down list and pick the field you want to base the symbolization on. Make any other settings you wish, then click OK.

At this point the graduated symbols probably make little sense without a base layer behind them as a reference.

7   Add any other layers to your map that you feel are necessary to make the map understandable.

TASK 2

Create an interpolated surface from point data

You can interpolate the asthma rates attached to the centroids to create a continuous surface map of childhood asthma hospitalization rates across the Bronx.

8   Insert a new data frame into the map. Name the new data frame **Asthma Interpolation**. Copy the Bx_zipcentroid layer from the ZIP Centroids data frame and paste it into the Asthma Interpolation data frame. (The join between the asthma table and Bx_zipcentroid will be preserved.)

9   From the Geostatistical Analyst menu, click Geostatistical Wizard. In the Dataset 1 tab, set the Input to Bx_zipcentroid and the Attribute to Asthma.RatePer1000. Click Finish, and click OK on the Output Layer Information dialog box.

The Geostatistical Wizard used the hospitalization rates stored with each point to create the Inverse Distance Weighting (IDW) Prediction Map of hospitalization rates across the entire Bronx.

10   Add the Shorelines layer to the Asthma Interpolation data frame, and with the Display tab of the TOC active, drag it above the IDW Prediction Map.

The boundaries of the IDW Prediction Map do not precisely match the boundaries of the Shorelines layer. You can resolve this issue by changing settings in the map's Properties.

11   Open the Properties of the IDW Prediction Map and click the Extent tab. From the Set the extent to drop-down list, choose the rectangular extent of Shorelines. Click OK, then zoom to the full extent of the map.

The boundaries of the IDW Prediction Map extend to match the bounding coordinates of the Shorelines layer. Now that the IDW Prediction Map covers the entire extent of the Bronx, you can use the data frame properties to make its extent match perfectly with the boundary of the Shorelines layer.

12   Open the Properties of the Asthma Interpolation data frame and click the Data Frame tab. In the Clip to Shape frame, check the Enable box, then click Specify Shape. In the Data Frame Clipping dialog box, choose the Outline of Features option, and pick Shorelines from the Layer drop-down list. Click OK in the two open dialog boxes.

At this point the IDW Prediction Map fits neatly within the borders of the Shoreline layer. It's an attractive layer, but can you safely assume that it accurately predicts the asthma hospitalization rates of children across the Bronx? Do you feel comfortable that it fairly shows the spatial trends of childhood asthma cases in the Bronx?

ADVANCED CHALLENGE

Experiment with interpolation methods

AC 1   Use the Geostatistical Wizard to experiment with other methods of interpolation, such as kriging, and determine which method is most illustrative of the data. If you have the ArcGIS Spatial Analyst extension available to you, use it to create surface maps from the asthma data and then convert these surfaces to contour maps of the asthma hospitalization rates.

TASK 3

Data exploration with histograms and scatter plots

13 After finalizing your interpolated surface, experiment with the Explore Data option in the Geostatistical Analyst to create histograms, Voronoi diagrams (Thiessen polygons), variograms, and scatterplots of the asthma rates data stored with the Bx_centroids layer. Add any of the graphs or maps you create during the data exploration phase by clicking the Add to Layout button. If you do add these graphs to your layout, it's helpful to add some explanatory text about them.

TASK 4

Prepare a final layout

14 Based on this initial data exploration and visualization, decide what further analyses would likely be necessary. Think about how these would be done. Then, using the maps showing interpolated surfaces or contours created in task 2, and the data exploration graphs created in task 3, prepare a final layout that presents your preliminary findings about asthma hospitalization rates in the Bronx.

15 When you are satisfied with your layout, save your map and print the layout. Close ArcMap.

Advanced Layout Techniques

You have already learned how to make basic presentation layouts in previous exercises. In this assignment, you will learn how to make your layout cleaner, more readable, and more effective. You will use standard map elements such as neatlines and scale bars, learn about legends and scale bar creation, and discover how to arrange elements on the page.

TASK 1

Create new data frames

1   Open ArcMap, then open the **BronxAsthma** map document that you created in lab exercise 10. Switch to Data View if necessary. Use the Save As command to save the map as **BronxAsthma_Layout.mxd** in your **UrbanGIS\ MyData** folder.

At the end of this lab exercise you will have a layout containing three data frames showing the following content: a choropleth map of asthma hospitalization rates by ZIP Code in the Bronx, a map showing percentage of Bronx population under age 18, and an inset map of the under-18 population of the South Bronx.

2   Insert three new data frames into your map document, and name them as follows, from top to bottom in the TOC: **Asthma Rate by ZIP Code, Under 18 by Census Tract**, and **South Bronx Inset**.

In the next step you will populate the data frames with the appropriate datasets. Keep in mind that in many cases the layer that you need already exists in one of the existing data frames in your map. In these cases, you can just copy and paste these layers into the new data frame as they are needed.

3   Activate the Asthma Rate by ZIP Code data frame. Inside this data frame, create a choropleth map showing by ZIP Code the childhood asthma hospitalization rates in the Bronx. Turn on any labels you wish and add any other layers you think are necessary to the data frame.

4   Activate the Under 18 by Census Tract data frame. Inside this data frame, create a choropleth map showing by census tract the percentage of the population that is under 18 years old. (If necessary, use the METNAMES text file that's in your **UrbanGIS\SourceData** folder to determine which field in Bx_demog contains the values you need to create this map.)

5   Activate the South Bronx Inset data frame. In this data frame, create the same choropleth map that you created in the Under 18 by Census Tract data frame, then zoom in to the South Bronx.

6   Review the contents of each of the three new data frames. For each data frame, add any other layers that you feel are necessary to support the data frame's theme.

TASK 2

Give your map some geographic context

To prevent your map from making the Bronx look like some isolated island in the Pacific Ocean, you should add another layer to the map to give the Bronx a positional reference within the other boroughs and counties in the New York City area.

7 Unless you have already done so, add the **nyczip** shapefile from your **UrbanGIS\SourceData** folder to each of the three new data frames. Once it's in the data frames, place it below the other layers and make its outline color and fill color the same for each frame. You should also choose a muted color for nyczip so that it will blend in with the background, not distract map readers.

TASK 3

Lay out your map, use Zoom to Layer, and set bookmarks

In this task you will arrange the data frames on your layout and define the scale and extent shown in each of the data frames.

By default, ArcMap stacks all the data frames in the map document on the center of the layout page. If there are data frames you don't want to show in your final map, you must drag them off or remove them from the map.

8 Remove all the data frames from your map document except for the three that you created in task 1. (To remove a data frame, right-click the name in the TOC and choose Remove.)

9 Switch to Layout View. Click the stack of data frames with the Select Elements tool and drag the selected data frame off the stack, and repeat for the other two data frames. Once you have separated the data frames, move the Asthma Rate by ZIP Code data frame to the top center portion of the layout. Then move the Under 18 by Census Tract data frame to the lower left corner, and the South Bronx Inset data frame to the lower right corner. (If you click the Toggle Draft Mode button on the Layout toolbar, the data frames will be labeled with their names.) Once the data frames are where you want them on the page, resize them accordingly.

While in Layout View, you can change the extent and scale of the data in the frames by using the Zoom and Pan tools from the Tools toolbar. If you want to zoom in on elements on the layout page without changing their scale or extent, use the Zoom and Pan tools on the Layout toolbar.

10   Activate the Asthma Rate by ZIP Code data frame by clicking it on the virtual page using the Select Elements tool. If necessary, center the Bronx within the data frame by right-clicking the layer in the TOC and clicking Zoom To Layer.

Now that you have the data in the Asthma Rate by ZIP Code data frame positioned, you can preserve its extent and scale using the Bookmark function. After setting a bookmark for the data frame, you can quickly return to the bookmarked extent if, for example, you accidentally change the spatial extent shown in the data frame.

11   Verify that the Asthma Rate by ZIP Code data frame is active and that it's set to your desired spatial extent. From the View menu, point to Bookmarks, then click Create. Name the bookmark **Bronx Asthma** and click OK.

If necessary, you can return the Asthma Rate by ZIP Code data frame back to the bookmarked extent by choosing Bookmarks from the View menu and selecting Bronx Asthma. You can only access a data frame's bookmarks if it's active.

12   Activate the Under 18 by Census Tract data frame. If necessary, zoom to and center the Bronx within the data frame, then create a bookmark for the current extent named Bronx Under 18.

13   Activate the South Bronx Inset data frame. Using the Zoom In tool from the Tools toolbar, zoom to the portion of the South Bronx that you want as the focus for the data frame. Once you have the extent of this data frame defined, create a bookmark named **Bronx Inset**.

Keep in mind that bookmarks are also useful for general navigation purposes. You can use them to zoom to a particular study area or block without having to guess each time where it is on the larger map.

TASK 4

Add an extent rectangle

Your layout currently has three data frames, two of them showing the complete extent of the Bronx, and another zoomed to the South Bronx. To help the map reader see what portion of the Bronx the South Bronx Inset map represents, you will create a box in the Under 18 by Census Tract data frame that reflects the spatial extent shown in the South Bronx Inset data frame.

14   Open the Properties of the Under 18 by Census Tract data frame. In the Data Frame Properties dialog box, click the Extent Rectangles tab. In the Other data frames box, click the South Bronx Inset, then click the right pointing arrow. Click the Frame button. In the Frame Properties dialog box define a Border style and click OK. If you want a leader—a line connecting the extent rectangle to the data frame it represents—click the Show Leader checkbox. You can then click the Leader Symbol button to choose a style and color for the leader line. Click OK in all of the open dialog boxes and close the Data Frame Properties.

A convenience of using extent rectangles is that they update automatically each time you change the extent of the data frame they represent. You can also use several extent rectangles per layout in order to show several different areas in detail on a single data frame.

TASK 5

Define the symbology of the layers in your data frames

ArcMap automatically assigns a random color to a layer when it's added to the map.

You can quickly change the color of a layer by right-clicking its color chip in the TOC, then selecting a new color from the drop-down color palette. To change more advanced symbology settings, such as changing the outline width and color of a fill symbol, you can left-click a layer's color chip in the TOC to open the Symbol Selector. You can make the most advanced changes to layer symbology, such as creating graduated symbols or a dot density map, by using the Symbology tab of the Layer Properties dialog box. By this point, you have probably used all of these methods to change the symbology of a layer.

If you create customized colors or styles, you can save these to be used again. This is useful when you create a new color (for example, a certain shade of green for park areas) that you want to use in future projects. After creating or updating a symbol, click the Save button in the Symbol Selector, then give the symbol a name and define its category. From that point on the symbol will be available to you in the left-hand pane of the Symbol Selector. To show only the symbols in the categories you created, click the Category drop-down list and choose the category in which you saved your symbols. (When saving custom symbols, it's a good idea to put them in well-defined categories for easier access.)

Here are a few points to remember about color and symbology:

+ Make sure you choose colors that are relatively light, not bright or dark. Printers often print darker than what you see on the screen. ArcMap contains preset colors that you can choose, and in many cases the lighter ones are the most effective.

+ If the map is intended for publication on the Internet, be aware that there will almost always be a color shift. The range of colors you can see on your screen is usually more limited than what others can see on theirs, and much more limited than what a good printer can show.

+ Try to avoid using green for anything other than parkland or open space, or blue for anything other than water. Blue and green have strong connections in people's minds to water and vegetation that are difficult to overcome. You don't want someone to mistake your carefully selected blue tax lot for a rectangular lake.

+ Be careful about using too many lines and shapes in one map. If you have on your map a street grid, tax lots, development zones, census tracts, and ZIP Codes, no one is going to able to tell one from another even if the colors of each element are very different. You are better off using more then one map to tell different stories than to try to differentiate elements by color.

+ If you're symbolizing ranges of data, it's best to use color ramps (from light to dark) instead of a group of disparate colors. Color ramps more effectively convey range data without the map reader having to keep referring to the legend to figure out which color is associated with what value.

15　To the best of your abilities, define the symbology of all the layers in your map document in a way that is cartographically functional and aesthetically pleasing.

You can modify ArcMap's preset color ramps to make the colors more distinguishable. Often, if you're using more than four or five classes in your data, the individual colors are very close together and it becomes difficult to tell one color value from another. To modify a color ramp, in the Symbology tab of the Layer Properties dialog box, change the color of the symbol listed for the lowest value, then change the color of the symbol assigned to the highest value. Right-click one of the symbols and choose Ramp Colors. ArcMap will define a new range of colors based on the symbol colors you defined. Once the ramp is created you can save it by right-clicking on the Color Ramp drop-down list and choosing Save to Style.

TASK 6

Add and arrange elements on the layout

In this task you will add elements such as legends, scale bars, and north arrows. You will begin by inserting a legend.

Before you insert a legend you should seriously consider changing layer names that may be difficult to understand to something descriptive and easily discernable to the map reader.

You can rename a layer using the General tab of the Layer Properties dialog, or by clicking a layer name in the TOC to select it, then clicking its name again to activate a text cursor to type in a new name. You can also change the text that appears in the legend below the layer name. The legend you add to your layout is dynamic, so if you change a layer name after the legend is inserted, the legend will automatically update.

16   If necessary, rename the layers in your map, giving them descriptive names, and change any legend text in the TOC. For example, in the Under 18 by Census Tract data frame, the KIDPC (or the UNDER18) field name appears just below the layer name. You could either remove this name from the legend or rename it to something more descriptive, such as Percentage of Persons < Age 18.

17   Activate the Asthma Rate by ZIP Code data frame. From the Insert menu, click Legend.

The first panel of the Legend Wizard contains two boxes: the Map Layers box lists all the layers in the active data frame, while the Legend Items box lists all the layers that will appear in the legend. By default, all of the layers in the data frame will appear in the Legend Items box. You can remove items from the legend by clicking the Layer in the Legend Items box, and clicking the left pointing arrow. You can also change the order of the items in the legend by clicking a layer in the Legend Items pane and clicking the up or down arrow.

18   Remove any layers that you do not want and define their draw order. Click Next.

The second panel of the Legend Wizard is used to define the legend title and its style. By default the title is Legend. This is not particularly helpful; most people who see a legend on a map will know it's a legend.

19   It's recommended that you delete the word "Legend" and leave the Legend Title box blank. Click Next.

The third panel is used to create a legend frame. By default, no legend frame is defined, but if you want one, within this panel you can define the line style of the border, the background color, and add a drop shadow effect. You can also change the distance between the legend elements and the border by changing the Gap number. You can also change the square border to a rounded border by putting a number in the Rounding box; a value of 10 to 20 works well.

20   If you want a frame, define the Legend Frame settings, then click Next.

The fourth panel lets you change the properties of the symbols (called patches) used in the legend. For example, you can use the Area drop-down list in the Patch box to change the symbols used for the Asthma layer from rectangles to ellipses.

21   Accept the default symbol patches assigned to your legend items, or change them to suit your preferences. When finished defining the patch styles, click Next.

The fifth panel lets you review and change the spacing among all the parts of your legend.

22   Accept the spacing values, and click Finish. After the legend appears on the layout page, use the Select Elements tool to move it to the location where you want it to appear on the final map.

Like legends, scale bars are dynamic features on the layout that change in accordance with changes in the scale of their associated data frame. When adding a scale bar to a layout with multiple data frames, it's very important to make sure the data frame for which you are creating a scale bar is active; otherwise, you might unknowingly place a scale bar on the layout with a data frame that it does not represent.

23   Make sure the Asthma Rate by ZIP Code data frame is active. From the Insert menu, choose Scale Bar. Scroll through the list of scale bars and click the one you like most. While the scale bar is selected in the list, click the Properties button. In the Scale and Units tab of the Scale Bar dialog box, click the When resizing drop-down arrow and choose Adjust width. (This setting preserves the division value and the number of divisions in the scale bar, and adjusts the scale bar width if the map scale changes.) In the Units frame, click the Division Units drop-down list and choose Miles. In the Scale box, set the Division Value to 5 and the Number of Divisions to 2. Click OK in the two open dialogs. If necessary, move the scale bar to a location just below the Asthma Rate by ZIP Code data frame.

Based on the setting you just made, the scale bar will represent a distance of ten miles using two five-mile divisions. If you change the map scale, the scale bar will preserve the distance it represents and the number of divisions it contains by automatically changing sizes. Feel free to experiment with other properties.

It's better to set the properties for the scale bar after you place it on the map. That allows you to make changes to the properties of the scale bar while also previewing it. To open the properties of a scale bar, right-click the scale bar with the Select Elements tool and choose Properties. Once the Properties dialog box is open, move it to where it does not block the scale bar. Each time you make changes you can click the Apply button to preview the effect of your changes.

24 From the Inset menu, choose North Arrow. From the list of arrows, click the one you want and click OK. Then use the Select Elements tool to move it where you want it on the Layout and to resize it.

25 At this point, finalize the position, and size of the legend, north arrow, and scale bar. They should all be placed in an area on the page that does not interfere with any important map information.

Keep in mind that when you have more than one data frame on a layout, it must be very clear which map element refers to which data frame. For smaller (larger-scale) maps showing the same data as the main map, an inset map for example, you might not need a separate scale bar and you would need only a separate north arrow when the orientation of the two maps is different.

The final map element you will add in this task is a neatline. Adding a neatline creates a nice, clean border around your map page. Your map frames will have their own borders by default, but they probably won't take up the whole page. And, if you add a neatline you should consider removing the border from your main map frame on the layout. Having too many large enclosing boxes will make the map look heavy and confusing.

26   From the Insert menu, click Neatline. Set the properties for the neatline, then click OK.

Each printer has different margin dimensions that can affect how large the neatline is when added to the layout. If you change printer options after adding a neatline, you can select the neatline and change its size by dragging one of the nodes at its corner.

TASK 7

Finish the layout

At this point you should have several elements on your layout including three data frames, a scale bar, a north arrow, and a legend.

27　In layout view, arrange the data frames with the main data frame taking up the most room in the center of the page. Arrange the smaller data frames so they don't cover up any important data in the main map. Make sure the north arrow and the scale bar are placed far enough away from the smaller maps so they clearly show they refer to the main map and not the smaller insets. It's a good idea to spread out the elements enough so they don't interfere with one another. Use the pan and zoom tools on the Tools toolbar to focus each data frame on the area of interest, and make sure there is not too much extra space within your data frames—that is, to show the features in the data frames at the largest scale possible.

The last element to add is a title.

28　From the Insert menu, choose Title. The default title is the name of the map document. Change the title to something more fitting than the map document name, then move the title to an appropriate location on the layout. (If there is an old title in the layout, delete or edit it.)

You can select the title with the Select Elements tool, right-click, and choose Properties. In the Text tab of the Properties dialog, you can edit the title's text, and change the font type and font style.

You also need to add some text describing what's in each of the smaller data frames on your layout. For these you can use the Text tool found on the Draw toolbar. Turn on this toolbar if necessary.

29　On the toolbar, click New Text tool **A** . Click above the data frame you want to label and type an appropriate title.

To set off the title so it's more visible, use the New Rectangle tool on the Draw toolbar to draw a rectangle around, and slightly larger than, the title text. Open the properties of the rectangle by double-clicking it with the Select Elements tool. Use the Properties dialog box to change the rectangle to a light color. If needed, place the text on top of the rectangle by moving the rectangle, then right-clicking on the text and choosing Order, then Bring to Front. Drag the text back on top of the text box. You can also align the text within the rectangle by selecting both the text and the box, then right-clicking and choosing the appropriate Align option.

TASK 8

Print the layout

You are now ready to print your layout.

30   Save your map document, then use the Print Preview dialog box to make sure the page is set up the way you want. Click the Print button, then click OK in the Print dialog box.

31   After a successful printing, save your map document again, then close ArcMap.

Afterword

At the end of part 1, we touched on advanced GIS functions and on new GIS data sources that are becoming increasingly available to GIS users at reasonable cost. As is always the case with GIS, the effective use of advanced GIS functionality depends on the data. These new data sources, coming mainly from the development and expansion of other technologies—such as remote sensing, Global Positioning Systems, and the Internet— promise to put advanced GIS analysis techniques into the hands of even those municipalities that have limited budgets and small GIS staffs. Three-dimensional GIS analysis is especially useful in the dense urban environment, where buildings and infrastructure share the same horizontal space at different levels, and where the aesthetics and environmental impact of proposed high-rise development need to be understood. The potential of three-dimensional GIS for examining both the built and natural environments is just beginning to be appreciated by planners.

Although this book has focused mainly on a city in the more developed world, these new techniques and data sources should prove to be especially useful to the planners and administrators of rapidly growing urban areas. As the world becomes predominately urban, and cities such as Nairobi, Shanghai, Mexico City, and Istanbul grow at unprecedented rates, the need for GIS, and for skillful and ethical GIS users, will keep pace with this phenomenon.

This book has attempted to provide a solid grounding in the current uses of GIS for the urban environment, outlining basic concepts and commonly used analytical techniques. More advanced techniques have just been touched upon here, and deserve their own book. Although the subject of GIS programming is beyond the scope of this book, it should be noted that programming languages such as Microsoft Visual Basic and Sun Microsystems Java make it possible to customize GIS applications for nontechnical user groups to meet specialized urban needs. For example, nontechnical staff may need to access GIS information for operational or planning purposes, and a GIS can be customized to allow them to do so with minimal GIS training. As a GIS user–analyst for a government agency, nonprofit organization, or consulting firm, you may be called upon to program such simple applications.

The authors believe that, due to the interdisciplinary character of GIS and the evolving nature of the technology, the GIS learning process is ongoing, and this is part of its appeal. As you begin to apply GIS to urban problems, you will quickly find that no book, however comprehensive, can cover the myriad possible GIS functions and analytical needs that may present themselves. In fact, many GIS users and analysts will find themselves in the position of creating their own GIS techniques and methods, often developing novel solutions in the process. It is not unusual for even beginning GIS users and analysts by necessity to invent new ways to analyze urban spatial issues. As mentioned in the Introduction, this need for innovation by the everyday user is part of the adventure of GIS.

About the authors

Juliana A. Maantay, PhD, MUP, M Phil, MA, BSc

Juliana Maantay is Associate Professor of Environmental and Urban Geography at City University of New York/Lehman College's Department of Geology and Geography, where she has been teaching since 1997. As director of the Geographic Information Science (GISc) Program at Lehman, she developed six different GIS courses at the undergraduate, graduate, and doctoral levels, and established an active internship program for GIS students, as well as a GIS certificate program. She has also taught graduate-level GIS and planning courses at Pratt Institute's Graduate Center for Planning and the Environment, and Hunter College's Department of Urban Affairs, and is a research scientist with the National Oceanic and Atmospheric Administration's (NOAA) Cooperative Center for Remote Sensing Science and Technology (CREST) at City University of New York.

Dr. Maantay's experience includes more than 20 years as an environmental analyst and urban planner for a variety of governmental, nonprofit, and consulting firms, such as the New York City Department of Environmental Protection, the Department of City Planning, and the Center for a Sustainable Urban Environment. She also works extensively with community-based planning groups.

Dr. Maantay's main research interests are environmental justice, urban health, risk assessment, and sustainable development, and she views GIS as an excellent vehicle for problem solving in these areas. Her work has been published in major journals, including the *American Journal of Public Health; Environmental Health Perspectives; The Journal of Law, Medicine, and Ethics; Projections: the Planning Journal of Massachusetts Institute of Technology (MIT);* and *Planners Network.* She has also contributed to chapters in books such as *World Minds: Geographical Perspectives on 100 Problems;* and *Health and Social Justice: A Reader on Politics, Ideology and Inequity in the Distribution of Disease.*

Juliana Maantay holds a PhD and Master of Philosophy degree in Environmental and Urban Geography from Rutgers University, as well as a Master of Urban Planning degree from New York University, a Master of Arts in Geographic Information Systems from Hunter College, and a Bachelor of Science in Environmental Analysis from Cornell University. Dr. Maantay is a native New Yorker who, despite traveling widely, is always happy to return to her urban roots.

John C. Ziegler, AICP, RA, MUP, M Arch, BA

Mr. Ziegler is the founder and president of Space Track Inc., a New York City-based GIS consulting company. His interest in Geographic Information Systems (GIS) developed from his experience in urban planning and architecture, his government work with emergency management programs, and his professional practice in architecture. As president of Space Track, Mr. Ziegler has worked with government, nonprofit, and private-sector organizations since 1989, and has overseen the development of Space Track's own municipal GIS applications. Space Track is the developer of LotInfo, and is a business partner of ESRI, Terrapoint, and other GIS companies. Space Track was a part of the disaster response team that worked with New York City's Office of Emergency Management in the weeks following the 9/11 attack on the World Trade Center, using GIS to produce mapping for emergency responders.

Prior to founding Space Track, Mr. Ziegler was in private practice as an architect and planner in New York City, and served as the New York regional director of the Federal Insurance Administration (FIA), and deputy regional director of the Federal Emergency Management Agency (FEMA). As the FIA's New York regional director, he was responsible for the National Flood Insurance Program in New York, New Jersey, Puerto Rico, and the Virgin Islands, and implemented the program to map riverine and coastal floodplains in these areas. When the FIA became a part of FEMA, he oversaw the full range of federal disaster response and hazard mitigation programs in the New York region.

Mr. Ziegler is an adjunct professor of planning at New York University's Wagner Graduate School of Public Service, where he taught urban planning, data visualization, and GIS from 1989 to 2001.

He holds a Master of Urban Planning degree from New York University, a Master of Architecture from Columbia University, and a Bachelor of Arts from Harvard College. He is a charter member of the American Institute of Certified Planners (AICP), and a Registered Architect in New York.

Data credits

Data source and image/illustration credits

All graphics and illustrations without attribution to third-party copyright holders were created by the authors, with the assistance of Karen Kaplan and Andrew Maroko of the Lehman College Urban GISc Lab, and Kenneth Mack, formerly of Space Track Inc. Data sources for illustrations and maps, and authorial responsibilities, are given below. Vector drawings not otherwise credited were redrawn by Jennifer Galloway of ESRI Press, using Adobe Illustrator.

Chapter 2

2.2, 2.3 Kenneth Mack

2.4, 2.9 LotInfo (Juliana Maantay), Source: Bytes of the Big Apple™ Tax Block & Lot Files ©1995, 1997, 1998, 2002, 2004, 2005 and Pluto™© 2003, 2004, 2005. NYC Department of City Planning. All rights reserved.

2.5 LotInfo (Juliana Maantay), Source: Bytes of the Big Apple™ Tax Block & Lot Files ©1995, 1997, 1998, 2002, 2004, 2005 and Pluto™ © 2003, 2004, 2005. NYC Department of City Planning. All rights reserved. Bronx Data Center: Lehman College, City University of New York. © William Bosworth, Bronx Data Center. U.S. Census Bureau 2000; USGS

2.7 LotInfo (Juliana Maantay), Source: Bytes of the Big Apple™ Tax Block & Lot Files ©1995, 1997, 1998, 2002, 2004, 2005 and Pluto™ © 2003, 2004, 2005. NYC Department of City Planning. All rights reserved. ESRI Data & Maps 2002. Bronx Data Center: Lehman College, City University of New York. © William Bosworth, Bronx Data Center. (Juliana Maantay)

2.8 ESRI Data & Maps 2002; Bronx Data Center: Lehman College, City University of New York. © William Bosworth, Bronx Data Center. U.S. Census 2000

2.13, 2.15–2.19 ESRI Data & Maps 2002 (Juliana Maantay)

Chapter 3

3.4, 3.6, 3.9, 3.10 LotInfo (Kenneth Mack), Source: Bytes of the Big Apple™ Tax Block & Lot Files ©1995, 1997, 1998, 2002, 2004, 2005 and Pluto™ © 2003, 2004, 2005. NYC Department of City Planning. All rights reserved.

3.7 LotInfo (Karen Kaplan), Bytes of the Big Apple™ Tax Block & Lot Files ©1995, 1997, 1998, 2002, 2004, 2005 and Pluto™ © 2003, 2004, 2005. NYC Department of City Planning. All rights reserved.

3.8 ESRI Data & Maps 2002 (Juliana Maantay)

3.10 LotInfo (Kenneth Mack), Source: Bytes of the Big Apple™ Tax Block & Lot Files © 1995, 1997, 1998, 2002, 2004, 2005 and Pluto™ © 2003, 2004, 2005. NYC Department of City Planning. All rights reserved.

3.11, 3.18 U.S. Census Bureau 2000 (Kenneth Mack)

3.12 National Health Statistics, Centers for Disease Control and Prevention: Division of Tuberculosis Elimination, Surveillance Reports 2000 (Juliana Maantay)

3.13–3.17 Bronx Data Center: Lehman College, City University of New York. © William Bosworth, Bronx Data Center. U.S. Census Bureau 2000 (Juliana Maantay)

3.19 Bronx Data Center: Lehman College, City University of New York. © William Bosworth, Bronx Data Center. U.S. Census Bureau 2000; New York State Department of Health. Garg R., Karpati A., Leighton J., Perrin M., Shah M. Asthma Facts, Second Edition. New York City Department of Health and Mental Hygiene, May 2003. (Juliana Maantay). U.S. EPA

3.20   Bronx Data Center: Lehman College, City University of New York. © William Bosworth, Bronx Data Center. New York State Department of Health. Garg R., Karpati A., Leighton J., Perrin M., Shah M. Asthma Facts, Second Edition. New York City Department of Health and Mental Hygiene, May 2003

3.22   USGS (Kenneth Mack)

3.31   Bronx Data Center: Lehman College, City University of New York. © William Bosworth, Bronx Data Center. New York State Department of Health. Garg R., Karpati A., Leighton J., Perrin M., Shah M. Asthma Facts, Second Edition. New York City Department of Health and Mental Hygiene, May 2003. (Juliana Maantay)

Chapter 4

4.1   Bronx Data Center: Lehman College, City University of New York. © William Bosworth, Bronx Data Center. U.S. Census Bureau 2000 (Juliana Maantay)

4.2   LotInfo (Juliana Maantay), Source: Bytes of the Big Apple™ Tax Block & Lot Files ©1995, 1997, 1998, 2002, 2004, 2005 and Pluto™ © 2003, 2004, 2005. NYC Department of City Planning. All rights reserved. ESRI Data & Maps (Juliana Maantay)

4.3   Bronx Borough President's Office, Bronx Solid Waste Management Plan, 1997 (Juliana Maantay)

4.4, 4.5, 4.16, 4.17, 4.19   Bronx Data Center: Lehman College, City University of New York. © William Bosworth, Bronx Data Center. U.S. Census Bureau 2000 (Juliana Maantay)

4.6   ESRI Data & Maps (Juliana Maantay)

4.8–4.13   U.S. Department of Health and Human Services, Centers for Disease Control and Prevention. Chronic diseases and their Risk Factors: The Nation's Leading Causes of Death, 1999

4.14   Montefiore Medical Center

4.18   U.S. Department of Health and Human Services, Centers for Disease Control and Prevention. Chronic Diseases and their Risk Factors: The Nation's Leading Causes of Death, 1999.

4.19   Bronx Data Center: Lehman College, City University of New York. c William Bosworth, Bronx Data Center. U.S. Census Bureau 2000 (Juliana Maantay). Source: Bytes of the Big Apple™ Tax Block & Lot Files ©1995, 1997, 1998, 2002, 2004, 2005 and Pluto™ © 2003, 2004, 2005. NYC Department of City Planning. All rights reserved.

Chapter 5

5.2, 5.3, 5.7, 5.8, 5.9, 5.11, 5.12   LotInfo (Karen Kaplan), Source: Bytes of the Big Apple™ Tax Block & Lot Files ©1995, 1997, 1998, 2002, 2004, 2005 and Pluto™ © 2003, 2004, 2005. NYC Department of City Planning. All rights reserved.

5.4, 5.14, 5.21, 5.27   Bronx Data Center: Lehman College, City University of New York. © William Bosworth, Bronx Data Center. U.S. Census Bureau 2000 (Juliana Maantay)

5.10   Bronx Data Center: Lehman College, City University of New York. © William Bosworth, Bronx Data Center. U.S. Census Bureau 2000. EPA 2000 (Juliana Maantay)

5.13   Bronx Data Center: Lehman College, City University of New York. © William Bosworth, Bronx Data Center. Source: Bytes of the Big Apple™ Tax Block & Lot Files ©1995, 1997, 1998, 2002, 2004, 2005 and Pluto™ © 2003, 2004, 2005. NYC Department of City Planning. All rights reserved. New York City Department of Sanitation 2002 (Juliana Maantay)

5.15   U.S. Department of Health and Human Services, Centers for Disease Control and Prevention. Chronic Diseases and their Risk Factors: The Nation's Leading Causes of Death, 1999

5.17, 5.20, 5.23   ESRI Data & Maps 2002 (Juliana Maantay)

5.18   LotInfo (Karen Kaplan), Source: Bytes of the Big Apple™ Tax Block & Lot Files ©1995, 1997, 1998, 2002, 2004, 2005 and Pluto™ © 2003, 2004, 2005. NYC Department of City Planning. All rights reserved.

5.22   New York State Department of Health. Garg R., Karpati A., Leighton J., Perrin M., Shah M. Asthma Facts, Second Edition. New York City Department of Health and Mental Hygiene, May 2003. Bronx Data Center: Lehman College, City University of New York. © William Bosworth, Bronx Data Center. (Juliana Maantay)

Chapter 6

6.1, 6.2, 6.9   U.S. Census Bureau 2000 (Kenneth Mack)

Chapter 7

7.2   U.S. Census Bureau 2000 (Kenneth Mack)

7.3   LotInfo (Kenneth Mack), Source: Bytes of the Big Apple™ Tax Block & Lot Files ©1995, 1997, 1998, 2002, 2004, 2005 and Pluto™ © 2003, 2004, 2005. NYC Department of City Planning. All rights reserved.

Chapter 8

8.1   LotInfo (John Ziegler), Source: Bytes of the Big Apple™ Tax Block & Lot Files ©1995, 1997, 1998, 2002, 2004, 2005 and Pluto™ © 2003, 2004, 2005. NYC Department of City Planning. All rights reserved.

8.3   LotInfo (Kenneth Mack), Source: Bytes of the Big Apple™ Tax Block & Lot Files ©1995, 1997, 1998, 2002, 2004, 2005 and Pluto™ © 2003, 2004, 2005. NYC Department of City Planning. All rights reserved.

8.4   LotInfo (Andrew Maroko), Source: Bytes of the Big Apple™ Tax Block & Lot Files ©1995, 1997, 1998, 2002, 2004, 2005 and Pluto™ © 2003, 2004, 2005. NYC Department of City Planning. All rights reserved.

Tables 8.1–8.8   The City of New York Department of Finance

Table 8.9   The City of New York Department of Finance; Source: Bytes of the Big Apple™ Tax Block & Lot Files ©1995, 1997, 1998, 2002, 2004, 2005 and Pluto™ © 2003, 2004, 2005. NYC Department of City Planning. All rights reserved.

Chapter 9

9.1, 9.2, 9.6   ESRI Data & Maps (Juliana Maantay)

9.3, 9.4   LotInfo (Kenneth Mack), Source: Bytes of the Big Apple™ Tax Block & Lot Files ©1995, 1997, 1998, 2002, 2004, 2005 and Pluto™ © 2003, 2004, 2005. NYC Department of City Planning. All rights reserved.

9.5   Juliana Maantay

9.7–9.11   Andrew Maroko

9.12   LotInfo (Juliana Maantay), Source: Bytes of the Big Apple™ Tax Block & Lot Files ©1995, 1997, 1998, 2002, 2004, 2005 and Pluto™ © 2003, 2004, 2005. NYC Department of City Planning. All rights reserved. U.S. EPA

9.13–9.15, 9.17, 9.19   Andrew Maroko; Juliana Maantay

9.21   New York State Department of Health. Garg R., Karpati A., Leighton J., Perrin M., Shah M. Asthma Facts, Second Edition. New York City Department of Health and Mental Hygiene, May 2003. Bronx Data Center: Lehman College, City University of New York. © William Bosworth, Bronx Data Center. (Juliana Maantay)

9.22, 9.23   Bronx Data Center: Lehman College, City University of New York. © William Bosworth, Bronx Data Center. Montefiore Medical Center

9.25   Bronx Data Center: Lehman College, City University of New York. © William Bosworth (Karen Kaplan, Juliana Maantay)

9.26   New Jersey Department of Environmental Protection (Karen Kaplan, Juliana Maantay)

Chapter 12

12.3b, 12.4   LotInfo (Kenneth Mack), Source: Bytes of the Big Apple™ Tax Block & Lot Files ©1995, 1997, 1998, 2002, 2004, 2005 and Pluto™ © 2003, 2004, 2005. NYC Department of City Planning. All rights reserved.

Case Study 1

1   New York City Fire Department; U.S. Census 2000 (Micaela Birmingham)

2   Municipal Art Society; LotInfo (Micaela Birmingham), Source: Bytes of the Big Apple™ Tax Block & Lot Files ©1995, 1997, 1998, 2002, 2005 and Pluto™ © 2003, 2004, 2005. NYC Department of City Planning. All rights reserved.

3–5, 8, 9   Municipal Art Society; LotInfo (Micaela Birmingham), Source: Bytes of the Big Apple™ Tax Block & Lot Files ©1995, 1997, 1998, 2002, 2005 and Pluto™ © 2003, 2004, 2005. NYC Department of City Planning. All rights reserved.

6a   Municipal Art Society (Micaela Birmingham)

6b   Municipal Art Society (Greg Studwell)

7   Municipal Art Society; U.S. Census Bureau 2000 (Micaela Birmingham)

Case Study 2

1   University Neighborhood Housing Program and Fordham Bedford Housing Corporation; Source: Bytes of the Big Apple™ Tax Block & Lot Files ©1995, 1997, 1998, 2002, 2004, 2005 and Pluto™ © 2003, 2004, 2005. NYC Department of City Planning. All rights reserved.

2   Enterprise Foundation

3   The Enterprise Foundation. LotInfo, Source: Bytes of the Big Apple™ Tax Block & Lot Files ©1995, 1997, 1998, 2002, 2004, 2005 and Pluto™ © 2003, 2004, 2005. NYC Department of City Planning. All rights reserved.

Case Study 3

1, 2   Author's survey © Zvia Segal Naphtali. LotInfo. InfoShare

3, 4, 7, 8   InfoShare. LotInfo

5, 6   U.S. Census Bureau 2000. InfoShare

Case Study 4

1–5   Regional Plan Association (Jennifer Cox)

Case Study 5

1–3   LotInfo (Karen Kaplan), Bytes of the Big Apple™ Tax Block & Lot Files ©1995, 1997, 1998, 2002, 2004, 2005 and Pluto™ © 2003, 2004, 2005. NYC Department of City Planning. All rights reserved. Bronx Data Center: Lehman College, City University of New York. © William Bosworth

Case Study 7

1, 2   New York City Department of Environmental Protection

Case Study 8

1–4   LotInfo. Source: Bytes of the Big Apple™ Tax Block & Lot Files ©1995, 1997, 1998, 2002, 2005 and Pluto™ © 2003, 2004, 2005. NYC Department of City Planning. All rights reserved. Courtesy of New York City Landmarks Preservation Commission (Kenneth Mack)

5   LotInfo. Source: Bytes of the Big Apple™ Tax Block & Lot Files ©1995, 1997, 1998, 2002, 2005 and Pluto™ © 2003, 2004, 2005. NYC Department of City Planning. All rights reserved. Courtesy of New York City Landmarks Preservation Commission; Robinson 1885 Manhattan plate; David Rumsey.com (Kenneth Mack)

6   Courtesy of New York City Landmarks Preservation Commission; Robinson 1885 Manhattan plate; David Rumsey.com (Kenneth Mack)

Case Study 9

1   Photographs by Juan Carlos Saborio

2   Bronx Data Center: Lehman College, City University of New York. © William Bosworth. U.S. Census Bureau 2000; U.S. EPA; New York State Department of Health. Garg R., Karpati A., Leighton J., Perrin M., Shah M. Asthma Facts, Second Edition. New York City Department of Health and Mental Hygiene, May 2003 (Juliana Maantay)

3   Bronx Data Center: Lehman College, City University of New York. © William Bosworth. U.S. Census Bureau 2000. U.S. EPA (Holly Porter-Morgan and Juliana Maantay)

4   Bronx Data Center: Lehman College, City University of New York. © William Bosworth. U.S. Census Bureau 2000. New York State Department of Health. Garg R., Karpati A., Leighton J., Perrin M., Shah M. Asthma Facts, Second Edition. New York City Department of Health and Mental Hygiene, May 2003. (Juliana Maantay)

5   Andrew Maroko

6   Bronx Data Center: Lehman College, City University of New York. © William Bosworth. U.S. Census Bureau 2000. Montefiore Medical Center; New York State Department of Health. Garg R., Karpati A., Leighton J., Perrin M., Shah M. Asthma Facts, Second Edition. New York City Department of Health and Mental Hygiene, May 2003. (Juliana Maantay)

7   Bronx Data Center: Lehman College, City University of New York. © William Bosworth. U.S. Census Bureau 2000. Montefiore Medical Center; New York City Department of Mental Health, New York Department of Health, Statewide Planning and Research Cooperative System (SPARCS), with permission from the New York State Department of Health's Data Protection Review Board; U.S. EPA (Juliana Maantay)

8   Bronx Data Center: Lehman College, City University of New York. © William Bosworth. U.S. Census Bureau 2000 (Juan Carlos Saborio)

9   Bronx Data Center: Lehman College, City University of New York. © William Bosworth. U.S. Census Bureau 2000. U.S. EPA (Juliana Maantay)

10   Montefiore Medical Center; New York State Department of Health. Garg R., Karpati A., Leighton J., Perrin M., Shah M. Asthma Facts, Second Edition. New York City Department of Health and Mental Hygiene, May 2003. (Juan Carlos Saborio)

11   Montefiore Medical Center; New York State Department of Health. Garg R., Karpati A., Leighton J., Perrin M., Shah M. Asthma Facts, Second Edition. New York City Department of Health and Mental Hygiene, May 2003. (Holly Porter-Morgan and Juliana Maantay)

Case Study 10

1, 3   New York Police Department (Herrmann, Maroko)

2   New York Police Department; U.S. Census Bureau 2000 (Herrmann, Maroko)

Case Study 11

1, 2   Author's survey; LotInfo (Thomas Paino), Source: Bytes of the Big Apple™ Tax Block & Lot Files ©1995, 1997, 1998, 2002, 2005 and Pluto™ © 2003, 2004, 2005. NYC Department of City Planning. All rights reserved.

3   Photographs by Thomas Paino

4– 7   Author's survey; LotInfo (Thomas Paino), Source: Bytes of the Big Apple™ Tax Block & Lot Files ©1995, 1997, 1998, 2002, 2005 and Pluto™ © 2003, 2004, 2005. NYC Department of City Planning. All rights reserved.

Case Study 12

1–3   Metropolitan Waterfront Alliance © 2001

Data license agreement

Important:
Read carefully before opening the sealed media package

Environmental Systems Research Institute, Inc. (ESRI), is willing to license the enclosed data and related materials to you only upon the condition that you accept all of the terms and conditions contained in this license agreement. Please read the terms and conditions carefully before opening the sealed media package. By opening the sealed media package, you are indicating your acceptance of the ESRI License Agreement. If you do not agree to the terms and conditions as stated, then ESRI is unwilling to license the data and related materials to you. In such event, you should return the media package with the seal unbroken and all other components to ESRI.

ESRI License Agreement

This is a license agreement, and not an agreement for sale, between you (Licensee) and Environmental Systems Research Institute, Inc. (ESRI). This ESRI License Agreement (Agreement) gives Licensee certain limited rights to use the data and related materials (Data and Related Materials). All rights not specifically granted in this Agreement are reserved to ESRI and its Licensors.

Reservation of Ownership and Grant of License: ESRI and its Licensors retain exclusive rights, title, and ownership to the copy of the Data and Related Materials licensed under this Agreement and, hereby, grant to Licensee a personal, nonexclusive, nontransferable, royalty-free, worldwide license to use the Data and Related Materials based on the terms and conditions of this Agreement. Licensee agrees to use reasonable effort to protect the Data and Related Materials from unauthorized use, reproduction, distribution, or publication.

Proprietary Rights and Copyright: Licensee acknowledges that the Data and Related Materials are proprietary and confidential property of ESRI and its Licensors and are protected by United States copyright laws and applicable international copyright treaties and/or conventions.

Permitted Uses: Licensee may install the Data and Related Materials onto permanent storage device(s) for Licensee's own internal use.

Licensee may make only one (1) copy of the original Data and Related Materials for archival purposes during the term of this Agreement unless the right to make additional copies is granted to Licensee in writing by ESRI.

Licensee may internally use the Data and Related Materials provided by ESRI for the stated purpose of GIS training and education.

Uses Not Permitted: Licensee shall not sell, rent, lease, sublicense, lend, assign, time-share, or transfer, in whole or in part, or provide unlicensed Third Parties access to the Data and Related Materials or portions of the Data and Related Materials, any updates, or Licensee's rights under this Agreement.

Licensee shall not remove or obscure any copyright or trademark notices of ESRI or its Licensors.

Term and Termination: The license granted to Licensee by this Agreement shall commence upon the acceptance of this Agreement and shall continue until such time that Licensee elects in writing to discontinue use of the Data or Related Materials and terminates this Agreement. The Agreement shall automatically terminate without notice if Licensee fails to comply with any provision of this Agreement. Licensee shall then return to ESRI the Data and Related Materials. The parties hereby agree that all provisions that operate to protect the rights of ESRI and its Licensors shall remain in force should breach occur.

Disclaimer of Warranty: THE DATA AND RELATED MATERIALS CONTAINED HEREIN ARE PROVIDED "AS-IS," WITHOUT WARRANTY OF ANY KIND, EITHER EXPRESS OR IMPLIED, INCLUDING, BUT NOT LIMITED TO, THE IMPLIED WARRANTIES OF MERCHANTABILITY, FITNESS FOR A PARTICULAR PURPOSE, OR NONINFRINGEMENT. ESRI does not warrant that the Data and Related Materials will meet Licensee's needs or expectations, that the use of the Data and Related Materials will be uninterrupted, or that all nonconformities, defects, or errors can or will be corrected. ESRI is not inviting reliance on the Data or Related Materials for commercial planning or analysis purposes, and Licensee should always check actual data.

Data Disclaimer: The Data used herein has been derived from actual spatial or tabular information. In some cases, ESRI has manipulated and applied certain assumptions, analyses, and opinions to the Data solely for educational training purposes. Assumptions, analyses, opinions applied, and actual outcomes may vary. Again, ESRI is not inviting reliance on this Data, and the Licensee should always verify actual Data and exercise their own professional judgment when interpreting any outcomes.

Limitation of Liability: ESRI shall not be liable for direct, indirect, special, incidental, or consequential damages related to Licensee's use of the Data and Related Materials, even if ESRI is advised of the possibility of such damage.

No Implied Waivers: No failure or delay by ESRI or its Licensors in enforcing any right or remedy under this Agreement shall be construed as a waiver of any future or other exercise of such right or remedy by ESRI or its Licensors.

Order for Precedence: Any conflict between the terms of this Agreement and any FAR, DFAR, purchase order, or other terms shall be resolved in favor of the terms expressed in this Agreement, subject to the government's minimum rights unless agreed otherwise.

Export Regulation: Licensee acknowledges that this Agreement and the performance thereof are subject to compliance with any and all applicable United States laws, regulations, or orders relating to the export of data thereto. Licensee agrees to comply with all laws, regulations, and orders of the United States in regard to any export of such technical data.

Severability: If any provision(s) of this Agreement shall be held to be invalid, illegal, or unenforceable by a court or other tribunal of competent jurisdiction, the validity, legality, and enforceability of the remaining provisions shall not in any way be affected or impaired thereby.

Governing Law: This Agreement, entered into in the County of San Bernardino, shall be construed and enforced in accordance with and be governed by the laws of the United States of America and the State of California without reference to conflict of laws principles. The parties hereby consent to the personal jurisdiction of the courts of this county and waive their rights to change venue.

Entire Agreement: The parties agree that this Agreement constitutes the sole and entire agreement of the parties as to the matter set forth herein and supersedes any previous agreements, understandings, and arrangements between the parties relating hereto.

Installing the data

GIS for the Urban Environment includes one CD that contains the exercise data. The exercise data takes up about 30 megabytes of hard-disk space. The data installation process takes about five minutes.

Installing the data

Put the data CD in your computer's CD drive. Click the "Install Lab Exercise Data for GIS for the Urban Environment" button on the splash screen. Follow the installation procedures and note the location where the data will be written.

Uninstalling the exercise data

To uninstall the exercise data from your computer, open your operating system's control panel and double-click the Add/Remove Programs icon. In the Add/Remove Programs dialog box, select the "GIS for the Urban Environment" entry to remove it.

Index

Numbers

2.5-meter resolution satellite image of urban area, 302
3D GIS: expressing attribute values with, 298; using, 294–298
3D GIS model of buildings in NYC, 297
3D image of midtown Manhattan at lot level, 297
3D models, displaying, 19
3D points, coordinates in, 294
3D representation of nonspatial attributes of geographic entities, 298
7.5-minute quadrangles, overview of, 172
9/11 attacks: GIS response to, 368; memorialization of, 324
30×30 meter DEM, use of, 294
911 emergency response programs, hit rates in, 184
911 updating, performing, 183

A

abandoned buildings, producing thematic map of, 185
absolute method of scaling, using, 132
access privileges, defining for databases, 200
Access to health care in Greenpoint–Williamsburg, Brooklyn, New York case study, 341–347
Add Field dialog box, opening, 474, 477, 486
address locators, creating, 507–509
address matching: explanation of, 171; performing, 183–184
address styles, selecting, 507–508
addresses: correcting and rematching unmatched addresses, 511–514; geocoding TRI facility addresses, 510–511; locating with TIGER/Line files, 184; point geocoding by, 183–184; reviewing, 513; in TRI facilities table, 506
administrative boundaries, significance of, 38
Advocacy Planning and Public Information case study, 424–430
advocacy purposes, using GIS for, 275–277
aerial photographs: availability of, 304; use in Ground Zero case study, 369
The Agency for Toxic Substances and Disease Registry (Center for Disease Control): Improving public health through GIS Web site, 22
AIDS case data, aggregation of, 37
AIDS population diffusion during 1970s and 1980s map, 77
AIDS prevalence rates in New York City by ZIP Code, 110
air passenger travel in U.S. map, 78
air pollutant quality proportional symbol map, 73
air pollution sources in Bronx map, 398
Al Idrisi map (1154), 5
Albers Equal Area projection, use of, 47–48
Alias values, listing and overwriting, 502–503

aliases, creating and verifying, 503
American Indian Population cartogram and proportional symbol map, 82
Amherst, NY needs-assessment survey, 254
analog models, using, 224
and Boolean operation: explanation of, 222; using, 223
apparent-magnitude method of scaling, using, 132
ArcCatalog: components of, 460–461; starting, 460; using, 460
ArcGIS, using in auto thefts case study, 411
ArcGIS 8.3, using in Landmark Preservation Commission (NYC) case study, 384
ArcGIS 9 Query Builder screen, 197
archaeology GIS project for the NYC Landmark Preservation Commission case study, 382–391
ArcMap: adding data layers in, 501; adding fields in, 489; closing, 457; launching, 462; query capabilities of, 390; saving work in, 457; starting, 444; toggling layer visibility in, 445; using in Land Preservation Commission (NYC) case study, 382–391; viewing maps in, 445
ArcMap components, reviewing, 445–447
ArcMap interface, components of, 446
ArcToolbox: closing, 533; opening, 532
arithmetic line graphs, using, 128
arithmetic means: considering in interval and ratio scales, 103–105; using, 100
Army of the Pure poster (Pakistan), 271
arrays, definition of, 98
arrows, selecting, 563
ASCII files, overview of, 192
assessed land values, tracking, 198
assessment methodology, applying to natural habitat and open-space assessment, 350–354
asthma and environmental hazards in the Bronx case study, 393–407
asthma data, shortcomings of, 406
Asthma Hospitalization Rates in Bronx maps, 147, 236, 394–397, 398, 400, 403, 547
Asthma Rates data frame, creating, 542
asthma rates, interpolating, 551–552
asthma statistics: basing maps on, 546; joining to ZIP Code centroids, 550–551
Asthma table: adding fields to, 542; joining to Attributes of Bx_zip, 546; joining to Bx_zipcentroid, 550; populating, 543
atlases, reference maps in, 61

attribute data: accuracy and currency of, 163–164; collecting from various sources, 160; examining with buffers, 215; explanation of, 17; integrating and processing, 260; organizing, 192; reclassifying, 212–213; versus spatial data, 192; use in Ground Zero case study, 372–375

attribute data, importing, 15

attribute databases: mapping, 182; querying, 17

attribute tables: examining, 447; importing into GIS, 192; records in, 464; unmapped attribute tables, 185

attribute values, expressing with 3D, 298

attributes: generating in reclassification, 212; reclassifying, 213; viewing for country layer, 451

auto thefts in Bronx, using GIS for, 409–413

averages, using, 100

axes in line graphs, starting, 128

azimuthal projection: example of, 45; explanation of, 44

B

Babylonian map, 4

backups of databases, maintaining, 261

Baltimore County, Maryland Web site, 440

bar graph symbol of deaths due to stroke, 115

bar graphs: creating, 488; using, 130. *See also* histograms

basemap feature labels, including in maps, 140

basemap layer, using in Landmark Preservation Commission case study, 385

basemaps: use in *Imagine New York* project, 325; using in green roofs case study, 360; using in Long Island Sound case study, 355; using in thematic maps, 63

batch geocoding, explanation of, 182

BBL (borough-block-lot) numbers, using in Landmark Preservation Commission case study, 385–387

Bergen County, NJ GIS problem-solving example, 242–244

binary models, using, 226–227

Birmingham, Micaéla: Public outreach: *Imagine New York,* The Lower Manhattan Preservation Fund, The campaign for community-based planning: Olympics 2012 case study, 322–333

birth rates by country, representing with choropleth maps, 70–71

bivariate data, mapping, 114

blank space, proportioning in maps, 144

blocks layer, joining DOF summary table to, 523–524

blue, connotation of, 560

board-presentation maps, creation Enterprise Foundation, 339

bookmarks: naming, 452; using, 558

Boolean operators, using, 222–224

Boolean queries, examples of, 224

borders, adding around maps, 563

borough geographic unit, using for NYC, 161

breast cancer mortality rates integrated map and graph, 134 significance of, 329–330

British Columbia Special Locations maps, 276

British Isles map projections, 48

Bronx (Hunts Point), thematic map, 62

Bronx, NY: aggregating by land-use class, 359; land-use data for, 360; locating TRI facilities in, 500; mapping asthma hospitalization rates in, 547; mapping environmental hazards and asthma in, 392–407; transportation and density map, 141. *See also* South Bronx

The Bronx, NY: Aggregated by Land-Use Class map, 359

Bronx auto thefts using GIS case study, 408–413

Bronx census tracts dot density map, 67

Bronx Inset bookmark, creating, 558

Bronx Streets maps, line weights in, 146

Bronx ZIP Code districts, accessing, 543

BronxAsthma map document, opening, 550, 556

BronxBuffers map document, opening, 542

Bronx.mdb database: adding feature classes to, 462; changing colors in, 463

BronxPopLayout map document, naming, 484

BronxPop map document: naming, 473; opening, 484

BronxTRI map document: naming, 502; opening, 506; opening and saving as new map, 532

Brooklyn, NY: Greenpoint–Williamsburg case study, 341–347; land values expressed as contour map, 76

Brooklyn assessed total value of lots 3D representation, 298

buffer analysis, using clipping in, 216

buffer distances, uncertainty in asthma and environmental hazards in Bronx case study, 406

buffers: creation of, 215; generating, 532–533; selecting census tracts within, 534–535; using in asthma and environmental hazards in Bronx case study, 401, 403

Buffers Used in Estimating Impact Extent of Air Pollution Sources map, 402

Buildings That Matter map, 326

business analysis, overview of, 300

Bx_blocks feature class, adding, 520

Bx_blocks table, opening attributes of, 524

Bx_demog layer in Bronx.mdb: explanation of, 462; highlighting in TOC, 462; loading, 473; opening, 479, 480

Bx_demog table, identifying field names in, 484

Bx_streets feature class: loading, 506; opening attributes of, 513

Bx_zip file, adding to map, 544

Bx_zipcentroid, joining Asthma table to, 550

bx-DOF table: displaying in TOC, 520; opening, 521

C

CAD: connecting to GIS, 18–19; creating simulation models in, 296

CAD images, use in Ground Zero case study, 372

cadastral maps: example of, 7; as reference maps, 61

Canada, land platting in, 52

car theft case study, using GIS in, 408–413

Carrier Phase Tracking application, description of, 307

cartogram and proportional symbol map of American Indian population, 82

cartograms: contiguous versus noncontiguous cartograms, 81; of emissions of greenhouse gases, 80; of gross national products of countries, 79; of U.S. population, 81

cartographic conventions, following, 137–138

cartographic license, explanation of, 35

cartographic models, using, 224–225

Catalog tree in ArcCatalog, description of, 461

CBSAs (core based statistical areas), significance of, 170

CDP (Census designated place), significance of, 168

census block, explanation of, 168–169

Census Bureau designations of urban areas: CDP (Census designated place), 168; census block, 168–169; census block group, 168; census tract, 168; consolidated city, 168

Census Bureau inventory units: congressional, state, and local districts, 169; metropolitan areas, 169; school districts, 169; TAZs (traffic analysis zones), 169; ZCTAs (ZIP Code Tabulation Areas), 169

Census hierarchy for Maryland and D.C., 170

Census table, opening attributes of, 464

census tract boundaries in NYC boroughs, 176

census tracts: aggregating, 485–486; designation of, explanation of, 168; intersection with TRI buffers, 535; selecting within buffers, 534–535

central tendencies, summarizing data with, 100

centroids of polygons, point geocoding to, 183

change detection, description of, 305

character string, relationship to attribute data, 192

charts, types of, 126

choropleth and dot density map, 83

choropleth maps: creating, 479; creating for asthma hospitalization rates in Bronx, 547; creating with dot density theme on top, 481; of deaths due to stroke, 115; of hate groups, 84; using, 70–72; using in graffiti case study, 421; using unclassed choropleth maps, 112

citation information FGDC metadata category, explanation of, 166

CitiScan Project (Hartford, CT), significance of, 269

city block lots table, 193

city points, resizing, 456–457

city populations, creating proportional symbol map of, 456–457

class breaks: labeling, 479; manipulating, 111

class formation from same data and different classification methods, 109

class range labels, changing, 479

classification: issues related to, 111–112; using as method of data exploration, 114

classification maps, description of, 305

classification methods: changing, 479–480; equal interval method, 106; natural breaks method, 107; overview of, 105; quantile method, 106–107, 112; standard deviation, 108–109, 112

Classify button, using, 479

clearinghouses, finding GIS data in, 176

clients, role in GIS projects, 248

Clip tool, converting shapefiles with, 543–544

clipping, using, 216

clustering methodology, using in auto thefts case study, 411–412

CMAP (Community Mapping Assistance Project), role in MWA case study, 425, 427

CMSAs (Consolidated Metropolitan Statistical Areas), significance of, 169

color and symbology, guidelines for use of, 560

color contrast, considering in map designs, 150

color ramps: modifying, 560; using in standard deviation classification, 109

colors: changing for country features, 454; changing for layers, 559–560; changing in Bronx.mdb database, 463; considering in map designs, 148-150; conventions for, 149–150; value, hue, and saturation of, 150

column graphs. *See* bar graphs

columns, joining to produce unique ID fields, 199

combination thematic maps, using, 83–84

community, defining relative to PGIS, 273–274

Community Board 2, Queens, NY (graffiti case study): neighborhoods and land-use types in, 416–418

Community Mapping Assistance Project for the Metropolitan Waterfront Alliance Open Accessible Space Information Systems Internet Map Site case study, 424–430

community names, significance of, 89

Community-Based Economic Development Recommendations & New York City Median Income map, 330

community-based GIS: access to hardware, software, and skills, 280; components of, 278

Community-Based Open Space Recommendations: Creating a Citywide Open Space Network map, 331

community-based organizations, GIS applications for, 272–273

community-based planning: campaign for, 328; case study, 415–423; GIS analysis of recommendations, 330–333

community-based plans: compilation of, 329; implementation of, 333

Compstat process, overview of, 409

concatenated field, explanation of, 199

conceptual models, using, 224

confidentiality, ethical issues related to, 285–286

conformal projection, definition of, 47

congressional inventory units, significance of, 169

conical projection: example of, 45; explanation of, 44

Connect to Folder button, using, 461

connectivity, representing, 17

consolidated city Census designation, explanation of, 168

contact information FGDC metadata category, explanation of, 166

containment, representing, 17

contiguity, representing, 17

continuous versus discrete data, 28–29

contour intervals, using with USGS topographic maps, 294–295

contour maps: of Brooklyn, NY land values, 76; generating with 30×30 meter DEM, 294; using, 75–77

control segment of GPS, description of, 306

coordinate systems: application of, 40; SPC (State Plane Coordinate) system, 50; types of, 48; UTM (Universal Transverse Mercator) system, 49

coordinates: defining points by, 162; point geocoding by, 182

coordinating committee: empowering, 250; establishing, 249; role in GIS projects, 248

Corridors of Concern map, 327

counter-mapping, significance of, 275–277

counties in U.S., number of, 167

country features, changing colors for, 454

country layer, viewing attributes of, 451

country names: identifying, 450; viewing alphabetically, 452

Cox, Jennifer R.: Long Island Sound Stewardship System case study, 348–355

Create Graph option, using, 488

Create New Address Locator option, choosing, 507

CRG (Community Resource Guide), significance of, 336

crime pattern analysis case study, 409–413

CrimeStat III software: kernel functions in, 413; role in crime thefts case study, 411–412

Crino, Michael et al: GIS mapping of the New York City sewer and storm drain system case study, 376–381

criteria maps, grouping in Long Island Sound case study, 350

CSV files, downloading, 500

Cultural Resources criteria, using in Long Island Sound case study, 353

cylindrical equidistant projection, explanation of, 47

cylindrical projection: example of, 45; explanation of, 44

D

data: adding to data frames, 473; classifying, 479; combining from multiple tables, 201; combining with overlay operations, 214; generating from different maps, 109; grouping, 197–200; importing from external database tables, 202; reclassifying, 479–480; summarizing with statistics, 100–105; symbolizing, 454; viewing and sorting in tables, 451–452

data access: ethical issues related to, 282; restricting with views, 196

data acquisition, explanation of, 14

data aggregation: determining for choropleth maps, 72; explanation of, 36

data availability, surveying for GIS projects, 256

data classification exercise, 118

data clearinghouses, obtaining data from, 176

data collection, potential complications with, 256–257

data conversion, outsourcing, 260

data dictionary from LotInfo GIS product, 258–259

data display: and interaction, 14; reordering, 444

data dissemination, ethical issues related to, 283–285

data exploration, performing, 112–115

data frames: creating, 526, 556; defining symbology of layers in, 559–560; explanation of, 472; populating with datasets, 556; removing, 557; toggling between, 473; viewing, 472

data integration, implementing in RDBMS, 205

data layers: adding, 501; masking, 467; preparing for symbolization, 465–466. *See also* layers

data liability, ethical issues related to, 283–285

data ownership, ethical issues related to, 283

data patterns, revealing, 481

data quality FGDC metadata category, explanation of, 165

data quality, setting standards for, 205

data sources, including in maps, 139

data storage and retrieval, explanation of, 14

data survey report, contents of, 256, 257

data synthesis and presentation, explanation of, 15

Data View, displaying ArcMaps in, 445

data visualization, significance of, 135–137

database administrator, role in GIS projects, 248

database design and administration, technical qualifications for, 258

database design principles, following, 200

database fields, data types for, 194

database maintenance staff, role in GIS projects, 248

database management, explanation of, 14

database tables: creating, 194; importing data from, 202; managing, 194–196; selecting fields from, 195. *See also* tables

databases: adding spatial dimension to, 203–204; expandability of, 200; incorporating local knowledge into, 278–279; planning, designing, and constructing, 258–259; querying spatially, 203

datasets: in asthma and environmental hazards in Bronx case study, 396; calculating modes for, 103; deriving range of values in, 106; differentiating with quartiles, 102–103; locations of, 461; in Long Island Sound case study, 350; populating data frames with, 556; producing with union function, 217; reviewing statistics for, 550; saving as layer files, 467–468

DBMS (database management systems): calculating summary statistics in, 197; managing derived fields in, 198; working with external systems, 194

D.C. (District of Columbia), Census hierarchy for, 170

deaths due to strokes, data exploration of, 115

decennial census. *See* U.S. decennial census hierarchy

decimal degree, definition of, 43

decision trees, using in rule-based reasoning, 231

decision-making tool, GIS as, 14–15

definition query, creating for DOF table, 520

degree of longitude, size of, 43

degrees, calculating on earth's surface, 41

degrees of latitude: dividing into minutes, 41; size of, 43

Delete key, using with caution, 492

DEM (digital elevation models): overview of, 173; significance of, 294

demographic analysis, performing, 536–537

demographic comparison data, creating graph of, 537

demographic data, creating maps with, 481

demographic information, source of, 167–170

demographics, analysis of, 324

demographics, designating, 284

density methodology, using in auto thefts case study, 411–412

DEP (Department of Environmental Protection), role in NYC sewer and storm drain system case study, 378–379

derived data fields; occurrence of, 212; managing, 198

derived variables, use of, 97

descriptive statistics exercise, 117

design and application development, performing for GIS projects, 255

designer, role in GIS projects, 248

Detroit maps, 268

DGPS (Differential GPS), overview of, 307

digital mapping: components of, 162; sources of, 174

digital sensors, using with satellite imagery, 301

DIME (Dual Independent Map Encoding), significance of, 171

disaster response and emergency management case study, 366–375

discrete versus continuous data, 28–29

dissolve function, using, 216

distribution FGDC metadata category, explanation of, 166

documentation, maintaining, 261

DOF summary table, joining to blocks layer, 523–524

DOF table: creating definition query for, 520; generalizing, 521–522

DOF_sum table, opening, 523

DOQQ (digital orthophoto quadrangles), overview of, 173

dot density maps: creating, 469; of deaths due to stroke, 115; designing, 67; plotting, 465–466; problem with dot placement on, 467; of reported auto thefts in Bronx, NY, 411; using, 65–69

dot density theme, creating choropleth map with, 481

dot matrix graphs, using in graffiti case study, 421

dot size, varying in dot density maps, 68

DotDensity.mxd file: naming, 467; opening, 472

DR-1391-NY New York City BoO Locations—September 17 map, 374

drawing order, controlling, 494

drawing toolbar in ArcMap, identifying, 446

DTM (digital terrain models), explanation of, 294

E

early-warning system, creating for housing, 337

Earth, imaginary lines on, 42

earth's surface: calculating degrees on, 41; locating points on, 41

eco-effectiveness, explanation of, 358

ecological fallacy, explanation of, 284

edge matching software, description of, 305

Edit sessions, starting, 514

Editor toolbar, opening, 487, 543

electric utility poles, geocoding, 182

EMC (Emergency Mapping Center), role in responding to 9/11 attacks, 368

emergency management and disaster response case study, 366–375

emissions of greenhouse gases cartogram, 80

Enterprise Foundation, maps created by, 338–339

entity and attribute information FGDC metadata category, explanation of, 165

environmental hazards and asthma in the Bronx case study, 392–407

Environmental Justice Alliance map, 275

environmental simulation, performing, 296–297

EPA (Environmental Protection Agency): adding fields to table from, 501; Web site, 20, 500–501

equal area projection, definition of, 47

equal interval classification: of adults without HS diploma, 122; using, 106

equations, checking in Field Calculator, 476

equator: degree of longitude at, 43; representing, 41

equator locations, using Mollweide projection for, 47

equatorial map projection, example of, 45

equatorial plane, explanation of, 41

equidistant projection, definition of, 47

equivalent projection, definition of, 47

Erie County, New York Web site, 441

ESRI GIS Dictionary, consulting, xii

ESRI Web site, 20

ethical issues, considering, 268

EVRAP (Extensive Vegetative Roof Acreage Potential), significance of, 357

EVRAP: Estimated Roof Area Analysis map, 363

Executive order 12906, significance of, 165

"Experience of Place," mapping, 277

expert systems, overview of, 231–232

exploration, role of maps in, 6

Explore Data option, using, 553

Exploring Bronx auto thefts using GIS case study, 408–413

Export Data dialog box, opening, 525

expressions, creating in Field Calculator, 476

extent rectangles, adding, 559

external database tables, importing data from, 202

F

Facility_ID field, adding, 514

false easting, explanation of, 50

false northing, explanation of, 50

Farag, Magdi et al: GIS Mapping of the New York City Sewer and Storm Drain System case study, 376–381

FBHC (Fordham Bedford Housing Corporation) case study, 335–339

FCAP (Fordham Community Action Plan), significance of, 336

FCAP Focus Area map, 337

FDNY September 11th Fatalities map, 324

feature classes: adding to Bronx.mdb database, 462; exporting selected records to, 525–526

features: buffering, 215; identifying, 449; labeling, 454–455; selecting, 450–451; selecting interactively, 536; selecting with attribute queries, 454

ferry stops, adding in MWA case study, 426–427

FGDC (Federal Geographic Data Committee), significance of, 162

FGDC metadata categories, overview of, 165

FGDC-STD-001-1998 (metadata content standard), 165

Field Calculator: accessing, 474; checking equations in, 476; creating expressions in, 476; opening for Percent_Minority field, 478; using, 475

field names, identifying in Bx_demog table, 484

field values: aggregating for graphing, 485–486; identifying, 464

fields: adding in ArcMap, 489; adding to Asthma table, 542; adding to table from EPA Web site, 501; adding to tables, 486–487; creating and calculating values for, 473–478; definition of, 98; deleting, 546; displaying summary statistics for, 485; formatting for table joins, 545–546; managing derived fields, 198; naming, 474; populating, 474; selecting from database tables, 195

figure-ground relationships, considering in map designs, 146–148

Figures: 2.5-meter resolution satellite image of urban area, 302; 3D GIS, 19; 3D GIS model of buildings in NYC, 297; 3D image of midtown Manhattan at lot level, 297; 3D representation of nonspatial attributes of geographic entities, 298; 1999 Odds Ratios by Buffer Type (asthma case study in Bronx), 405; AIDS population diffusion during 1970s and 1980s, 77; AIDS prevalence rates in New York City by ZIP Code, 110; air pollution sources in Bronx map, 398; Al Idrisi map (1154), 5; ArcGIS 9 Query Builder screen, 197; arithmetic line graph, 128; Army of the Pure poster (Pakistan), 271; Asthma Hospitalization Cases in Bronx County, 403; Asthma Hospitalization Cases in Bronx map, 398; asthma hospitalization rates for children, 147; Asthma Hospitalization Rates in Bronx map, 400; Asthma Hospitalization Rates map, 397; Asthma Hospitalizations in Bronx map, 398; attribute and spatial queries combined, 204; axes in line graphs, 127; bar graphs, 131; bar or column graph, 130; basemap, thematic overlay, and thematic map, 63; basemap for priority areas methodology (Long Island Sound case study), 355; binary model flow chart, 226; The Bronx, NY: Aggregated by Land-Use Class, 359; Brooklyn, NY land values expressed as contour map, 76; buffering point, line, and polygon features, 215; Buffers Used in Estimating Impact Extent of Air Pollution Sources map, 402; buffers used individually, 215; buildings damaged in 9/11 attacks, 372; Buildings That Matter, 326; CAD drawings incorporated into GIS, 18; cadastral map, 7; cartogram and proportional symbol map of American Indian population, 82; cartogram of emissions of greenhouse gases, 80; cartogram of GNP (gross national products) of countries, 79; cartograms of U.S. population, 81; Census hierarchy for Maryland and D.C., 170; Census tract boundaries in NYC boroughs, 176; choropleth an dot density map, 83; choropleth map of hate groups, 84; choropleth maps, 70, 71; class formation from same data and different classification methods, 109; classification methods, 112; clip function, 216; Combinatorial GIS Model: Kernel Density and Deviational Ellipse Bronx, New York (2000), 413; Community Board 2, Queens, NY (graffiti case study) map, 416; Community-Based Economic Development Recommendations & New York City Median Income map, 330; Community-Based Open Space Recommendations: Creating a Citywide Open Space Network map, 331; Corridors of Concern map, 327; data dictionary from LotInfo GIS product, 258–259; data exploration of deaths due to stroke, 115; design and application development, 255; Detroit maps, 268; dissolve function, 216; dot density map, 65; dot density map of Bronx census tracts, 67; dot density maps, 66, 69; dot size variations, 68; DR-1391-NY New York City BoO Locations - September 17 map, 374; Enterprise Foundation housing map, 338; Environmental Justice Alliance map, 275; equal interval classification of Hispanic deaths by heart disease, 106; EVRAP: Estimated Roof Area Analysis map, 363; expert system flow chart, 232; FCAP Focus Area Map, 337; F.D.NY September 11th Fatalities, 324; figure-ground relationships, 147; First Nations (British Columbia) Bioregion Mapping Project, 277; flat map, 39; flow map and proportional symbol map, 78; font sizes, 151; generalizing map features, 34; geocoded electric utility poles, 182; geocoded sites in Land Preservation Commission (NYC) case study, 388; geographic extent, 36; geographic profiling, 239; Geography Network, 177; georegistering historic maps, 391; GIS applications list from needs-assessment survey, 254; GIS components, 9; GIS development cycle, 261; GIS development process, 251; GIS Tree, 9; graffiti types (graffiti case study), 418; graph components, 126; graticule components, 42; Greenpoint–Williamsburg transportation network, 343; Harbor Loop Ferry map

(MWA case study), 426; health care coverage in 2003 (Greenpoint–Williamsburg case study), 346; health-care access, 342; histogram of land-use classes, 99; hospital admissions from Williamsburg in 2004, 344; Hunters Point Pilot Area (Industrial/Residential) map (graffiti case study), 421; IDW interpolation, 236; IDW versus spline interpolation, 237; information layers, 12; inset maps, 141; integrated map and graph, 134; intersect function, 217; isarithmic map of advance of gentrification, 77; isolines showing elevation in map of Staten Island, 75; Kiepersol, Eastern Transvaal, South Africa map, 272; kriging interpolation, 238; Landmark Preservation Commission (NYC) mapping discrepancy, 387; large-scale versus small-scale maps, 33; latitude–longitude coordinate system, 41; lead poisoning center flow chart, 219–221; lidar data, 308–310; line graph, 126; line graph with dependent and independent variables, 127; line weights in Bronx Streets maps, 146; line weights varied and same, 148; Long Island City Pilot Area map (graffiti case study), 421; lots color coded by land use, 195; map elements, 138; map of abandoned buildings in Manhattan, 185; map projections, 45; map projections of British Isles, 48; map projections with spatial properties, 46; map typology, 58; *mappae mundi*, 7; map-scale representation, 30; Mark Monmonier's maps with class breaks, 111; master data list of needs assessment, 253; merge function, 216; multiple line graph, 129; multivariate cartogram, 85; multivariate map of Los Angeles (1971), 85; multivariate map of quality of life, South Carolina (1992), 86; multivariate map with segmented symbols, 86; natural breaks classification of Hispanic rates of death from heart disease, 108; New York City maps with varying details, 35; nominal maps of locations of qualitative features, 95; NYC2012 Proposed Olympic Plan for New York City map, 332; oldest world map, 4; Open Space Resources map (Long Island Sound case study), 353; ordinal measurement scaling map, 96; overlay operations, 214; Percentage of Population Below Poverty, within Buffers map, 401; pie graph, 131; Pilot Study Area Context (Western Half) map (graffiti case study), 417; pollution sources in Bronx, 394; population by census tract (Greenpoint–Williamsburg case study), 345; population density at different levels of aggregation, 72; population density by ZIP Code in NYC, 160; process model flow chart, 231; propaganda map symbols, 270; proportional symbol graph, 132; proportional symbol map, 74; proportional symbol pie graph as thematic overlay, 135; proportional symbols for air pollutants, 73; Public Access map (Long Island Sound case study), 351; qualitative map, 64; qualitative map with named categories of data, 94; quality of life map, 269; quantile classification of Hispanic rates of death from heart disease, 107; quantitative map, 64; quantitative map with numerical categories of data, 94; raster format based on grid cells, 27; raster image of USGS 7.5-minute quadrangle map, 173; ratio measurement scaling, 97; reclassification of zoning categories, 213; Recreational Need map (Long Island Sound case study), 354; reference map of streets and subway lines, 61; relative frequency histogram, 133; Reported Auto Theft Locations, Bronx, NY (2000), 411; Reported Auto Thefts in Bronx, NY, 410; RGDC metadata content standard, 166; Sanborn map, 175; scatter diagram, 132; selection by attribute, 210; selection by location, 211; semilog line graph, 129; serif and sans serif typefaces, 152; sewer cross section (Manhattan, 1888), 380; sewer map (Queens, 1945), 381; site plans created by MAS, 328; site repopulation in Land Preservation Commission (NYC) case study, 389–390; size difference between proportional symbols, 74; space perception influenced by identity, 59; spatial features in vector formats, 26; spatial join, 218; spatial query

results, 203; SPC (State Plane Coordinate) system, 50; spherical globe transformed into flat plane, 39; standard deviation classification of Hispanic rates of death from heart disease, 108; state plane zones, 51; street address located with TIGER/Line file, 184; surface graph, 130; table components, 98; thematic maps, 62; thermal image at Ground Zero, 369; Thiessen polygons, 234; TIN (Triangulated Irregular Network), 295; TIN model of shaded terrain surface, 296; TLC version of Manhattan neighborhoods, 60; topological relationships, 17; tour map by Enterprise Foundation, 339; type faces (normal, gold, and italic), 151; unclassified map of population density in U.S., 113; Uniform Land Use Review Procedure, 225; union function, 217; units of data aggregation in New York City, 38; U.S. bulls-eye map in wake of 9/11, 271; U.S. Census geographic entities standard hierarchy, 171; USGS 7.5-minute quadrangle map, 61; USGS DOQ (digital orthophoto quadrangle), 173; UTM (Universal Transverse Mercator) coordinate system, 49; vector map and raster map combined, 187; vector versus raster format, 29; Venn diagrams of Boolean operators, 223; Virginia map (1585), 6; visual balance, 144; visual contrast optimization, 145; Wards Island WPCP Drainage Basin, 361; Water Resource map (Long Island Sound case study), 352; Waterfront Issues: NYC Council District 33 map (MWA case study), 427; weighted model, 229; workshops across Tri-State area (NY, CT, NJ), 325. *See also* maps

FILE styles, using with shapefiles, 507
First Nations (British Columbia) Bioregion Mapping Project, 277
fixed radius method, using with IDW, 236
Fixed Zoom In button, identifying, 447
fixed-length file, explanation of, 192
flat maps, making, 39
flood-prone parcels, alleviating problem of, 228
flow map and proportional symbol maps: community-based planning in NYC, 329; Number of Graffiti per Surveyed Lot map (Western Half) map, 420; NYC population density by county, 161
flow maps, using, 78
fonts and size, considering in map designs, 151–153
Format Labels option, using, 480
frequency distribution, definition of, 99
Full Extent button, using, 452
future scenarios, using process models with, 230–231

G
Gall-Peters projection: example of, 46; explanation of, 47
Gap number, changing for legends, 562
GDB styles, using with geodatabases, 507
generalization techniques, 34–35
gentrification isarithmic map, 77
geocentricity, persistence of, 5
Geocode Addresses dialog box, opening, 510
geocoded electric utility poles, 182
geocoded sites in Land Preservation Commission (NYC) case study, 388
geocoding TRI facility addresses, 510–511
geodatabase files, using GDB styles with, 507
geographic context, adding to maps, 557
geographic data: generalizing, 34; importing, 15
geographic datasets, querying, 17
geographic demographics, designating, 284
geographic extents: bookmarking, 452; explanation of, 36; selecting, 37
geographic profiling example, 239

Geographic Profiling: One Example in Canada excerpt, 239
geographic unit of analysis, explanation of, 36
Geography Network: obtaining clearinghouse information from, 176–177; Web site, 176
geoprocessing: buffers in, 215; join processes in, 217–218; overlay operations, 214; overview of, 213
geoprocessing tools: clip function, 216; dissolve function, 216; intersect function, 217; merge function, 216; union function, 217
georeferencing: creating vector mapping with, 186; explanation of, 28
Georeferencing tools, using in Land Preservation Commission (NYC) case study, 390
geospatial accuracy standards, explanation of, 163
geostatistical analysis, overview of, 299
Geostatistical Analyst, using, 550–551
Geostatistical Wizard, using, 551–552
Geosupport database, role in Landmark Preservation Commission (NYC) case study, 385–386
GIS (geographic information system): for advocacy purposes, 275–277; analyzing reported auto thefts with, 410–411; applications of, 9; approaching as bottom-up technology, 281; conceptualizing, 8; connecting to CAD, 18–19; as decision-making tool, 14–15; democratization of, 272–274, 281; importing attribute tables into, 192; join processes in, 217–218; proposing, 248; role in WTC (World Trade Center) attacks, 367; surface modeling with, 295–296; technical foundations of, 9; in urban planning, 11; using for problem solving, 242–244; using in community-based organizations, 272–273; using in planning functions, 10; using in urban planning, 9; vernacularization of, 281–282
GIS analysis: adding time dimension, 299; business analysis, 300; of community-based planning recommendations, 330–333; conclusion in community-based planning, 333; geostatistical analysis, 299; network analysis, 299
GIS application, development phase of, 250–251
GIS components, 9
GIS data: availability from state and local government, 174; determining availability and cost of, 256–257; finding in clearinghouses, 176; naming conventions for, 462; quality requirements of, 164. *See also* urban data sources
GIS development cycle components: availability and cost of GIS data, 256–257; choice of GIS hardware and software, 257; database planning, design, and construction, 258–259; design and application development, 255; GIS use and maintenance, 261; metadata and data quality, 257; needs assessment, 252; pilot study, 260; survey of available data, 256; systems integration, 260
GIS development process, drivers of, 248
GIS for Housing and Urban Development report, significance of, xvii
GIS for the Urban Environment: audience for, xi; classroom use of, xiii–xiv; focus of, xviii; geographic extent of, xiv; goals of, xiv–xv; organization of, xii; scope of, xi
GIS hardware, choosing, 257
GIS images, availability of, 305
GIS layers: choosing geographic units for, 160–161; overlapping at differing levels of detail, 161; using in Ground Zero case study, 369; using in Long Island Sound case study, 350
GIS mapping of the New York City sewer and storm drain system case study, 376–381
GIS process, steps in, 13–14
GIS projects: end result of, 15; estimating costs and benefits of, 250; initiating, 249–250; management perspective of, 262; roles in, 248
GIS software: choosing, 257; using LotInfo with, 417
GIS tasks, examples of, 15

GIS technology, adoption and popularity of, 13
GIS Tree, 9
GIS users, importance of, 8–9
GISMO (GIS and Mapping Operations), significance of, 368
GNP (gross national products) of countries cartogram, 79
Go Back to Previous Extent button, identifying, 448
Go to Next Extent button, identifying, 448
GPS (Global Positioning System): Carrier Aided Tracking and surveying, 307; components of, 306; DGPS (Differential GPS), 307; segments of, 306; use of, 187
GPS readings, achieving high degree of accuracy in, 307
graduated color maps. *See* choropleth maps
graduated point maps. *See* proportional symbol maps
graduated symbol maps. *See* proportional symbol maps: creating, 550-551
Graffiti in Community District Two, Queens, New York: Private Expression in the Public Realm case study, 414–423
graffiti types, examples of, 418
graph scales, using, 127–128
Graph Wizard, accessing, 488
Graph_Stats table, opening, 486
graphed values, creating tables for storage of, 486
graphic design components: color, 148–150; figure-ground relationships, 146–148; hierarchies, 144; line weights, 145; overview of, 142–143; placement of map elements, 145; typography, 150–153; visual balance, 143–144; visual contrast, 145
graphic scale: drawing, 29; uses of, 31
graphs: bar graphs, 130, 488; creating for demographic comparison data, 537; integrating with maps, 134, 140; line graphs, 126–128; pie graphs, 131, 488–489; saving, 490; surface graphs, 130; types of, 126; using dot matrix graphs in graffiti case study, 421; using proportional symbols in, 132
graticular network, representing, 41
graticules: components of, 42; explanation of, 41; including in maps, 140; placing over maps, 148
grayscale versus color in map designs, 149
Great Britain Historical GIS Project Web site, 21
green, connotation of, 560
green roofs: benefits of, 358; uses of, 357
greenhouse gases emissions cartogram, 80
Greenpoint–Williamsburg (NY) health-care services case study, 341–347
grid cells: basing raster format on, 27; lateral dimension of, 303; use with satellite sensors, 302
Ground Truth, significance of, xvii
Ground Zero case study, 366–375
GSD (ground sample distance), relationship to satellite grid cells, 303

H
Harbor Loop Ferry map (MWA case study), 426
hardware: choosing for GIS projects, 257; obtaining for community-based GIS, 280
Hartford, CT CitiScan Project, significance of, 269
hate groups choropleth map, 84
heads-up digitizing, explanation of, 28
health care, access to (Greenpoint–Williamsburg case study), 346
"Health Disparities in New York City" study, findings of, 341
health insurance, access to (Greenpoint–Williamsburg case study), 346
health service delivery, planning for, 345
health-care access map, 342

health-care services case study, 341–347

heart disease mortality rates integrated map and graph, 134

helicopter-based lidar, use of, 308

Hell's Kitchen (New York), reputation of, 89

Hermann, Christopher et al: Exploring Bronx auto thefts using GIS case study, 408–413

hierarchical clustering routine, using in auto thefts case study, 412

hierarchy of map elements, considering, 144

high-rise building locations, comparing with TRI facilities, 527

Hi-rise Buildings data frame, renaming, 526

Hispanic deaths: due to strokes, 115; by heart disease equal interval classification, 106

Hispanic population: as percentage of total population by state, 97

Hispanic population by state map, ordinal scaling used in, 96

Hispanic rates of death from heart disease: class formation from same data and different classification methods, 109; natural breaks classification, 108; quantile classification, 107; standard deviation classification, 108

histograms: creating, 553; definition of, 99; displaying next to maps, 133; including in maps, 140; using, 133. *See also* bar graphs

historic maps, georegistering in Land Preservation Commission (NYC) case study, 390

historic structures in lower Manhattan, mapping of, 326–327

hit rates: in 911 emergency response programs, 184; determining in address matching, 184; improving, 184

hospital admission patterns (Greenpoint–Williamsburg case study), 344

hospital admissions from Williamsburg in 2004 map, 344

hospital usage in Greenpoint–Williamsburg, overview of, 342

hospitalization for asthma in Bronx case study, 392–407

hospitals, transportation to (Greenpoint–Williamsburg case study), 343

"hot spot" maps, using in auto thefts case study, 411, 412

hourly wage related to rent example, 120–121

household income choropleth map, 71

household income table, 104

housing, creating early-warning system for, 337

housing affordability, using GIS maps for, 337

housing initiatives case study, 335–339

housing map files, creation by Enterprise Foundation, 338–339

HUD homes, mapping data related to, 337

hue, definition of, 150

Hunters Point Pilot Area (Industrial/Residential) map (graffiti case study), 421

Hunts Point (Bronx): asthma hospitalizations in, 394; thematic map, 62; waste-related facilities, 275

Hunts Point root versus land area: bar or column graph, 130; multiple line graph, 129; pie graph, 131; proportional symbol graph, 132; surface graph, 130

hyperspectral sensors, explanation of, 301

I

ID fields, producing from joined columns, 199

ID number of census block groups, identifying in asthma and environmental hazards in Bronx case study, 400

ID numbers, labeling TRI facilities with, 514–515

identification FGDC metadata category, explanation of, 165

Identify Results window, displaying, 449

Identify tool, using, 449

identity, influence on maps, 59

Identity Site tool, using in Land Preservation Commission (NYC) case study, 389

IDW (inverse distance weighted) interpolation, using, 236

IDW Prediction Map, using with asthma case study, 551–552

IF>THEN>ELSE rules, using with decision trees, 231–232

IKONOS satellite system, use in Ground Zero case study, 369

Imagine New York project, overview of, 324–325

income by household choropleth map, 71

income by household table, 104

information layers in GIS, 12

inset maps, including, 141, 516

insets, including in maps, 140

Integrated Data and Workflow Management for Registrars of Voters Using GIS Web site, 21

Interactive Review dialog box, using, 512

international date line, definition of, 43

Internet mapping software, overview of, 311

interpolated surface, creating from point data, 551–552

interpolation: IDW (inverse distance weighted) method of, 236; kriging method of, 238–239; performing, 233; process of, 232; sample selection in, 233; spline method of, 237; Thiessen polygon method of, 234–236

interpolation methods, experimenting with, 552

interquartile range, explanation of, 103

intersect function, using, 217

interval scales: arithmetic mean and standard deviation considerations, 103–105; using, 96

Ireland data visualization experiment, 136

isarithmic interval, choosing for maps, 76

isarithmic maps: of Brooklyn, NY land values, 76; generating with 30×30 meter DEM, 294; using, 75–78

isobar, definition of, 75

isobath, definition of, 75

isohyets, definition of, 75

isolines: isobaths. showing in topographic map of Staten Island, 75; types of, 75

isopleth maps, using, 77

isospecies, definition of, 75

isotherm, definition of, 75

IT staff, involving in GIS projects, 249

Ithaca, New York Web site, 441

J

Jenks Optimization algorithm, using, 107

Join Data dialog box, opening, 523

join operations, performing on tables, 545–546

join processes, types of, 217–218

joined columns, producing unique ID fields from, 199

Jost, Gregory et al: FBHC (Fordham Bedford Housing Corporation) case study, 335–339

jurisdictions, concerns associated with, 38

K

Kaplan, Karen S.: Land-use Classification for Extensive Vegetative Roof Acreage Potential in the Bronx case study, 356–364

kernel smoothing, using in auto thefts case study, 412–413

key fields, including in tables, 200–201

Kiepersol, Eastern Transvall, South Africa map, 272

Knight Spielman, Christy et al: Community Mapping Assistance Project for the Metropolitan Waterfront Alliance Open Accessible Space Information Systems Internet Map Site case study, 424–430

kriging interpolation, using, 238–239

L

Label Features option, using, 454

label properties, assigning to TRI facilities, 514–515

labels, placement of, 153

Lake Tappan, applying and implementing category 1 water purity standards to, 242–244

Lambert Azimuthal projection, use of, 48

Lambert Conformal Conic projection, use of, 47–48

LAN36005 layer in Bronx.mdb, explanation of, 462

land partitioning systems, examples of, 48, 51–52

Landsat project, satellites of, 300

landscape orientation, deciding on, 491

land-use class, aggregating The Bronx, NY by, 359

Land-use Classification for Extensive Vegetative Roof Acreage Potential in the Bronx case study, 356–364

land-use data for Bronx, using in green roofs case study, 360

land-use decision-making process in NYC, flow chart of, 225

large-scale versus small-scale maps, 32–35

latitude, size of degree of, 43

lat-long coordinate system: explanation of, 40–41; variation on, 43

Launch ArcMap button, using, 462

layer files: adding, 468; repairing connections for, 468; saving datasets as, 467–468

layer properties, storing, 468

layer symbolization, editing, 454–455

layer visibility, toggling, 445

layers: adding, 520; adding street data layer, 506; changing colors of, 559–560; definition of, 444; deriving new information from, 214; including all features in, 216; looking at individually, 463; masking, 467; order of, 444; organizing mapped information into, 16; overlaying with intersect function, 217; preparing for symbolization, 465–466; renaming, 561; renaming in map document, 462; storing in data frames, 472. See also data layers

Layers drop-down list, locating, 449

layers in data frames, defining symbology of, 559–560

layout composition, finalizing, 495–496

layout elements, finalizing, 495–496

layout format, planning, 491–492

Layout toolbar, distinguishing from Tools toolbar, 493

Layout View: displaying ArcMaps in, 445; using, 557–558; working in, 491

layouts: adding and arranging elements on, 561–563; adding map elements to, 494–495; adding north arrows to, 494; adding scale bars to, 494, 562–563; adding text elements to, 494; adding titles to, 495; composing, 492–493; creating, 537–538, 547, 557–558; creating and printing, 528; displaying, 496; finalizing and printing, 517; finishing, 564; incorporating maps into, 491; preparing, 490; preparing for asthma case study, 553; printing, 496, 565; visual structure of, 495

lead poisoning center: exposure submodel for, 220; optimal location for, 222; performing location analysis for, 219; risk submodel for, 219; vulnerability submodel for, 221

legend frames, creating, 562

Legend option, choosing, 493

Legend Wizard, using, 561

legends: inserting, 139, 561; using symbols in, 562

legibility: impact of color on, 150; of type, 152–153

Lehman College (Bronx), coordinates of, 43, 49

Lehman College GISc team, role in asthma and environmental hazards in Bronx case study, 395–396

lettering, considering in map designs, 150–153

lidar (Light Detection and Ranging) technology: overview of, 308–309; use in Ground Zero case study, 369

line buffer, example of, 215

line graphs: arithmetic line graphs, 128; axes in, 127; logarithmic line graphs, 128; multiple line graphs, 129; using, 126–128; variables on, 127

line objects, mapping street segments as, 183

line of tangency, representing, 45

line segments, defining, 26

line weights, considering in map designs, 145

linear features, labeling, 153

lines: smoothness of, 26; using, 560

lines of latitude and longitude, representing, 42

LION digital map, use in Ground Zero case study, 372–373

LION street centerline data layer, using in Landmark Preservation Commission (NYC) case study, 386–387

local electoral district inventory units, significance of, 169

local knowledge, incorporating into databases, 278–279

location analysis, overview of, 218–222

locations: selecting by, 534–535; selection by, 211

log scales, using, 128

Logan, Patrick et al: FBHC (Fordham Bedford Housing Corporation) case study, 335–339

logarithmic line graphs, using, 128

logical operators, examples of, 17

Long Integer type, using, 486

Long Island City Pilot Area map (graffiti case study), 421

Long Island Sound Stewardship System case study, 348–355

longitude, size of degree of, 43

Los Angeles (1971) multivariate map, 85

lot area, averaging for land-use type, 197–200

lot level data, viewing citywide, 199

LotInfo software, use in graffiti case study, 417–420

lots, breaking down by land-use category, 197–200

lots in city block table, 193

lots with residential land uses, displaying, 196–197

Lower East Side (Manhattan) isarithmic map of advance gentrification, 77

Lower Manhattan Preservation Fund, overview of, 326–327

LPC (Landmark Preservation Commission) case study, 382–391

lung cancer mortality rates integrated map and graph, 134

.lyr extension, explanation of, 468

M

Maantay, Juliana: Mapping Environmental Hazards and Asthma in the Bronx case study, 392–407

Mack, Kenneth: Archaeology GIS Project for NYC Landmark Preservation Commission case study, 382–391

main menu in ArcCatalog, description of, 461

mainframe file formats, overview of, 193

management, involving in GIS projects, 249

Manhattan: lot-level 3D image of midtown, 297; map of abandoned buildings in, 185; mapping historic structures in, 326–327; neighborhoods map, 60; sewer cross section map (1888), 380

Manhattan's Lower East Side isarithmic map of advance gentrification, 77

map accuracy standards, availability of, 163

map author information, including in maps, 139

map development, importance of, 6

map display in ArcMap, identifying, 446

map documents: renaming layers in, 462; saving, 467, 480; saving graphs in, 490; saving to, 484

map elements: accessing properties of, 492; adding to layouts, 494–495; controlling drawing order of, 494; hierarchy of, 142; overview of, 139–142; placement of, 145

map extents, restoring, 447

map features, generalizing, 34–35

map images, including in maps, 139

map projections: of British Isles, 48; choosing, 47–48; and coordinate systems, 39; definition of, 39; divisions of, 47; reconciling, 16; types of, 44

map resolution, explanation of, 36

map scale factors, examples of, 31

map scales: definition of, 29; determining map detail from, 32; examples of, 30; expressing, 29; including, 139; reconciling, 16; reducing, 34; significance of, 29–32; temporal scale, 34

map types: mental maps, 58–60; tangible maps, 60

mapmaking, 137–138; geocentricity of, 5

mappae mundi, significance of, 7

mapped demographic information, source of, 167–170

mapped information, organizing into layers, 16

mapped tables, merging unmapped attribute tables with, 185

mapping: democratization of, 8; origins of, 5; subject matter treated by, 4

Mapping environmental hazards and asthma in the Bronx case study, 392–407

maps: accuracy of, 162–163; adding borders to, 563; adding detail to, 35; adding geographic context to, 557; basing on asthma statistics, 546; creating, 16; creating graduated symbol maps, 550–551; creating inset maps, 516; creating with demographic data, 481; definition of, 4; designating point segments in, 26; displaying histograms next to, 133; by Enterprise Foundation, 338–339; generating from same data, 109; geographic extent of, 36; graphic design of, 142–148; grouping in Long Island Sound case study, 350; impact of identity on, 59; information contained in, 15; integrating graphs with, 134; large-scale versus small-scale maps, 32–34; laying out, 557–558; linking to other images, 18; preparing for presentation, 515, 527; as propaganda, 270–271; reference maps, 61–62; saving, 467, 490; as secretive documents, 6; subjectivity of, 268–269; symbolic and religious significance of, 7; thematic maps, 61–62; use of paper maps in Ground Zero case study, 373–375; using in collection of taxes, duties, and revenues, 7; viewing in ArcMap, 445. *See also* Figures; thematic maps

marketing zones, delineation of, 284

Maroko, Andrew et al: Exploring Bronx Auto Thefts Using GIS case study, 408–413

Maryland, Census hierarchy for, 170

MAS (Municipal Art Society): description of, 323; involvement in *Imagine New York* project, 324; use of GIS in Olympics 2012 bid, 333

MAS community-based planning campaign, goals of, 328

MAS site map, example of, 328

masking data layers, 467

master data list, including in needs assessment report, 253

Match Automatically option, using, 511

Match Interactively option, using, 511–512

mathematical models, using, 224

MCDs (minor civil divisions), classification of, 167

MCE (multi-criteria evaluation), significance of, 219

McGreevey seeks Purity Standards for 2 Bergen Reservoirs article, 241

McGuinness, Carol (Northern Ireland data visualization experiment), 136

McHarg, Ian (introduction of layering system), 16

measure of dispersion, definition of, 100

measurement scales: interval scaling, 96; nominal scaling, 95; ordinal scaling, 96; ratio scaling, 97

median, relationship to ordinal scale, 101–102

median household income choropleth map, 71

mental mapping: exercise, 90; impact of, 89

mental maps: initiating planning with, 60; of South Bronx, 89; using, 58–60

menu bar in ArcMap, identifying, 446

Mercator projection: example of, 46; explanation of, 47

merge function, using, 216

merging images software, description of, 305

meridian, definition of, 43

meridians of longitude, representing, 41

Mesopotamian map, origin of, 5

metadata: maintaining, 261; organizing, 165–166; overview of, 165; quality considerations in GIS projects, 257

metadata files, content standard for, 165

metadata mode in ArcCatalog, description of, 461

metadata reference FGDC metadata category, explanation of, 166

metes-and-bounds surveying, explanation of, 51

metropolitan area inventory units, significance of, 169

metropolitan divisions, identifying, 170

Metropolitan Statistical Areas, identifying, 170

metropolitan statistical units, redefinition of, 169

Miami–Dade County, Florida Web site, 440

Miller Cylindrical projection: example of, 46; explanation of, 47

minorities in census tracts, calculating numbers of, 476

minority population bar graphs by percent, 131

minority population by percent choropleth map, 71

minority population by percent, displaying for TRI theme, 516

minutes, dividing degrees of latitude into, 41

mobile mapping, overview of, 310

mode measurement, definition of, 100

modeling: raster model, 27; vector model, 26–27

models: analog models, 224; binary models, 226–227; cartographic models, 224–225; conceptual models, 224; mathematical models, 224; process models, 230–231; ranking models, 228; rating models, 228; spatial models, 224–225; weighted models, 229–230

modes, calculating for datasets, 103

Mollweide projection, use of, 47

Monmonier, Mark: definition of maps, 4; maps with class breaks, 111

Mott Have (Bronx), asthma hospitalizations in, 394

Multifamily Foreclosure Prevention Clearinghouse project, significance of, 337

multiple line graphs, using, 129

multispectral imagery: definition of, 300; use in Ground Zero case study, 371

multivariate cartogram, 85

multivariate data, mapping, 114

multivariate maps: of Los Angeles (1971), 85; of quality of life in South Carolina (1992), 86; with segmented symbols, 86; using, 85–86

MWA (Metropolitan Waterfront Alliance), advocacy GIS mapping for, 425–428

.mxd file extension, explanation of, 457

N

Naphtali, Zvia Segal: Access to Health Care in Greenpoint–Williamsburg, Brooklyn, New York case study, 340–347

national map accuracy standards, availability of, 163

natural breaks classification: of adults without H.S. diploma, 122; using, 107

natural habitat and open space assessment case study, 350–355

neatlines, adding, 140, 563

NECMAs (New England County metropolitan areas), significance of, 169

NECTAs (New England city and town areas), significance of, 170

needs assessment, performing for GIS projects, 252–254

neighborhood demographics, analysis of, 324

neighborhood names, significance of, 89

neighborhood-assessment maps, creation Enterprise Foundation, 339

neighborhoods with common local issues, depicting, 331

network analysis, overview of, 299

New Jersey (Bergen County) GIS problem-solving example, 242–244

New Jersey Department of Environmental Protection Maps for Mayors Web site, 244

New Map File button, using, 468

New Rectangle tool, using, 564

New York City. See NYC (New York City)

New York State, axes in, 50

Nnh (nearest neighbor hierarchical) spatial clustering, using in auto thefts case study, 412

NOAA (National Oceanic and Atmospheric Administration) Web site, 20

nominal scale, mode and variation ratio considerations, 100

nominal scaling, using, 95

nonspatial data. See attribute data

nonspatial information: example of, 15; explanation of, 17

normal map projection, example of, 45

North Arrow option, using, 563

north arrows, adding to layouts, 494

north pole, representing in UTM coordinate system, 49

Northern Ireland data visualization experiment, 136

Northwest British Columbia Special Locations maps, 276

not Boolean operation, explanation of, 222

NRC (National Research Council) report, significance of, xvii

NSDI (National Spatial Data Infrastructure), development of, 162

NSDI Geospatial Data Clearinghouse: significance of, 176; Web site, 176

numerical data, representing, 132

NYC (New York City): administrative boundaries of, 38; New York City AIDS prevalence rates by ZIP Code, 110; auto thefts in, 410–411; community-based planning in, 329; Hell's Kitchen section of, 89; land-use decision-making process in, 225; population density by ZIP Code in, 160; relative frequency histogram of residents living below poverty level, 133; sewer and storm drain system case study, 376–381; units of data aggregation in, 38

NYC 3D GIS model of buildings, 297

NYC boroughs, Census tract boundaries in, 176

NYC communities, scatter diagram of adults with H.S. diplomas and recycling rates, 132

NYC data sources, using in Landmark Preservation Commission case study, 385–386

NYC Landmark Preservation Commission case study, 382–391

NYC 2012 Proposed Olympic Plan for New York City map, 332

NYCMap project: role in sewer and storm drain system case study (NYC), 380–381; significance of, 368

nyczip shapefile, contents of, 543

NYP.D.'s Compstat process, overview of, 409

NYPIRG (New York Public Interest Research Group) Web site, 21

O

OASIS (Open Accessible Space Information System): role in MWA case study, 428–430; significance of, 428–430; Web site, 440

oblique map projection, example of, 45

ODBC connectivity, using with external DBMS, 194

Olympics 2012, New York's bid for, 333

on-screen data conversion, creating vector mapping with, 186

on-screen editing of spatial databases, creating vector mapping with, 186

Open Space Resources map (Long Island Sound case study), 353

open-space and waterfront access points, identifying relationships between, 331

operations, examples of, 12

or Boolean operation, explanation of, 222

ordinal scale: median and percentile considerations, 101–103; using, 96

organizations, examining for GIS projects, 249

orientation: choosing for maps, 491; including in maps, 139

Orlando, Florida Web site, 441

overlay analysis, buffers as variation of, 215

overlay operations, overview of, 214

P

Paino, Thomas: Graffiti in Community District Two, Queens, New York: Private expression in the public realm case study, 414–423

Pakistan Army of the Pure poster, 271

Pan tool, identifying, 447

panchromatic satellite images, definition of, 300

paper maps, use in Ground Zero case study, 373–375

parallels of latitude, representing, 41

parcel and zoning inquiry by resident data flow, 255

parcels of land, digitizing and geocoding, 331

Parks and Public Lands Asthma Hospitalization Cases combination map, 83

passive electro-optical sensor, alternative to, 305

patches, using in legends, 562

patient locations, mapping in asthma and environmental hazards in Bronx case study, 398, 402

patterns in data, revealing, 481

percent, definition of, 98

percent minority population bar graphs, 131

Percent_Minority field: naming, 477; opening Field Calculator for, 478

Percentage of Population Below Poverty, within Buffers map, 401

percentage values, calculating, 477

percentiles, examples of, 102–103

perceptual method of scaling, using, 132

pesticide spraying expert system, 232

PGIS (Participatory GIS): criticisms of, 273; issues related to, 272–274; significance of, 10

phenomenon-based search, explanation of, 210

physical models, using, 224

pictograms, sizing, 132

pie graphs: creating, 488–489; using, 131

pilot studies, performing for GIS projects, 260

Pilot Study Area Context (Western Half) map (graffiti case study), 417 planar projection: example of, 45; explanation of, 44

planners, use of GIS by, 9–11. *See also* urban planning

planning: explanation of, 12; initiating with mental maps, 60. *See also* urban planning

Planning for All New Yorkers: A Briefing Book of Community-Based Plans, significance of, 329–330

planning functions, using GIS in, 10

Plate Carrée projection: example of, 46; explanation of, 47

PMSAs (prime metropolitan statistical areas), significance of, 169

point buffer, example of, 215

point data: creating interpolated surface from, 551–552; placement of, 444

point geocoding: to centroids of polygons, 183; by coordinates, 182; by street address, 183–184

point segments, designating in maps, 26

points, defining by coordinates, 162

points in type styles, choosing, 151

points of tangency, definition of, 44

polar map projection, example of, 45

polar regions, using Lambert Azimuthal projection for, 48

policy makers, maps for education of, 427

political control, relationship to maps, 6

political propaganda, using maps as, 270

pollution sources in Bronx, 394

polygon boundaries, eliminating with dissolve function, 216

polygon buffer, example of, 215

polygon features, components of, 26

Polynesian oceanic explorers, maps created by, 5

Poor Workers Finding Modest Housing Unaffordable, Study Says article, 119–120

population maps: by block group dot density, 69; by census tract (Greenpoint–Williamsburg case study), 345; by county dot density map, 69; by ZIP Code dot density map, 69

population density: calculating, 475; at different levels of aggregation choropleth map, 72; in U.S. (unclassified) map, 113

population distribution (Greenpoint–Williamsburg case study), 345

population per square mile by county maps, 112

populations, representing in dot density maps, 66–67

portrait orientation, deciding on, 491

poverty levels of NYC residents, 133

PPGIS (Public Participation GIS), significance of, 10

PPS (Precise Positioning Service), significance of, 306

presentation layout, creating, 547

preview mode in ArcCatalog, description of, 461

prime meridian: definition of, 43; representing, 41

printing layouts, 496, 517, 565

privacy, ethical issues related to, 285–286

private sector, use of GIS in, 11

process models, using, 230–231

project investment committee package maps, creation Enterprise Foundation, 339

projection process, conceptualizing, 44

projections: of British Isles, 48; choosing, 47–48; and coordinate systems, 39; definition of, 39; divisions of, 47; reconciling, 16; types of, 44

projects: end result of, 15; estimating costs and benefits of, 250; initiating, 249–250; management perspective of, 262; roles in, 248

propaganda, maps as, 270–271

property boundaries, recording in United States, 51

proportion, definition of, 98

proportional symbol map and cartogram of American Indian population, 82

proportional symbol maps: creating, 456–457; of deaths due to stroke, 115; using, 73–74

proportional symbols: for air pollutants map, 73; displaying residential units as, 526; using in graphs, 132

proposer, role in GIS projects, 248

proximal search. *See* buffers

proximity analysis. *See* buffers

Public Access to the Water category, using in Long Island Sound case study, 351

public information, significance of, 13

Public Outreach: *Imagine New York,* The Lower Manhattan Preservation Fund, The Campaign for Community-based Planning Olympics 2012 case study, 323–333

public outreach, implementing in *Imagine New York* project, 324

public relation maps, creation Enterprise Foundation, 338

Q

qualitative information, displaying in thematic maps, 94

qualitative map with named categories of data, 94

qualitative thematic maps, using, 64

quality of life map, 269

quantile, classification by, 112

quantile classification: of adults without H.S. diploma, 122; using, 106–107

quantitative information, displaying in thematic maps, 94

quantitative map with numerical categories of data, 94

quantitative thematic maps, using, 64

quartile deviation, explanation of, 103

quartiles, differentiating datasets with, 102–103

Queens, NY: sewer map (1945), 381; graffiti case study, 414–423

queries: capabilities in ArcMap, 390; creating definition query for DOF table, 520; overview of, 195; performing, 453–454; simple queries, 210–211

Query Builder screen in ArcGIS 9, 197

query dialog box, example of, 196–197

querying databases spatially, 203

R

Race Contours (Racial/ethnic changes in the United States) Web site, 21

ranking models, using, 228

raster data, use in Ground Zero case study, 374

raster image file formats, examples of, 27

raster image of USGS 7.5-minute quadrangle map, 173

raster images: problem associated with, 28; uses of, 27

raster maps, combining vector maps with, 187

raster model: combining with vector model, 28; overview of, 27

raster-based analysis, conducting in Long Island Sound case study, 350

rating models, using, 228

ratio, definition of, 97

ratio scales: arithmetic mean and standard deviation considerations, 103–105; using, 97

RDBMS (relational database management systems): combining data from multiple tables in, 201; and data integration, 205; designing, 200; joining tables with unequal number of rows in, 202; key fields in, 201

RDBMS structure, documenting, 200

reclassification, performing on attribute and spatial data, 212–213 records: adding to tables, 486–487; in attribute table, 464; definition of, 98; exporting to new feature classes, 525–526; matching, 513–514

records in fields, calculating values for, 475

Recreation Resources criteria, using in Long Island Sound case study, 353

Recreational Need category, using in Long Island Sound case study, 354

Recreational Need map (Long Island Sound case study), 354

rectangles: adding extent rectangles, 559; opening properties of, 564

reference maps: design of, 63; using, 61–62

references and further reading: for Attribute Data and Relational Database Management Systems, 207; for Basics of Mapping and GIS, 23; Community-based Planning, 423; for Crime Pattern Analysis, 413; for Data Classification Methods and Date Exploration, 123; for Data Visualization and Map Design, 155; for Delivering Health-care Services to an Urban Population, 347; for Ethical Issues in GIS and Urban Planning, 289–290; for Urban Environmental Planning, 364; for GIS Project Development and Institutional Issues, 264; for Health and Environmental Justice, 407; for Mapping Databases, 189; for Methods of Spatial Data Analysis, 245; for Other Geotechnologies and Recent Developments in GIS, 314; for Sources of Urban Data, 179; for Spatial Data and Basic Mapping Concepts, 55; for Thematic Mapping, 91

reflectance, measurement of, 301

relative frequency, definition of, 99

relative frequency histogram of NYC residents below poverty level, 133

relative frequency histograms, usefulness of, 133

Relief Web: Emergencies, natural and environmental disasters, armed conflicts, and humanitarian crises Web site, 21

Rematch Criteria frame, accessing, 511

remote sensing data, processing, 305

remote sensing datasets, examples of, 158

remote sensing, definition of, 300

Renting at Minimum Wage map, 120–121

residential units, displaying as proportional symbols, 526

ResidentialHighRise map, naming, 520

resolution: of datasets, 37–38; explanation of, 36

RF (representative fraction) method: advantages of, 31–32; explanation of, 29

risk classes, establishing in ranking models, 228

Robinson projection: example of, 46; explanation of, 47

Romalewski, Steven et al: Community Mapping Assistance Project for the Metropolitan Waterfront Alliance and Open Accessible Space Information Systems Internet Map Site case study, 424–430

roof acreage, determining availability of, 362

roofs, urban uses of, 357

rows, grouping, 197–200

Royal Observatory (Greenwich, United Kingdom), significance of, 5

RPA (Regional Plan Association), significance of, 349–350

rubber-sheeting capability, explanation of, 28

rule-based reasoning, overview of, 231–232

Rutberg, Karen et al: GIS mapping of the New York City sewer and storm drain system case study, 376–381

S

Sanborn maps, uses for, 175

SanGIS Web site, 440

sans serif typeface, example of, 152

SAR (Synthetic Aperture Radar), overview of, 305

satellite imagery: availability of, 300; spatial resolution of, 302–303; spatial versus spectral types of, 300–301; temporal resolution of, 303–304

satellite systems, use in Ground Zero case study, 369

saturation, definition of, 150

Save As Layer File option, using, 468

Save button, using, 480

scale bars: adding to layouts, 494, 562–563; drawing, 29

scale factors, examples of, 31

scales. *See* map scales

scaling pictograms, 132

scanned images and digital photographs, creating vector mapping with, 186

scatter diagrams: of adults with H.S. diplomas and community recycling rates, 132; using, 133

scatter plots: including in maps, 140; using with natural breaks classification, 107

school district inventory units, significance of, 169

scrubbing process, using in sewer and storm drain system case study (NYC), 380

seconds, dividing minutes into, 41

Select By Attributes dialog box, opening, 453

Select By Location dialog box, using, 534–535

Select Elements tool, using, 492, 493, 563

Select Features tool, using, 450–451, 456

selected records, exporting to new feature classes, 525–526

selection by location, performing, 211

selections, clearing, 478

semilog graphs, lines in, 128–129

sensors, using with satellite imagery, 301

September 11, memorialization of, 324

serif typeface, example of, 152

sewer (Queens, 1945) map, 381

sewer and storm drain system case study (NYC), 376–381

sewer cross section (Manhattan, 1888), 380

sewer mapping project, phases of, 379

shapefiles: converting with Clip tool, 543–544; using FILE styles with, 507

shapes, using, 560

Show/Hide ArcToolbox Window button, using, 532

SHR36005 layer in Bronx.mdb, explanation of, 462

simple queries: selection by location in, 211; using, 210–211

simulation models, creating for built environments, 296–297

Sinusoidal projection: example of, 46; explanation of, 47

site map example (Municipal Art Society), 328

site records, matching in Landmark Preservation Commission (NYC) case study, 386-388

skills, obtaining for community-based GIS, 280

small-scale versus large-scale maps, 32–35

sociodemographic factors in multivariate map, 85

software: choosing for GIS projects, 257; obtaining for community-based GIS, 280

sorting selected data in tables, 451–452

South Bronx. connotations of, 89; mental mapping of, 89. *See also* Bronx, NY

South Bronx article, 87–88

South Carolina (1992) multivariate map of quality of life, 86

space segment of GPS, description of, 306

SPARCS (Statewide Planning and Research Cooperative System) data: limitations of, 404–405; using in asthma and environmental hazards in Bronx case study, 398

spatial aggregation, exploring data with, 114

spatial analysis: explanation of, 14–15, 17; in green roofs case study, 362; types of, 210; using simple queries in, 210–211

spatial and statistical methodologies, combining in auto thefts case study, 411

spatial data: versus attribute data, 192; continuous versus discrete data, 28–29; examining with buffers, 215; explanation of, 26; reclassifying, 212–213

spatial data organization FGDC metadata category, explanation of, 165

spatial databases, linking tables to, 503

spatial density, describing variation in, 65–69

spatial dimension, adding to databases, 203–204

spatial extent, reducing with clipping, 216

spatial features, representing in vector formats, 26

spatial information, example of, 15

spatial joins, explanation of, 218

spatial models, using, 224–225

spatial patterns, identifying in crime analysis, 410

spatial perceptions, reflecting with mental maps, 58

spatial properties, examples of, 47

spatial reclassification: explanation of, 212; performing, 213

spatial reference organization FGDC metadata category, explanation of, 165

spatial resolution: choosing for satellite images, 302; relationship to satellite grid cells, 302; versus temporal resolution, 304

spatial satellite imagery, overview of, 300–301

Spatial Statistics Toolbox, using in auto thefts case study, 411

SPC (State Plane Coordinate) system, overview of, 50

spectral satellite imagery, overview of, 300–301

Spelling Sensitivity, setting, 511

spheres, flattening, 39

spline interpolation, using, 237

SPOT IMAGE satellite system, use in Ground Zero case study, 369

spreadsheets, organization of, 192

SPS (Standard Positioning Service), significance of, 306

SPS (Stationary Point Sources), mapping, 73

SPS buffers, using in asthma and environmental hazards in Bronx case study, 403–404

SQL (Structured Query Language), significance of, 195

square root method of scaling, using, 132

standard deviation: calculating, 104; classification by, 112; considering in interval and ratio scales, 103–105

standard deviation classification: of adults without H.S. diploma, 122; using, 108–109

standard toolbar in ArcMap, identifying, 446

state inventory units, significance of, 169

Staten Island isolines map, 75

states, multiple zones in, 50

statistical and spatial methodologies, combining in auto thefts case study, 411

statistics: goal of, 97; reviewing for datasets, 550; summarizing data with, 100–105

Statistics dialog box, opening, 485

status bar in ArcMap, identifying, 446

Stop Editing option, choosing, 487

storm drain and sewer system case study (NYC), 376–381

street addresses: locating with TIGER/Line files, 184; point geocoding by, 183–184

street data layer, adding, 506

streets, reference map of, 61

stroke deaths, mapping, 115

styles, customizing and saving, 560

subway lines, reference map of, 61

suitability analysis, overview of, 218–222

Summarize dialog box, opening, 522

summary statistics, calculating in DBMS, 197

surface graphs, using, 130

surface map, creating for asthma rates attached to centroids, 551–552

surface modeling, performing with GIS, 295–296

surveillance, ethical issues related to, 285–286

symbol size, perception of, 74

symbolization, preparing data layers for, 465–466

symbology: and color guidelines, 560; defining for layers in data frames, 559–560

symbols: using in graphs, 132; using in legends, 562; using in proportional symbol maps, 73

system integration, requirements for, 260

T

table digitizing, creating vector mapping with, 186

table joins: explanation of, 217; formatting fields for, 545-546; performing, 546

table view, example of, 196

Tables: aggregate data summarized for green roofs case study, 362; asthma hospitalizations of children in Bronx, 403; attribute data, 195; attribute database table, 193; combining data from multiple tables in RDBMS, 201; derived fields, 198; grouping land-use data, 197; image resolution correspondence to scale of paper map, 304; importing data from external database tables, 202; joining columns to produce unique ID fields, 199; mean values with different standard deviations, 104; median values with different ranges, 102; odds ratios of increased risks of Asthma hospitalizations in Bronx case study, 404; rating model priority factors, 228; relative frequencies of land use by property lots, 99; remote sensing data used in Ground Zero case study, 370–371; roof areas case study, 362

tables: adding fields and records to, 486–487; combining data from, 201; components of, 98; creating, 542; creating to store graphed values, 486; exporting, 200; including key fields in, 200; limiting to single categories, 200; linking to spatial databases, 503; viewing and sorting selected data in, 451–452. *See also* database tables

tables with unequal number of rows, joining, 202

tangible maps, using, 60

tax lot maps, using in Landmark Preservation Commission (NYC) case study, 385

taxes, relationship to maps, 7

TAZ (traffic analysis zone) inventory units, significance of, 169

technology, considerations related to community-based GIS, 279–280

Tel Aviv, Israel, 2.5-meter resolution of satellite image, 302

temporal scale, specifying, 34

temporal versus spatial resolution, 304

terrain elevations, representing, 294

text elements, adding, 494

text files, overview of, 192

Text tool, using, 564

thematic data, importance of, 63

thematic map description exercise, 116

thematic maps: cartograms, 79; choropleth maps, 70–72; classification of, 105; classification themes in, 111; combination thematic maps, 83–84; components of, 63; dot density maps, 65–69; flow maps, 78; isarithmic maps, 75–78; measurement scales in, 95–97; multivariate maps, 85–86; proportional symbol maps, 73–74; quantitative and qualitative information in, 94; quantitative versus qualitative maps, 64; using, 61–62; using colors in, 149. *See also* maps

THEN branch of decision tree, significance of, 231

thermal imagery: definition of, 300; use in Ground Zero case study, 371

Thiessen polygon interpolation, performing, 234–236

TIGER mapping system, significance of, 171

TIGER/Line files: contents of, 171; downloading in ASCII format, 171; locating street addresses with, 184

T-in-O map, example of, 7

time dimension, adding in GIS analysis, 299

time period FGDC metadata category, explanation of, 166

TIN (Triangulated Irregular Network) example, 295

TIN versus wire frame models, 295–296

titles, adding to layouts, 495, 564

titles, including in maps, 139

TOC (table of contents): displaying bx-DOF table in, 520; identifying in ArcMap, 446

toolbars in ArcCatalog, description of, 461

Tools toolbar: description of, 446; distinguishing from Layout toolbar, 493

topographic mapping, overview of, 294–295

topology: explanation of, 17; in vector model, 27

tour maps, creation by Enterprise Foundation, 338

townships, formation of, 53

trade, role of maps in, 6

Transportation and Population Density in The Bronx, NY map, 141

transverse map projection, example of, 45

TRI (Toxic Release Inventory) facilities: geocoding to street address location, 218; mapping, 73–74

TRI buffers: intersection with census tracts, 535; using in asthma and environmental hazards in Bronx case study, 403–404

TRI facilities: comparing locations of high-rise buildings with, 527; downloading information on, 500–501; labeling, 514–515

TRI facilities table, examining addresses in, 506

TRI facility addresses, geocoding, 510–511

TRI Locations data frame, renaming, 503

TRI table: adding, 501–502; opening and reviewing contents of, 502

TRI theme, displaying with percent minority theme, 516

TRI_Buffers file, saving, 532

TRI.csv file, opening, 501

tuberculosis by racial and ethnic group in U.S. maps, 66

type, legibility of, 152–153

typeface styles, choosing, 152

typography, considering in map designs, 150–153

U

Undo button, using, 492

unemployment rates by ZIP Code, 102

UNHCR (United Nations High Commissioner for Refugees) Web site, 21

UNHP (University Neighborhood Housing Program): significance of, 335–339; Web site of, 336

Uniform Land Use Review Procedure example, 225

union function, using, 217

unique ID fields, producing from joined columns, 199

unique ID numbers, labeling TRI facilities with, 514–515

unit of analysis, selecting, 37

United States. *See* U.S. (United States)

univariate map, definition of, 83

Universal Soil Loss Equation map, 231

University of Arkansas: CAST (Center for Advanced Spatial Technologies) Web site, 21

University of California at Santa Barbara: Project Gigalapolis Web site, 21

University of Iowa: Department of Public Health Administration Web site, 22

unmapped attribute tables, overview of, 185

unmatched addresses, correcting and rematching, 511–514

updating responsibilities, defining for databases, 200

urban, connotation of, 11

urban applications of GIS, categories of, 12–13

urban area, definition of, 167

urban data: choosing, 159–160; considering relevance of, 159; increased availability of, 158; level of detail in, 160–161; quality of, 162

Urban Data Solutions: 3D urban mapping Web site, 21

urban data sources: commercial sources, 174–175; data clearinghouses, 176; federal agencies, 174; Internet, 176; U.S. decennial census, 167–169; U.S. postal zone mapping, 172; USGS (U.S. Geological Survey) mapping, 172. *See also* GIS data

urban environmental planning case study, 356–364

urban GIS data, accessing, 177

urban planning: requirements for, 9; using GIS in, 9, 11. *See also* planning

urban services case study, 335–339

urbanized areas in U.S., identification by Census, 167

U.S. (United States): division into 7.5-minute quadrangles, 172; land partitioning in, 51–53; land platting in, 52; number of counties in, 167; recording property boundaries in, 51

U.S. air passenger travel map, 78

U.S. bulls-eye map in wake of 9/11, 271

U.S. Census TIGER mapping system, significance of, 171

U.S. decennial census hierarchy: census division, 167; census region, 167; county, 167; county subdivisions and places, 167; state, 167

U.S. Department of Transportation: Various transportation planning projects Web site, 22

U.S. population cartograms, 81

U.S. population density map (unclassified), 113

U.S. postal zone mapping, overview of, 172

user segment of GPS, description of, 306

users, role in GIS projects, 248–249

USF (Universal Site File), using in Landmark Preservation Commission (NYC) case study, 384–385, 389

USGS (U.S. Geologic Survey) Web site, 20

USGS mapping: 7.5-minute quadrangles, 172; DEM (digital elevation models), 173; DOQQ (digital orthophoto quadrangles), 173

USGS 7.5-minute quadrangle map, 61

USGS DEM products, using, 294

Using GIS at Ground Zero case study, 366–375

Using GIS to map crime victim services Web site, 22

USPLS (U.S. Land Partitioning Survey) system, explanation of, 52–53

UTM (Universal Transverse Mercator) system, overview of, 49

V

value of color, definition of, 150

value-by-area maps. *See* cartograms

values, definition of, 98
variables: definition of, 98; plotting on line graphs, 127
variation ratio: calculating, 101; explanation of, 100
vector data, use in Ground Zero case study, 374
vector mapping methods: georeferencing, 186; on-screen data conversion, 186; on-screen editing of spatial databases, 186; scanned images and digital photographs, 187; table digitizing, 186
vector maps: combining with raster maps, 187; foundation of, 26
vector model: combining with raster model, 28; flexibility of, 26; overview of, 26–27; topology considerations, 27
vegetative roofs, benefits of, 358
Venn diagrams: of Boolean operators, 223; definition of, 222; of optimal location analysis of lead poisoning center, 223
verbal scale, example of, 29
viewing selected data in tables, 451–452
views: exporting, 200; restricting data access with, 196
Virginia map (1585), 6
ViSC (Visualization in Scientific Computing), significance of, 135–137
visual balance, considering in graphic design of maps, 143–144
visual comparisons, performing with overlay operations, 214
visual contrast, considering in map designs, 145
Voronoi diagrams. *See* Thiessen polygon interpolation

W

WAAS (Wide Area Augmentation System), significance of, 306–307
WANs (Wide Area Networks), distributing GIS data on, 311
Wards Island WPCP Drainage Basin map, 361
Washing 'South' Out of Bronx Mouths article, 87
waste-related facilities in Hunts Point (Bronx), 275
Water Resources Protection category, using in Long Island Sound case study, 352
waterfront and open-space access points, identifying relationships between, 331
Waterfront Issues: NYC Council District 33 map (MWA case study), 427
WBD36005 layer in Bronx.mdb, explanation of, 462
Web sites: The Agency for Toxic Substances and Disease Registry (Center for Disease Control): Improving public health through GIS, 22; Baltimore County, Maryland, 440; EPA (Environmental Protection Agency), 20; Erie County, New York, 441; ESRI, 20; ESRI GIS Dictionary, xii; Geography Network, 176; Great Britain Historical GIS Project, 21; Integrated Data and Workflow Management for Registrars of Voters Using GIS, 21; Ithaca, New York, 441; Miami-Dade County, Florida, 440; New Jersey Department of Environmental Protection Maps for Mayors, 244; NOAA (National Oceanic and Atmospheric Administration), 20; NSDI Geospatial Data Clearinghouse, 176; NYPIRG (New York Public Interest Research Group), 21; OASIS (Open Accessible Space Information System), 428, 440; Orlando, Florida, 441; Race Contours (Racial/ethnic changes in the United States), 21; Relief Web: Emergencies, natural and environmental disasters, armed conflicts, and humanitarian crises, 21; SanGIS, 440; UNHCR (United Nations High Commissioner for Refugees), 21; UNHP (University Neighborhood Housing Program), 336; University of Arkansas: CAST (Center for Advanced Spatial Technologies), 21; University of California at Santa Barbara: Project Gigalapolis, 21; University of Iowa: Department of Public Health Administration, 22; Urban Data Solutions: 3D urban mapping, 21; U.S. Department of Transportation: Various transportation planning projects, 22; USGS (United States Geologic Survey), 20; Using GIS to map crime victim services, 22; World Resources Institute: Where are the poor?, 21

weighted models, using, 229–230
Where Do the Elderly Live? exercise, 154
Whetten, Michelle et al: FBHC (Fordham Bedford Housing Corporation) case study, 335–339
White, John (Virginia map, 1585), 6
white space, proportioning in maps, 144
wire frame versus TIN models, 295–296
Woodcliff Lake, applying and implementing category 1 water purity standards to, 242–244
word statement: description of, 31; example of, 29
workshops across Tri-State area (NY, CT, NJ), 325
world maps, early examples of, 4–5
World Resources Institute: Where are the poor? Web site, 21
World.mdb database, locating, 444
WTC (World Trade Center) attacks: memorialization of, 324; role of GIS in, 367

X

x-axis: in histograms, 133; plotting independent variables on, 127
xor Boolean operation, explanation of, 222

Y

y-axis: in histograms, 133; plotting dependent variables on, 127

Z

z-axis, expressing nonspatial attributes of geographic entities with, 298
z-coordinate, using with variables, 98
ZCTA (ZIP Code Tabulation Area) inventory units, significance of, 169
Ziegler, John: Using GIS at Ground Zero case study, 366–375
ZIP Centroids data frame, naming, 550
ZIP Code area maps, use by UNHP, 336
ZIP Code areas for NYC, accessing, 543
ZIP Code centroids, joining asthma statistics to, 550–551
ZIP Codes, number of, 172
ZIP+4 mapping, availability of, 172
zoning categories, reclassifying, 213
Zoom In tool, identifying, 448
zooming, 447

Books from

ESRI
Press

Advanced Spatial Analysis: The CASA Book of GIS 1-58948-073-2

ArcGIS and the Digital City: A Hands-on Approach for Local Government 1-58948-074-0

ArcView GIS Means Business 1-879102-51-X

A System for Survival: GIS and Sustainable Development 1-58948-052-X

A to Z GIS: An Illustrated Dictionary of Geographic Information Systems 1-58948-140-2

Beyond Maps: GIS and Decision Making in Local Government 1-879102-79-X

Cartographica Extraordinaire: The Historical Map Transformed 1-58948-044-9

Cartographies of Disease: Maps, Mapping, and Medicine 1-58948-120-8

Charting the Unknown: How Computer Mapping at Harvard Became GIS 1-58948-118-6

Children Map the World: Selections from the Barbara Petchenik Children's World Map Competition 1-58948-125-9

Community Geography: GIS in Action 1-58948-023-6

Community Geography: GIS in Action Teacher's Guide 1-58948-051-1

Confronting Catastrophe: A GIS Handbook 1-58948-040-6

Conservation Geography: Case Studies in GIS, Computer Mapping, and Activism 1-58948-024-4

Designing Better Maps: A Guide for GIS Users 1-58948-089-9

Designing Geodatabases: Case Studies in GIS Data Modeling 1-58948-021-X

Disaster Response: GIS for Public Safety 1-879102-88-9

Extending ArcView GIS (version 3.x edition) 1-879102-05-6

Fun with GPS 1-58948-087-2

Getting to Know ArcGIS Desktop, Second Edition Updated for ArcGIS 9 1-58948-083-X

Getting to Know ArcObjects: Programming ArcGIS with VBA 1-58948-018-X

Getting to Know ArcView GIS (version 3.x edition) 1-879102-46-3

GIS and Land Records: The ArcGIS Parcel Data Model 1-58948-077-5

GIS for Environmental Management 1-58948-142-9

GIS for Everyone, Third Edition 1-58948-056-2

GIS for Health Organizations 1-879102-65-X

GIS for Landscape Architects 1-879102-64-1

GIS for the Urban Environment 1-58948-082-1

GIS for Water Management in Europe 1-58948-076-7

GIS in Public Policy: Using Geographic Information for More Effective Government 1-879102-66-8

GIS in Telecommunications 1-879102-86-2

GIS Means Business, Volume II 1-58948-033-3

GIS, Spatial Analysis, and Modeling 1-58948-130-5

GIS Tutorial for Health 1-58948-148-8

GIS Tutorial: Workbook for ArcView 9 1-58948-127-5

GIS Worlds: Creating Spatial Data Infrastructures 1-58948-122-4

Hydrologic and Hydraulic Modeling Support with Geographic Information Systems 1-879102-80-3

Continued on next page

When ordering, please mention book title and ISBN (number that follows each title)

Books from ESRI Press (continued)

Integrating GIS and the Global Positioning System *1-879102-81-1*

Making Community Connections: The Orton Family Foundation Community Mapping Program *1-58948-071-6*

Managing Natural Resources with GIS *1-879102-53-6*

Mapping Census 2000: The Geography of U.S. Diversity *1-58948-014-7*

Mapping Global Cities: GIS Methods in Urban Analysis *1-58948-143-7*

Mapping Our World: GIS Lessons for Educators, ArcGIS Desktop Edition *1-58948-121-6*

Mapping Our World: GIS Lessons for Educators, ArcView GIS 3.x Edition *1-58948-022-8*

Mapping the Future of America's National Parks: Stewardship through Geographic Information Systems *1-58948-080-5*

Mapping the News: Case Studies in GIS and Journalism *1-58948-072-4*

Marine Geography: GIS for the Oceans and Seas *1-58948-045-7*

Measuring Up: The Business Case for GIS *1-58948-088-0*

Modeling Our World: The ESRI Guide to Geodatabase Design *1-879102-62-5*

Past Time, Past Place: GIS for History *1-58948-032-5*

Remote Sensing for GIS Managers *1-58948-081-3*

Salton Sea Atlas *1-58948-043-0*

Spatial Portals: Gateways to Geographic Information *1-58948-131-3*

The ESRI Guide to GIS Analysis, Volume 1: Geographic Patterns and Relationships *1-879102-06-4*

The ESRI Guide to GIS Analysis, Volume 2: Spatial Measurements and Statistics *1-58948-116-X*

The GIS Guide for Local Government Officials *1-58948-141-0*

Think Globally, Act Regionally: GIS and Data Visualization for Social Science and Public Policy Research *1-58948-124-0*

Thinking About GIS: Geographic Information System Planning for Managers (paperback edition) *1-58948-119-4*

Transportation GIS *1-879102-47-1*

Undersea with GIS *1-58948-016-3*

Unlocking the Census with GIS *1-58948-113-5*

Zeroing In: Geographic Information Systems at Work in the Community *1-879102-50-1*

Ask for ESRI Press titles at your local bookstore or order by calling 1-800-447-9778. You can also shop online at www.esri.com/esripress. Outside the United States, contact your local ESRI distributor.

ESRI Press titles are distributed to the trade by the following:

In North America, South America, Asia, and Australia:
Independent Publishers Group (IPG)
Telephone (United States): 1-800-888-4741 • Telephone (international): 312-337-0747
E-mail: frontdesk@ipgbook.com

In the United Kingdom, Europe, and the Middle East:
Transatlantic Publishers Group Ltd.
Telephone: 44 20 7373 2515 • Fax: 44 20 7244 1018 • E-mail: richard@tpgltd.co.uk

ESRI Press • 380 New York Street • Redlands, California 92373-8100 • www.esri.com/esripress